FOUNDATIONS OF TOPOLOGY

Foundations of Topology

An Approach to Convenient Topology

by

Gerhard Preuss
Institut für Mathematik I,
Freie Universität Berlin,
Berlin, Germany

Springer-Science+Business Media, B.V.

A C.I.P. Catalogue record for this book is available from the Library of Congress.

ISBN 978-94-010-3940-6 **ISBN 978-94-010-0489-3 (eBook)**
DOI 10.1007/978-94-010-0489-3

Printed on acid-free paper

Dedicated to the memory of

Felix Hausdorff

Miroslav Katětov

and

André Weil

Contents

Preface

This book is based on lectures the author has given at the Free University of Berlin over many years starting in 1996. Furthermore, he presented parts of the material during several international conferences (Oberwolfach 1991, Berlin 1992, L'Aquila 1994, Oberwolfach 1995, Prague 1996, Antwerp 1998, Kent/Ohio 1998, ICM 98, Bremen 2000). Up to now there was no textbook combining topological and uniform concepts on a convenient level. The course is written for graduate students and interested mathematicians who know already the basic facts on topological spaces. These ones are summarized (without proof) in chapter 0 so that everybody can check his knowledge. It is *not* expected that the reader is familiar with the theory of uniform spaces which is entirely developed here. This book is self-contained with the exception of the last chapter where some (repeated) facts on nearness spaces are useful.

After some preliminary remarks on Set Theory, General Topology and Category Theory (chapter 0) topological constructs (chapter 1) are introduced together with numerous examples most of them are fundamental for the following. In order to describe the interactions between various topological constructs the categorical theory of reflections and coreflections (chapter 2) is extremely useful. But the language of category theory is also highly appreciated to formulate convenient properties for topological constructs (chapter 3).

From now on semiuniform convergence spaces are studied extensively since they form the best known topological construct making a topologist's life easier. In chapter 4 a completion theory for semiuniform convergence spaces is developed with applications to Hausdorff compactifications of topological spaces. The theory of connection and disconnection profits from the better behaviour of quotients in the realm of semiuniform convergence spaces (chapter 5). In chapter 6 function spaces are studied, first in the framework of classical General Topology and then in the construct **SUConv** of semiuniform convergence spaces. Last but not least the classical Ascoli theorem results from the characterization of compactness in the natural function spaces in **SUConv**. The last chapter is devoted to the fact that subspaces of topological spaces behave better when they are formed in **SUConv**. Normality, paracompactness and dimension theory profit from this better behaviour. Concerning the appendix it is assumed that the reader has some knowledge on cohomology theory and homotopy theory of topological spaces. For a quick overview an implication scheme, a table and a diagram are included at the end of the book. Since it is worthwile to have some "rules" for filters in mind, these are summarized under 0.2.3.15 as well as in the exercises 17) and 18).

The present status of Convenient Topology is not the end of the theory. Much more can (and must) be done in the future. Thus, I hope that the presented material will help to create further new results.

I have dedicated the present book to the memory of F. Hausdorff, M. Katětov and A. Weil because topological and uniform spaces (created by F. Hausdorff and A. Weil respectively) as well as filter spaces (invented by M. Katětov) form leading examples of semiuniform convergence spaces. I am very grateful to my friend and colleague Horst Herrlich for developing various categorical methods which are applicable to Convenient Topology and for encouraging me to continue my investigations of semiuniform convergence spaces (immediately after presenting him my first results during my invited talk in one of his CatMAT seminars in Bremen). I thank too Dr. J. Schröder (South Africa) for proofreading. But the text would not have been appeared in printed form without the help of Mrs. Annette Mühlenfeld, who prepared the TEX file of my manuscript. I thank her very much. I thank also Mrs. Barbara Wengel and Mrs. Simone Aubram for assisting her.

Finally, I thank Dr. Elisabeth Mol from Kluwer Academic Publishers for her current interest in the progress of this book and Mr. Attila Kimpan for technical help.

Berlin, March 2002

Gerhard Preuß

Introduction

In 1906 Fréchet [47] introduced metric spaces, a very fruitful concept for many purposes in analysis. But unfortunately, the class of metric spaces is not big enough in order to describe *pointwise convergence* in function spaces as the following example shows: On the set X of all maps from the set $\mathbb{R}$ of real numbers into itself there is no metric d describing pointwise convergence, i.e. such that for each $f \in X$ and each sequence (f_n) in X the following are equivalent:

(a) (f_n) converges pointwise to f, i.e. $(f_n(x))$ converges to $f(x)$ for each $x \in X$ with respect to the usual metric on $\mathbb{R}$.

(b) (f_n) converges to f in (X, d).

(Namely, let A be the set of all continuous maps from $\mathbb{R}$ to $\mathbb{R}$ and (r_n) a sequence of *all* rational numbers. For each $n \in \mathbb{N}$ define $f_n : \mathbb{R} \longrightarrow \mathbb{R}$ by

$$f_n(x) = \begin{cases} 1 & \text{if } x \in \{r_1, \ldots, r_n\} \\ 0 & \text{otherwise} \end{cases}$$

and let $f : \mathbb{R} \longrightarrow \mathbb{R}$ be defined by

$$f(x) = \begin{cases} 1 & \text{if } x \in \mathbb{Q} \\ 0 & \text{otherwise} \end{cases} .$$

Then (f_n) converges pointwise to f. If there would be a metric d on X describing pointwise convergence, then the following would be valid (provided $\overline{A}$ denotes the closure of A in (X, d)):

(α) $f_n \in \overline{A}$ for each $n \in \mathbb{N}$

(β) $f \in \overline{\overline{A}} = \overline{A}$

(γ) $f \notin \overline{A}$, i.e. f is not the pointwise limit of a sequence in A, because $U = \{x \in \mathbb{R} : f \text{ is not continuous at } x\}$ is not first category in $\mathbb{R}$ [i.e. U is not a union of a sequence of subsets of $\mathbb{R}$, whose closures do not contain a non–empty open subset of $\mathbb{R}$].).

As is well–known in the realm of topological spaces, introduced by Hausdorff [58] in 1914 (nowadays called Hausdorff spaces) and (in the usual meaning) by Kuratowski [91] in 1922, pointwise convergence can be desribed. But the category **Top** of topological spaces (and continuous maps) has other deficiencies, which will be considered in the following.

In 1921 Hahn [56] introduced 'continuous convergence' in his book "Theorie der reellen Funktionen". According to Carathéodory [25] uniform convergence should be substituted by continuous convergence in Function Theory whenever this is possible because it is much easier to handle. In the realm of topological spaces continuous convergence can be defined as follows: If $C(\mathbf{X},\mathbf{Y})$ denotes the set of all continous maps from a topological space $\mathbf{X} = (X, \mathcal{X})$ into a topological space $\mathbf{Y} = (Y, \mathcal{Y})$, then the evaluation map $e_{X,Y} : X \times C(\mathbf{X},\mathbf{Y}) \longrightarrow Y$ is defined by $e_{X,Y}((x, f)) = f(x)$ for each $(x, f) \in X \times C(\mathbf{X},\mathbf{Y})$. A filter $\mathcal{F}$ on $C(\mathbf{X},\mathbf{Y})$ *converges continuously* to $f \in C(\mathbf{X},\mathbf{Y})$ whenever for each $x \in X$ and each filter $\mathcal{G}$ on X converging to x in $\mathbf{X}$ the filter $e_{X,Y}(\mathcal{G} \times \mathcal{F})$ converges to $f(x)$ in $\mathbf{Y}$, where $\mathcal{G} \times \mathcal{F}$ denotes the product filter (i.e. $\mathcal{G} \times \mathcal{F}$ is generated by the filter base $\{G \times F : G \in \mathcal{G} \text{ and } F \in \mathcal{F}\}$). On the other hand the following is valid:

① a) *On the set $C(\mathbf{X},\mathbf{Y})$ there is in general no topology $\mathcal{C}$ describing continuous convergence.*

Namely, if $\mathcal{C}$ were such a topology, it would be the coarsest topology such that the evaluation map is continuous:

(A) The evaluation map $e_{X,Y} : \mathbf{X} \times (C(\mathbf{X},\mathbf{Y}), \mathcal{C}) \longrightarrow \mathbf{Y}$ is continuous.
(Let $\mathcal{H}$ be a filter on $X \times C(\mathbf{X},\mathbf{Y})$ converging to some $(x, f) \in X \times C(\mathbf{X},\mathbf{Y})$. Then $p_1(\mathcal{H})$ converges to x and $p_2(\mathcal{H})$ converges continuously to f provided p_1 (resp. p_2) denotes the first (resp. second) projection map.
Thus $e_{X,Y}(p_1(\mathcal{H}) \times p_2(\mathcal{H})) \subset e_{X,Y}(\mathcal{H})$ converges to $f(x)$ and consequently $e_{X,Y}(\mathcal{H})$ converges to $e_{X,Y}((x, f)) = f(x)$.

(B) If $\mathcal{C}'$ is a topology on $C(\mathbf{X},\mathbf{Y})$ such that the evaluation map is continuous, then $\mathcal{C} \subset \mathcal{C}'$.
(Let $\mathcal{U}_{\mathcal{Z}}(z)$ be the neighborhood filter of $z \in Z$ for every topological space $(Z, \mathcal{Z})$. If $\mathcal{F}$ is a filter on $C(\mathbf{X},\mathbf{Y})$ such that $\mathcal{F} \supset \mathcal{U}_{\mathcal{C}'}(f)$ for some $f \in C(\mathbf{X},\mathbf{Y})$, then $\mathcal{F} \supset \mathcal{U}_{\mathcal{C}}(f)$, because it follows from the continuity of $e_{X,Y} : \mathbf{X} \times (C(\mathbf{X},\mathbf{Y}), \mathcal{C}') \longrightarrow \mathbf{Y}$ that $e_{X,Y}(\mathcal{G} \times \mathcal{F})$ converges to $f(x)$ for each filter $\mathcal{G}$ on X converging to x and, since $\mathcal{C}$ is the topology of continuous convergence, $\mathcal{F}$ converges to f with respect to $\mathcal{C}$. But then $\mathcal{C} \subset \mathcal{C}'$ because it follows from $O \in \mathcal{C}$ and $f \in O$ that $O \in \mathcal{F} = \mathcal{U}_{\mathcal{C}'}(f)$ since $O \in \mathcal{U}_{\mathcal{C}}(f) \subset \mathcal{F}$.)

Already in 1946 Arens proved that e.g. on $C(\mathbb{R}^{\mathbb{N}}, [0,1])$ there is no coarsest topology such that the evaluation map is continuous (cf. [4; theorem 3]). In other words:
For infinite-dimensional analysis continuous convergence cannot be described in the realm of topological spaces.

If the topology of continuous convergence would always exist, then the following would also be valid:

(C) For each topological space $\mathbf{Z} = (Z, \mathcal{Z})$ and each continuous map $h : \mathbf{X} \times \mathbf{Z} \longrightarrow \mathbf{Y}$ the map $h^* : \mathbf{Z} \longrightarrow (C(\mathbf{X},\mathbf{Y}), \mathcal{C})$ defined by $(h^*(z))(x) = h(x, z)$ is continuous.
(Let $\mathcal{G}$ be a a filter on Z converging to some $z \in Z$. In order to prove that $h^*(\mathcal{G})$ converges to $h^*(z)$ it must be shown that $e_{X,Y}(\mathcal{F} \times h^*(\mathcal{G}))$ converges to

$e_{X,Y}(x, h^*(z)) = h(x,z)$ for each $x \in X$ and each filter $\mathcal{F}$ on X such that $\mathcal{F} \supset \mathcal{U}_X(x)$. Since $e_{X,Y}(\mathcal{F} \times h^*(\mathcal{G})) = h(\mathcal{F} \times \mathcal{G})$, this follows from the continuity of h.)

Thus, if the topology of continuous convergence would always exist, the category **Top** of topological spaces (and continuous maps) would be *cartesian closed*, i.e. for each pair $(\mathbf{X}, \mathbf{Y})$ of topological spaces there would exist a topology $\mathcal{C}$ on $C(\mathbf{X}, \mathbf{Y})$ such that (A) and (C) are satisfied, in other words: **Top** would have *natural function spaces.* Then the map $* : C(\mathbf{X} \times \mathbf{Z}, \mathbf{Y}) \longrightarrow C(\mathbf{Z}, (C(\mathbf{X}, \mathbf{Y}), \mathcal{C}))$ defined by $*(f) = f^*$ (cf. (C)) would be bijective (use (A) and (C)), i.e.

(D) $C(\mathbf{X} \times \mathbf{Z}, \mathbf{Y}) \cong C(\mathbf{Z}, (C(\mathbf{X}, \mathbf{Y}), \mathcal{C}))$

in the category **Set** of sets (and maps).
Property (D) is highly appreciated by mathematicians working in the field of *Functional Analysis (Duality Theory), Algebraic Topology (Homotopy Theory) or Topological Algebra (quotients).* But unfortunately the following is valid:

① b) *The category* **Top** *is not cartesian closed.*
We have already proved that ① b) implies ① a). We will see later that if **Top** were cartesian closed, then the product of two quotient maps would be a quotient map. But this is not true in **Top** as the following example under ② shows.

② *In* **Top** *products of quotients need not be quotients,* i.e. if $f_i : (\mathbf{X}_i \longrightarrow \mathbf{Y}_i)_{i \in I}$ is any family of quotient maps in **Top**, then $\Pi f_i : \Pi_{i \in I} \mathbf{X}_i \longrightarrow \Pi_{i \in I} \mathbf{Y}_i$ defined by $\Pi f_i((x_i)) = (f_i(x_i))$, is a surjective continuous map but no quotient map in general in contrast to the situation for the category **Set** [or **Group** of groups (and homomorphisms)]. In order to prove ② let us consider the following *example:* Let π be an equivalence relation on $\mathbb{R}$ defined by $x_1 \pi x_2$ iff $x_1 = x_2$ or $\{x_1, x_2\} \subset \mathbb{Z}$. Let $\omega : \mathbb{R} \longrightarrow \mathbb{R}/\pi$ be the natural map. Then the map $\omega \times 1_{\mathbb{Q}} : \mathbb{R} \times \mathbb{Q} \longrightarrow \mathbb{R}/\pi \times \mathbb{Q}$ is surjective and continuous. If it were a quotient map, it would map closed saturated sets to closed sets (note: If $f : X \to Y$ is a quotient map, then $A \subset X$ is saturated iff $A = f^{-1}[B]$ for some $B \subset Y$). Now let $(a_n)_{n \in \mathbb{N}}$ be a sequence of irrational numbers converging to 0. For each n let $(r_{n,m})_{m \in \mathbb{N}}$ be a sequence of rational numbers converging to a_n. Put

$$A = \{(n + \frac{1}{m}, r_{n,m}) : n, m \in \mathbb{N} \text{ and } m > 1\}.$$

A is closed and saturated in $\mathbb{R} \times \mathbb{Q}$ but $A = (\omega \times 1_{\mathbb{Q}})[A]$ is not closed in $(\mathbb{R}/\pi) \times \mathbb{Q}$ (note: $(\omega(0), 0) \notin A$ but each neigborhood of $(\omega(0), 0)$ in $(\mathbb{R}/\pi) \times \mathbb{Q}$ meets A).

③ *In* **Top** *quotients are not hereditary,* i.e. if $f : (X, \mathcal{X}) \longrightarrow (Y, \mathcal{Y})$ is a quotient map in **Top** and $A \subset Y$, then $(f|f^{-1}[A])' : f^{-1}[A] \longrightarrow A$ need not be a quotient map in **Top** provided that $f^{-1}[A]$ (resp. A) is endowed with the relative topology, in contrast to the situation for the category **Set** (or **Group**) in which quotient maps are always hereditary. According to Arhangel'skii [5] many irregularities in Topology depend on the fact that quotients are not hereditary. ③ is proved by means of the following example:
Consider the topological spaces X and Y as well as the subspaces A of Y and B of X, whose non–empty open sets are illustrated in the following picture:

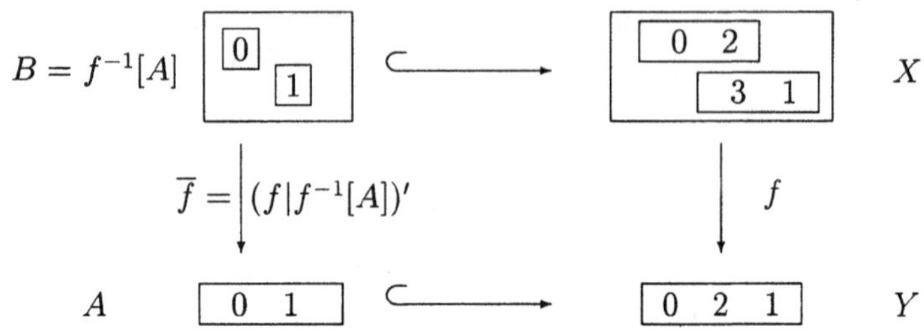

Then f is a quotient map, but $\overline{f}$ is not a quotient map.

④ *Uniform concepts such as uniform continuity, uniform convergence, Cauchy sequences (or Cauchy filters) and completeness are not available in* **Top**.

What are the reasons?
First let us consider the concept of Cauchy sequence in Analysis: A sequence $(x_n)_{n \in \mathbb{N}}$ of real numbers is called a *Cauchy sequence* iff for each $\varepsilon > 0$ there is some $N(\varepsilon) \in \mathbb{N}$ such that

$$(*) \qquad |x_n - x_m| < \varepsilon$$

whenever $n, m \geq N(\varepsilon)$.

$(*)$ may be interpreted as follows: x_n belongs to an ε–neighborhood of x_m (= ε–sphere about x_m) and x_m belongs to an ε–neigborhood of x_n, but x_n and x_m are distinct elements of $\mathbb{R}$ in general. ε–neighborhood of distinct points may be considered to have the same size. In a topological space $(X, \mathcal{X})$ there is assigned a neighborhood system $\mathcal{U}(x)$ to each $x \in X$ such that certain axioms are satisfied (according to Hausdorff's original definition of topological spaces) but there is no possibility to compare neighborhoods of different points with respect to their size. Thus, whenever it is possible to compare neighborhoods of different points with respect to their size (e.g. in metric spaces), Cauchy sequences can be defined. The same argument may be used to explain why uniform continuity and uniform convergence are not available in the realm of topological spaces. Last not least, we cannot define completeness in topological spaces because of the absence of the concept of Cauchy filter.

Since the class of metric spaces is too small (cf. the beginning of this introduction), the question arises, whether there is a bigger class than the class of metric spaces such that uniform concepts are available. This question has been solved by Weil [147] in 1937. He introduced so–called *uniform spaces* and constructed a completion for each uniform space generalizing Hausdorff's completion of metric spaces [58], where the latter one is a generalization of Cantor's construction of the real numbers from the rational numbers. We will see later on that the category **Unif** of uniform spaces is not cartesian closed. Thus having all the points ① – ④ in mind the following question arises:

What is the right concept of space in Topology?

Obviously, such a concept should remedy all the deficiencies of **Top** mentioned under ① b), ②, ③ and ④. Furthermore, all results on topological and uniform spaces should remain important special cases. Last but not least, the definition of such a concept of space should be easy. In this book we will present a solution by introducing **semiuniform convergence spaces**. Once having done this, we get further nice results which cannot be obtained in topological and (or) uniform spaces, namely:

⑤ The localization of the concept of compactness leads to a cartesian closed category and 'locally compact' is equivalent to 'compactly generated' (the Hausdorff–axiom is not required!).

⑥ The localization of the concept of precompactness (= total boundedness) leads to a cartesian closed category in which quotients are hereditary and 'locally precompact' is equivalent to 'precompactly generated'.

⑦ The localization of the concept of connectedness leads to a cartesian closed category.

⑧ The (Hausdorff) completion of a separated uniform space can be easily constructed by means of function spaces.

⑨ Not only the structure of pointwise convergence and uniform convergence but also the structure of continuous convergence can be derived from the natural function space structure in the realm of semiuniform convergence spaces.

Furthermore, let us consider the formation of subspaces of topological spaces. Though there is a difference of a "topological" nature between the removal of a point and the removal of a closed interval of length one from the usual topological space $I\!R_t$ of real numbers, the obtained topological spaces are homeomorphic. But if subspaces are formed in the realm of semiuniform convergence spaces and $I\!R_t$ is considered to be a semiuniform convergence space, then we obtain non–isomorphic semiuniform convergence spaces by means of the above procedure. Consequently, semiuniform convergence spaces behave well w.r.t. the formation of subspaces. The following enlargements of classical topological results can be obtained, whenever subspaces of symmetric topological spaces are formed in the framework of semiuniform convergence spaces (where they need not be necessarily topological unless they are closed):

⑩ Subspaces of fully normal (normal) symmetric topological spaces are fully normal (normal).

⑪ Dense subspaces of connected (locally connected) symmetric topological spaces are connected (locally connected).

⑫ There is a dimension function for semiuniform convergence spaces such that the dimension of each subspace of a symmetric topological space $\mathbf{X}$ is less than or equal to the dimension of $\mathbf{X}$, where for symmetric topological spaces this dimension function coincides with the (Lebesgue) covering dimension dim.

Let me end this introduction by citing Hausdorff [58; p. 211] because his argumentation concerning topological spaces is also valid for semiuniform convergence spaces:

"Und zwar ist der Gewinn an Allgemeinheit nicht etwa mit einer erhöhten Komplikation, sondern gerade umgekehrt mit einer erheblichen Vereinfachung verbunden, indem wir, wenigstens für die Grundzüge der Theorie, nur von ganz wenigen und einfachen Voraussetzungen (Axiomen) Gebrauch zu machen haben."

[English translation: "In particular the gain of generality is not connected with much more complication but vice versa with a considerable simplification by using, at least for the foundations of the theory, only a few and simple assumptions (axioms)."]

Chapter 0

Preliminaries

Though it is advisible to have some basic knowledge of Set Theory and General Topology, in the following some fundamental definitions and results concerning sets and topological spaces are repeated as far as they are useful for a better understanding of this book. Since proofs are omitted, hints to the literature are given. Furthermore, some categorical concepts are introduced.

0.1 Set theoretical concepts

0.1.1 Remark

In Cantor's *naive set theory* every collection of objects specified by some property was called a set. As is well–known this approach leads to contradictions, e.g. to the antinomy of the set R of all sets not members of themselves [Using the terminology under 0.1.2. ① we obtain

$$R \in R \text{ if and only if } R \notin R$$

provided that R is a set.]. In order to resolve this contradiction we introduce two types of collections: classes and sets. Then a *class* is a collection of objects specified by some property, whereas a *set* is a class which is a member of some class. Thus, R is not a set but a class, i.e. a proper class, and also the concept of the class of all sets makes sense.

The *axiomatic set theory* tries to avoid further antinomies. The axiomatic approach due to Gödel, Bernays and von Neumann is suitable to handle classes and sets. For further details the interested reader is referred to Kelley [82], although it is not necessary for understanding this book to know all the details.

0.1.2 Sets

① The objects making up a set are called *elements* (*members, points*) of the set, where the membership relation is denoted by $\in$, e.g. $x \in A$ means that x is an element of A (or A *contains* the element x), and we write $x \notin A$ if x is not an element of A. Let X be a set and P a "property". Then the set of all members of X that have the property P is denoted by $\{x \in X : P(x)\}$, e.g. if $I\!R$ is the set of real numbers, the closed unit interval $[0,1]$ is the set $\{x \in I\!R : 0 \leq x \leq 1\}$. If a set X has exactly the elements $x_1, \ldots, x_n$, we write $X = \{x_1, \ldots, x_n\}$; in case $n = 1$ we call X a *singleton.*

② If A and B are sets, then A is a *subset* of B iff each member of A is a member of B. In this case we say also that A is *contained* in B (or B *contains* A), and we write $A \subset B$ (or $B \supset A$). Furthermore, we say sets A and B are *equal,* $A = B$, iff they have the same elements. Evidently, $A = B$ iff $A \subset B$ and $B \subset A$.

③ The *empty set* $\emptyset$ is the set having no elements; it is a subset of every set A.

④ If A is a set, then the set $\mathcal{P}(A)$ of all subsets of A is called the *power set* of A. Subsets of $\mathcal{P}(A)$ are usually denoted by capital script letters.

⑤ Let $\mathcal{B} \subset \mathcal{P}(A)$ for some set A:
1) a) The set $\{x \in A : x \in B$ for each $B \in \mathcal{B}\}$ is called the *intersection* of all $B \in \mathcal{B}$, denoted by $\bigcap_{B\in\mathcal{B}} B$, or $\bigcap\{B : B \in \mathcal{B}\}$.
b) The set $\{x \in A : x \in B$ for some $B \in \mathcal{B}\}$ is called the *union* of all $B \in \mathcal{B}$, denoted by $\bigcup_{B\in\mathcal{B}} B$, or $\bigcup\{B : B \in \mathcal{B}\}$.
c) If $\mathcal{B} = \emptyset$, then $\bigcup_{B\in\mathcal{B}} B = \emptyset$ and $\bigcap_{B\in\mathcal{B}} B = A$.
2) If $\mathcal{B} = \{C, D\}$ one writes $C \cap D$ instead of $\bigcap_{B\in\mathcal{B}} B$, and $C \cup D$ instead of $\bigcup_{B\in\mathcal{B}} B$.

⑥ 1) Let X be a set and $A, B, C \in \mathcal{P}(X)$. Then
a) *Associative laws:*
α) $A \cup (B \cup C) = (A \cup B) \cup C$,
β) $A \cap (B \cap C) = (A \cap B) \cap C$.
b) *Distributive laws:*
α) $A \cup (B \cap C) = (A \cup B) \cap (A \cup C)$,
β) $A \cap (B \cup C) = (A \cap B) \cup (A \cap C)$.
2) The above laws are also valid for arbitrary unions and intersections.

⑦ Two subsets A and B of a set X are called *disjoint* iff $A \cap B = \emptyset$; we say A *meets* B iff $A \cap B \neq \emptyset$.

⑧ Let A be a subset of a set X. Then the set $\{x \in X : x \notin A\}$ is denoted by $X \backslash A$ and called the *complement* of A in X. We have
a) $X\backslash(X\backslash A) = A$, b) $X\backslash X = \emptyset$, c) $X\backslash\emptyset = X$.

⑨ If A and B are subsets of a set X, then
a) $X\backslash(A \cup B) = (X\backslash A) \cap (X\backslash B)$
b) $X\backslash(A \cap B) = (X\backslash A) \cup (X\backslash B)$
c) $A \subset B$ iff $(X\backslash A) \supset (X\backslash B)$.

⑩ Let X and Y be sets and $x \in X$, $y \in Y$. Then the set $\{\{x\}, \{x, y\}\}$ is denoted by (x, y) and called an *ordered pair.* If $x' \in X$ and $y' \in Y$, then $(x, y) = (x', y')$ iff $x = x'$ and $y = y'$. The *cartesian product* $X \times Y$ of X and Y is defined to be the set of all ordered pairs (x, y) such that $x \in X$ and $y \in Y$.

0.1.3 Relations and maps

① Any subset R of the cartesian product $X \times Y$ is called a *relation* between X and Y; one writes also $x\, R\, y$ instead of $(x, y) \in R$ and says x is *R–related* to y. If $X = Y$, we say that R is a relation on X.

② Let $R, S \subset X \times Y$. Then R is called *finer* than S, or S *coarser* than R, provided that $R \subset S$.

③ A relation f between X and Y is called a *map* from X to (or into) Y, denoted by $f : X \longrightarrow Y$, provided that for each $x \in X$, there is exactly one $y \in Y$ such that $(x, y) \in f$. X is called the *domain* of f and Y the *range* (or *codomain*) of f. One writes also $y = f(x)$ instead of $(x, y) \in f$ and calls $f(x)$ the *value* of f at x, or the *image* of x under f. Intuitively speaking, a map $f : X \longrightarrow Y$ *assigns* to each $x \in X$ the value $f(x)$, or *sends* x to $f(x)$. The terms map, *function* and *operator* are synonymous.

Remark. More exactly, a map f from a set X to set Y is a triple[1] (X, F, Y), where $F \subset X \times Y$ such that for each $x \in X$, there is exactly one $y \in Y$ with $(x, y) \in F$. F is the *graph* of the map f.

④ Let $f : X \longrightarrow Y$ be a map, $A \subset X$ and $B \subset Y$. Then
a) $f[A] = \{y \in Y : y = f(x) \text{ for some } x \in A\}$ is the *image* of A under f, and
b) $f^{-1}[B] = \{x \in X : f(x) \in B\}$ is the *inverse image* of B under f.

⑤ 1) A map $f : X \longrightarrow Y$ is called
a) *surjective* provided that $f[X] = Y$,
b) *injective* provided that for each $(x, x') \in X \times X$, $f(x) = f(x')$ implies $x = x'$,
c) *bijective* provided that it is injective and surjective.

2) A bijective map $f : X \longrightarrow Y$ is said to define a *one–one correspondence* between the elements of the set X and the elements of the set Y.

3) If $f : X \longrightarrow Y$ is a bijective map, then the map $g : Y \longrightarrow X$, defined by $g(y) = x \in X$ for each $y = f(x) \in Y$, is called the *inverse map* and is denoted by f^{-1}.

⑥ a) Let $f : X \longrightarrow Y$ be a map and $A \subset X$. Then a map $f|A : A \longrightarrow Y$ is defined by $(f|A)(x) = f(x)$ for each $x \in A$, called the *restriction* of f to A.
b) Let $g : A \longrightarrow Y$ be a map and $A \subset X$. Then a map $f : X \longrightarrow Y$ is called an *extension* of g provided that $f|A = g$.

⑦ 1) Let $A, B \subset X$ and $C, D \subset Y$. If $f : X \longrightarrow Y$ is a map, then
a) $f[A \cap B] \subset f[A] \cap f[B]$,
b) $f[A \cup B] = f[A] \cup f[B]$,
c) $f^{-1}[A \cap B] = f^{-1}[A] \cap f^{-1}[B]$,
d) $f^{-1}[A \cup B] = f^{-1}[A] \cup f^{-1}[B]$.

[1] The triple (x, y, z) is defined to be the ordered pair $((x, y), z)$.

2) The rules 1) a) – d) are also valid for infinite unions and intersections.

⑧ If $f : X \longrightarrow Y$, $A \subset X$ and $B \subset Y$, then
a) $f^{-1}[Y \backslash B] = X \backslash f^{-1}[B]$,
b) α) $f^{-1}[f[A]] \supset A$,
β) $f^{-1}[f[A]] = A$ provided that f is injective,
c) α) $f[f^{-1}[B]] \subset B$,
β) $f[f^{-1}[B]] = B$ provided that f is surjective.

⑨ Let $f : X \longrightarrow Y$ and $g : Y \longrightarrow Z$ be maps. Then a map $g \circ f : X \longrightarrow Z$ is defined by $(g \circ f)(x) = g(f(x))$ for each $x \in X$, called the *composition* of f and g.

Remark. If $f : X \longrightarrow Y$ is a map and $A \subset X$, then the restriction $f|A$ is equal to the composition $f \circ i$, where $i : A \longrightarrow X$ denotes the *inclusion map*, defined by $i(x) = x$ for each $x \in A$.

⑩ If several maps are considered, we often use *diagrams*, e.g.

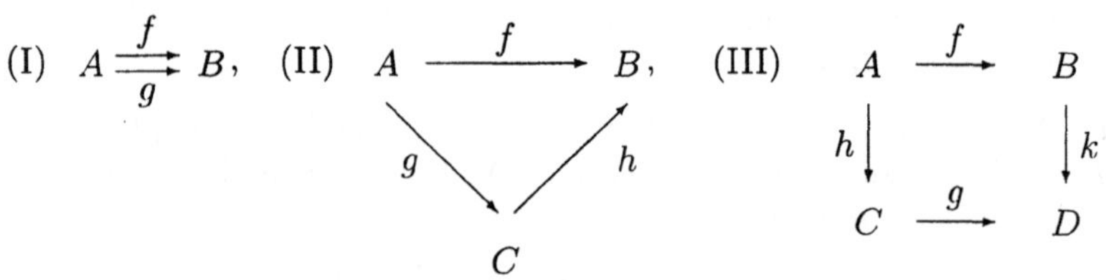

The diagram (I) is called commutative iff $f = g$.
The diagram (II) is called commutative iff $h \circ g = f$.
The diagram (III) is called commutative iff $g \circ h = k \circ f$.

A diagram being composed of the basic types (I), (II) and (III) is called *commutative* provided that each basic type is commutative. Sometimes, we say "a diagram commutes" instead of "a diagram is commutative".

0.1.4 Concepts related to maps

① Let X and Y be sets:
a) We say that X is *equipotent* to Y, and we write $X \sim Y$ iff there is a bijective map $f : X \longrightarrow Y$.
b) If $\mathbb{N}$ denotes the set of positive integers, then X is called *finite* provided that there is some $m \in \mathbb{N}$ such that $X \sim \{n \in \mathbb{N} : n \leq m\}$, or $X = \emptyset$.
c) X is called *infinite* provided that X is not finite.
d) X is *countably infinite* iff X and $\mathbb{N}$ are equipotent.
e) X is *countable* iff it is either finite or countably infinite.

② A *family* $(x_i)_{i \in I}$ of elements of a set X is a map $f : I \longrightarrow X$ sending $i \in I$ to $f(i) = x_i$. I is called the *index set* of the family $(x_i)_{i \in I}$; we say also that the family $(x_i)_{i \in I}$ is indexed by the set I. If $I = \mathbb{N}$, the family $(x_i)_{i \in I}$ is called a *sequence* in X; we write $(x_n)_{n \in \mathbb{N}}$ or (x_n).

Remark. Every set $\mathcal{N}$ of subsets of a set X may be regarded as a family of elements of $\mathcal{P}(X)$; namely, choose $I = \mathcal{N}$ and $f = 1_I$, where $1_I : I \longrightarrow I$ is defined by $1_I(i) = i$ for each $i \in I$ and is called the *identity map*.

0.1.5 Equivalence relations

① 1) Let X be a set and R a relation on X. Then R is
a) *reflexive* iff $x\,R\,x$ for each $x \in X$,
b) *symmetric* iff $x\,R\,y$ implies $y\,R\,x$ for all $x, y \in X$,
c) *transitive* iff $x\,R\,y$ and $y\,R\,z$ imply $x\,R\,z$ for all $x, y, z \in X$.
2) An *equivalence relation* on X is a reflexive, symmetric and transitive relation on X.

② Let R be an equivalence relation on a set X:
a) The *equivalence class* of $x \in X$ is the set $[x]_R = \{y \in X : x\,R\,y\}$. Each $y \in [x]_R$ is called a *representative* of this class.
b) If $x, y \in X$, then $[x]_R = [y]_R$, i.e. x is R-related to y, or $[x]_R \cap [y]_R = \emptyset$, i.e. x is not R-related to y.
c) The set X/R whose elements are the equivalence classes $[x]_R$ for each $x \in X$ is called the *quotient set* of X by R.
d) The map $\omega : X \longrightarrow X/R$ defined by $\omega(x) = [x]_R$ for each $x \in X$ is called the *natural map*; it is surjective.

③ If X is a set, then a subset $\mathcal{M}$ of $\mathcal{P}(X)$ is called a *decomposition* (*partition*) of X provided that the following are satisfied:
1) If $M, N \in \mathcal{M}$, then $M = N$ or $M \cap N = \emptyset$, i.e. $\mathcal{M}$ is *pairwise disjoint*,
2) $\bigcup_{M \in \mathcal{M}} M = X$, i.e. $\mathcal{M}$ is a *cover* of X.

Remarks. 1) The quotient set X/R of X by R (cf. ② c)) is a partition of X.
2) Every partition $\mathcal{M}$ of a set X defines an equivalence relation R on X as follows:

$$x\,R\,y \quad \text{iff} \quad \{x, y\} \subset M \quad \text{for some} \quad M \in \mathcal{M}.$$

3) For $A \subset X$, let R_A be the equivalence relation $(A \times A) \cup \{(x, x) : x \in X\}$. The quotient set X/R_A is *the set X with A identified to a point* and is written X/A.

④ Let $f : X \longrightarrow Y$ be a map. An equivalence relation π_f on X is defined by $x\,\pi_f\,y$ iff $f(x) = f(y)$ for all $x, y \in X$. Furthermore, a map $s : X/\pi_f \longrightarrow Y$ is defined by the property that it makes the diagram

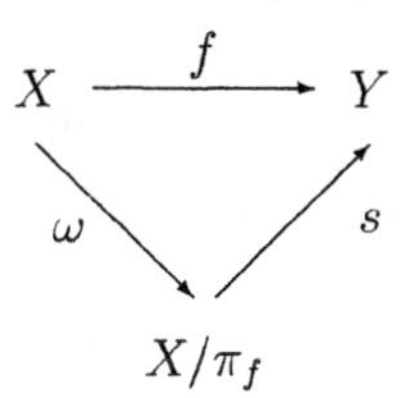

commutative (i.e. $s(\omega(x)) = f(x)$ for each $x \in X$), where $\omega : X \longrightarrow X/\pi_f$ denotes the natural map. The map s is injective, and if f is surjective, it is bijective (*Set theoretical analogue of the homomorphism theorem in Group Theory*).

⑤ Let R_X be an equivalence on a set X, R_Y an equivalence relation on a set Y, and $f : X \longrightarrow Y$ a map. If $\omega_X : X \longrightarrow X/R_X$ and $\omega_Y : Y \longrightarrow Y/R_Y$ are the natural maps, then a map f^*, induced by f, exists which is defined by the property that it makes the diagram

$$\begin{array}{ccc} X & \xrightarrow{f} & Y \\ {\scriptstyle\omega_X}\big\downarrow & & \big\downarrow{\scriptstyle\omega_Y} \\ X/R_X & \xrightarrow{f^*} & Y/R_Y \end{array}$$

commutative provided that for each $x \in X$,

$$f[\omega_X(x)] \subset \omega_Y(f(x)).$$

⑥ Just as an equivalence relation on a set is defined, one can define an equivalence relation on a class, e.g. if $\mathcal{U}$ denotes the class of all sets, then $\sim$ as defined under 0.1.4. ① a) is an equivalence relation on $\mathcal{U}$. For each set X, the equivalence class of X with respect to $\sim$ is called the *cardinal number*, or the *cardinality*, of X, denoted by $|X|$ (or card (X)). Thus, for all $X, Y \in \mathcal{U}$, $|X| = |Y|$ iff X is equipotent to Y. The cardinality of $\mathbb{N}$ is denoted by $\aleph_0$ (aleph–zero), whereas the cardinality of a finite set is denoted by the number of its elements, e.g. card $(\emptyset) = 0$ and card $(\{\emptyset\}) = 1$.

0.1.6 Order relations

① A relation R on a set X is *antisymmetric* iff $x\,R\,y$ and $y\,R\,x$ imply $x = y$ for all $x, y \in X$. A *partial order* on a set is a reflexive, antisymmetric and transitive relation on it. A *partially ordered set* is a pair $(X, \prec)$, where X is a set and $\prec$ a partial order on X. If $(X, \prec)$ is a partially ordered set such that for any $x, y \in X$, $x \prec y$ or $y \prec x$, then $(X, \prec)$ is called a *totally ordered* (*linearly ordered*) *set* or a *chain*.

Examples. 1) If $\leq$ is the usual relation on the set $\mathbb{R}$ of real numbers, then $(\mathbb{R}, \leq)$ is a totally odered set.
2) $(\mathcal{P}(X), \subset)$ is a partially ordered set.

② Let $(X, \prec)$ and $(X', \prec')$ be partially ordered sets. Then a map $f : X \longrightarrow X'$ is called *isotone* (*monotone, order–preserving*) iff $f(x) \prec' f(x')$ whenever $x \prec x'$.

③ Let $(X, \prec)$ be a partially ordered set and $A \subset X$:
1) a) $x \in X$ is an *upper bound* of A iff for each $y \in A$, $y \prec x$.
   b) $x \in X$ is a *lower bound* of A iff for each $y \in A$, $x \prec y$.
2) a) $s \in X$ is a *least upper bound* or *supremum* of A, denoted by $s = \text{lub}\, A$ or $s = \sup A$, iff it is an upper bound of A and for each upper bound x of A, $s \prec x$.

b) $i \in X$ is a *greatest lower bound* or *infimum* of A, denoted by $i = \text{glb } A$ or $i = \inf A$, iff it is a lower bound of A and for each lower bound x of A, $x \prec i$.
3) a) An element m of X is called *maximal* provided that $x \in X$ and $m \prec x$ imply $x = m$.
b) An element m of X is called *minimal* provided that $x \in X$ and $x \prec m$ imply $x = m$.

Remark. If $(X, \prec)$ is a totally ordered set, there is at most one maximal element and at most one minimal element. The maximal element, if it exists, is called the *smallest element*, whereas the minimal element, if it exists, is called the *largest element.*

④ A totally ordered set $(X, \prec)$ is *well-ordered* iff each non-empty subset A of X has a smallest element; e.g. $(\mathbb{N}, \leq)$ is well-ordered, whereas $(\mathbb{R}, \leq)$ is not well-ordered.
If $(X, \prec)$ is a well-ordered set, then the *principle of transfinite induction* is valid:

For each $x \in X$, let $S(x)$ be a statement that is true or false. If
a) $S(x_1)$ is true, where x_1 denotes the smallest element of X,
b) $S(y)$ is true for all $y \prec x$, $y \neq x$, implies $S(x)$ is true,
then $S(x)$ is true for each $x \in X$.

⑤ The following are equivalent (cf. [60]):
1) (*Zermelo's Theorem*): Every set can be well-ordered.
2) (*Axiom of choice*): For each surjective map $f : X \longrightarrow Y$ between sets, there is a map $g : Y \longrightarrow X$ such that $f \circ g = 1_Y$.
3) (*Zorn's Lemma*): If each chain in a non-empty partially ordered set $(X, \prec)$ has an upper bound (resp. a lower bound), then there is a maximal (resp. minimal) element of X. Hence, for each $x \in X$, there is even a maximal (resp. minimal) element $m \in X$ such that $x \prec m$ (resp. $x \succ m$).

0.1.7 Some modifications and supplements

① $\{\cdots : \cdots\}$ or $\{\cdots | \cdots\}$ denotes the class of all $\cdots$ such that $\cdots$, whereas $\in$ stands for membership and is read 'is an element of' or 'belongs to'.

② In order to form classes (or sets) of cardinal numbers, it is useful to call the equivalence classes, introduced under 0.1.5. ⑥, *equipotence classes* instead of cardinal numbers. Then the *cardinal number* $|X|$, or card (X), of a set X is a suitable representative of the equipotence class of X, which can be uniquely determined by means of the theory of ordinal numbers (cf. [38; chapter II, 7.5]).[2] The cardinal number of a set X is called less than or equal to the cardinal number of a set Y, denoted by $|X| \leq |Y|$, iff there is an injective map $f : X \longrightarrow Y$. We write $|X| < |Y|$ iff $|X| \leq |Y|$ and $|X| \neq |Y|$; it is also usual to write $|X| > |Y|$ iff $|Y| < |X|$. Thus, $|X| > \aleph_0 = |\mathbb{N}|$ means that the set X is not countable. Since obviously, order relations on classes can be defined analogously to order relations

[2] There is no misunderstanding, if the reader has an arbitrary representative in mind.

on sets, one obtains that $\leq$ is a partial order on the class of all cardinal numbers. In order to prove that this class is well–ordered by $\leq$, one needs the theory of ordinal numbers, which is not developed in this book (cf. eg. [38]).

③ *Maps between classes* can be defined analogously to maps between sets. If I is a class, then *a family* $(X_i)_{i\in I}$ of sets is a map $X : I \longrightarrow \mathcal{U}$ sending $i \in I$ to $X(i) = X_i$. We say also that the family $(X_i)_{i\in I}$ is *indexed* by I. If I is a set, then the following sets can be formed:
1) the *union* $\bigcup_{i\in I} X_i = \{x : x \in X_i \text{ for some } i \in I\}$,
2) the *intersection* [3] $\bigcap_{i\in I} X_i = \{x : x \in X_i \text{ for each } i \in I\}$, provided that $I \neq \emptyset$,
3) the *disjoint union* $\dot\bigcup_{i\in I} X_i = \bigcup_{i\in I} X_i \times \{i\}$,
4) the *cartesian product* $\prod_{i\in I} X_i = \{x : I \longrightarrow \bigcup_{i\in I} X_i \mid x(i) \in X_i \text{ for each } i \in I\}$, where we write x_i instead of $x(i)$ and also $(x_i)_{i\in I}$, or shortly (x_i), instead of x provided that $x \in \prod_{i\in I} X_i$. Furthermore, if X is a set and $X_i = X$ for each $i \in I$, $\prod_{i\in I} X_i$ is denoted by X^I, and in case $I = \{1, \ldots, n\}$ by X^n.

④ 1) For each $i \in I$, the map $p_i : \prod_{i\in I} X_i \longrightarrow X_i$, defined by $p_i(x) = x(i)$ for each $x \in \prod_{i\in I} X_i$, is called the i–th *projection*. If $\prod_{i\in I} X_i \neq \emptyset$, then for each $i \in I$, the i–th projection is surjective.

2) The following are equivalent:
a) If $(X_i)_{i\in I}$ is a family of non–empty sets, i.e. $X_i \neq \emptyset$ for each $i \in I$, then $\prod_{i\in I} X_i \neq \emptyset$.
b) If $(X_i)_{i\in I}$ is a family of non–empty sets, then there is a map $f : I \longrightarrow \bigcup_{i\in I} X_i$ such that $f(i) \in X_i$ for each $i \in I$.
c) If $(X_i)_{i\in I}$ is a family of non–empty sets, which is *pairwise disjoint* (i.e. $X_i \cap X_j = \emptyset$, whenever $i, j \in I$ and $i \neq j$), then there is a map $f : I \longrightarrow \bigcup_{i\in I} X_i$ such that $f(i) \in X_i$ for each $i \in I$.
d) The axiom of choice.

3) In this book we will use sometimes the axiom of choice, e.g. if we want to apply Zorn's lemma or 2) a).

⑤ In the realm of the Gödel–Bernays–von Neumann set theory, mentioned under 0.1.1., it is postulated that *the elements of classes are sets.*

0.2 Topological structures

The proofs of the statements in this section can be found in any book dealing with topological spaces (cf. e.g. [43], [67], [108] or [149]).

0.2.1 Metric and pseudometric spaces

0.2.1.1 Definitions. 1) A *metric space* is a pair (X, d), where X is a set and $d : X \times X \longrightarrow \mathbb{R}$ a map satisfying, for $x, y, z \in X$:
M_1) α) $d(x, x) = 0$; β) $d(x, y) = 0$ implies $x = y$,

[3] If $I = \emptyset$, $\bigcap_{i\in I} X_i = \mathcal{U}$ is not a set, since $R \subset \mathcal{U}$ is not a set.

M_2) $d(x,y) = d(y,x)$,
M_3) $d(x,y) + d(y,z) \geq d(x,z)$ (*triangle inequality*).

2) A *pseudometric space* is a pair (X,d), where X is a set and $d : X \times X \longrightarrow \mathbb{R}$ a map satisfying M_1) α), M_2) and M_3) for $x, y, z \in X$.

3) If (X,d) is a (pseudo)metric space, then d is called a *(pseudo)metric* on X, and for each $(x,y) \in X \times X$, $d(x,y)$ is called the *distance* between x and y.

0.2.1.2 Remark. If we substitute z by x in M_3), it follows by means of M_1) α) and M_2), $d(x,y) \geq 0$, i.e. the distance between x and y is non-negative.

0.2.1.3 Examples of metric spaces. 1) Let X be a set. Define $d : X \times X \longrightarrow \mathbb{R}$ by $d(x,y) = 1$ for $x \neq y$ and $d(x,x) = 0$.

2) Let $X = \mathbb{R}^n$. Define $d : X \times X \longrightarrow \mathbb{R}$ by

$$d((x_1, \cdots, x_n), (y_1, \cdots, y_n)) = \sqrt{\sum_{i=1}^{n} (x_i - y_i)^2}$$

(Euclidean metric or usual metric). Then $(\mathbb{R}^n, d)$ is called *Euclidean n–space.*

3) Let X be the set of all sequences $(x_n)_{n \in \mathbb{N}}$ of real numbers such that the series $\sum_{n=1}^{\infty} x_n^2$ is convergent. Define $d : X \times X \longrightarrow \mathbb{R}$ by $d((x_n),(y_n)) = \sqrt{\sum_{n=1}^{\infty} (x_n - y_n)^2}$. Then (X,d) is called *Hilbert space.*

4) Let X be the set of all continuous maps from the closed unit interval $[0,1]$ into $\mathbb{R}$. Define $d : X \times X \longrightarrow \mathbb{R}$ by

$$d(f,g) = \max\{|f(x) - g(x)| : x \in [0,1]\}.$$

5) Let X be as under 4). Define $d : X \times X \longrightarrow \mathbb{R}$ by

$$d(f,g) = \left[\int_0^1 (f(t) - g(t))^2 dt\right]^{\frac{1}{2}}.$$

0.2.1.4 Remark. If X is the set of all Riemann integrable functions from $[0,1]$ to $\mathbb{R}$ and $d : X \times X \longrightarrow \mathbb{R}$ is defined as under 0.2.1.3. 5), then (X,d) is *a pseudometric space which is not a metric space.*

0.2.1.5 Definitions. Let (X,d) be a (pseudo)metric space.

1) For each $x \in X$ and each real number $\varepsilon > 0$, the *ε–sphere* about x is the set $V_\varepsilon(x) = \{y \in X : d(x,y) > \varepsilon\}$.

2) $O \subset X$ is *open* iff for each $x \in O$, there is some $\varepsilon > 0$ such that $V_\varepsilon(x) \subset O$.

0.2.1.6 Remark. If (X,d) is a (pseudo)metric space, then for each $x \in X$, each ε–sphere about x is open.

0.2.1.7 Proposition. *Let (X,d) be a (pseudo)metric space. Then the set $\mathcal{X}_d$ of all open subsets of X has the following properties:*

1) $\emptyset \in \mathcal{X}_d$; $X \in \mathcal{X}_d$,

2) $O_1, O_2 \in \mathcal{X}_d$ *imply* $O_1 \cap O_2 \in \mathcal{X}_d$,
3) $\mathcal{U} \subset \mathcal{X}_d$ *implies* $\bigcup_{U \in \mathcal{U}} U \in \mathcal{X}_d$.

0.2.1.8 Remark. Arbitrary intersections of open subsets in a (pseudo)metric space need not be open: Consider the intervals $(-\frac{1}{n}, \frac{1}{n})$ for all $n \in I\!N$ in the Euclidean 1–space, i.e. the $\frac{1}{n}$–spheres about $0 \in I\!R$. Then $\bigcap_{n \in I\!N}(-\frac{1}{n}, \frac{1}{n}) = \{0\}$ is not open.

0.2.2 Topological spaces and continuous maps

0.2.2.1 Definition. A *topological space* is a pair $(X, \mathcal{X})$, where X is a set and $\mathcal{X}$ a subset of $\mathcal{P}(X)$ such that the following are satisfied:
1) $\emptyset \in \mathcal{X}$; $X \in \mathcal{X}$,
2) $O_1 \in \mathcal{X}$ and $O_2 \in \mathcal{X}$ imply $O_1 \cap O_2 \in \mathcal{X}$,
3) $\mathcal{S} \subset \mathcal{X}$ implies $\bigcup_{S \in \mathcal{S}} S \in \mathcal{X}$.

If $(X, \mathcal{X})$ is a topological space, then $\mathcal{X}$ is called a *topology* (or *topological structure*) on X and the elements of $\mathcal{X}$ are called *open sets* (of the topological space $(X, \mathcal{X})$). Sometimes, we write X instead of $(X, \mathcal{X})$ provided that there is no confusion about the topology on X.

0.2.2.2 Examples. 1) If (X, d) is a (pseudo)metric space, then $(X, \mathcal{X}_d)$ is a topological space (cf. 0.2.1.7.); $\mathcal{X}_d$ is called *the topology* on X *induced by the (pseudo)metric d*. The topology induced by the Euclidean metric on the set $I\!R$ of real numbers is called the *usual topology*; it is the only topology employed in classical Analysis. In order to emphasize that $I\!R$ carries the usual topology we write also $I\!R_t$.

2) Let X be a set and $\mathcal{D} = \mathcal{P}(X)$. Then $(X, \mathcal{D})$ is a topological space, and $\mathcal{D}$ is called the *discrete* topology on X.

3) Let X be a set and $\mathcal{I} = \{\emptyset, X\}$. Then $(X, \mathcal{I})$ is a topological space, and $\mathcal{I}$ is called the *indiscrete* topology on X.

4) a) Let X be a set with $|X| \geq \aleph_0$. Then the *topology of finite complements* (or *cofinite topology*) $\mathcal{X}_f$ on X is defined by $O \in \mathcal{X}_f$ iff $X \backslash O$ is finite or $O = \emptyset$.
b) Let X be a set with $|X| > \aleph_0$. Then the *topology of countable complements* $\mathcal{X}_c$ on X is defined by $O \in \mathcal{X}_c$ iff $X \backslash O$ is countable or $O = \emptyset$.

0.2.2.3 Definition. If $(X, \mathcal{X})$ is a topological space, then a subset A of X is *closed* iff $X \backslash A \in \mathcal{X}$.

0.2.2.4 Proposition. *Let $(X, \mathcal{X})$ be a topological space and $\mathcal{A} = \{A \subset X : A$ is closed$\}$. Then the following are valid:*
1) $\emptyset \in \mathcal{A}$; $X \in \mathcal{A}$,
2) $A_1 \in \mathcal{A}$ *and* $A_2 \in \mathcal{A}$ *imply* $A_1 \cup A_2 \in \mathcal{A}$,
3) $\mathcal{B} \subset \mathcal{A}$ *implies* $\bigcap_{B \in \mathcal{B}} B \in \mathcal{A}$.

0.2.2.5 Definition. If $(X, \mathcal{X})$ is a topological space and $x \in X$, then each $O \in \mathcal{X}$ with $x \in O$ is called an *open neighborhood* of x; a subset U of X is a *neighborhood* of x iff it contains an open neighborhood of x, and the *neighborhood system* of x is the set of all neighborhoods of x, denoted by $\mathcal{U}_{\mathcal{X}}(x)$ (or shortly: $\mathcal{U}(x)$).

0.2.2.6 Proposition. *A subset S of a topological space X is open iff it is a neighborhood of each of its points.*

0.2.2.7 Proposition. *Let $(X, \mathcal{X})$ be a topological space. Then for each $x \in X$, the neighborhood system $\mathcal{U}(x)$ fulfills the following:*
1) *$\mathcal{U}(x) \neq \emptyset$, and $x \in U$ for each $U \in \mathcal{U}(x)$,*
2) *$V \in \mathcal{U}(x)$ whenever $V \supset U$ for some $U \in \mathcal{U}(x)$,*
3) *$U \in \mathcal{U}(x)$ and $V \in \mathcal{U}(x)$ imply $U \cap V \in \mathcal{U}(x)$,*
4) *If $U \in \mathcal{U}(x)$, then there is some $V \in \mathcal{U}(x)$ such that for each $y \in V$, $U \in \mathcal{U}(y)$.*

0.2.2.8 Definitions. Let $(X, \mathcal{X})$ be a topological space and $A \subset X$. Then

a) $A^0 = \text{int}_{\mathcal{X}} A = \bigcup\{O : O \subset A$ and $O \in \mathcal{X}\}$ is called the *interior* of A. The elements of A^0 are called *interior points* of A.

b) $\overline{A} = cl_{\mathcal{X}} A = \bigcap\{B : B \supset A$ and $X \setminus B \in \mathcal{X}\}$ is called the *closure* of A. The elements of $\overline{A}$ are called *adherent points* of A.

0.2.2.9 Remark. Obviously, if $(X, \mathcal{X})$ is a topological space and $A \subset X$, then A is open (resp. closed) iff $A = A^0$ (resp. $A = \overline{A}$); furthermore, $x \in X$ is an interior point (resp. adherent point) of A iff A contains some neighborhood of x (resp. each neighborhood of x contains some point of A).

0.2.2.10 Proposition. *Let $(X, \mathcal{X})$ be a topological space and $A \subset X$. Then $X \setminus (A^0) = \overline{(X \setminus A)}$ and $X \setminus \overline{A} = (X \setminus A)^0$.*

0.2.2.11 Theorem. *If $(X, \mathcal{X})$ is a topological space and A, B are subsets of X, then the following are satisfied:*

I_1) $X^0 = X$,
I_2) $A^0 \subset A$,
I_3) $A^{00} = A^0$,
I_4) $(A \cap B)^0 = A^0 \cap B^0$.

H_1) $\overline{\emptyset} = \emptyset$,
H_2) $A \subset \overline{A}$,
H_3) $\overline{\overline{A}} = \overline{A}$,
H_4) $\overline{(A \cup B)} = \overline{A} \cup \overline{B}$.

Corollaries.
1) *$A \subset B$ implies $A^0 \subset B^0$*
2) *$A^0 \cup B^0 \subset (A \cup B)^0$*

1′) *$A \subset B$ implies $\overline{A} \subset \overline{B}$*
2′) *$\overline{A \cap B} \subset \overline{A} \cap \overline{B}$*

0.2.2.12 Remarks. 1) In order to obtain alternative descriptions of topological spaces, each of the concepts 'closed set', 'neighborhood', 'interior' and

'closure' can be axiomatized, where the properties 0.2.2.4. 1) – 3), 0.2.2.7. 1) – 4), 0.2.2.11. I_1) – I_4) and 0.2.2.11. H_1) – H_4) respectively can be used as axioms.
2) If $(X, \mathcal{X})$ is a topological space, then a subset $\mathcal{B}$ of $\mathcal{X}$ is a *base* for $\mathcal{X}$ iff each $O \in \mathcal{X}$ is the union of elements of $\mathcal{B}$, whereas $\mathcal{S} \subset \mathcal{X}$ is a *subbase* for $\mathcal{X}$ iff the set of all finite intersections of elements of $\mathcal{S}$ is a base for $\mathcal{X}$. Obviously, every non–empty set $\mathcal{S}$ of subsets of a set X is a subbase for a topology on X. If (X, d) is a (pseudo)metric space, then the topology $\mathcal{X}_d$ induced by d has the set of all ε–spheres about the points of X as a base. The Euclidean n–space $(\mathbb{R}^n, d)$ has even a countable base $\mathcal{B}$:

$$\mathcal{B} = \{V_\varepsilon(x) : x \in \mathbb{Q}^n \text{ and } \varepsilon > 0 \text{ rational}\},$$

where $\mathbb{Q}$ denotes the set of rational numbers. A topological space having a countable base is said to fulfill the *second axiom of countability.*

0.2.2.13 Definition. A *continuous map* $f : (X, \mathcal{X}) \longrightarrow (Y, \mathcal{Y})$ between topological spaces $(X, \mathcal{X})$ and $(Y, \mathcal{Y})$ is a map $f : X \longrightarrow Y$ such that $f^{-1}[O] \in \mathcal{X}$ for each $O \in \mathcal{Y}$.

0.2.2.14 Remark. Evidently, if $(X, \mathcal{X})$ and $(Y, \mathcal{Y})$ are topological spaces, then a map $f : X \longrightarrow Y$ is continuous iff $f^{-1}[O] \in \mathcal{X}$ for each $O \in \mathcal{S}$, where $\mathcal{S}$ is a subbase for $\mathcal{Y}$.

0.2.2.15 Theorem. *Let $(X, \mathcal{X})$ and $(Y, \mathcal{Y})$ be topological spaces and $f : X \longrightarrow Y$ a map. Then the following are equivalent:*
(1) $f : (X, \mathcal{X}) \longrightarrow (Y, \mathcal{Y})$ *is a continuous map,*
(2) $f^{-1}[A]$ *is closed in X for each closed subset A of Y,*
(3) $f^{-1}[A^0] \subset (f^{-1}[A])^0$ *for each subset A of Y,*
(4) $f[\overline{A}] \subset \overline{f[A]}$ *for each subset A of X.*

0.2.2.16 Definition. Let $(X, \mathcal{X})$, $(Y, \mathcal{Y})$ be topological spaces and $x_0 \in X$. Then a map $f : X \longrightarrow Y$ is called *continuous at* x_0 provided that for each $V \in \mathcal{U}(f(x_0))$, there is some $U \in \mathcal{U}(x_0)$ such that $f[U] \subset V$.

0.2.2.17 Proposition. *If $(X, \mathcal{X})$ and $(Y, \mathcal{Y})$ are topological spaces, then a map $f : X \longrightarrow Y$ is continuous iff it is continuous at each $x_0 \in X$.*

0.2.2.18 Examples of continuous maps.
1) All continuous maps considered in Analysis.
2) Constant maps between topological spaces.
3) For each topological space $(X, \mathcal{X})$, the identity map $1_X : X \longrightarrow X$.
4) Each map from a discrete topological space to an arbitrary topological space.
5) Each map from an arbitrary topological space to an indiscrete topological space.

0.2.2.19 Remark. If $\mathcal{X}$ and $\mathcal{X}'$ are topologies on the same set X, then $\mathcal{X}$ is called *finer* than $\mathcal{X}'$ (or $\mathcal{X}'$ *coarser* than $\mathcal{X}$) iff $1_X : (X, \mathcal{X}) \longrightarrow (X, \mathcal{X}')$ is continuous [equivalently: $\mathcal{X}' \subset \mathcal{X}$].

0.2.2.20 Proposition. *Let $(X, \mathcal{X})$, $(Y, \mathcal{Y})$ and $(Z, \mathcal{Z})$ be topological spaces, $f : X \longrightarrow Y$ a map which is continuous at $x_0 \in X$ and $g : Y \longrightarrow Z$ a map which is continuous at $f(x_0)$. Then the composite $g \circ f : X \longrightarrow Z$ is continuous at x_0.*

Corollary. *If $f : (X, \mathcal{X}) \longrightarrow (Y, \mathcal{Y})$ and $g : (Y, \mathcal{Y}) \longrightarrow (Z, \mathcal{Z})$ are continuous maps, then $g \circ f : (X, \mathcal{X}) \longrightarrow (Z, \mathcal{Z})$ is a continuous map.*

0.2.3 Filters and convergence

0.2.3.1 Remark. A sequence $(x_n)_{n \in I\!N}$ of points in a topological space $(X, \mathcal{X})$ is said to converge to $x_0 \in X$, and we write $x_n \longrightarrow x_0$, iff for each neighborhood U of x_0, there is some $n_0 \in I\!N$ such that $x_n \in U$ for all $n \geq n_0$. In contrast to the situation in Analysis, it is not true that a continuous map $f : (X, \mathcal{X}) \longrightarrow (Y, \mathcal{Y})$ between topological spaces $(X, \mathcal{X})$ and $(Y, \mathcal{Y})$ is continuous at $x_0 \in X$ iff for each sequence $(x_n)_{n \in I\!N}$ in $(X, \mathcal{X})$ converging to $x_0 \in X$, the sequence $(f(x_n))_{n \in I\!N}$ converges to $f(x_0)$, as the following *example* shows: Let X be a set with $|X| > \aleph_0$ and $\mathcal{X}_c$ the topology of countable complements on X. Then a sequence $(x_n)_{n \in I\!N}$ in $(X, \mathcal{X}_c)$ converges to $x_0 \in X$ iff there is some $n_0 \in I\!N$ such that $x_n = x_0$ for all $n \geq n_0$. Furthermore, the map $1_X : (X, \mathcal{X}_c) \longrightarrow (X, \mathcal{P}(X))$ is not continuous, but each convergent sequence in $(X, \mathcal{X}_c)$ is convergent in $(X, \mathcal{P}(X))$.
Thus, in the following the concept of sequence will be substituted by the wider concept of filter.

0.2.3.2 Definitions. 1) Let X be a set. A non–empty set $\mathcal{F}$ of subsets of X is called a *filter* on X provided that the following are satisfied:
F_1) $\emptyset \notin \mathcal{F}$,
F_2) $F_1 \in \mathcal{F}$ and $\mathcal{F}_2 \in \mathcal{F}$ imply $F_1 \cap F_2 \in \mathcal{F}$,
F_3) $F \in \mathcal{F}$ and $F \subset F' \subset X$ imply $F' \in \mathcal{F}$.
2) The set of all filters on a set X is denoted by $F(X)$.

0.2.3.3 Examples. 1) Let X be a non–empty set:
a) $\mathcal{F} = \{X\}$ is a filter on X.
b) If $A \subset X$ is non–empty, then $\mathcal{F} = \{U \subset X : A \subset U\}$ is a filter on X; it is denoted by (A). If $A = \{x\}$, we write $\dot{x}$ instead of $(\{x\})$.

2) Let $(X, \mathcal{X})$ be a topological space, $x \in X$ and A a non–empty subset of X:
a) The neighborhood system $\mathcal{U}(x)$ is a filter on X, which is sometimes called the *neighborhood filter of* x.
b) $\mathcal{U}(A) = \{U \subset X : A \subset O \subset U$ for some $O \in \mathcal{X}\}$ is filter on X, called the *neighborhood filter of* A; its elements are called *neighborhoods of* A.

3) Let $(x_n)_{n \in I\!N}$ be a sequence in a set X. Then $\mathcal{F}_e = \{U \subset X : x_n \in U$ for all but finitely many $n \in I\!N\}$ is a filter on X, called the *elementary filter of the sequence* $(x_n)_{n \in I\!N}$.

4) Let X be a set with $|X| \geq \aleph_0$ and k a cardinal number with $\aleph_0 \leq k \leq |X|$.

Then $\mathcal{F} = \{U \subset X : |X \backslash U| < k\}$ is a filter on X. If $k = \aleph_0$, this filter is called the *filter of finite complements*.

0.2.3.4 Definition. Let X be a set. A non–empty set $\mathcal{B}$ of subsets of X is called a *filter base* on X provided that the following are satisfied:
FB_1) $\emptyset \notin \mathcal{B}$,
FB_2) If $B_1 \in \mathcal{B}$ and $B_2 \in \mathcal{B}$, then there is some $B_3 \in \mathcal{B}$ such that $B_3 \subset B_1 \cap B_2$.

0.2.3.5 Examples. 1) Let A be a non–empty subset of a set X. Then $\mathcal{B} = \{A\}$ is a filter base on X.
2) Let $(X, \mathcal{X})$ be a topological space, $x \in X$ and A a non–empty subset of X:
a) $\overset{\circ}{\mathcal{U}}(x) = \{O \in \mathcal{X} : x \in O\}$ is a filter base on X.
b) $\overset{\circ}{\mathcal{U}}(A) = \{O \in \mathcal{X} : A \subset O\}$ is a filter base on X.
3) Let $(x_n)_{n \in \mathbb{N}}$ be a sequence in a set X and $E_m = \{x_n : n \geq m\}$ for each $m \in \mathbb{N}$. Then $\mathcal{B} = \{E_m : m \in \mathbb{N}\}$ is a filter base on X.

0.2.3.6 Remarks. 1) If $\mathcal{B}$ is a filter base on a set X, then $(\mathcal{B}) = \{U \subset X : B \subset U$ for some $B \in \mathcal{B}\}$ is a filter on X, called *the filter generated by the filter base* $\mathcal{B}$. Evidently, the filters under 0.2.3.3. 1), 2) and 3) are generated by the filter bases under 0.2.3.5. 1), 2) and 3) respectively.
2) A filter can be generated by different filter bases, e.g. if $(X, \mathcal{X})$ is a *(pseudo)metrizable topological space*, i.e. a topological space whose topology is induced by a (pseudo)metric, then for each $x \in X$, the neighborhood filter $\mathcal{U}(x)$ is generated by the following filter bases:
a) $\mathcal{B}_0 = \overset{\circ}{\mathcal{U}}(x)$,
b) $\mathcal{B}_1 = \{V_\varepsilon(x) : \varepsilon > 0$ is a real number$\}$,
c) $\mathcal{B}_2 = \{V_{\frac{1}{n}}(x) : n \in \mathbb{N}\}$.

Obviously, $\mathcal{B}_2$ is countable. By the way, a filter base generating the neighborhood filter at a point x in a topological space $(X, \mathcal{X})$ is often called a *neighborhood base* at x. A topological space $(X, \mathcal{X})$ is said to fulfill the *first axiom of countability* iff for each $x \in X$, there is a countable neighborhood base at x. Using this terminology, *every (pseudo)metrizable topological space fulfills the first axiom of countability*. Necessary and sufficient conditions for the metrizability of a topological space have been found in the midst of the 20th century by R.H. Bing, J. Nagata and Ju.M. Smirnov (cf. e.g. [43;4.4.7 and 4.4.8]).

0.2.3.7 Definitions. 1) If $\mathcal{F}_1$ and $\mathcal{F}_2$ are filters on a set X, then $\mathcal{F}_2$ is *finer* than $\mathcal{F}_1$ (or $\mathcal{F}_1$ is *coarser* than $\mathcal{F}_2$) iff $\mathcal{F}_1 \subset \mathcal{F}_2$.
2) Let $(X, \mathcal{X})$ be a topological space and $x \in X$. Then a filter $\mathcal{F}$ on X *converges to* x, denoted by $\mathcal{F} \longrightarrow x$, iff $\mathcal{F}$ is finer than the neighborhood filter of x, i.e. $\mathcal{F} \supset \mathcal{U}(x)$. If $\mathcal{F} \longrightarrow x$, then x is called a *limit* of $\mathcal{F}$.
3) A sequence $(x_n)_{n \in \mathbb{N}}$ of points in a topological space $(X, \mathcal{X})$ *converges* to $x \in X$ iff the elementary filter $\mathcal{F}_e$ of $(x_n)_{n \in \mathbb{N}}$ converges to x; in this case we write $x_n \longrightarrow x$ and call x a *limit* of the sequence $(x_n)_{n \in \mathbb{N}}$.

0.2.3.8 Remarks. 1) The above definition of the convergence of a sequence is equivalent to the corresponding definition given under 0.2.3.1. In particular, if we consider the set $I\!R$ of real numbers, endowed with the usual topology, we obtain the well-known definition of convergence for a sequence of real numbers.
2) a) If X is an indiscrete topological space, then every filter on X converges to every point of X.
b) If X is a discrete topological space, then a filter $\mathcal{F}$ on X converges to $x \in X$ iff $\mathcal{F} = \dot{x}$.
3) a) If X is a topological space and $x \in X$, then a filter $\mathcal{F}$ on X converges to x whenever there is some filter $\mathcal{G}$ on X converging to x such that $\mathcal{G} \subset \mathcal{F}$.
b) Let $(x_n)_{n \in I\!N}$ be a sequence on a set X and $(j_n)_{n \in I\!N}$ a sequence in $I\!N$ such that $j_n < j_{n+1}$ for all $n \in I\!N$. Then $(x_{j_n})_{n \in I\!N}$ is called a *subsequence* of $(x_n)_{n I\!N}$ (or more exactly: an infinite subsequence). Evidently, the elementary filter of $(x_{j_n})_{n \in I\!N}$ is finer than the elementary filter of $(x_n)_{n \in I\!N}$, Thus, it follows from a) that each subsequence of a convergent sequence is convergent.

0.2.3.9 Definition. Let $(X, \mathcal{X})$ be a topological space and $x \in X$:
1) If $\mathcal{F}$ is a filter on X, then x is called an *adherence point* (or accumulation point) of $\mathcal{F}$ iff it is an adherent point of each $F \in \mathcal{F}$.
2) If $(x_n)_{n \in I\!N}$ is a sequence in X, then x is called an *accumulation point* of $(x_n)_{n \in I\!N}$ iff it is an adherence point of the elementary filter of $(x_n)_{n \in I\!N}$.

0.2.3.10 Proposition. *Let $(x_n)_{n \in I\!N}$ be a sequence in a topological space $(X, \mathcal{X})$ and $x \in X$. Then $(x_n)_{n \in I\!N}$ has x as an accumulation point iff for each neighborhood $U \in \mathcal{U}(x)$ there are infinitely many $i \in I\!N$ such that $x_i \in U$.*

0.2.3.11 Proposition. 1) *A filter $\mathcal{F}$ on a topological space $(X, \mathcal{X})$ has a point $x \in X$ as an adherence point iff there is a filter $\mathcal{G} \in F(X)$ finer than $\mathcal{F}$ which converges to x.*
2) *If $(X, \mathcal{X})$ is a topological space which fulfills the first axiom of countability, then a sequence $(x_n)_{n \in I\!N}$ in X has a point $x \in X$ as an accumulation point iff there is a subsequence of $(x_n)_{n \in I\!N}$ converging to x.*

0.2.3.12 Definition. Let $\mathcal{F}$ be a filter on a set X and $f : X \longrightarrow Y$ a map from X into a set Y. Then the filter $f(\mathcal{F})$ generated by the filter base $\{f[F] : F \in \mathcal{F}\}$ is called the *image of $\mathcal{F}$ under f*.

0.2.3.13 Proposition. 1) *A map $f : (X, \mathcal{X}) \longrightarrow (Y, \mathcal{Y})$ is continuous at $x_0 \in X$ iff for each filter $\mathcal{F} \in F(X)$ converging to x_0, the filter $f(\mathcal{F})$ converges to $f(x_0)$.*
2) *A map $f : (X, \mathcal{X}) \longrightarrow (Y, \mathcal{Y})$ between a topological space $(X, \mathcal{X})$ fulfilling the first axiom of countability and an arbitrary topological space $(Y, \mathcal{Y})$ is continuous at $x_0 \in X$ iff for each sequence $(x_n)_{n \in I\!N}$ in X converging to x_0, the sequence $(f(x_n))_{n \in I\!N}$ converges to $f(x_0)$.*

0.2.3.14 Definition. Let $\mathcal{F}$ be a filter on a set Y and $f : X \longrightarrow Y$ a map from a set X into Y. If $\mathcal{B} = \{f^{-1}[F] : F \in \mathcal{F}\}$ is a filter base on X, then the filter $(\mathcal{B})$, denoted by $f^{-1}(\mathcal{F})$, is called the *inverse image of $\mathcal{F}$ under f*.

0.2.3.15 Remarks. 1) The inverse image of a filter $\mathcal{F}$ under a map f exists iff $f^{-1}[F] \neq \emptyset$ for each $F \in \mathcal{F}$. If $\mathcal{F}$ is a filter on a set X, A a subset of X and $i : A \longrightarrow X$ the inclusion map, then the inverse image of $\mathcal{F}$ under i, if it exists (i.e. $A \cap F \neq \emptyset$ for each $F \in \mathcal{F}$), is called the *trace* of $\mathcal{F}$ on A, denoted by $\mathcal{F}_A$. Obviously, $\mathcal{F}_A = \{F \cap A : F \in \mathcal{F}\}$. Furthermore, $A \in \mathcal{F}$ implies $i(\mathcal{F}_A) = \mathcal{F}$.
2) If $f : X \longrightarrow Y$ is a map, $\mathcal{F}$ a filter on X and $\mathcal{G}$ a filter on Y, then it follows from 0.1.3. ⑧ b) and c) that the following are satisfied:

a)	α) $f^{-1}(f(\mathcal{F})) \subset \mathcal{F}$, β) $f^{-1}(f(\mathcal{F})) = \mathcal{F}$ provided that f is injective.
b)	If $f^{-1}(\mathcal{G})$ exists, then α) $f(f^{-1}(\mathcal{G})) \supset \mathcal{G}$, β) $f(f^{-1}(\mathcal{G})) = \mathcal{G}$ provided that f is surjective.

0.2.3.16 Proposition. *Let $(X, \mathcal{X})$ be a topological space, A a subset of X and $x \in X$. Then the following are equivalent:*
(1) $x \in \overline{A}$.
(2) *$\mathcal{U}(x)$ has a trace on A.*
(3) *There is a filter $\mathcal{F}$ on X such that $\mathcal{F} \longrightarrow x$ and $A \in \mathcal{F}$.*

0.2.3.17 Remark. If $(X, \mathcal{X})$ is a topological space fulfilling the first axiom of countability, then a point $x \in X$ is an adherent point of a subset A of X iff there exists a sequence of points of A converging to x.

0.2.3.18 Proposition. *Let $(\mathcal{F}_i)_{i \in I}$ be a family of filters on a non-empty set X. Then $\mathcal{M} = \{\mathcal{F}_i : i \in I\}$ has a supremum in the partially ordered set $(F(X), \subset)$, i.e. there is a coarsest filter on X which finer than each filter belonging to $\mathcal{M}$, iff the following condition is satisfied: If $\mathcal{F}_{i_1}, \cdots, \mathcal{F}_{i_n}$ are finitely many elements of $\mathcal{M}$ and $F_{i_j} \in \mathcal{F}_{i_j}$ for each $j \in J = \{1, \ldots, n\}$, then $\bigcap_{j \in J} F_{i_j} \neq \emptyset$.
If sup $\mathcal{M}$ exists, then sup $\mathcal{M}$ = $\{\bigcap_{k \in K} F_k$: K is any finite subset of I, and $F_k \in \mathcal{F}_k$ for each $k \in K\}$.*

0.2.3.19 Remark. If $(\mathcal{F}_i)_{i \in I}$ is a non-empty family of filters on a set X, then $\mathcal{F} = \bigcap_{i \in I} \mathcal{F}_i$ is a filter on X. Evidently, $\mathcal{F}$ is the *infimum* of the set $\mathcal{M} = \{\mathcal{F}_i : i \in I\}$ in the partially ordered set $(F(X), \subset)$, i.e. the finest filter on X which is coarser than each $\mathcal{F}_i$, and $\mathcal{F} = \inf \mathcal{M} = \{\bigcup_{i \in I} F_i : F_i \in \mathcal{F}_i$ for each $i \in I\}$.

0.2.3.20 Definition. A filter $\mathcal{U}$ on a set X is called an *ultrafilter* provided that for each filter $\mathcal{F}$ on X, $\mathcal{F} \supset \mathcal{U}$ implies $\mathcal{F} = \mathcal{U}$, i.e. iff it is a maximal element in the partially ordered set $(F(X), \subset)$.

0.2.3.21. Zorn's lemma is needed for the proof of the following

Theorem. *For each filter $\mathcal{F}$ on a set X, there is an ultrafilter $\mathcal{U}$ on X such that $\mathcal{F} \subset \mathcal{U}$.*

Corollary. *Let $\mathcal{F}$ be a filter on a set X. Then $\mathcal{F} = \bigcap\{\mathcal{U} \in F(X) : \mathcal{U}$ is an ultrafilter and $\mathcal{U} \supset \mathcal{F}\}$.*

0.2.3.22 Remark. The above theorem is an existence theorem for ultrafilters. For each set X, the only explicitly constructed ultrafilters on X are the filters $\dot{x}$ for each $x \in X$.

0.2.3.23 Proposition. *Let $\mathcal{U}$ be a filter on a set X. Then the following are equivalent:*
(1) *$\mathcal{U}$ is an ultrafilter,*
(2) *$A \cup B \in \mathcal{U}$ implies $A \in \mathcal{U}$ or $B \subset \mathcal{U}$,*
(3) *If A is a subset of X, then either $A \in \mathcal{U}$ or $X \setminus A \in \mathcal{U}$,*
(4) *$A \cap F \neq \emptyset$, for all $F \in \mathcal{U}$, implies $A \in \mathcal{U}$.*

0.2.3.24 Proposition. *Let $(X, \mathcal{X})$ be a topological space, $\mathcal{U}$ an ultrafilter on X, and $x \in X$. Then $\mathcal{U}$ converges to x iff x is an adherence point of $\mathcal{U}$.*

0.2.3.25 Proposition. *Let $f : X \longrightarrow Y$ be a map and $\mathcal{U}$ an ultrafilter on X. Then $f(\mathcal{U})$ is an ultrafilter on Y.*

0.2.3.26 Proposition. *Let $f : X \longrightarrow Y$ be a map, $\mathcal{F}$ a filter on X and $\mathcal{V}$ an ultrafilter on Y such that $\mathcal{V} \supset f(\mathcal{F})$. Then there exists an ultrafilter $\mathcal{U}$ on X such that $f(\mathcal{U}) = \mathcal{V}$.*

0.2.4 Initial and final topological structures

0.2.4.1 Proposition. *Let $((X_i, \mathcal{X}_i))_{i \in I}$ be a family of topological spaces, indexed by some set (or class) I, X a set and*

a) $(f_i : X \longrightarrow X_i)_{i \in I}$ | b) $(f_i : X_i \longrightarrow X)_{i \in I}$

a family of maps. Then

a) *$\mathcal{S} = \{f_i^{-1}[O_i] : i \in I,\ O_i \in \mathcal{X}_i\}$ is a subbase for the coarsest topology $\mathcal{X}$ on X making all f_i continuous. It is the* **initial topology** *with respect to the family $(f_i)_{i \in I}$, i.e. for any topological space $(Y, \mathcal{Y})$, a map $g : (Y, \mathcal{Y}) \longrightarrow (X, \mathcal{X})$ is continuous iff for every $i \in I$, $f_i \circ g : (Y, \mathcal{Y}) \longrightarrow (X_i, \mathcal{X}_i)$ is continuous.*

b) *$\mathcal{X} = \{O \subset X : f_i^{-1}[O] \in \mathcal{X}_i$ for each $i \in I\}$ is the finest topology on X making all f_i continuous. It is the* **final topology** *with respect to the family $(f_i)_{i \in I}$, i.e. for any topological space $(Y, \mathcal{Y})$, a map $g : (X, \mathcal{X}) \longrightarrow (Y, \mathcal{Y})$ is continuous iff for every $i \in I$, $g \circ f_i : (X_i, \mathcal{X}_i) \longrightarrow (Y, \mathcal{Y})$ is continuous.*

0.2.4.2 Examples. 1) Let $(X, \mathcal{X})$ and $(Y, \mathcal{Y})$ be topological spaces. Then

a) $(Y, \mathcal{Y})$ is called a *subspace* of $(X, \mathcal{X})$ provided that there is an *embedding* $f : (Y, \mathcal{Y}) \longrightarrow (X, \mathcal{X})$, i.e. an injective map $f : X \longrightarrow Y$ such that $\mathcal{Y}$ is the initial topology on A w.r.t. f. In this case, $\mathcal{Y} = \{f^{-1}[O] : O \in \mathcal{X}\}$ is called the *subspace topology* w.r.t. f.

If $A \subset X$, the subspace topology on A w.r.t. the inclusion map $i : A \longrightarrow X$ is denoted by $\mathcal{X}_A$ and called the *relative topology*; obviously, $\mathcal{X}_A = \{O \cap A : O \in \mathcal{X}\}$. $(A, \mathcal{X}_A)$ is called the *subspace of* $(X, \mathcal{X})$ *determined by* A.

b) $(Y, \mathcal{Y})$ is called a *quotient space* of $(X, \mathcal{X})$ provided that there is a *quotient map* $f : (X, \mathcal{X}) \longrightarrow (Y, \mathcal{Y})$, i.e. a surjective map $f : X \longrightarrow Y$ such that $\mathcal{Y}$ is the final topology on Y w.r.t. f. In this case, $\mathcal{Y} = \{O \subset Y : f^{-1}[O] \in \mathcal{X}\}$ is called the *quotient topology* w.r.t. f.

If R is an equivalence relation on X, then the quotient topology on X/R w.r.t. the natural map $\omega : X \longrightarrow X/R$ is denoted by $\mathcal{X}_R$, and $(X/R, \mathcal{X}_R)$ is called the *quotient space of* $(X, \mathcal{X})$ *by* R.

2) Let $((X_i, \mathcal{X}_i))_{i \in I}$ be a family of topological spaces indexed by some set I. Then

a) the initial topology $\mathcal{X}$ on $\prod_{i \in I} X_i$ w.r.t. the family $(p_i)_{i \in I}$ of projections $p_i : \prod_{i \in I} X_i \to X_i$ is called the *product topology*; $(\prod_{i \in I} X_i, \mathcal{X})$ is called the *product space* of $((X_i, \mathcal{X}_i))_{i \in I}$. Evidently, $\mathcal{B} = \{\prod_{i \in I} O_i : O \in \mathcal{X}_i$ for each $i \in I$ and $O_i = X_i$ for all but finitely many $i \in I\}$ is a base for the product topology.

b) the final topology $\mathcal{X}$ on $\bigcup_{i \in I} X_i \times \{i\}$ w.r.t. the family $(j_i)_{i \in I}$ of injections $j_i : X_i \longrightarrow \bigcup_{i \in I} X_i \times \{i\}$, defined by $j_i(x_i) = (x_i, i)$ for each $x_i \in X_i$, is called the *sum topology*; $(\bigcup_{i \in I} X_i \times \{i\}, \mathcal{X})$ is called the *sum space* of $((X_i, \mathcal{X}_i))_{i \in I}$.

3) Let $\mathcal{M} = \{\mathcal{X}_i : i \in I\}$ be a set of topologies on a set X and $\mathcal{T}_X$ the set of all topologies on X. Then

a) $\mathcal{M}$ has a *least upper bound* in the partially ordered set $(\mathcal{T}_X, \subset)$, i.e. there is a coarsest topology $\mathcal{X}$ on X finer than each $\mathcal{X}_i \in \mathcal{M}$, namely the initial topology on X w.r.t. the family $(1_X^i : X \longrightarrow (X, \mathcal{X}_i))_{i \in I}$ of identity maps.

b) $\mathcal{M}$ has a *greatest lower bound* in the partially ordered set $(\mathcal{T}_X, \subset)$, i.e. there is a finest topology $\mathcal{X}$ on X coarser than each $\mathcal{X}_i \in \mathcal{M}$, namely the final topology on X w.r.t. the family $(1_X^i : (X, \mathcal{X}_i) \longrightarrow X)_{i \in I}$ of identity maps.

0.2.4.3 Remarks. 1) A partially ordered set $(X, \prec)$ is called a *lattice* provided that every finite subset of X has an infimum and a supremum; a lattice $(X, \prec)$ is called *complete* provided that every subset of X has an infimum and a supremum. By 0.2.4.2. 3), $(\mathcal{T}_X, \subset)$ is a complete lattice for each set X.

2) A continuous map $f : (X, \mathcal{X}) \longrightarrow (Y, \mathcal{Y})$ between topological spaces is a *homeomorphism* (or *isomorphism*) iff it is bijective and the inverse map $f^{-1} : (Y, \mathcal{Y}) \longrightarrow (X, \mathcal{X})$ is continuous. If there is a homeomorphism between the

topological spaces $(X,\mathcal{X})$ and $(Y,\mathcal{Y})$, we say that $(X,\mathcal{X})$ is *homeomorphic* (or *isomorphic*) to $(Y,\mathcal{Y})$ and we write $(X,\mathcal{X}) \cong (Y,\mathcal{Y})$. If $f:(X,\mathcal{X}) \longrightarrow (Y,\mathcal{Y})$ is an embedding, then the map $f':(X,\mathcal{X}) \longrightarrow (f[X],\mathcal{Y}_{f[X]})$ defined by $f'(x)=f(x)$ for each $x \in X$, is a homeomorphism, i.e. *a subspace $(X,\mathcal{X})$ of a topological space $(Y,\mathcal{Y})$ is homeomorphic to a subspace of $(Y,\mathcal{Y})$ determined by some subset of Y.* If $f:(X,\mathcal{X}) \longrightarrow (Y,\mathcal{Y})$ is a quotient map, then the map $s:(X/\pi_f,\mathcal{X}_{\pi_f}) \longrightarrow (Y,\mathcal{Y})$ defined under 0.1.5. ④ is a homeomorphism, i.e. a *quotient space $(Y,\mathcal{Y})$ of a topological space $(X,\mathcal{X})$ is homeomorphic to a quotient space of $(X,\mathcal{X})$ by an equivalence relation on X.*

3) Usually, we do not distinguish between homeomorphic topological spaces. The reason lies in the fact that in the realm of topological structures we are only interested in the study of *topological invariants*, i.e. properties which when possessed by a topological space $(X,\mathcal{X})$ are also possessed by each topological space $(Y,\mathcal{Y})$ being homeomorphic to $(X,\mathcal{X})$. A topological invariant P is
a) *hereditary* iff each subspace of a topological space with P has P,
b) *productive* iff the product space of each family of topological spaces enjoying P has P,
c) *divisible* iff the quotient space of each topological space with P has P,
d) *summable* if the sum space of each family of topological spaces with P has P,
e) *initially closed* iff for any family $((X_i,\mathcal{X}_i))_{i\in I}$ of topological spaces with P, any set X and any family $(f_i : X \longrightarrow X_i)_{i\in I}$ of maps, $(X,\mathcal{X})$ has P, where $\mathcal{X}$ is the initial topology on X w.r.t. $(f_i)_{i\in I}$,
f) *finally closed* iff for any family $((X_i,\mathcal{X}_i))_{i\in I}$ of topological spaces with P, any set X and any family $(f_i : X_i \longrightarrow X)_{i\in I}$ of maps, $(X,\mathcal{X})$ has P, where $\mathcal{X}$ is the final topology on X w.r.t. $(f_i)_{i\in I}$.

In the following some useful topological invariants other than the first and second axiom of countability are defined:

0.2.4.4 Definitions. I) A topological space $(X,\mathcal{X})$ is
1) (*separation axioms*)
a) a *T_0–space* iff one of the following two equivalent conditions is satisfied:
α) For any two distinct points $x,y \in X$, there exists a neighborhood of one of these points not containing the other.
β) For any two distinct points $x,y \in X$, $\overline{\{x\}} \neq \overline{\{y\}}$;
b) a *T_1–space* iff one of the following two equivalent conditions is satisfied:
α) For any two distinct points $x,y \in X$, there exist a neighborhood U_x of x and a neighborhood U_y of y such that $x \notin U_y$ and $y \notin U_x$,
β) For each $x \in X$, $\{x\} = \overline{\{x\}}$;
c) a *Hausdorff space* (or T_2–space) iff one of the following two equivalent conditions is satisfied:
α) For any two distinct points $x,y \in X$, there exist neighborhoods U_x of x and U_y of y such that $U_x \cap U_y = \emptyset$.
β) Each filter on X has at most one limit.
d) *regular* iff one of the following two equivalent conditions is satisfied:

α) For every closed subset A of X and every point $x \in X \backslash A$, there exist neighborhoods U_A of A and U_x of x such that $U_A \cap U_x = \emptyset$,
β) For every $x \in X$ and every $U_x \in \mathcal{U}(x)$, there exists a $V_x \in \mathcal{U}(x)$ such that $\overline{V_x} \subset U_x$;

e) *completely regular* iff for every closed subset A of X and every point $x \in X \backslash A$, there is a continuous map $f : X \longrightarrow [0,1]$ such that $f[A] \subset \{1\}$ and $f(x) = 0$, where the topology on $[0,1]$ is the relative topology of the usual topology on $I\!R$;

f) *normal* iff for any two disjoint closed subsets of X, there exist neighborhoods U_A of A and U_B of B such that $U_A \cap U_B = \emptyset$;

2) a) *connected* iff one of the following three equivalent conditions is satisfied:
α) X is not the union of two non–empty, open and disjoint subsets of X,
β) There are no clopen (= closed and open) subsets of X other than $\emptyset$ and X,
γ) Every continuous map $f : (X, \mathcal{X}) \longrightarrow D_2$ is constant, where $D_2 = (\{0,1\}, \{\emptyset, \{0\}, \{1\}, \{0,1\}\})$ is the two–point discrete topological space;

b) *path connected* iff for any two points $x, y \in X$ there is a path in X running from x to y, i.e. a continuous map $f : [0,1] \longrightarrow X$ such that $f(0) = x$ and $f(1) = y$ (the topology on $[0,1]$ is the relative topology of the usual topology on $I\!R$);

3) *compact* iff one of the following three equivalent conditions is satisfied:
α) Every open cover $\mathcal{U}$ of X (i.e. every cover $\mathcal{U}$ of X each member of which is open in X) has a finite subcover $\mathcal{V}$ (i.e. $\mathcal{V}$ is a finite subset of $\mathcal{U}$ and a cover of X),
β) Every filter on X has an adherence point,
γ) Every ultrafilter on X has a limit.

II) 1) If $(X, \mathcal{X})$ is a topological space, A a subset of X and P a topological invariant, then we say A has P provided that the subspace of $(X, \mathcal{X})$ determined by A has P.
2) A topological space $(X, \mathcal{X})$ is
a) *locally connected* (*locally path connected*) iff each $x \in X$ has a neighborhood base consisting of connected (path connected) sets;
b) *locally compact* iff each $x \in X$ has a compact neighborhood.

0.2.4.5 Table

topological invariants	hereditary	productive	summable	divisible	initially closed	finally closed
1st axiom of countability	+	−	+	−	−	−
2nd axiom of countability	+	−	−	−	−	−
T_0	+	+	+	−	−	−
T_1	+	+	+	−	−	−
Hausdorff	+	+	+	−	−	−
regular	+	+	+	−	+	−
completely regular	+	+	+	−	+	−
normal	−	−	+	−	−	−
connected	−	+	−	+	−	−
path connected	−	+	−	+	−	−
locally connected	−	−	+	+	−	+
locally path connected	−	−	+	+	−	+
compact	−	+	−	+	−	−
locally compact	−	−	+	−	−	−

0.2.4.6 Remarks. 1) The connected subsets of the topological space $I\!R_t$ of realnumbers are exactly the empty set, all singletons and all intervals, whereas the compact subsets of $I\!R_t$ are all closed and bounded subsets of $I\!R$ (in particular, all closed intervals).

2) A topological invariant P is *closed under formation of continuous images* iff for every topological space $(X, \mathcal{X})$ with P, every topological space $(Y, \mathcal{Y})$, for which there is a surjective continuous map $f : (X, \mathcal{X}) \longrightarrow (Y, \mathcal{Y})$, has P; e.g. the properties 1^0 'connected' and 2^0 'compact' are closed under formation of continuous images, which implies for real–valued continuous functions the *intermediate value theorem* in the first case and the theorem that a continuous function $f : (X, \mathcal{X}) \longrightarrow I\!R_t$ from a non–empty compact topological space $(X, \mathcal{X})$ to $I\!R_t$ has a *maximum and minimum value* in the second case (use 1)), both known from Analysis.

3) A completely regular Hausdorff space is also called a *Tychonoff space.* Tychonoff spaces are exactly those spaces that can be densely embedded into compact Hausdorff spaces, where an embedding $f : (X, \mathcal{X}) \longrightarrow (Y, \mathcal{Y})$ is called dense provided that $\overline{f[X]} = Y$: Since subspaces of compact Hausdorff spaces are Tychonoff spaces and subspaces of Tychonoff spaces are again Tychonoff spaces, it remains to verify that for each Tychonoff space $(X, \mathcal{X})$, there is a compact Hausdorff space $(Y, \mathcal{Y})$ and a dense embedding $f : (X, \mathcal{X}) \longrightarrow (Y, \mathcal{Y})$. Let I be the set of all continuous maps from $(X, \mathcal{X})$ into the closed unit interval $[0, 1]$, considered as a subspace of $I\!R_t$. Then the map $e : X \longrightarrow [0, 1]^I$ defined by $p_i(e(x)) = i(x)$ for each $i \in I$ and each $x \in X$ is an embedding, i.e. the map $e' : X \longrightarrow e[X]$, defined by $e'(x) = e(x)$ for each $x \in X$, is a homeomorphism. Since closed subspaces of compact topological spaces are compact, $\beta(X) = \overline{e[X]}$ is a compact Hausdorff space. If $i : e[X] \longrightarrow \overline{e[X]}$ denotes the inclusion map, then

$\beta_X = i \circ e' : X \longrightarrow \beta(X)$ is a dense embedding. $\beta(X)$ is called the *Stone–Čech compactification* of the Tychonoff space X. It is characterized by the following

Theorem (Stone, Čech). *If X is a Tychonoff space, then for each compact Hausdorff space Y and each continuous map $f : X \longrightarrow Y$, there is a unique continuous map $\overline{f} : \beta(X) \longrightarrow Y$ such that the diagram*

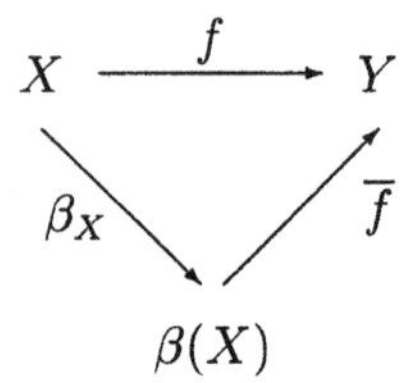

commutes.

Corollary. *If $\alpha_X : X \longrightarrow \alpha(X)$ is a continuous map from a Tychonoff X into a compact Hausdorff space $\alpha(X)$ such that for each compact Hausdorff space Y and each continuous map $f : X \longrightarrow Y$, there is a unique continuous map $\overline{f} : \alpha(X) \longrightarrow Y$ with $\overline{f} \circ \alpha_X = f$, then there is a homeomorphism $h : \beta(X) \longrightarrow \alpha(X)$ such that $h \circ \beta_X = \alpha_X$, in other words: the Stone–Čech compactification $\beta(X)$ is uniquely determined (up to homeomorphism) by the property stated in the theorem of Stone and Čech.*

0.3 Some categorical concepts

For each mathematical discipline we define at first objects and then suitable maps for describing the objects. This procedure is formalized by the concept 'category'.
0.3.1 Definition. A *category* $\mathcal{C}$ consists of
(1) a class $|\mathcal{C}|$ of *objects* (which are denoted by $A, B, C, \dots$),
(2) a class of pairwise disjoint sets $[A, B]_{\mathcal{C}}$ for each pair (A, B) of objects (the members of $[A, B]_{\mathcal{C}}$ are called *morphisms* from A to B), and
(3) a *composition* of morphisms, i.e. for each triple (A, B, C) of objects, there is a map from $[A, B]_{\mathcal{C}} \times [B, C]_{\mathcal{C}}$ to $[A, C]_{\mathcal{C}}$ sending $(f, g) \in [A, B]_{\mathcal{C}} \times [B, C]_{\mathcal{C}}$ to $g \circ f \in [A, C]_{\mathcal{C}}$ such that the following axioms are satisfied:
Cat_1) (*Associativity*). If $f \in [A, B]_{\mathcal{C}}$, $g \in [B, C]_{\mathcal{C}}$ and $h \in [C, D]_{\mathcal{C}}$, then $h \circ (g \circ f) = (h \circ g) \circ f$.
Cat_2) (*Existence of identities*). For each $A \in |\mathcal{C}|$, there is an *identity* (*morphism*) $1_A \in [A, A]_{\mathcal{C}}$ such that for all $B, C \in |\mathcal{C}|$, all $f \in [A, B]_{\mathcal{C}}$ and all $g \in [C, A]_{\mathcal{C}}$, $f \circ 1_A = f$ and $1_A \circ g = g$.

0.3.2 Remarks. 1) We write also $f : A \longrightarrow B$ or $A \xrightarrow{f} B$ instead of $f \in [A, B]_{\mathcal{C}}$. A (resp. B) is called the *domain* (resp. *codomain*) of f.
2) a) The identity 1_A is uniquely determined by A.
b) If $A, A' \in |\mathcal{C}|$ with $A \neq A'$, then $1_A \neq 1_{A'}$, because $[A, A]_{\mathcal{C}} \cap [A', A']_{\mathcal{C}} = \emptyset$.
3) The class $\bigcup_{(A,B) \in |\mathcal{C}| \times |\mathcal{C}|} [A, B]_{\mathcal{C}}$, i.e. the class of all morphisms of $\mathcal{C}$, is denoted

by Mor $\mathcal{C}$: its elements are called *$\mathcal{C}$–morphisms.*
4) The requirement of the pairwise disjointness of the morphism sets is not restrictive, since it is always fulfilled provided that $[A, B]_{\mathcal{C}}$ is replaced (when necessary) by $[A, B]'_{\mathcal{C}} = \{(A, \alpha, B) : \alpha \in [A, B]_{\mathcal{C}}\}$.
5) Sometimes we write $[A, B]$ instead of $[A, B]_{\mathcal{C}}$ when no confusion can result about $\mathcal{C}$.

0.3.3 Examples of categories.
① The category **Set** of sets (and maps): $|\mathbf{Set}| = \mathcal{U}$, i.e. the class of all sets; for each $(A, B) \in \mathcal{U} \times \mathcal{U}$, $[A, B]_{\mathbf{Set}}$ is the set of all maps from A to B; and the composition is the usual composition of maps, where identities are the identity maps.

Remark. In order to obtain the pairwise disjointness of the morphism sets, it is necessary to define maps as triples (cf. the remark under 0.1.3. ③).

② For each of the following categories, the composition of morphisms is the usual composition of maps, where the identities are the identity maps as usual:
a) The category **Top** of topological spaces (and continuous maps), i.e. objects are all topological spaces and morphisms are all continuous maps between them.
b) The category **Group** of groups (and homomorphisms), i.e. objects are all groups and morphisms are all homomorphisms between them.
c) The category **Pos** of partially ordered sets (and isotone maps), i.e. objects are all partially ordered sets and morphisms are all order preserving maps between them.

③ If $(S, \prec)$ is a partially ordered set, then a category $\mathcal{C}$ is defined as follows:

$$|\mathcal{C}| = S, \quad [x, y]_{\mathcal{C}} = \begin{cases} \{(x, y)\} & \text{if } x \prec y \\ \emptyset & \text{otherwise} \end{cases}, (y, z) \circ (x, y) = (x, z)$$

and $1_x = (x, x)$.

④ Let $\mathcal{C}$ be a category. Then the *dual category* $\mathcal{C}^*$ is defined as follows:
(1) $|\mathcal{C}^*| = |\mathcal{C}|$,
(2) $[A, B]_{\mathcal{C}^*} = [B, A]_{\mathcal{C}}$ for each $(A, B) \in |\mathcal{C}| \times |\mathcal{C}|$,
(3) The composition $\alpha \circ \beta$ in $\mathcal{C}^*$ is defined to be the composition $\beta \circ \alpha$ in $\mathcal{C}$.

Convention. If a $\mathcal{C}$–morphism f is regarded as a $\mathcal{C}^*$–morphism, we write f^* instead of f.

Remarks. 1) $(\mathcal{C}^*)^* = \mathcal{C}$.
2) For each statement in a category $\mathcal{C}$, there is a dual statement, namely the corresponding statement in $\mathcal{C}^*$ phrased as a statement in $\mathcal{C}$ (by reversing all arrows by means of which morphisms are symbolized).
3) Though the objects in a category $\mathcal{C}$ are sets (cf. 0.1.7. ④), the morphisms between them need not be maps as the example $\mathbf{Top}^*$ shows.

⑤ Let $\mathcal{C}_1$ and $\mathcal{C}_2$ be categories. Then the *product category* $\mathcal{C}_1 \times \mathcal{C}_2$ is defined by

$$|\mathcal{C}_1 \times \mathcal{C}_2| = |\mathcal{C}_1| \times |\mathcal{C}_2| = \{(X_1, X_2) : X_1 \in |\mathcal{C}_1|, X_2 \in |\mathcal{C}_2|\}$$

and

$$[(X_1, X_2), (Y_1, Y_2)]_{C_1 \times C_2} = [X_1, Y_1]_{C_1} \times [X_2, Y_2]_{C_2},$$

where the composition is given by

$$(f_1, g_1) \circ (f_2, g_2) = (f_1 \circ f_2, g_1 \circ g_2)$$

(provided that the compositions $f_1 \circ f_2$ and $g_1 \circ g_2$ exist).
Obviously, $(C_1 \times C_2)^* = C_1^* \times C_2^*$

0.3.4 Definition. A morphism $f : A \longrightarrow B$ in a category C is called an *isomorphism* provided that there is a C–morphism $g : B \longrightarrow A$ such that $g \circ f = 1_A$ and $f \circ g = 1_B$. Objects A and B in a category C are *isomorphic*, $A \cong B$, iff there exists an isomorphism $f \in [A, B]_C$.

0.3.5 Remarks. 1) In 0.3.4., g is uniquely determined by f (if $g' \in [B, A]_C$ with $g' \circ f = 1_A$ and $f \circ g' = 1_B$, then $g = g \circ 1_B = g \circ (f \circ g') = (g \circ f) \circ g' = 1_A \circ g' = g'$) and called the *inverse* of f, denoted by f^{-1}.
2) Obviously, an isomorphism in **Set** is a bijective map (and vice versa), while an isomorphism in **Top** is a homeomorphism (and vice versa).
3) For every category C, the identity $1_X : X \longrightarrow X$ is an isomorphism for each $X \in |C|$. If $f : X \longrightarrow Y$ is an isomorphism in C, then $f^{-1} : Y \longrightarrow X$ is also an isomorphism. Additionally, the composition of two isomorphisms in C is again an isomorphism. Thus, $\cong \subset |C| \times |C|$ is an equivalence relation on $|C|$; the corresponding equivalence classes are called *isomorphism classes.* A property P for the objects of C is called a C–*invariant* provided that the following is satisfied: If an object X of C has the property P, then all objects of the isomorphism class of X have the property P.
Example: **Top**–invariant means topological invariant.

0.3.6 Definitions. Let C be a category. A C–morphism $f : A \longrightarrow B$ is called

1) a *monomorphism* provided that for all pairs (α, β) of C–morphisms with codomain A such that $f \circ \alpha = f \circ \beta$, it follows that $\alpha = \beta$.

1′) an *epimorphism* provided that f^* is a monomorphism in the dual category C^* (i.e. for all pairs (α, β) of C–morphisms with domain B such that $\alpha \circ f = \beta \circ f$, it follows that $\alpha = \beta$).

2) an *extremal monomorphism* provided that the following are satified:
(1) f is a monomorphism.
(2) If $f = h \circ g$, where g is an epimorpism, then g must be an isomorphism.

2′) an *extremal epimorphism* provided that f^* is an extremal monomorphism in the dual category C^* (i.e. the following are satisfied:
(1′) f is an epimorphism.
(2′) If $f = g \circ h$, where g is a monomorphism, then g must be an isomorphism.).

0.3.7 Proposition. *Let C be a category and $f : A \longrightarrow B$ a C–morphism. then the following are equivalent:*

(1) *f is an isomorphism,*
(2) *f is an epimorphism and an extremal monomorphism,*
(3) *f is a monomorphism and an extremal epimorphism.*

Proof. (1) $\Longrightarrow$ (2). a) Let α, β be $\mathcal{C}$–morphisms such that $\alpha \circ f = \beta \circ f$. Then $\alpha \circ f \circ f^{-1} = \beta \circ f \circ f^{-1}$, i.e. $\alpha \circ 1_B = \alpha = \beta \circ 1_B = \beta$. Hence, f is an epimorphism.
b) α) Since f is an isomorphism, f is also a monomorphism (analogously to a)).
β) Let $f = h \circ g$, where g is an epimorphism, i.e. the diagram

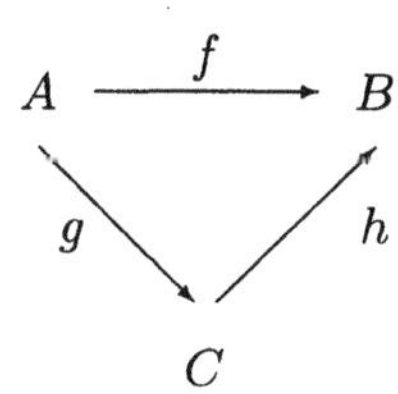

commutes. Then $1_A = f^{-1} \circ f = f^{-1} \circ (h \circ g) = (f^{-1} \circ h) \circ g$. Furthermore, $(g \circ (f^{-1} \circ h)) \circ g = g \circ ((f^{-1} \circ h) \circ g) = g \circ 1_A = g = 1_C \circ g$ which implies $g \circ (f^{-1} \circ h) = 1_C$ since g is an epimorphism. Therefore, g is an isomorphism.
(2) $\Longrightarrow$ (1). Since f is an epimorphism and an extremal monomorphism, it follows immediately from $f = 1_B \circ f$ that f is an isomorphism.
Thus, the proposition is proved because (2) and (3) are dual statements and (1) is self–dual [more exactly: f isomorphism $\Longleftrightarrow$ f^* isomorphism $\Longleftrightarrow$ f^* epimorphism and extremal monomorphism $\Longleftrightarrow$ f monomorphism and extremal epimorphism].

0.3.8 Remark. A morphism in a category $\mathcal{C}$ is called a *bimorphism* provided that it is an epimorphism and a monomorphism. Categories in which each bimorphism is already an isomorphism are called *balanced*; e.g. the categories **Set** and **Group** are balanced (note, that for a bijective homomorphism between groups, the inverse is automatically a homomorphism), whereas the category **Top** is not balanced (e.g. if $X = \{0,1\}$, then the map $1_X : (X, \mathcal{P}(X)) \longrightarrow (X, \{\emptyset, X\}$ is bijective and continuous, but its inverse is not continuous). Obviously, for a balanced category $\mathcal{C}$, there is no distinction between 'epimorphism' (resp. 'monomorphism') and 'extremal epimorphism' (resp. 'extremal monomorphism').

0.3.9 Definition. Let $\mathcal{C}$ be a category, I a set and $(A_i)_{i\in I}$ a family of objects in $\mathcal{C}$ (shortly: a family of $\mathcal{C}$–objects).

A pair $(P, (p_i)_{i\in I})$ with $P \in |\mathcal{C}|$ and $p_i \in [P, A_i]_{\mathcal{C}}$ for each $i \in I$ is called a *product* of the family $(A_i)_{i\in I}$ provided that for each pair $(Q, (q_i)_{i\in I})$ with $Q \in |\mathcal{C}|$ and $q_i \in [Q, A_i]_{\mathcal{C}}$ for each $i \in I$ there exists a unique $\mathcal{C}$–morphism q such that the diagram

A pair $(C, (j_i)_{i\in I})$ with $C \in |\mathcal{C}|$ and $j_i \in [A_i, C]_{\mathcal{C}}$ for each $i \in I$ is called a *coproduct* of the family $(A_i)_{i\in I}$ provided that $(C, (j_i^*)_{i\in I})$ is a product of $(A_i)_{i\in I}$ in the dual category $\mathcal{C}^*$ (i.e. for each pair $(D, (k_i)_{i\in I})$ with $D \in |\mathcal{C}|$ and $k_i \in [A_i, D]_{\mathcal{C}}$ for each $i \in I$ there exists

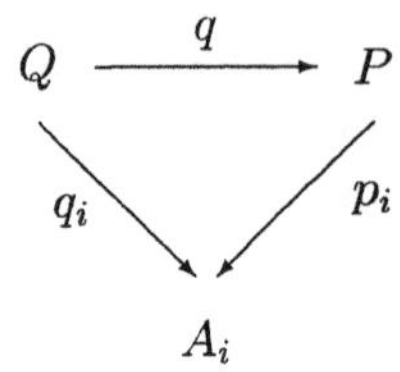

is commutative (i.e. $p_i \circ q = q_i$) for every $i \in I$.

We write $\prod_{\in I} A_i$ instead of P (cf. the following proposition); p_i is called the ith *projection*. Sometimes $\prod_{i\in I} A_i$ is already called the product of the family $(A_i)_{i\in I}$.

a unique $\mathcal{C}$–morphism k such that the diagram

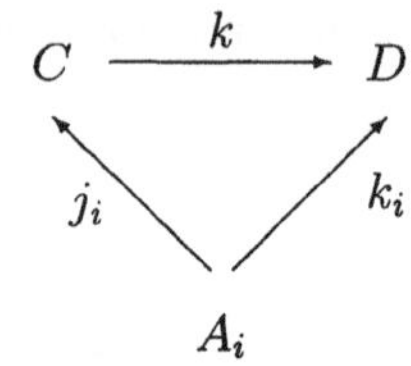

is commutative for each $i \in I$).
We write $\coprod_{i\in I} A_i$ instead of C (cf. the following proposition); j_i is called the ith *injection*. Sometimes $\coprod_{i\in I} A_i$ is already called the coproduct of the family $(A_i)_{i\in I}$.

0.3.10 Proposition. *Let $\mathcal{C}$ be a category, I a set and $(A_i)_{i\in I}$ a family of $\mathcal{C}$-objects.*

a) *If each of $(P, (p_i)_{i\in I})$ and $(P', (p'_i)_{i\in I})$ is a product of $(A_i)_{i\in I}$ in $\mathcal{C}$, then there is a unique isomorphism $k \in [P, P']_{\mathcal{C}}$ such that $p'_i \circ k = p_i$ for each $i \in I$.*

b) *If each of $(C, (j_i)_{i\in I})$ and $(C', (j'_i)_{i\in I})$ is a coproduct of $(A_i)_{i\in I}$ in $\mathcal{C}$, then there is a unique isomorphism $j \in [C', C]_{\mathcal{C}}$ such that $j \circ j'_i = j_i$ for each $i \in I$.*

Proof. It suffices to prove a): Since $(P', (p'_i)_{i\in I})$ is a product of $(A_i)_{i\in I}$, there is a unique $k \in [P, P']_{\mathcal{C}}$ such that $p'_i \circ k = p_i$ for each $i \in I$. Furthermore, since $(P, (p_i)_{i\in I})$ is a product of $(A_i)_{i\in I}$, there is a unique $h \in [P', P]_{\mathcal{C}}$ such that $p_i \circ h = p'_i$ for each $i \in I$. That is, the diagram

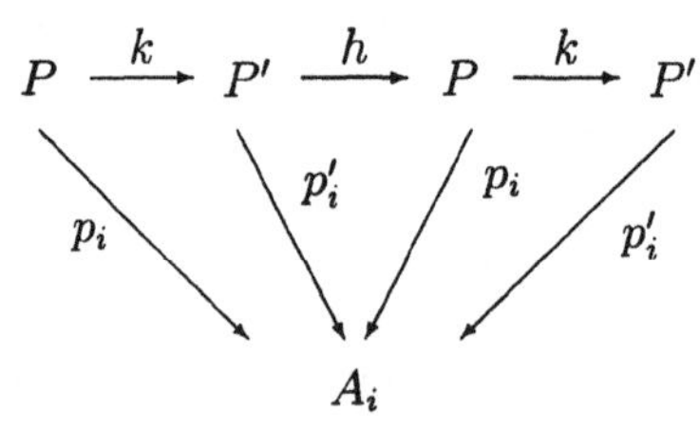

is commutative for all $i \in I$. There is a unique morphism from P to P such the triangle formed by the two left triangles in the above diagram commutes (note that $(P, (p_i)_{i\in I})$ is a product!). Thus, $h \circ k = 1_P$. Similarly, we may conclude that $k \circ h = 1_{P'}$. Therefore, k is an isomorphism.

0.3.11 Examples. The products in **Set** are the cartesian products, whereas the coproducts in **Set** are the disjoint unions. In **Top** the products are the usual (topological) products and the coproducts are the usual (topological) sums. Usually, in **Group** the products are called direct products and the coproducts are called direct sums.

0.3.12 Remark. Sometimes it is useful to consider also products of morphisms in a category $\mathcal{C}$: If $(f_i : A_i \longrightarrow B_i)_{i \in I}$ is a family of $\mathcal{C}$–morphisms indexed by some set I and if $(\prod_{i \in I} A_i, (p_i)_{i \in I})$ and $(\prod_{i \in I} B_i, (p'_i)_{i \in I})$ are products in $\mathcal{C}$, then the unique morphism $f : \prod_{i \in I} A_i \longrightarrow \prod_{i \in I} B_i$ that makes the diagram

$$\begin{array}{ccc} \prod_{i \in I} A_i & \xrightarrow{f} & \prod_{i \in I} B_i \\ {\scriptstyle p_i}\downarrow & & \downarrow{\scriptstyle p'_i} \\ A_i & \xrightarrow{f_i} & B_i \end{array}$$

commutative for each $i \in I$ is denoted by $\prod f_i$ and is called the *product* of the family $(f_i)_{i \in I}$. If $I = \{1, \ldots, n\}$, then we write $f_1 \times \cdots \times f_n$ instead of $\prod f_i$.

The *coproduct* $\coprod f_i$ of the family $(f_i)_{i \in I}$ is defined dually.

Chapter 1

Topological Constructs

In order to handle problems of a topological nature topologists have created not only topological spaces but also uniform spaces, filter spaces, convergence spaces and so on. Since constructions in the corresponding concrete categories of these spaces have striking similarities the question arises whether it is possible to postulate axioms for a construct (= concrete category) which may be regarded as topological. Thus, the problem consists in looking for one or more properties which are independent of the special structure of the considered objects in a construct (i.e. properties essentially characterized by morphisms) and which are not satisfied by "algebraic" constructs. This claim is fulfilled by the initial structures in the sense of N. Bourbaki [18] provided their unrestricted existence is required. In the category **Group** for instance there do not exist arbitrary initial structures, e.g. not every subset of a group is a subgroup. Further conditions may be added for getting the concept "topological construct" but they are of a more "technical" nature. In order to obtain final structures simultaneously it is useful (in constrast to N. Bourbaki) to require the existence of initial structures for families of maps which are indexed by a class (instead of a set). After the definition of a topological construct and numerous examples (up to measure theory and algebraic topology) the categorical properties of topological constructs are studied in this chapter.

The definition of a topological construct $\mathcal{C}$ used here is essentially due to H. Herrlich [63], where the additional requirement that there is exactly one $\mathcal{C}$–structure on the empty set has been proposed by O. Wyler [152].

1.1 Definition and examples

1.1.1 By a *construct* we mean a category $\mathcal{C}$ whose objects are structured sets, i.e. pairs (X, ξ) where X is a set and ξ a $\mathcal{C}$–structure on X, whose morhisms $f : (X, \xi) \longrightarrow (Y, \eta)$ are suitable maps between X and Y and whose composition law is the usual composition of maps (cf. 2.1.2. ③b) for a more advanced definition of a construct).

1.1.2 Definition. A construct $\mathcal{C}$ is called *topological* iff it satisfies the following conditions:

(1) *Existence of initial structures:*
For any set X, any family $((X_i, \xi_i))_{i\in I}$ of $\mathcal{C}$–objects indexed by a class I and any family $(f_i : X \longrightarrow X_i)_{i\in I}$ of maps indexed by I there exists a unique $\mathcal{C}$–structure ξ on X which is *initial* with respect to $(X, f_i, (X_i, \xi_i), I)$, i.e. such that for any $\mathcal{C}$–object (Y, η) a map $g : (Y, \eta) \longrightarrow (X, \xi)$ is a $\mathcal{C}$–morphism iff for every $i \in I$ the composite map $f_i \circ g : (Y, \eta) \longrightarrow (X_i, \xi_i)$ is a $\mathcal{C}$–morphism.

(2) For any set X, the class $\{(Y, \eta) \in |\mathcal{C}| : X = Y\}$ of all $\mathcal{C}$–objects with underlying set X is a set.

(3) For any set X with cardinality at most one, there exists exactly one $\mathcal{C}$–object with underlying set X (i.e. there exists exactly one $\mathcal{C}$–structure on X).

1.1.3 Remarks. 1) Let ξ be the initial structure on X with respect to $(X, f_i, (X_i, \xi_i), I)$. Then $f_i : (X, \xi) \longrightarrow (X_i, \xi_i)$ is a $\mathcal{C}$–morphism for each $i \in I$ (Hint. Let $(Y, \eta) = (X, \xi)$ and $g = 1_X$ in (1)).

2) In a topological construct $\mathcal{C}$ the following holds: If X is a set and ξ, η are $\mathcal{C}$–stuctures on X such that $1_X : (X, \xi) \longrightarrow (X, \eta)$ and $1_X : (X, \eta) \longrightarrow (X, \xi)$ are $\mathcal{C}$–morphisms, then $\xi = \eta$ (this follows immediately from the uniqueness of initial structures required in (1)).

1.1.4 Definition. Let $\mathcal{C}$ be a topological construct, let X be a set and let ξ, η be $\mathcal{C}$–structures on X. The $\mathcal{C}$–structure ξ is called *finer* than η (or η *coarser* than ξ), denoted by $\xi \leq \eta$, iff $1_X : (X, \xi) \longrightarrow (X, \eta)$ is a $\mathcal{C}$–morphism.

1.1.5 Proposition. *The initial structure ξ on a set X with respect to $(X, f_i, (X_i, \xi_i), I)$ in a topological construct $\mathcal{C}$ is the coarsest $\mathcal{C}$–structure on X such that f_i is a $\mathcal{C}$–morphism for each $i \in I$.*

Proof. By 1.1.3. 1) ξ is a $\mathcal{C}$–structure on X such that all $f_i : (X, \xi) \longrightarrow (X_i, \xi_i)$ are $\mathcal{C}$ morphisms. Let η be a $\mathcal{C}$–structure on X such that all $f_i : (X, \eta) \longrightarrow (X_i, \xi_i)$ are $\mathcal{C}$–morphisms. Since all $f_i \circ 1_X : (X, \eta) \longrightarrow (X_i, \xi_i)$ are $\mathcal{C}$–morphisms, $1_X : (X, \eta) \longrightarrow (X, \xi)$ is a $\mathcal{C}$–morphism, i.e. $\eta \leq \xi$.

1.1.6 Examples of topological constructs.

① The category **Top** of topological spaces (and continuous maps).

② The category **Unif** of uniform spaces (and uniformly continuous maps):

a) A *uniform space* is a pair $(X, \mathcal{W})$ where X is a set and $\mathcal{W}$ a non–empty subset of $\mathcal{P}(X \times X)$ such that the conditions F_2) and F_3) for a filter are satisfied as well as the following:

U_1) $W \in \mathcal{W}$ implies $\Delta = \{(x, x) : x \in X\} \subset W$.

U_2) $W \in \mathcal{W}$ implies $W^{-1} = \{(x, y) : (y, x) \in W\} \in \mathcal{W}$.

U$_3$) For each $W \in \mathcal{W}$ there is some $W^* \in \mathcal{W}$ such that $W^{*^2} = \{(x,y) \in X \times X$: there is some $z \in X$ with $(x,z) \in W^*$ and $(z,y) \in W^*\} \subset W$.

Obviously, if X is non–empty, $\mathcal{W}$ is a filter on $X \times X$. If $(X, \mathcal{W})$ is a uniform space, then $\mathcal{W}$ is called a *uniformity* on X and the elements of $\mathcal{W}$ are called *entourages*. Further, $\mathcal{B} \subset \mathcal{P}(X \times X)$ is called a *base* for the uniformity $\mathcal{W}$ provided that $\{W \subset X \times X : W \supset B$ for some $B \in \mathcal{B}\} = \mathcal{W}$.

b) A map $f : (X, \mathcal{W}) \longrightarrow (X', \mathcal{W}')$ between uniform spaces is called *uniformly continuous* provided that one of the following two equivalent conditions is satisfied:

(1) For each $W' \in \mathcal{W}'$ there is some $W \in \mathcal{W}$ such that $(f(x), f(y)) \in W'$ for each $(x,y) \in W$.

(2) $(f \times f)^{-1}[W'] \in \mathcal{W}$ for each $W' \in \mathcal{W}'$, where $f \times f : X \times X \longrightarrow Y \times Y$ is defined by $(f \times f)(x,y) = (f(x), f(y))$ for each $(x,y) \in X \times X$.

(Note: If (X,d) is a metric space [resp. pseudometric space] and $V_\varepsilon = \{(x,y) : d(x,y) < \varepsilon\}$ for each $\varepsilon > 0$, then $\mathcal{B} = \{V_\varepsilon : \varepsilon > 0\}$ is a base for a uniformity $\mathcal{W}_d$ on X; $\mathcal{W}_d$ is called the *uniformity induced by the (pseudo)metric d*. For a map $f : (X,d) \longrightarrow (X', d')$ between (pseudo)metric spaces the following are equivalent:

(1) $f : (X, \mathcal{W}_d) \longrightarrow (X', \mathcal{W}_{d'})$ is uniformly continuous.
(2) For each $\varepsilon > 0$ there is some $\delta > 0$ such that $d'(f(x), f(y)) < \varepsilon$ for each $(x,y) \in X \times X$ with $d(x,y) < \delta$.)

Let X be a set, $((X_i, \mathcal{W}_i))_{i \in I}$ a family of uniform spaces and $(f_i : X \longrightarrow X_i)_{i \in I}$ a family of maps. Put $g_i = f_i \times f_i$ for each $i \in I$. Then all finite intersections of elements of $\{g_i^{-1}[W_i] : W_i \in \mathcal{W}_i, i \in I\}$ form a base for a uniformity $\mathcal{W}$ on X which is initial with respect to $(X, f_i, (X_i, \mathcal{W}_i), I)$.
Much more will be said about uniform spaces later on.

③ a) The category **GConv** of generalized convergence spaces (and continuous maps):
α) For each set X let $F(X)$ be the set of all filters on X. Then a *generalized convergence space* is a pair (X, q), where X is a set and $q \subset F(X) \times X$ such that the following are satisfied:

C$_1$) $(\dot{x}, x) \in q$ for each $x \in X$, where $\dot{x} = \{A \subset X : x \in A\}$,
C$_2$) $(\mathcal{G}, x) \in q$ whenever $(\mathcal{F}, x) \in q$ and $\mathcal{G} \supset \mathcal{F}$.

If (X, q) is a generalized convergence space, then we write sometimes $\mathcal{F} \xrightarrow{q} x$ or shortly $\mathcal{F} \longrightarrow x$ instead of $(\mathcal{F}, x) \in q$ and say that $\mathcal{F}$ *converges* to x.

β) A map $f : (X, q) \longrightarrow (X', q')$ between generalized convergence spaces is called *continuous* provided that $(f(\mathcal{F}), f(x)) \in q'$ for each $(\mathcal{F}, x) \in q$.
(Note: If $(X, \mathcal{X})$ is a topological space and $q_\mathcal{X} \subset F(X) \times X$ is defined by $(\mathcal{F}, x) \in q_\mathcal{X}$ iff $\mathcal{F} \supset \mathcal{U}_\mathcal{X}(x)$ where $\mathcal{U}_\mathcal{X}(x)$ denotes the neighborhood filter of $x \in X$ in $(X, \mathcal{X})$, then $(X, q_\mathcal{X})$ is a generalized convergence space and $q_\mathcal{X}$ is called *the generalized convergence structure induced by the topology $\mathcal{X}$*. As is well–known the following are equivalent for each map $f : (X, \mathcal{X}) \longrightarrow (Y, \mathcal{Y})$ between topological spaces:

(1) $f : (X, \mathcal{X}) \longrightarrow (Y, \mathcal{Y})$ is continuous (in the usual sense),
(2) $f : (X, q_{\mathcal{X}}) \longrightarrow (X', q_{\mathcal{X}'})$ is continuous (in the sense above).)

Let X be a set, $((X_i, q_i))_{i\in I}$ a family of generalized convergence spaces and $(f_i : X \longrightarrow X_i)_{i\in I}$ a family of maps, then $q = \{(\mathcal{F}, x) \in F(X) \times X : (f_i(\mathcal{F}), f_i(x)) \in q_i$ for each $i \in I\}$ is a **GConv**–structure on X which is initial with respect to the given data.

b) The category **KConv** of Kent convergence spaces (and continuous maps):
A generalized convergence space (X, q) is called a *Kent convergence space* provided that the following is satisfied:

C_3) $(\mathcal{F} \cap \dot{x}, x) \in q$ whenever $(\mathcal{F}, x) \in q$.

The initial structures in **KConv** are formed as in **GConv.**
(Note: If $(X, \mathcal{X})$ is a topological space, then the generalized convergence structure $q_{\mathcal{X}}$ induced by $\mathcal{X}$ is a **KConv**–structure.)

④ The category **Fil** of filter spaces (and Cauchy continuous maps):
a) A *filter space* is pair (X, γ) where X is a set and γ a set of filters on X such that the following are satisfied:

(1) $\dot{x} \in \gamma$ for each $x \in X$,
(2) $\mathcal{G} \in \gamma$ whenever $\mathcal{F} \in \gamma$ and $\mathcal{F} \subset \mathcal{G}$.

If (X, γ) is a filter space, then the elements of γ are called *Cauchy filters.*

b) A map $f : (X, \gamma) \longrightarrow (X', \gamma')$ between filter spaces is called *Cauchy continuous* provided that $f(\mathcal{F}) \in \gamma'$ for each $\mathcal{F} \in \gamma$.

If X is a set, $((X_i, \gamma_i))_{i\in I}$ a family of filter spaces and $(f_i : X \longrightarrow X_i)_{i\in I}$ a family of maps, then

$$\gamma = \{\mathcal{F} \in F(X) : f_i(\mathcal{F}) \in \gamma_i \text{ for each } i \in I\}$$

is the initial **Fil**–structure on X with respect to the given data.
(Note: If $(X, \mathcal{W})$ is a uniform space, then

$\gamma_{\mathcal{W}} = \{\mathcal{F} \in F(X) :$ for each $W \in \mathcal{W}$ there is some $F \in \mathcal{F}$ such that $F \times F \subset W\}$

is a **Fil**–structure on X, called the *filter space structure induced by the uniformity* $\mathcal{W}$. By a *Cauchy filter on a uniform space* $(X, \mathcal{W})$ we always mean a filter $\mathcal{F}$ on X such that $\mathcal{F} \in \gamma_{\mathcal{W}}$, in other words: a Cauchy filter on a uniform space $(X, \mathcal{W})$ is a filter containing arbitrarily small sets [remember that in a metric space (X, d) the set $\{V_\varepsilon : \varepsilon > 0\}$ is a base for the uniformity $\mathcal{W}_d$ induced by d]. A sequence $(x_n)_{n\in I\!N}$ in a uniform space $(X, \mathcal{W})$ is called a *Cauchy sequence* iff the elementary filter $\mathcal{F}_e$ of the sequence (x_n), i.e. the filter generated by $\{E_m : m \in I\!N\}$ with $E_m = \{x_n : n \geq m\}$, is a Cauchy filter. In a metric space (X, d) the following are equivalent for each sequence (x_n) in X:

(1) (x_n) is a Cauchy sequence in $(X, \mathcal{W}_d)$,
(2) For each $\varepsilon > 0$ there is some $N(\varepsilon) \in I\!N$ such that $d(x_n, x_m) < \varepsilon$ whenever $n, m \geq N(\varepsilon)$.)

⑤ The category **SUConv** of semiuniform convergence spaces (and uniformly continuous maps):
a) A *semiuniform convergence space* is a pair $(X, \mathcal{J}_X)$, where X is a set and $\mathcal{J}_X$

a set of filters on $X \times X$ such that the following are satisfied:

UC$_1$) The filter generated by $\{(x,x)\}$, i.e. $\dot{x} \times \dot{x}$, belongs to $\mathcal{J}_X$ for each $x \in X$,
UC$_2$) $\mathcal{G} \in \mathcal{J}_X$ whenever $\mathcal{F} \in \mathcal{J}_X$ and $\mathcal{F} \subset \mathcal{G}$,
UC$_3$) $\mathcal{F} \in \mathcal{J}_X$ implies $\mathcal{F}^{-1} = \{F^{-1} : F \in \mathcal{F}\} \in \mathcal{J}_X$, where $F^{-1} = \{(y,x) : (x,y) \in F\}$.

If $(X, \mathcal{J}_X)$ is a semiuniform convergence space, then the elements of $\mathcal{J}_X$ are called *uniform filters.*
(Note: If $(X, \mathcal{W})$ is a uniform space and $[\mathcal{W}] = \{\mathcal{F} \in F(X \times X) : \mathcal{F} \supset \mathcal{W}\}$ then $(X, [\mathcal{W}])$ is a semiuniform convergence space. Thus, if X is non–empty, the uniformity $\mathcal{W}$ is a uniform filter.)

b) A map $f : (X, \mathcal{J}_X) \longrightarrow (Y, \mathcal{J}_Y)$ between semiuniform convergence spaces is called *uniformly continuous* provided that $(f \times f)(\mathcal{F}) \in \mathcal{J}_Y$ for each $\mathcal{F} \in \mathcal{J}_X$, shortly: $(f \times f)(\mathcal{J}_X) \subset \mathcal{J}_Y$.
(Note: If $(X, \mathcal{W})$, $(Y, \mathcal{R})$ are uniform spaces, then the following are equivalent for each map $f : (X, \mathcal{W}) \longrightarrow (Y, \mathcal{R})$:

(1) $f : (X, [\mathcal{W}]) \longrightarrow (Y, [\mathcal{R}])$ is uniformly continuous in **SUConv**,
(2) $f : (X, \mathcal{W}) \longrightarrow (Y, \mathcal{R})$ is uniformly continuous in **Unif.**)

If X is a set, $((X_i, \mathcal{J}_{X_i}))_{i \in I}$ a family of semiuniform convergence spaces, $(f_i : X \longrightarrow X_i)_{i \in I}$ a family of maps, then $\mathcal{J}_X = \{\mathcal{F} \in F(X \times X) : (f_i \times f_i)(\mathcal{F}) \in \mathcal{J}_{X_i}$ for each $i \in I\}$ is the *initial* **SUConv**–*structure* on X with respect to the given data.

⑥ The category **Meas** of measurable spaces (and measurable maps):
a) A *measurable space* is a pair $(X, \mathcal{A})$ where X is a set and $\mathcal{A} \subset \mathcal{P}(X)$ a σ–algebra on X, i.e. such that $\mathcal{A}$ satisfies the following conditions:

1) $X \in \mathcal{A}$,
2) $A \in \mathcal{A}$ implies $X \backslash A \in \mathcal{A}$,
3) $\bigcup_{n \in \mathbb{N}} A_n \in \mathcal{A}$ whenever $(A_n)_{n \in \mathbb{N}}$ is a sequence in $\mathcal{A}$.

b) A map $f : (X, \mathcal{A}) \longrightarrow (X', \mathcal{A}')$ between measurable spaces is called *measurable* provided that $f^{-1}[A'] \in \mathcal{A}$ for each $A' \in \mathcal{A}'$.

If X is a set, $((X_i, \mathcal{A}_i))_{i \in I}$ a family of measurable spaces and $(f_i : X \longrightarrow X_i)_{i \in I}$ a family of maps, then the intersection of all σ–algebras on X containing $\bigcup_{i \in I} f_i^{-1}\mathcal{A}_i$ with $f_i^{-1}\mathcal{A}_i = \{f_i^{-1}[A_i] : A_i \in \mathcal{A}_i\}$ is the initial σ–algebra on X with respect to the given data.

⑦ The category **Born** of bornological spaces (and bounded maps):
a) A *bornological space* is a pair $(X, \mathcal{B})$, where X is a set and $\mathcal{B}$ a subset of $\mathcal{P}(X)$ satisfying

1) $A, B \in \mathcal{B}$ imply $A \cup B \in \mathcal{B}$,
2) $B \in \mathcal{B}$ and $A \subset B$ imply $A \in \mathcal{B}$,
3) Each finite subset B of X belongs to $\mathcal{B}$.

If $(X, \mathcal{B})$ is a bornological space, then the elements of $\mathcal{B}$ are called *bounded sets.*

b) A map $f : (X, \mathcal{B}) \longrightarrow (X', \mathcal{B}')$ between bornological spaces is called *bounded* provided that $f[B] \in \mathcal{B}'$ for each $B \in \mathcal{B}$.

If X is a set, $((X_i, \mathcal{B}_i))_{i\in I}$ a family of bornological spaces and $(f_i : X \longrightarrow X_i)_{i\in I}$ a family of maps, then $\mathcal{B} = \{B \subset X : f_i[B] \in \mathcal{B}_i$ for each $i \in I\}$ is the initial **Born**–structure on X with respect to the given data.

⑧ The category **Rere** of reflexive relations (and relation preserving maps):
a) Objects of **Rere** are pairs (X, ρ), where X is a set and ρ a reflexive relation on X.
b) A map $f : (X, \rho) \longrightarrow (X', \rho')$ between **Rere**–objects is a **Rere**–morphism provided that $(f(x), f(y)) \in \rho'$ for each $(x, y) \in \rho$.

If X is a set, $((X_i, \rho_i))_{i\in I}$ a family of **Rere**–objects and $(f_i : X \longrightarrow X_i)_{i\in I}$ a family of maps, then $\rho = \{(x, y) \in X \times X : (f_i(x), f_i(y)) \in \rho_i$ for each $i \in I\}$ is the initial **Rere**–structure on X with respect to the given data.

⑨ The category **Simp** of simplicial complexes (and simplicial maps):
a) A *simplicial complex* is a pair $(K, \mathcal{K})$, where K is a set and $\mathcal{K}$ a subset of $\mathcal{P}(K)$ satisfying

Simp_1) $\{k\} \in \mathcal{K}$ for each $k \in K$,
Simp_2) $E \in \mathcal{K}$ implies that E is non–empty and finite,
Simp_3) $E \in \mathcal{K}$ and $F \subset E$ non–empty imply $F \in \mathcal{K}$.

If $(K, \mathcal{K})$ is a simplicial complex, then the elements of $\mathcal{K}$ (resp. K) are called *simplexes* (resp. *vertices*).

b) A map $f : (K, \mathcal{K}) \longrightarrow (K', \mathcal{K}')$ between simplicial complexes is called *simplicial* provided that $f[E] \in \mathcal{K}'$ for each $E \in \mathcal{K}$.
If K is a set, $((K_i, \mathcal{K}_i))_{i\in I}$ a family of simplicial complexes and $(f_i : K \longrightarrow K_i)_{i\in I}$ a family of maps, then
$\mathcal{K} = \{E \subset K : E$ is non–empty and finite such that $f_i[E] \in \mathcal{K}_i$ for each $i \in I\}$
is the initial **Simp**–structure on K with respect to the given data.

⑩ The category **Reg** (resp. **CReg**) of regular (resp. completely regular) topological spaces [and continuous maps.]
(Hint. |**Reg**| (resp. |**CReg**|) is closed under formation of initial structures in **Top**.)

⑪ The category **LCon** (resp. **LPCon**) of locally connected (resp. locally path-connected) topological spaces [and continuous maps].
(Hint. Determine the final structures and compare with the following theorem 1.2.1.1.)

1.2 Special categorical properties of topological constructs

1.2.1 Completeness and cocompleteness

1.2.1.1 Theorem. *Let $\mathcal{C}$ be a construct. Then the following are equivalent:*
(a) $\mathcal{C}$ satisfies (1) in 1.1.2.

(b) For any set X, any family $((X_i, \xi_i))_{i\in I}$ of $\mathcal{C}$–objects indexed by some class I and any family $(f_i : X_i \longrightarrow X)_{i\in I}$ of maps indexed by I there exists a unique $\mathcal{C}$–structure ξ on X which is final with respect to $((X_i, \xi_i), f_i, X, I)$, i.e. such that for any $\mathcal{C}$–object (Y, η) a map $g : (X, \xi) \longrightarrow (Y, \eta)$ is a $\mathcal{C}$–morphisms iff for every $i \in I$ the composite map $g \circ f_i : (X_i, \xi_i) \longrightarrow (Y, \eta)$ is a $\mathcal{C}$–morphism.

Proof. A) (a) $\Longrightarrow$ (b).

α) Let (R_j, ρ_j) be a $\mathcal{C}$–object and let $h_j : X \longrightarrow R_j$ be a map such that $h_j \circ f_i : (X_i, \xi_i) \longrightarrow (R_j, \rho_j)$ is a $\mathcal{C}$–morphism for each $i \in I$.

Let $((R_j, \rho_j), h_j)_{j\in J}$ be the family of all pairs determined in the way above with the index class J. Let ξ be the initial structure with respect to $(X, h_j, (R_j, \rho_j), J)$. Then ξ is the final structure with respect to $((X_i, \xi_i), f_i, X, I)$:

1. Let $g : (X, \xi) \longrightarrow (Y, \eta)$ be a $\mathcal{C}$–morphism and $i \in I$. Since $h_j \circ f_i : (X_i, \xi_i) \longrightarrow (R_j, \rho_j)$ is a $\mathcal{C}$–morphism for each $j \in J$ and ξ is initial, $f_i : (X_i, \xi_i) \longrightarrow (X, \xi)$ is a $\mathcal{C}$–morphism. Thus, $g \circ f_i : (X_i, \xi_i) \longrightarrow (Y, \eta)$ is a $\mathcal{C}$–morphism as a composition of two $\mathcal{C}$–morphisms.

2. Suppose all $g \circ f_i : (X_i, \xi_i) \longrightarrow (Y, \eta)$ are $\mathcal{C}$–morphisms. Then there is some $j \in J$ such that $g = h_j$ and $(Y, \eta) = (R_j, \rho_j)$. Hence, since ξ is initial, g is a $\mathcal{C}$–morphism (cf. 1.1.3.1)).

β) Let ξ' be another $\mathcal{C}$–structure on X which is final with respect to $((X_i, \xi_i), f_i, X, I)$. Then $1_X : (X, \xi) \longrightarrow (X, \xi')$ and $1_X : (X, \xi') \longrightarrow (X, \xi)$ are $\mathcal{C}$–morphisms. Thus $\xi = \xi'$ (cf. 1.1.3.2)).

B) (b) $\Longrightarrow$ (a): analogously to A).

1.2.1.2 By the previous theorem in a topological construct exist abitrary initial and final structures. In particular, corresponding to 1.1.5, the following holds:

Proposition. *In a topological construct $\mathcal{C}$ the final structure ξ on a set X with respect to $((X_i, \xi_i), f_i, X, I)$ is the finest $\mathcal{C}$–structure on X such that all f_i are $\mathcal{C}$–morphisms.*

Proof. Let $(Y, \eta) = (X, \xi)$ and $g = 1_X$ in 1.2.1.1 (2). Then each $f_i : (X_i, \xi_i) \longrightarrow (X, \xi)$ is a $\mathcal{C}$–morphism. Let η be a $\mathcal{C}$–structure on X for which each $f_i : (X_i, \xi_i) \longrightarrow (X, \eta)$ is a $\mathcal{C}$–morphism. Since all $1_X \circ f_i : (X_i, \xi_i) \longrightarrow (X, \eta)$ are $\mathcal{C}$–morphisms, the map $1_X : (X, \xi) \longrightarrow (X, \eta)$ is a $\mathcal{C}$–morphism, i.e. $\xi \leq \eta$.

1.2.1.3 Initial and final structures in a topological construct can be used to define products and coproducts respectively. The corresponding constructions are similar to those in **Top** as the following theorem shows:

Theorem. *Let $\mathcal{C}$ be a topological construct, I a set and $((X_i, \xi_i))_{i\in I}$ a family of $\mathcal{C}$–objects.*

a) If $X = \prod_{i\in I} X_i$ is the cartesian product of the family $(X_i)_{i\in I}$, $p_i : X \longrightarrow X_i$ are the projection maps for each $i \in I$ and ξ is the initial $\mathcal{C}$-structure on X with respect to $(X, p_i, (X_i, \xi_i), I)$, then $((X, \xi), (p_i)_{i\in I})$ is the product of the family $((X_i, \xi_i))_{i\in I}$ in the category $\mathcal{C}$.

b) If $X = \bigcup_{i\in I} X_i \times \{i\}$, $j_i : X_i \longrightarrow X$ are the (canonical) injection maps for each $i \in I$ (i.e. $j_i(y) = (y, i)$ for each $y \in X_i$ and each $i \in I$) and ξ is the final $\mathcal{C}$-structure on X with respect to $((X_i, \xi_i), j_i, X, I)$, then $((j_i)_{i\in I}, (X, \xi))$ is the coproduct of the family $((X_i, \xi_i))_{i\in I}$ in the category $\mathcal{C}$.

Proof.

a) Let (Y, η) be a $\mathcal{C}$–object and $p'_i : (Y, \eta) \longrightarrow (X_i, \xi_i)$ a $\mathcal{C}$–morphism for each $i \in I$.
Then a map $p : (Y, \eta) \longrightarrow (\prod_{i\in I} X_i, \xi)$ is defined by $p_i(p(y)) = p'_i(y)$ for each $y \in Y$ and each $i \in I$.
Since $p_i \circ p = p'_i$ is a $\mathcal{C}$–morphism for each $i \in I$ and ξ is initial, p is a $\mathcal{C}$–morphism.
Obviously, any map $p' : Y \longrightarrow \prod_{i\in I} X_i$ satisfying $p_i \circ p' = p'_i$ for each $i \in I$ coincides with p.

b) Let (Y, η) be a $\mathcal{C}$–object and $j'_i : (X_i, \xi_i) \longrightarrow (Y, \eta)$ a $\mathcal{C}$–morphism for each $i \in I$.
Then for each $x \in X$ there exists a unique $x_i \in X_i$ such that $(x_i, i) = x$ and thus a map $j : (X, \xi) \longrightarrow (Y, \eta)$ is defined by $j(x) = j((x_i, i)) = j(j_i(x_i)) = j'_i(x_i) \in Y$. Since $j \circ j_i = j'_i$ is a $\mathcal{C}$–morphism for each $i \in I$ and ξ is final, j is a $\mathcal{C}$–morphism.
Obviously, any map $j' : X \longrightarrow Y$ satisfying $j' \circ j_i = j'_i$ for each $i \in I$ coincides with j.

1.2.1.4 Definition. *A category $\mathcal{C}$ has products (resp. coproducts) provided that for every set I, each family $(A_i)_{i\in I}$ of $\mathcal{C}$-objects has a product (resp. coproduct) in $\mathcal{C}$.*

1.2.1.5 Remark. By the preceding definition, 1.2.1.3 means that every topological construct has products and coproducts.

1.2.1.6 Definition. *Let $\mathcal{C}$ be a category and $f, g : A \longrightarrow B$ $\mathcal{C}$-morphisms.*

1) A $\mathcal{C}$–morphism

$$k : K \longrightarrow A$$

is called an *equalizer* of f and g provided that the following conditions hold:

(1) $f \circ k = g \circ k$,
(2) For any $D \in |\mathcal{C}|$ and any $h \in [D, A]_{\mathcal{C}}$ such that $f \circ h = g \circ h$ there exists a unique $h' \in [D, K]_{\mathcal{C}}$ such that the diagram

$$c : B \longrightarrow C$$

is called a *coequalizer* of f and g provided that the following conditions hold:

(1′) $c \circ f = c \circ g$,
(2′) For any $D \in |\mathcal{C}|$ and any $h \in [B, D]_{\mathcal{C}}$ such that $h \circ f = h \circ g$, there exists a unique $h' \in [C, D]_{\mathcal{C}}$ such that the diagram

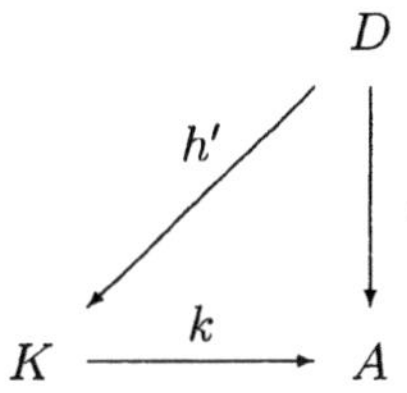

commutes (i.e. $h = k \circ h'$).
Instead of k we often write $E(f,g)$ (see the following proposition).

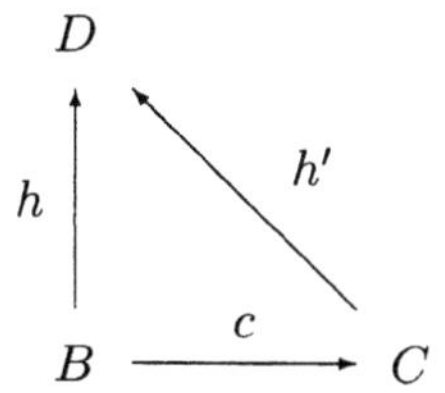

commutes (i.e. $h' \circ c = h$), i.e. c^* is an equalizer of f^* and g^* in the dual category $\mathcal{C}^*$. Instead of c we often write $CE(f,g)$ (see the following proposition).

2) A $\mathcal{C}$–object

$K \in |\mathcal{C}|$ is called an *equalizer of* $A \in |\mathcal{C}|$ provided that $[K,A]_\mathcal{C}$ contains an equalizer.

$C \in |\mathcal{C}|$ is called a *coequalizer of* $B \in |\mathcal{C}|$ provided that $[B,C]_\mathcal{C}$ contains a coequalizer (i.e. C is an equalizer of B in the dual category $\mathcal{C}^*$).

1.2.1.7 Proposition.

a) If $k : K \longrightarrow A$ and $k' : K' \longrightarrow A$ are equalizers of two $\mathcal{C}$–morphisms $f, g : A \longrightarrow B$ in a category $\mathcal{C}$, then there exists a unique $\mathcal{C}$–isomorphism $i : K \longrightarrow K'$ such that $k = k' \circ i$.

b) If $c : B \longrightarrow C$ and $c' : B \longrightarrow C'$ are coequalizers of two $\mathcal{C}$–morphisms $f, g : A \longrightarrow B$ in a category $\mathcal{C}$, then there exists a unique $\mathcal{C}$–isomorphism $i : C' \longrightarrow C$ such that $c = i \circ c'$.

Proof. Since a) and b) are dual, it suffices to prove a): Since k' is an equalizer of f and g, there is a unique $\mathcal{C}$–morphism $i : K \longrightarrow K'$ such that $k' \circ i = k$. Reversing the roles of k and k', there is a unique $\mathcal{C}$–morphism $i' : K' \longrightarrow K$ such that $k \circ i' = k'$. Hence $k \circ (i' \circ i) = (k \circ i') \circ i = k' \circ i = k = k \circ 1_K$. By 1.2.1.6 1) (2) there is a unique $h : K \longrightarrow K$ such that $k \circ h = k$. Thus $h = i' \circ i = 1_K$. Similarly, $i \circ i' = 1_{K'}$. Consequently, i is an isomorphism.

1.2.1.8 Theorem. *Let $\mathcal{C}$ be a topological construct. If (X,ξ), (Y,η) are $\mathcal{C}$–objects and $f, g : (X,\xi) \longrightarrow (Y,\eta)$ $\mathcal{C}$–morphisms, then the following is valid:*

a) Let $K = \{x \in X : f(x) = g(x)\}$ be endowed with the initial $\mathcal{C}$–structure ξ_K with respect to the inclusion map $i : K \longrightarrow X$. Then $i : (K,\xi_K) \longrightarrow (X,\xi)$ is the equalizer of f and g.

b) Let R be the finest equivalence relation on Y for which $f(x)$ and $g(x)$ are equivalent for each $x \in X$ (i.e. R is the intersection of all equivalence relations with this property). If $C = Y/R$ is endowed with the final $\mathcal{C}$–structure η_R with respect to the natural map $\omega : Y \longrightarrow C$, then $\omega : (Y,\eta) \longrightarrow (C,\eta_R)$ is the coequalizer of f and g.

Proof.

a) (1) $f \circ i = g \circ i$ is obvious because f and g coincide on K.

(2) Given a $\mathcal{C}$–morphism $h : (R, \rho) \longrightarrow (X, \xi)$ such that $f \circ h = g \circ h$. Then $h(y) \in K$ for each $y \in R$ because $f(h(y)) = g(h(y))$ for each $y \in R$.
Hence a map $h' : R \longrightarrow K$ is defined by $h'(y) = h(y)$ for each $y \in R$. Since $h = i \circ h'$ is a $\mathcal{C}$–morphism and ξ_K is initial, $h' : (R, \rho) \longrightarrow (K, \xi_K)$ is a $\mathcal{C}$–morphism. Obviously, any map $h'' : R \longrightarrow K$ such that $h = i \circ h''$ coincides with h'.

b) (1') $\omega \circ f = \omega \circ g$ is obvious because $f(x)$ and $g(x)$ are equivalent for each $x \in X$, i.e. $\omega(f(x)) = \omega(g(x))$ for each $x \in X$.

(2') Given a $\mathcal{C}$–morphism $h : (Y, \eta) \longrightarrow (R^*, \rho^*)$ such that $h \circ f = h \circ g$. An equivalence relation π_h defined by $(y, y') \in \pi_h$ iff $h(y) = h(y')$ is assigned to h. Then $f(x)$ and $g(x)$ are equivalent with respect to π_h for each $x \in X$. Thus $R \subset \pi_h$. Hence there exist two maps 1_Y^* and s such that the diagram

$$\begin{array}{ccccc} Y & \xrightarrow{1_Y} & Y & \xrightarrow{h} & R^* \\ {\scriptstyle\omega}\downarrow & & \downarrow{\scriptstyle\omega^*} & \nearrow{\scriptstyle s} & \\ Y/R & \xrightarrow{1_Y^*} & Y/\pi_h & & \end{array}$$

commutes (ω^*: natural map). Then $h = h' \circ \omega$ where $h' = s \circ 1_Y^*$. Since η_R is final with respect to ω, $h' : (Y/R, \eta_R) \longrightarrow (R^*, \rho^*)$ is a $\mathcal{C}$–morphism. Obviously, any map $h'' : Y/R \longrightarrow R^*$ such that $h = h'' \circ \omega$ coincides with h'.

1.2.1.9 Definitions. 1) A *category $\mathcal{C}$ has equalizers* (*resp. coequalizers*) provided that every pair (f, g) of $\mathcal{C}$–morphisms with common domain and common codomain has an equalizer (resp. coequalizer).
2) A category $\mathcal{C}$ is said to be
a) *complete* provided that $\mathcal{C}$ has products and equalizers,
b) *cocomplete* provided that $\mathcal{C}$ has coproducts and coequalizers, i.e. the dual category $\mathcal{C}^*$ is complete.

1.2.1.10 Theorem. *Every topological construct is complete and cocomplete.*

Proof. See 1.2.1.3 and 1.2.1.8.

1.2.2 Special objects and special morphisms

1.2.2.1 Definition. Let $\mathcal{C}$ be a topological construct and X a set. The initial $\mathcal{C}$–structure ξ_i (resp. final $\mathcal{C}$–structure ξ_d) on X with respect to the empty index class I is called *indiscrete* (resp. *discrete*).

1.2.2.2 Remarks. 1) If ξ_i is the indiscrete C–structure on X, then $f : (Y, \eta) \longrightarrow (X, \xi_i)$ is a C–morphism for every object (Y, η) and every map $f : Y \longrightarrow X$.
2) If ξ_d is the discrete C–structure, then $f : (X, \xi_d) \longrightarrow (Y, \eta)$ is a C–morphism for every C–object (Y, η) and every map $f : X \longrightarrow Y$.
3) If X is a set with cardinality at most one, then by 1.1.2 (3) there is exactly one C–structure on X. Hence $\xi_i = \xi_d$.

1.2.2.3 Proposition. *Let C be a topological construct and let (X, ξ), (Y, η) be C-objects. If $f : X \longrightarrow Y$ is a constant map, then $f : (X, \xi) \longrightarrow (Y, \eta)$ is a C-morphism.*

Proof. Since $f : X \longrightarrow Y$ is constant, $f[X]$ is a set with cardinality at most one. Suppose that μ is the initial C–structure on $f[X]$ with respect to the inclusion map $i : f[X] \longrightarrow Y$. Then by 1.2.2.2 3) μ is indiscrete. Hence $f' : (X, \xi) \longrightarrow (f[X], \mu)$ defined by $f'(x) = f(x)$ for each $x \in X$ is a C–morphism. Thus $f = i \circ f' : (X, \xi) \longrightarrow (Y, \eta)$ is a C–morphism as a composition of two C–morphisms.

1.2.2.4 Theorem. *In a topological construct C a C-morphism $f : (X, \xi) \longrightarrow (Y, \eta)$ is*

a) a monomorphism iff $f : X \longrightarrow Y$ is injective.

b) an epimorphism iff $f : X \longrightarrow Y$ is surjective.

Proof.

a) α) Let $x, y \in X$ such that $f(x) = f(y)$.
$\overline{x} : (X, \xi_d) \longrightarrow (X, \xi)$ defined by $\overline{x}(z) = x$ for each $z \in X$ and $\overline{y} : (X, \xi_d) \longrightarrow (X, \xi)$ defined $\overline{y}(z) = y$ for each $z \in X$ are C–morphisms (cf. 1.2.2.2 2)) such that $f \circ \overline{x} = f \circ \overline{y}$. Since f is a monomorphism, it follows $\overline{x} = \overline{y}$, i.e. $x = y$.

β) Let $\gamma, \delta : (X', \xi') \longrightarrow (X, \xi)$ be C–morphisms such that $f \circ \gamma = f \circ \delta$. Then $f(\gamma(x')) = f(\delta(x'))$ for each $x' \in X'$.
Since f is injective $\gamma(x') = \delta(x')$ for each $x' \in X'$. Thus $\gamma = \delta$.

b) α) (indirect). Suppose that f is not surjective. Then there is a $y' \in Y$ such that $y' \notin f[X]$. $\gamma, \delta : (Y, \eta) \longrightarrow (\{0, 1\}, \xi_i)$ defined by $\gamma(y) = 0$ for each $y \in Y$ and

$$\delta(y) = \begin{cases} 0 & \text{for each } y \in f[X] \\ 1 & \text{otherwise} \end{cases}$$

are C–morphisms (cf. 1.2.2.2. 1)) such that $\gamma \circ f = \delta \circ f$ and $\gamma \neq \delta$. Thus f is not an epimorphism.

β) If f is surjective, then for every $y \in Y$ there is an $x \in X$ such that $f(x) = y$. Hence $\alpha \circ f = \beta \circ f$ implies $\alpha(y) = \alpha(f(x)) = \beta(f(x)) = \beta(y)$ for each $y \in Y$, i.e. $\alpha = \beta$. Thus$f$ is an epimorphism.

1.2.2.5 Theorem. *In a topological construct C a C-morphism $f : (X, \xi) \longrightarrow$*

(Y, η) is an

a) extremal monomorphism if and only if it is an **embedding**, *i.e. one of the following two equivalent conditions is fulfilled:*
(1) $f' : (X, \xi) \longrightarrow (f[X], \eta_{f[X]})$ defined by $f'(x) = f(x)$ for each $x \in X$ is an isomorphism where $\eta_{f[X]}$ is the initial C-structure on $f[X]$ with respect to the inclusion map $i : f[X] \longrightarrow Y$.
(2) $f : X \longrightarrow Y$ is injective and ξ is the initial C-structure with respect to f.

b) extremal epimorphism if and only if f is a **quotient map**, *i.e. $f : X \longrightarrow Y$ is surjective and η is the final C-structure.*

Proof. First let us prove the equivalence of (1) and (2):

(1) $\Longrightarrow$ (2). Since $f = i \circ f'$, f is an initial monomorphism as a composition of two initial monomorphisms where a C-morphism $h : (R, \rho) \longrightarrow (S, \sigma)$ is called initial provided that ρ is the initial C-structure on R with respect to h.

(2) $\Longrightarrow$ (1). Obviously, $f' : (X, \xi) \longrightarrow (f[X], \eta_{f[X]})$ is bijective and a C-morphism since $f : (X, \xi) \longrightarrow (Y, \eta)$ is a C-morphism and $\eta_{f[X]}$ is initial with respect to i. Since ξ is initial with respect to f and $f \circ f'^{-1} = i$ is a C-morphism, f'^{-1} is a C-morphism. Thus f' is an isomorphism.

a) α) Suppose $f : (X, \xi) \longrightarrow (Y, \eta)$ is an extremal monomorphism.
The diagram

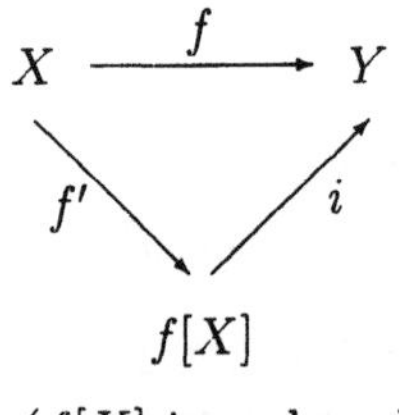

commutes ($f[X]$ is endowed with the initial C-structure w.r.t. the inclusion map $i : f[X] \longrightarrow Y$). Thus f' is a surjective C-morphism (see the definition of the initial C-structure), i.e. an epimorphism (see 1.2.2.4).

b) α) Suppose $f : (X, \xi) \longrightarrow (Y, \eta)$ is an extremal epimorphism. Then $f : X \longrightarrow Y$ is surjective. Let the diagram

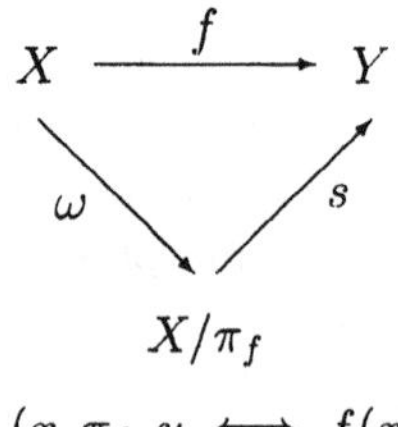

commute ($x \; \pi_f \; y \Longleftrightarrow f(x) = f(y)$; X/π_f is endowed with the final C-structure w.r.t. the natural map $\omega : X \longrightarrow X/\pi_f$). Thus s is an injective (bijective) C-morphism (see the definition of the final C-structure), i.e. s is a monomorphism (see 1.2.2.4.).

Since f is an extremal monomorphism, f' is an isomorphism.

β) Let $f : (X, \xi) \longrightarrow (Y, \eta)$ be an embedding. Furthermore, let the diagram in $\mathcal{C}$

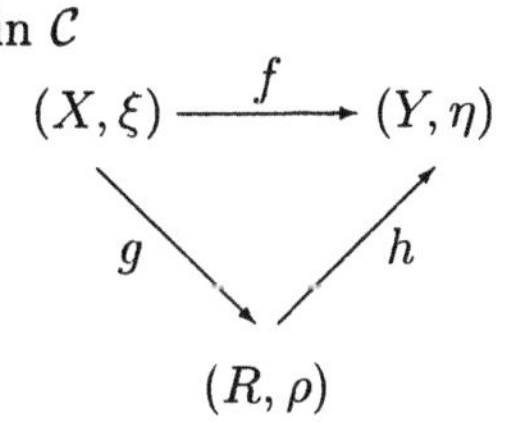

commute, where g is an epimorphism (i.e. a surjective $\mathcal{C}$–morphism). Since f is injective, g is bijective.
Moreover, $g^{-1} : (R, \rho) \longrightarrow (X, \xi)$ is a $\mathcal{C}$–morphism because $f \circ g^{-1} = h$ is a $\mathcal{C}$–morphism and ξ is initial w.r.t. f. Thus, g is an isomorphism. Consequently, f is an extremal monomorphism.

Since f is an extremal epimorphism, s is an isomorphism. It follows immediately that η is the final $\mathcal{C}$–structure w.r.t. f (because X/π_f is endowed with a final structure).

β) Let $f : (X, \xi) \longrightarrow (Y, \eta)$ be a quotient map. Furthermore, let the diagram in $\mathcal{C}$

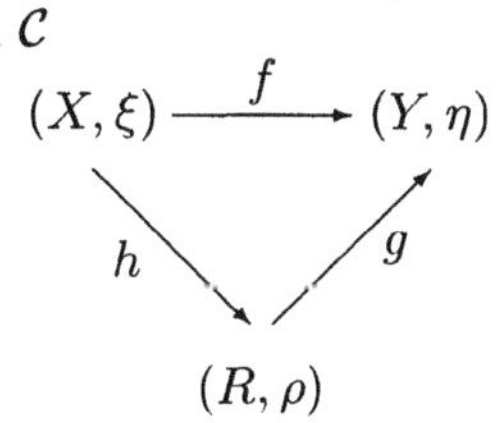

commute, where g is a monomorphism (i.e. an injective $\mathcal{C}$–morphism). Since f is surjective, g is bijective.
Moreover, $g^{-1} : (Y, \eta) \longrightarrow (R, \rho)$ is a $\mathcal{C}$–morphism because $g^{-1} \circ f = h$ is a $\mathcal{C}$–morphism and η is final w.r.t. f. Thus, g is an isomorphism. Consequently, f is an extremal epimorphism.

1.2.2.6 Remarks. 1) a) *In a topological construct $\mathcal{C}$ every coequalizer of two $\mathcal{C}$–morphisms is a quotient map* (cf. 1.2.1.8 b)) *and vice versa,* namely if $f : (X, \xi) \longrightarrow (Y, \eta)$ is a quotient map in $\mathcal{C}$, then f coincides with the natural map $\omega : (X, \xi) \longrightarrow (X/\pi_f, \eta)$ (up to an isomorphism) where $\pi_f \subset X \times X$ consists precisely of those pairs of points (x, y) for which $f(x) = f(y)$ and η is the final $\mathcal{C}$–structure w.r.t. ω. Let $X \times X$ be endowed with the initial $\mathcal{C}$–structure with respect to the projections $p_i : X \times X \longrightarrow X$ $(i = 1, 2)$ and π_f with the initial $\mathcal{C}$–structure with respect to the inclusion map $i : \pi_f \longrightarrow X \times X$. Then $\alpha = p_1|\pi_f$ and $\beta = p_2|\pi_f$ are $\mathcal{C}$–morphisms with codomain X and π_f is the finest equivalence relation on X for which $\alpha((x, y))$ and $\beta((x, y))$ are equivalent for each $(x, y) \in \pi_f$. Thus, by 1.2.1.8 b), ω is the coequalizer of α and β.

b) *In a topological construct $\mathcal{C}$ every equalizer of two $\mathcal{C}$–morphisms is an embedding* (cf. 1.2.1.8 a)) *and vice versa,* namely if $f : (X, \xi) \longrightarrow (Y, \eta)$ is an embedding in $\mathcal{C}$, then f coincides with the inclusion map $i : (U, \eta_U) \longrightarrow (Y, \eta)$ (up to an isomorphism) where $U = f[X]$ and η_U is the initial $\mathcal{C}$–structure on U w.r.t. i.

Put $Y_k = Y \times \{k\}$ for each $k \in \{1, 2\}$ and let $Y_1 \cup Y_2$ be endowed with the final $\mathcal{C}$–structure with respect to the injections $j_k : Y \longrightarrow Y_1 \cup Y_2$ $(k = 1, 2)$. Then an equivalence relation R on $Y_1 \cup Y_2$ is defined by

$$x\,R\,y \Longleftrightarrow \begin{cases} x = y & \text{or} \\ j_1^{-1}(x) = j_2^{-1}(y) \in U & \text{or} \\ j_2^{-1}(x) = j_1^{-1}(y) \in U \end{cases}$$

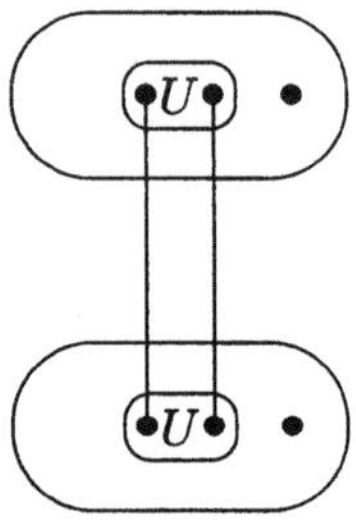

Let $Y_1 \cup Y_2/R$ be endowed with the final $\mathcal{C}$–structure w.r.t. the natural map $\omega : Y_1 \cup Y_2 \longrightarrow Y_1 \cup Y_2/R$. Put $\alpha = \omega \circ j_1$ and $\beta = \omega \circ j_2$. Then α and β are $\mathcal{C}$–morphisms such that $U = \{y \in Y : \alpha(y) = \beta(y)\}$. Thus, by 1.2.1.8 a), i is the equalizer of α and β.

2) In a category $\mathcal{C}$ extremal monomorphisms (resp. extremal epimorphisms) can be used to define subobjects (resp. quotient objects), namely a $\mathcal{C}$–object X is called a *subobject* (resp. *quotient object* [1]) of a $\mathcal{C}$–object Y provided that $[X, Y]_{\mathcal{C}}$ (resp. $[Y, X]_{\mathcal{C}}$) contains an extremal monomorphism (resp. extremal epimorphism). In a topological construct $\mathcal{C}$ we often say *subspace* (resp. *quotient space*) instead of subobject (resp. quotient object).

3) a) A category $\mathcal{C}$ is called (*epi, extremal mono*)*–factorizable* provided that for every $\mathcal{C}$–morphism f there are an epimorphism e and an extremal monomorphism m in $\mathcal{C}$ such that $f = m \circ e$. *Every topological construct $\mathcal{C}$ is* (*epi, extremal mono*)*–factorizable*, namely if $f : (X, \xi) \longrightarrow (Y, \eta)$ is a $\mathcal{C}$–morphism, then $f = i \circ f'$ is the desired factorization, where $f' : (X, \xi) \longrightarrow (f[X], \eta_{f[X]})$ is defined by $f'(x) = f(x)$ for each $x \in X$ and $i : (f[X], \eta_{f[X]}) \longrightarrow (Y, \eta)$ denotes the inclusion map provided that $\eta_{f[X]}$ is the initial $\mathcal{C}$–structure on $f[X]$.

b) A category $\mathcal{C}$ is called (extremal epi, mono)–factorizable provided that the dual category $\mathcal{C}^*$ is (epi, extremal mono)–factorizable, i.e. for every $\mathcal{C}$–morphism f there are an extremal epimorphism e and a monomorphism m in $\mathcal{C}$ such that $f = m \circ e$. *Every topological construct $\mathcal{C}$ is* (*extremal epi, mono*)*–factorizable*, namely if $f : (X, \xi) \longrightarrow (Y, \eta)$ is a $\mathcal{C}$–morphism, then $f = s \circ \omega$ is the desired factorization where $\omega : (X, \xi) \longrightarrow (X/\pi_f, \xi_{\pi_f})$ is the natural map ($x \pi_f x' \Longleftrightarrow f(x) = f(y)$) and $s : (X/\pi_f, \xi_{\pi_f}) \longrightarrow (Y, \eta)$ is defined by $f = s \circ \omega$ provided that ξ_{π_f} is the final $\mathcal{C}$–structure on X/π_f w.r.t. ω.

1.2.2.7 Proposition. *If (X, ξ) is an object in a topological construct $\mathcal{C}$ and*

[1] Sometimes, we say *quotient* instead of quotient object.

$f : X \longrightarrow Y$ a bijective map, then there exists a unique $\mathcal{C}$–structure η on Y such that $f : (X, \xi) \longrightarrow (Y, \eta)$ is an isomorphism, shortly: $\mathcal{C}$–structures are transportable.

Proof. Let η be the final $\mathcal{C}$–structure on Y w.r.t. f. Then $f : (X, \xi) \longrightarrow (Y, \eta)$ and $f^{-1} : (Y, \eta) \longrightarrow (X, \xi)$ are $\mathcal{C}$–morphisms such that $f^{-1} \circ f = 1_X$ and $f \circ f^{-1} = 1_Y$. Conseqently, f is an isomorphism. In order to prove the uniqueness of η let η' be a $\mathcal{C}$–structure on Y such that $f : (X, \xi) \longrightarrow (Y, \eta')$ is an isomorphism. Obviously, η' is the final $\mathcal{C}$–structure on Y w.r.t f. Since final structures are unique (see 1.2.1.1 (2)), we obtain $\eta = \eta'$.

1.2.2.8 Definition. A category $\mathcal{C}$ is called

a) *well–powered* provided that for $X \in |\mathcal{C}|$ there is a set $\{m_i : X_i \longrightarrow X\}$ of monomorphisms which is representative in the following sense:

For each monomorphism $m : Y \longrightarrow X$ there is an m_i and an isomorphism $h_m : Y \longrightarrow X_i$ satisfying $m = m_i \circ h_m$.

b) *co–well–powered* provided that $\mathcal{C}^*$ is well–powered, i.e. for every $X \in |\mathcal{C}|$ there is a set $\{e_i : X \longrightarrow X_i\}$ of epimorphisms which is representative in the following sense:

For each epimorphism $e : X \longrightarrow Y$ there is an e_i and an isomorphism $h_e : X_i \longrightarrow Y$ satisfying $e = h_e \circ e_i$.

1.2.2.9 Theorem. *Every topological construct $\mathcal{C}$ is well–powered and co–well–powered.*

Proof. A) Given a cardinal number k. Then there is a set $\mathcal{Q}$ of $\mathcal{C}$–objects such that every $\mathcal{C}$–object (Y, η) satisfying $|Y| \leq k$ is isomorphic to an object of $\mathcal{Q}$: Let Z be a set such that $|Z| = k$ and let (Y, η) be a $\mathcal{C}$–object such that $|Y| \leq k$. Then Y is equipotent with a subset X of Z, i.e. there is a bijective map $f : Y \longrightarrow X$. By 1.2.2.7 there exists a unique $\mathcal{C}$–structure ξ on X such that $f : (Y, \eta) \longrightarrow (X, \xi)$ is an isomorphism. Now let $\mathcal{Q} = \{(X, \xi) \in |\mathcal{C}| : X \subset Z\}$ and let $\mathcal{M}_X$ be the set of all $\mathcal{C}$–objects with underlying set X (cf. 1.1.2 (2)). Then $\mathcal{Q} = \bigcup_{X \in \mathcal{P}(Z)} \mathcal{M}_X$ is a set.

B) α) Let $f : (Y, \eta) \longrightarrow (X, \xi)$ be a mononormphism in $\mathcal{C}$. Then $|Y| \leq |X| = k$ because f is injective. Hence, using A), there is a representative set of monomorphisms with codomain (X, ξ). Thus, $\mathcal{C}$ is well–powered.

β) Let $f : (X, \xi) \longrightarrow (Y, \eta)$ be an epimorphism in $\mathcal{C}$, i.e. $f : X \longrightarrow Y$ is surjective. Then $|Y| \leq |X| = k$ (For each $y \in Y$ choose a unique $x \in f^{-1}(y)$. Then an injective map $h : Y \longrightarrow X$ is defined by $h(y) = x$ for each $y \in Y$.) Hence, using A), there is a representative set of epimorphisms with domain (X, ξ). Thus, $\mathcal{C}$ is co–well–powered.

Chapter 2

Reflections and Coreflections

As is well–known topological spaces can be related to each other by means of continuous maps. More generally, objects in a category can be related to each other by means of morphisms. There is an analogous relationship between categories via so–called functors. The classical definition of universal maps in the sense ofN. Bourbaki [18] corresponds to a categorical one which utilizes a functor. The existence of all universal maps with respect to a given functor $\mathcal{F}$ is related to a pair of adjoint functors $(\mathcal{G}, \mathcal{F})$, where $\mathcal{G}$ (resp. $\mathcal{F}$) is called a left adjoint (resp. right adoint). The relationships between these functors are described by means of natural transformations u and v (which occur as "maps" between functors). Thus, an adjoint situation $(\mathcal{G}, \mathcal{F}, u, v)$ is obtained. In the first part of this chapter adjoint situations are studied together with some examples. In the second part an important special case of adjoint situations $(\mathcal{G}, \mathcal{F}, u, v)$ is investigated, namely the case where $\mathcal{F}$ is an inclusion functor $\mathcal{I}$ from a subcategory $\mathcal{A}$ of a category $\mathcal{C}$ to $\mathcal{C}$ (the notion of inclusion functor corresponds to the notion of inclusion map in classical mathematics). Then $\mathcal{G}$ is called a reflector from $\mathcal{C}$ to $\mathcal{A}$ and $\mathcal{A}$ is called reflective. If the morphisms belonging to all universal maps with respect to $\mathcal{I}$ are epimorphisms, extremal epimorphisms or bimorphisms, then $\mathcal{G}$ is called an epireflector, extremal epireflector or bireflector respectively and we say epireflective, extremal epireflective or bireflective subcategory rather than reflective subcategory. The famous characterization theorem for epireflective (and extremal epireflective) subcategories is proved and the results are applied to bireflective subconstructs of topological constructs. All concepts are dualized and the role of bireflectors (dually: bicoreflectors) in the realm of topological constructs is clarified. Concerning the historical development of all these categorical concepts and results the interested reader in referred to an article of H. Herrlich and G.E. Strecker [73]. In the third part of the present chapter bireflectors and bicoreflectors are used to describe the relations between various types of convergence structures and uniform convergence structures as well as the linkage between them.

Besides topological spaces which have been introduced by F. Hausdorff [58] (nowadays called Hausdorff spaces) and which have been defined in its present form by C. Kuratowski [91] other important types of convergence structures have been

introduced by various authors, namely 1^0 pretopological spaces (= closure spaces in the terminology of E. Čech [29] or, together with the axiom T_1, "Gestufte Räume" in the sense of F. Hausdorff [59]) introduced by F. Riesz [125] under the name "Mathematisches Kontinuum" (where he assumed the T_1 axiom and a weakening of the T_2 axiom), 2^0 pseudotopological spaces by G. Choquet [30], 3^0 limit spaces by H.-J. Kowalsky [90] and H.R. Fischer [44] and 4^0 Kent convergence spaces by D.C. Kent [84]. Uniform spaces introduced by A. Weil [147] belong to classical uniform convergence structures and are considered in some detail. Generalizations of uniform spaces such as uniform limit spaces are due to C.H. Cook and H.R. Fischer [33] and are studied here in a slightly modified form proposed by O. Wyler [151], since in this case natural function spaces exist, which has been demonstrated by R. Lee [93].
Further generalizations of uniform limit spaces are obtained by omitting some of their defining axioms, e.g. semiuniform limit spaces and semiuniform convergence spaces (these names go back to O. Wyler [151]). Filter spaces originally introduced by M. Katětov [80] in the realm of merotopic spaces (cf. chapter 7 for the details) are the link between uniform convergence structures and convergence structures provided that the latter fulfill a certain less restrictive symmetry condition due to W.A. Robertson [127]. The class of filter spaces is nicely embedded into the class of semiuniform convergence spaces which form the common roof of the above mentioned uniform convergence structures and (symmetric) convergence structures and for which the concept of filter due to H. Cartan ([26] and [27]) is fundamental (filter bases have been introduced much earlier by L. Vietoris [145] under the name 'Kränze'). A first systematic treatment of the uniform part of semiuniform convergence structures goes back to A. Behling [9].

2.1 Universal maps and adjoint functors

2.1.1 Definition. Let $\mathcal{C}$ and $\mathcal{D}$ be categories, let $\mathcal{F}_1 : |\mathcal{C}| \longrightarrow |\mathcal{D}|$ and $\mathcal{F}_2 :$ Mor $\mathcal{C} \longrightarrow$ Mor $\mathcal{D}$ be maps. Then the quadruple $\mathcal{F} = (\mathcal{C}, \mathcal{D}, \mathcal{F}_1, \mathcal{F}_2)$ is called a *functor* from $\mathcal{C}$ to $\mathcal{D}$ or more exactly a *covariant functor* (denoted by $\mathcal{F} : \mathcal{C} \longrightarrow \mathcal{D}$) provided that the following are satisfied, where we write $\mathcal{F}(A)$ (resp. $\mathcal{F}(f)$) instead of $\mathcal{F}_1(A)$ (resp. $\mathcal{F}_2(f)$):

F_1) $f \in [A, B]_{\mathcal{C}}$ implies $\mathcal{F}(f) \in [\mathcal{F}(A), \mathcal{F}(B)]_{\mathcal{D}}$,
F_2) $\mathcal{F}(f \circ g) = \mathcal{F}(f) \circ \mathcal{F}(g)$ whenever $f \circ g$ is defined (i.e. the domain of f is equal to the codomain of g),
F_3) $\mathcal{F}(1_A) = 1_{\mathcal{F}(A)}$ for each $A \in |\mathcal{C}|$.

If F_1) and F_2) are replaced by

F_1') $f \in [A, B]_{\mathcal{C}}$ implies $\mathcal{F}(f) \in [\mathcal{F}(B), \mathcal{F}(A)]_{\mathcal{D}}$ and
F_2') $\mathcal{F}(f \circ g) = \mathcal{F}(g) \circ \mathcal{F}(f)$ (whenever $f \circ g$ is defined) respectively, then $\mathcal{F}$ is called a *contravariant functor* from $\mathcal{C}$ to $\mathcal{D}$ (which may also be defined as a covariant functor from $\mathcal{C}^*$ to $\mathcal{D}$).

2.1.2 Examples. ① The *identity functor* $\mathcal{I}_{\mathcal{C}} : \mathcal{C} \longrightarrow \mathcal{C}$ maps objects and morphisms identically to themselves (covariant functor!).

② *Constant functors:* Let $\mathcal{C}$ and $\mathcal{D}$ be arbitrary categories and let $X \in |\mathcal{D}|$. For every $A \in |\mathcal{C}|$ and every $f \in \text{Mor}\, \mathcal{C}$, put $\mathcal{F}(A) = X$ and $\mathcal{F}(f) = 1_X$ (co- and contravariant functor!)

③ *Forgetful* (or *underlying*) *functors:* a) Let $\mathcal{C}$ be a (topological) construct and let $\mathcal{F} : \mathcal{C} \longrightarrow \mathbf{Set}$ be defined by $\mathcal{F}((X,\xi)) = X$ and $\mathcal{F}(f) = f$ [= map between the underlying sets] (covariant functor!).
b) Let $\mathcal{C}$ and $\mathcal{X}$ be categories and $\mathcal{F} : \mathcal{C} \longrightarrow \mathcal{X}$ a functor which is *faithful*, i.e. for each pair $(A, B) \in |\mathcal{C}| \times |\mathcal{C}|$ the map $F_{A,B} : [A, B]_{\mathcal{C}} \longrightarrow [\mathcal{F}(A), \mathcal{F}(B)]_{\mathcal{D}}$ defined by $F_{A,B}(f) = \mathcal{F}(f)$ for each $f \in [A, B]_{\mathcal{C}}$ is injective. Then the pair $(\mathcal{C}, \mathcal{F})$ is called a *concrete category* over $\mathcal{X}$. Sometimes $\mathcal{F}$ is called the forgetful (or underlying) functor of the concrete category $(\mathcal{C}, \mathcal{F})$ and $\mathcal{X}$ is called the *base category* for $(\mathcal{C}, \mathcal{F})$. Thus, a *construct* may be defined as a concrete category over **Set**.

④ The *dualizing functor* $\mathcal{F} : \mathcal{C} \longrightarrow \mathcal{C}^*$ is defined by $\mathcal{F}(X) = X$ and $\mathcal{F}(f) = f^*$ (contravariant functor!).

⑤ *Inclusion functors:* Let $\mathcal{C}$ be a category, $\mathcal{A}$ a *subcategory*, i.e. $\mathcal{A}$ is a category such that

1. $|\mathcal{A}| \subset |\mathcal{C}|$,
2. $[A, B]_{\mathcal{A}} \subset [A, B]_{\mathcal{C}}$ for each $(A, B) \in |\mathcal{A}| \times |\mathcal{A}|$,
3. The composition of morphisms in $\mathcal{A}$ coincides with the composition of these morphisms in $\mathcal{C}$,
4. For each $A \in |\mathcal{A}|$ the identity 1_A is the same in $\mathcal{A}$ and in $\mathcal{C}$.

(If $[A, B]_{\mathcal{A}} = [A, B]_{\mathcal{C}}$ is satisfied instead of 2., then $\mathcal{A}$ is called *full*.)

The inclusion functor $\mathcal{F}_e : \mathcal{A} \longrightarrow \mathcal{C}$ is defined by $\mathcal{F}_e(A) = A$ for each $A \in |\mathcal{A}|$ and $\mathcal{F}_e(f) = f$ for each $f \in \text{Mor}\, \mathcal{A}$ (covariant functor!).

2.1.3 Remarks. 1) Because of the property F_2) *a functor preserves commutative diagrams and* by F_2) and F_3) it preserves *isomorphisms.*

2) A subcategory $\mathcal{A}$ of a construct $\mathcal{C}$ is also called a *subconstruct.*

2.1.4 Definition. Let $\mathcal{A}$ and $\mathcal{B}$ be categories, $\mathcal{F} : \mathcal{A} \longrightarrow \mathcal{B}$ a functor and $B \in |\mathcal{B}|$. A pair (u, A) with $A \in |\mathcal{A}|$ and $u : B \longrightarrow \mathcal{F}(A)$ is called a *universal map for B with respect to $\mathcal{F}$* provided that for each $A' \in |\mathcal{A}|$ and each $f : B \longrightarrow \mathcal{F}(A')$ there exists a unique $\mathcal{A}$–morphism $\overline{f} : A \longrightarrow A'$ such that the diagram

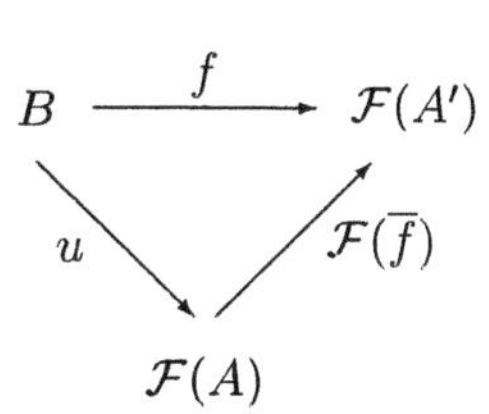

commutes.

2.1.5 Examples. ① Let $\mathbf{Top}_0$ be the category of all topological T_0–spaces (and continuous maps) and $\mathcal{F}_e : \mathbf{Top}_0 \longrightarrow \mathbf{Top}$ the inclusion functor. For each $(X, \mathcal{X}) \in |\mathbf{Top}|$ define an equivalence relation R on X as follows

$$x \, R \, y \Longleftrightarrow \overline{\{x\}} = \overline{\{y\}} \, .$$

Let $\omega : X \longrightarrow X/R$ be the natural map. If $\mathcal{X}_R$ denotes the final **Top**–structure w.r.t. ω, then $(X/R, \mathcal{X}_R)$ is a T_0–space and $(\omega, (X/R, \mathcal{X}_R))$ is a universal map for $(X, \mathcal{X})$ w.r.t. $\mathcal{F}_e$, i.e. for every T_0–space $(Y, \mathcal{Y})$ and every continuous map $f : (X, \mathcal{X}) \longrightarrow (Y, \mathcal{Y})$ there is a unique continuous map $\overline{f} : (X/R, \mathcal{X}_R) \longrightarrow (Y, \mathcal{Y})$ such that the diagram

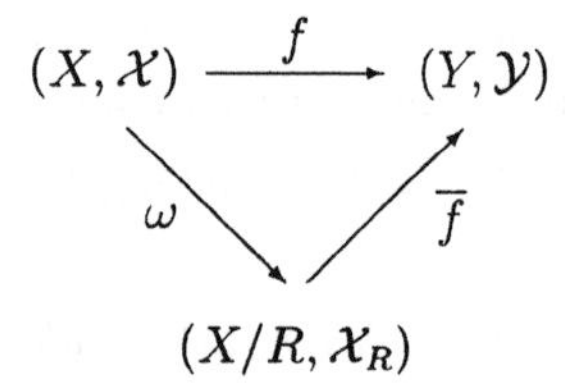

commutes.

② Let **Tych** be the category of Tychonoff spaces [= completely regular T_1–spaces = completely regular T_0–spaces] (and continuous maps) and let $\mathbf{CompT}_2$ be the category of compact Hausdorff spaces (and continuous maps). If the Stone–Čech compactification of $X \in |\mathbf{Tych}|$ is denoted by $\beta(X)$ and $\beta_X : X \longrightarrow \beta(X)$ is the (canonical) embedding, then $(\beta_X, \beta(X))$ is a universal map for X with respect to the inclusion functor $\mathcal{F}_e : \mathbf{CompT}_2 \longrightarrow \mathbf{Tych}$ (cf. e.g. [82; chapter 5, theorem 27]).

③ Let $\mathcal{C}$ be a topological construct and $\mathcal{F}_u : \mathcal{C} \longrightarrow \mathbf{Set}$ the forgetful functor. If X is a set and ξ_d the discrete $\mathcal{C}$–structure on X, then $(1_X, (X, \xi_d))$ is a universal map for X with respect to $\mathcal{F}_u$.

2.1.6 Proposition. *Let each of (u, A) and (u', A') be a universal map for some $B \in |\mathcal{B}|$ with respect to $\mathcal{F} : \mathcal{A} \longrightarrow \mathcal{B}$. Then there is an isomorphism $f : A \longrightarrow A'$ such that the diagram*

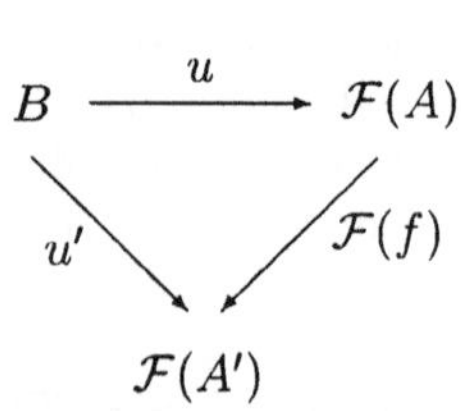

commutes.

Proof. Since (u, A) is a universal map, there is a unique morphism $f : A \longrightarrow A'$ such that the diagram

$$(\mathrm{D}_1)\qquad \begin{array}{ccc} B & \xrightarrow{u'} & \mathcal{F}(A') \\ & {\scriptstyle u}\searrow \quad \nearrow{\scriptstyle \mathcal{F}(f)} & \\ & \mathcal{F}(A) & \end{array}$$

commutes. Since (u', A') is a universal map there is a unique morphism $g : A' \longrightarrow A$ such that the diagram

$$(\mathrm{D}_2)\qquad \begin{array}{ccc} B & \xrightarrow{u} & \mathcal{F}(A) \\ & {\scriptstyle u'}\searrow \quad \nearrow{\scriptstyle \mathcal{F}(g)} & \\ & \mathcal{F}(A') & \end{array}$$

commutes. The diagrams (D_1) and (D_2) form the following commutative diagram

$$(\mathrm{D})\qquad \begin{array}{ccc} & & \mathcal{F}(A) \\ & {\scriptstyle u}\nearrow & \uparrow {\scriptstyle \mathcal{F}(g)} \\ B & \xrightarrow{u'} & \mathcal{F}(A') \\ & {\scriptstyle u}\searrow & \uparrow {\scriptstyle \mathcal{F}(f)} \\ & & \mathcal{F}(A) \end{array}$$

There is a unique morphism from A to A whose image under $\mathcal{F}$ makes the outer triangle of (D) commutative. Since $\mathcal{F}$ is a functor, $g \circ f$ and 1_A are two morphisms satisfying this property. Thus, $g \circ f = 1_A$. Similarly, one can show that $f \circ g = 1_{A'}$. Consequently, f is an isomorphism.

2.1.7. For every functor $\mathcal{H} : \mathcal{C} \longrightarrow \mathcal{D}$ there is (in a natural way) a *dual functor* (or *opposite functor*) $\mathcal{H}^* : \mathcal{C}^* \longrightarrow \mathcal{D}^*$. One obtains $\mathcal{H}^*$ by applying first the dualizing functor $\mathcal{C}^* \longrightarrow (\mathcal{C}^*)^* = \mathcal{C}$, then the functor $\mathcal{H} : \mathcal{C} \longrightarrow \mathcal{D}$ and finally the dualizing functor $\mathcal{D} \longrightarrow \mathcal{D}^*$. It holds $(\mathcal{H}^*)^* = \mathcal{H}$.

If $\mathcal{F} : \mathcal{A} \longrightarrow \mathcal{B}$ is a functor and $B \in |\mathcal{B}|$, then a pair (A, u) with $A \in |\mathcal{A}|$ and $u : \mathcal{F}(A) \longrightarrow B$ is called a *co-universal map for B with respect to $\mathcal{F}$*, provided that (u^*, A) is a universal map for B with respect to $\mathcal{F}^* : \mathcal{A}^* \longrightarrow \mathcal{B}^*$; i.e., provided that for each $A' \in |\mathcal{A}|$ and each $\mathcal{B}$-morphism $f : \mathcal{F}(A') \longrightarrow B$, there exists a unique $\mathcal{A}$-morphism $\overline{f} : A' \longrightarrow A$ such that the diagram

$$\begin{array}{ccc} \mathcal{F}(A) & \xrightarrow{u} & B \\ & {\scriptstyle \mathcal{F}(\overline{f})}\nwarrow \quad \nearrow{\scriptstyle f} & \\ & \mathcal{F}(A') & \end{array}$$

commutes.

2.1.8 Definition. Let $\mathcal{C}$ be a category. A family $(f_i : X \longrightarrow X_i)_{i\in I}$ of $\mathcal{C}$–morphisms indexed by some class I, shortly: a *source* in $\mathcal{C}$, is called a *mono–source* in $\mathcal{C}$ provided that for any pair $Y \overset{\alpha}{\underset{\beta}{\rightrightarrows}} X$ of C–morphisms such that $f_i \circ \alpha = f_i \circ \beta$ for each $i \in I$, it follows that $\alpha = \beta$.

Dually: A family $(f_i : X_i \longrightarrow X)_{i\in I}$ of $\mathcal{C}$–morphisms indexed by some class I, shortly: a *sink* in $\mathcal{C}$, is called an *epi–sink* in $\mathcal{C}$ provided that $(f_i^* : X \longrightarrow X_i)_{i\in I}$ is a mono–source in $\mathcal{C}^*$; i.e., provided that for any pair $X \overset{\alpha}{\underset{\beta}{\rightrightarrows}} Y$ of $\mathcal{C}$–morphisms such that $\alpha \circ f_i = \beta \circ f_i$ for each $i \in I$, it follows that $\alpha = \beta$.

2.1.9 Examples. In each of the following examples let $\mathcal{C}$ be a category:

① A single $\mathcal{C}$–morphism $f : X \longrightarrow Y$ considered as a one–element source (resp. sink) is a mono–source (resp. epi–sink) iff it is a monomorphism (resp. an epimorphism).

② If $(P, (p_i)_{i\in I})$ is the product of some family $(A_i)_{i\in I}$ of $\mathcal{C}$–objects, then $(p_i : P \longrightarrow A_i)_{i\in I}$ is a mono–source.

③ If $((j_i)_{i\in I}, C)$ is the coproduct of some family $(A_i)_{i\in I}$ of $\mathcal{C}$–objects, then $(j_i : A_i \longrightarrow C)_{i\in I}$ is an epi–sink.

2.1.10 Theorem. *Let $\mathcal{F} : \mathcal{A} \longrightarrow \mathcal{B}$ be a functor such that for each $B \in |\mathcal{B}|$ there exists a universal map (u_B, A_B) w.r.t. $\mathcal{F}$. Then the following are valid:*

(1) $\mathcal{F}$ preserves mono–sources, i.e. if $(f_i : A \longrightarrow A_i)_{i\in I}$ is a mono–source in $\mathcal{A}$, then $(\mathcal{F}(f_i) : \mathcal{F}(A) \longrightarrow \mathcal{F}(A_i))_{i\in I}$ is a mono–source in $\mathcal{B}$.
(2) $\mathcal{F}$ preserves products, i.e. if $(P, (p_i)_{i\in I})$ is a product in $\mathcal{A}$, then $(\mathcal{F}(P), (\mathcal{F}(p_i))_{i\in I})$ is a product in $\mathcal{B}$.
(3) $\mathcal{F}$ preserves equalizers, i.e. if $k = E(f, g)$ in $\mathcal{A}$, then $\mathcal{F}(k) = E(\mathcal{F}(f), \mathcal{F}(g))$ in $\mathcal{B}$.

Proof. (1) Let $(f_i : A \longrightarrow A_i)_{i\in I}$ be a mono–source in $\mathcal{A}$ and let $B \overset{\alpha}{\underset{\beta}{\rightrightarrows}} \mathcal{F}(A)$ be a pair of $\mathcal{B}$–morphisms such that $\mathcal{F}(f_i)\circ\alpha = \mathcal{F}(f_i)\circ\beta$ for each $i \in I$. If (u_B, A_B) is a universal map for B w.r.t. $\mathcal{F}$, then there exist $\mathcal{A}$–morphisms $A_B \overset{\overline{\alpha}}{\underset{\overline{\beta}}{\rightrightarrows}} A$ with $\mathcal{F}(\overline{\alpha}) \circ u_B = \alpha$ and $\mathcal{F}(\overline{\beta}) \circ u_B = \beta$. Thus, for each $i \in I$,

$$\begin{aligned}\mathcal{F}(f_i \circ \overline{\alpha}) \circ u_B &= \mathcal{F}(f_i) \circ \mathcal{F}(\overline{\alpha}) \circ u_B = \mathcal{F}(f_i) \circ \alpha = \mathcal{F}(f_i) \circ \beta \\ &= \mathcal{F}(f_i) \circ \mathcal{F}(\overline{\beta}) \circ u_B \\ &= \mathcal{F}(f_i \circ \overline{\beta}) \circ u_B \text{ and consequently, } f_i \circ \overline{\alpha} = f_i \circ \overline{\beta}\end{aligned}$$

for each $i \in I$. Since $(f_i : A \longrightarrow A_i)_{i\in I}$ is a mono–source, we obtain $\overline{\alpha} = \overline{\beta}$, and thus $\alpha = \beta$.

(2) Let $(P, (p_i)_{i\in I})$ be a product in $\mathcal{A}$ of a family $(A_i)_{i\in I}$ of $\mathcal{A}$–objects and let $(Q, (q_i)_{i\in I})$ be given with $Q \in |\mathcal{B}|$ and $q_i : Q \longrightarrow \mathcal{F}(A_i)$ for each $i \in I$. If (u_Q, A_Q)

is a universal map for Q for w.r.t. $\mathcal{F}$, then, for each $i \in I$, there is an $\mathcal{A}$–morphism $f_i : A_Q \longrightarrow A_i$ such that $\mathcal{F}(f_i) \circ u_Q = q_i$. Since $(P, (p_i)_{i \in I})$ is a product, there is a (unique) $\mathcal{A}$–morphism $p : A_Q \longrightarrow P$ such that $p_i \circ p = f_i$ for each $i \in I$. Thus, for each $i \in I$, the diagram

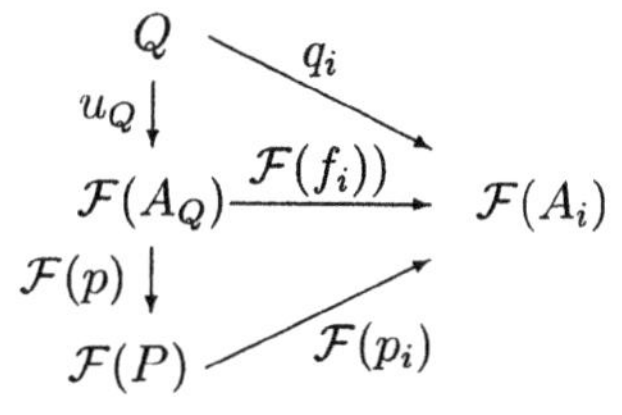

commutes. The uniqueness of the $\mathcal{B}$–morphism $h = \mathcal{F}(p) \circ u_Q$ follows from the fact that $(\mathcal{F}(p_i) : \mathcal{F}(P) \longrightarrow \mathcal{F}(p_i))_{i \in I}$ is a mono–source according to (1) (cf. 2.1.9. ②). Thus, $(\mathcal{F}(P), (\mathcal{F}(p_i))_{i \in I})$ is a product in $\mathcal{B}$.

(3) Let $k : K \longrightarrow A$ be an equalizer of two $\mathcal{A}$–morphisms $f, g : A \longrightarrow A'$. Further, let $\ell : B \longrightarrow \mathcal{F}(A)$ be a $\mathcal{B}$–morphism with $\mathcal{F}(f) \circ \ell = \mathcal{F}(g) \circ \ell$. If (u_B, A_B) is a universal map w.r.t. $\mathcal{F}$, then there is a (unique) $\mathcal{A}$–morphism $h : A_B \longrightarrow A$ such that $\mathcal{F}(h) \circ u_B = \ell$, which implies $\mathcal{F}(f \circ h) \circ u_B = \mathcal{F}(f) \circ \mathcal{F}(h) \circ u_B = \mathcal{F}(f) \circ \ell = \mathcal{F}(g) \circ \ell = \mathcal{F}(g) \circ \mathcal{F}(h) \circ u_B = \mathcal{F}(g \circ h) \circ u_B$, and consequently, $f \circ h = g \circ h$. Since $k : K \longrightarrow A$ is an equalizer, there is a (unique) $h' : A_B \longrightarrow K$ such that $k \circ h' = h$. Hence, the diagram

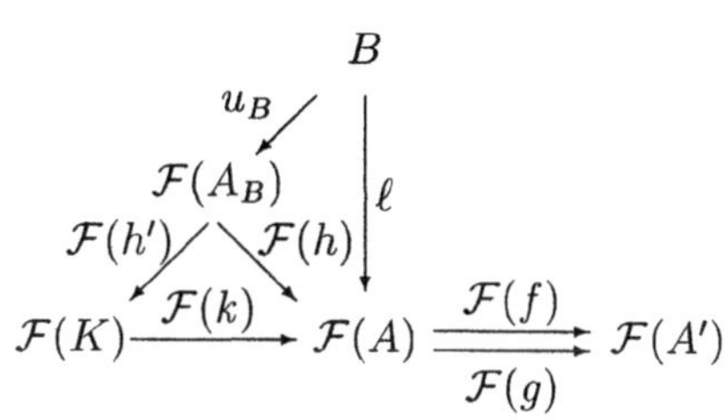

commutes. The uniqueness of $\ell' = \mathcal{F}(h') \circ u_B$ follows from the fact that $\mathcal{F}(k)$ is a monomorphism (because $k : K \longrightarrow A$ is a monomorphism and (1) is valid). Thus $\mathcal{F}(k) = E(\mathcal{F}(f), \mathcal{F}(g))$.

2.1.11. In 2.1.1 we have defined functors in order to describe categories. Now the question arises how we can describe functors. In other words: What are the right 'maps' between functors?
The answer is given by introducing the so–called natural transformations. We will also explain, when we do not distinguish between two functors, i.e. when they are isomorphic:

Definitions. Let $\mathcal{C}$ and $\mathcal{D}$ be categories and $\mathcal{F}, \mathcal{G} : \mathcal{C} \longrightarrow \mathcal{D}$ functors.

1) A family $\eta = (\eta_A)_{A \in |\mathcal{C}|}$ such that $\eta_A \in [\mathcal{F}(A), \mathcal{G}(A)]_{\mathcal{D}}$ for each $A \in |\mathcal{C}|$ is called a *natural transformation* (denoted by $\eta : \mathcal{F} \longrightarrow \mathcal{G}$) provided that for each pair $(A, B) \in |\mathcal{C}| \times |\mathcal{C}|$ and each $f \in [A, B]_{\mathcal{C}}$ the following diagram

$$\begin{array}{ccc} \mathcal{F}(A) & \xrightarrow{\eta_A} & \mathcal{G}(A) \\ {\scriptstyle \mathcal{F}(f)}\downarrow & & \downarrow{\scriptstyle \mathcal{G}(f)} \\ \mathcal{F}(B) & \xrightarrow{\eta_B} & \mathcal{G}(B) \end{array}$$

commutes.

2) A natural transformation $\eta : \mathcal{F} \longrightarrow \mathcal{G}$ is called a *natural equivalence* (or a *natural isomorphism*) provided that for each $A \in |\mathcal{C}|$, η_A is an isomorphism.

3) $\mathcal{F}$ and $\mathcal{G}$ are said to be *naturally equivalent* (or *naturally isomorphic*) [denoted by $\mathcal{F} \approx \mathcal{G}$] iff there exists a natural equivalence $\eta : \mathcal{F} \longrightarrow \mathcal{G}$.

2.1.12. Let $\mathcal{F} : \mathcal{A} \longrightarrow \mathcal{B}$ and $\mathcal{G} : \mathcal{B} \longrightarrow \mathcal{C}$ be functors, i.e. $\mathcal{F} = (\mathcal{A}, \mathcal{B}, \mathcal{F}_1, \mathcal{F}_2)$ and $\mathcal{G} = (\mathcal{B}, \mathcal{C}, \mathcal{G}_1, \mathcal{G}_2)$. Then a functor $\mathcal{G} \circ \mathcal{F} : \mathcal{A} \longrightarrow \mathcal{C}$ (the composition of $\mathcal{F}$ and $\mathcal{G}$) is defined by $\mathcal{G} \circ \mathcal{F} = (\mathcal{A}, \mathcal{C}, \mathcal{G}_1 \circ \mathcal{F}_1, \mathcal{G}_2 \circ \mathcal{F}_2)$, i.e. $(\mathcal{G} \circ \mathcal{F})(A) = \mathcal{G}(\mathcal{F}(A))$ for each $A \in |\mathcal{A}|$ and $(\mathcal{G} \circ \mathcal{F})(f) = \mathcal{G}(\mathcal{F}(f))$ for each $f \in \operatorname{Mor} \mathcal{A}$.
With this notation the following holds:

Theorem. *Let $\mathcal{F} : \mathcal{A} \longrightarrow \mathcal{B}$ be a functor such that, for each $B \in |\mathcal{B}|$, there is a universal map (u_B, A_B) with respect to $\mathcal{F}$. Then there exists a unique functor $\mathcal{G} : \mathcal{B} \longrightarrow \mathcal{A}$ such that the following are satisfied:*

(1) *$\mathcal{G}(B) = A_B$ for each $B \in |\mathcal{B}|$.*
(2) *$u = (u_B) : \mathcal{I}_{\mathcal{B}} \longrightarrow \mathcal{F} \circ \mathcal{G}$ is a natural transformation, where $\mathcal{I}_{\mathcal{B}} : \mathcal{B} \longrightarrow \mathcal{B}$ denotes the identity functor.*

Corollary. *There is a unique natural transformation $v = (v_A) : \mathcal{G} \circ \mathcal{F} \longrightarrow \mathcal{I}_{\mathcal{A}}$ ($\mathcal{I}_{\mathcal{A}} : \mathcal{A} \longrightarrow \mathcal{A}$ is the identity functor) such that the following are valid:*

(a) *$\mathcal{F}(v_A) \circ u_{\mathcal{F}(A)} = 1_{\mathcal{F}(A)}$ for each $A \in |\mathcal{A}|$.*
(b) *$v_{\mathcal{G}(B)} \circ \mathcal{G}(u_B) = 1_{\mathcal{G}(B)}$ for each $B \in |\mathcal{B}|$.*

Proof of the theorem. By (1) a map is defined from the object class of $\mathcal{B}$ to the object class of $\mathcal{A}$. We try to find a map between the corresponding morphism classes. If $f : B \longrightarrow B'$ is a $\mathcal{B}$–morphism, then there is a unique $\mathcal{A}$–morphism $\overline{f} : A_B \longrightarrow A_{B'}$, such that the diagram

(D₁)
$$\begin{array}{ccc} \mathcal{I}_{\mathcal{B}}(B) = B & \xrightarrow{u_B} & \mathcal{F}(A_B) = \mathcal{F}(\mathcal{G}(B)) \\ {\scriptstyle f}\downarrow & & \downarrow{\scriptstyle \mathcal{F}(\overline{f})} \\ \mathcal{I}_{\mathcal{B}}(B') = B' & \xrightarrow{u_{B'}} & \mathcal{F}(A_{B'}) = \mathcal{F}(\mathcal{G}(B')) \end{array}$$

commutes, because (u_B, A_B) is a universal map for B w.r.t. $\mathcal{F}$. For each $f \in \operatorname{Mor} \mathcal{B}$, put $\mathcal{G}(f) = \overline{f}$. Then $\mathcal{G}$ is a functor: If $f : B \longrightarrow B'$ and $g : B' \longrightarrow B''$ are $\mathcal{B}$–morphisms, then the following diagram

(D$_2$)
$$\begin{array}{ccc} B & \xrightarrow{u_B} & \mathcal{F}(A_B) \\ {\scriptstyle f}\downarrow & & \downarrow{\scriptstyle \mathcal{F}(\overline{f})} \\ B' & \xrightarrow{u_{B'}} & \mathcal{F}(A_{B'}) \\ {\scriptstyle g}\downarrow & & \downarrow{\scriptstyle \mathcal{F}(\overline{g})} \\ B'' & \xrightarrow{u_{B''}} & \mathcal{F}(A_{B''}) \end{array}$$

commutes. Hence $\overline{g} \circ \overline{f}$ is a morphism such that its image under $\mathcal{F}$ makes the outer square of (D$_2$) commutative (because $\mathcal{F}(\overline{g} \circ \overline{f}) = \mathcal{F}(\overline{g}) \circ \mathcal{F}(\overline{f})$). Since there is a unique morphism of this kind, namely $\overline{g \circ f}$, it follows that $\overline{g \circ f} = \overline{g} \circ \overline{f}$, i.e. $\mathcal{G}(g \circ f) = \mathcal{G}(g) \circ \mathcal{G}(f)$. For each $B \in |\mathcal{B}|$, $1_{A_B} : A_B \longrightarrow A_B$ is a morphism such that its image under $\mathcal{F}$ makes the diagram

(D$_3$)
$$\begin{array}{ccc} B & \xrightarrow{u_B} & \mathcal{F}(A_B) \\ {\scriptstyle 1_B}\downarrow & & \downarrow{\scriptstyle \mathcal{F}(1_{A_B})} \\ B & \xrightarrow{u_B} & \mathcal{F}(A_B) \end{array}$$

commutative (because $\mathcal{F}(1_{A_B}) = 1_{\mathcal{F}(A_B)}$). Since there is a unique morphism of this kind, namely $\overline{1_B}$, it follows that $\overline{1_B} = 1_{A_B}$, i.e. $\mathcal{G}(1_B) = 1_{\mathcal{G}(B)}$.

From the commutativity of (D$_1$) it follows that $u = (u_B) : \mathcal{I}_{\mathcal{B}} \longrightarrow \mathcal{F} \circ \mathcal{G}$ is a natural transformation.

Let $\mathcal{G}' : \mathcal{B} \longrightarrow \mathcal{A}$ be any functor satisfying (1) and (2). Then $\mathcal{G}'(B) = A_B = \mathcal{G}(B)$ by definition and $\mathcal{G}'(f) = \overline{f} = \mathcal{G}(f)$ by the uniqueness of $\overline{f}$. Thus, $\mathcal{G} = \mathcal{G}'$.

Proof of the corollary. For each $A \in |\mathcal{A}|$, $\mathcal{F}(A) \in |\mathcal{B}|$. Since $(u_{\mathcal{F}(A)}, \mathcal{G}(\mathcal{F}(A)))$ is a universal map for $\mathcal{F}(A)$ w.r.t. $\mathcal{F}$, there is a unique morphism $v_A : \mathcal{G}(\mathcal{F}(A)) \longrightarrow A$ such that the diagram

(D$'_1$)
$$\begin{array}{ccc} \mathcal{F}(A) & \xrightarrow{1_{\mathcal{F}(A)}} & \mathcal{F}(A) \\ & {\scriptstyle u_{\mathcal{F}(A)}}\searrow \quad \nearrow{\scriptstyle \mathcal{F}(v_A)} & \\ & \mathcal{F}(\mathcal{G}(\mathcal{F}(A))) & \end{array}$$

commutes, i.e. (a) is satisfied. Now we have to show that (b) is also fulfilled, i.e. that the diagram

(D$'_2$)
$$\begin{array}{ccc} \mathcal{G}(B) & \xrightarrow{1_{\mathcal{G}(B)}} & \mathcal{G}(B) \\ & {\scriptstyle \mathcal{G}(u_B)}\searrow \quad \nearrow{\scriptstyle v_{\mathcal{G}(B)}} & \\ & \mathcal{G}(\mathcal{F}(\mathcal{G}(B))) & \end{array}$$

commutes. Since $(u_B, \mathcal{G}(B))$ is a universal map for B with respect to $\mathcal{F}$, there is a unique morphism $h : \mathcal{G}(B) \longrightarrow \mathcal{G}(B)$ such that the diagram

(D$'_3$)
$$\begin{array}{ccc} B & \xrightarrow{u_B} & \mathcal{F}(\mathcal{G}(B)) \\ & {\scriptstyle u_B}\searrow \quad \nearrow{\scriptstyle \mathcal{F}(h)} & \\ & \mathcal{F}(\mathcal{G}(B)) & \end{array}$$

commutes. Since $\mathcal{F}$ is a functor, $\mathcal{F}(1_{\mathcal{G}(B)}) \circ u_B = 1_{\mathcal{F}(\mathcal{G}(B))} \circ u_B = u_B$. Further, $u : \mathcal{I}_B \longrightarrow \mathcal{F} \circ \mathcal{G}$ is a natural transformation, i.e. the diagram

(D$'_4$)
$$\begin{array}{ccc} B & \xrightarrow{u_B} & \mathcal{F}(\mathcal{G}(B)) \\ {\scriptstyle u_B}\downarrow & & \downarrow{\scriptstyle \mathcal{F}(\mathcal{G}(u_B))} \\ \mathcal{F}(\mathcal{G}(B)) & \xrightarrow{u_{\mathcal{F}(\mathcal{G}(B))}} & \mathcal{F}(\mathcal{G}(\mathcal{F}(\mathcal{G}(B)))) \end{array}$$

commutes, and additionally, (a) is valid. Thus, $\mathcal{F}(v_{\mathcal{G}(B)} \circ \mathcal{G}(u_B)) \circ u_B = \mathcal{F}(v_{\mathcal{G}(B)}) \circ \mathcal{F}(\mathcal{G}(u_B)) \circ u_B = \mathcal{F}(v_{\mathcal{G}(B)}) \circ u_{\mathcal{F}(\mathcal{G}(B))} \circ u_B = 1_{\mathcal{F}(\mathcal{G}(B))} \circ u_B = u_B$. Consequently, $h = v_{\mathcal{G}(B)} \circ \mathcal{G}(u_B) = 1_{\mathcal{G}(B)}$, i.e. (b) is fulfilled. It remains to show: $v = (v_A) : \mathcal{G} \circ \mathcal{F} \longrightarrow \mathcal{I}_\mathcal{A}$ is a natural transformation, i.e. for each pair $(A, A') \in |\mathcal{A}| \times |\mathcal{A}|$ and each $f \in [A, A']_\mathcal{A}$, the following diagram commutes:

(D$'_5$)
$$\begin{array}{ccc} \mathcal{G}(\mathcal{F}(A)) & \xrightarrow{v_A} & A \\ {\scriptstyle \mathcal{G}(\mathcal{F}(f))}\downarrow & & \downarrow{\scriptstyle f} \\ \mathcal{G}(\mathcal{F}(A')) & \xrightarrow{v_{A'}} & A' \end{array}$$

Since $(u_{\mathcal{F}(A)}, \mathcal{G}(\mathcal{F}(A)))$ is a universal map for $\mathcal{F}(A)$ w.r.t. $\mathcal{F}$, there is a unique morphism $h : \mathcal{G}(\mathcal{F}(A)) \longrightarrow A'$ such that the diagram

(D$'_6$)
$$\begin{array}{ccc} \mathcal{F}(A) & \xrightarrow{\mathcal{F}(f)} & \mathcal{F}(A') \\ & {\scriptstyle u_{\mathcal{F}(A)}}\searrow \quad \nearrow{\scriptstyle \mathcal{F}(h)} & \\ & \mathcal{F}(\mathcal{G}(\mathcal{F}(A))) & \end{array}$$

commutes. Since $\mathcal{F}$ is a functor and (a) is satisfied, $\mathcal{F}(f \circ v_A) \circ u_{\mathcal{F}(A)} = \mathcal{F}(f) \circ \mathcal{F}(v_A) \circ u_{\mathcal{F}(A)} = \mathcal{F}(f) \circ 1_{\mathcal{F}(A)} = \mathcal{F}(f)$. Further, $u = (u_B) : \mathcal{I}_\mathcal{B} \longrightarrow \mathcal{F} \circ \mathcal{G}$ is a natural transformation, i.e. the diagram

(D$'_7$)
$$\begin{array}{ccc} \mathcal{F}(A) & \xrightarrow{u_{\mathcal{F}(A)}} & \mathcal{F}(\mathcal{G}(\mathcal{F}(A))) \\ {\scriptstyle \mathcal{F}(f)}\downarrow & & \downarrow{\scriptstyle \mathcal{F}(\mathcal{G}(\mathcal{F}(f)))} \\ \mathcal{F}(A') & \xrightarrow{u_{\mathcal{F}(A')}} & \mathcal{F}(\mathcal{G}(\mathcal{F}(A'))) \end{array}$$

commutes, and additionally, (a) is valid. Thus, $\mathcal{F}(v_{A'} \circ \mathcal{G}(\mathcal{F}(f))) \circ u_{\mathcal{F}(A)} = \mathcal{F}(v_{A'}) \circ \mathcal{F}(\mathcal{G}(\mathcal{F}(f))) \circ u_{\mathcal{F}(A)} = \mathcal{F}(v_{A'}) \circ u_{\mathcal{F}(A')} \circ \mathcal{F}(f) = 1_{\mathcal{F}(A')} \circ \mathcal{F}(f) = \mathcal{F}(f)$. Consequently, $h = f \circ v_A = v_{A'} \circ \mathcal{G}(\mathcal{F}(f))$, i.e. ($D_5'$) is commutative. Therefore, the corollary is proved.

2.1.13 Remark. If for every $B \in |\mathcal{B}|$ a universal map (u'_B, A'_B) w.r.t. $\mathcal{F}$ is chosen, then by the preceding theorem, there is a functor $\mathcal{G}' : \mathcal{B} \longrightarrow \mathcal{A}$ with properties corresponding to $\mathcal{G}$. Since (u_B, A_B) and (u'_B, A'_B) are isomorphic in the sense of 2.1.6, $\mathcal{G}$ is naturally equivalent to $\mathcal{G}'$:

For every $B \in |\mathcal{B}|$, let $i_B : A_B = \mathcal{G}(B) \longrightarrow A'_B = \mathcal{G}'(B)$ be the isomorphism existing by 2.1.6 with $\mathcal{F}(i_B) \circ u_B = u'_B$. Then $i = (i_B) : \mathcal{G} \longrightarrow \mathcal{G}'$ is the desired natural equivalence, because for every $(B, B') \in |\mathcal{B}| \times |\mathcal{B}|$ and every $f \in [B, B']_{\mathcal{B}}$, the partial diagrams (I), (III), (IV) and the outer square of the diagram

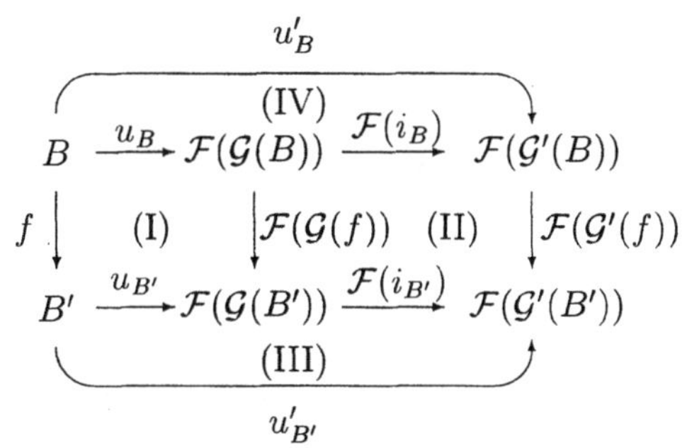

are commutative. Since there is a unique morphism $h : \mathcal{G}(B) \longrightarrow \mathcal{G}'(B')$ such that the diagram

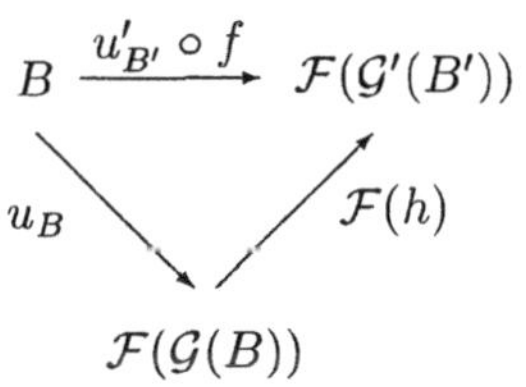

commutes ($(u_B, \mathcal{G}(B))$ is a universal map for B w.r.t. $\mathcal{F}$), then $h = (\mathcal{G}'(f) \circ i_B = i_{B'} \circ \mathcal{G}(f)$ (so the whole diagram above is commutative, because $\mathcal{F}$ is a functor).

2.1.14 Definition. If $\mathcal{F} : \mathcal{A} \longrightarrow \mathcal{B}$ and $\mathcal{G} : \mathcal{B} \longrightarrow \mathcal{A}$ are functors and $u = (u_B) : \mathcal{I}_{\mathcal{B}} \longrightarrow \mathcal{F} \circ \mathcal{G}$ and $v = (v_A) : \mathcal{G} \circ \mathcal{F} \longrightarrow \mathcal{I}_{\mathcal{A}}$ are natural transformations such that

(1) $\mathcal{F}(v_A) \circ u_{\mathcal{F}(A)} = 1_{\mathcal{F}(A)}$ for each $A \in |\mathcal{A}|$ and
(2) $v_{\mathcal{G}(B)} \circ \mathcal{G}(u_B) = 1_{\mathcal{G}(B)}$ for each $B \in |\mathcal{B}|$,

then $\mathcal{G}$ is said to be a *left adjoint* of $\mathcal{F}$, $\mathcal{F}$ is said to be a *right adjoint* of $\mathcal{G}$ and $(\mathcal{G}, \mathcal{F})$ is called a *pair of adjoint functors*.

2.1.15. We have proved already that if $\mathcal{F} : \mathcal{A} \longrightarrow \mathcal{B}$ is a functor and each

$B \in |\mathcal{B}|$ has a universal map $(u_B, \mathcal{G}(B))$ with respect to $\mathcal{F}$, then there is a functor $\mathcal{G}$ which is a left adjoint of $\mathcal{F}$ (cf. the preceding theorem together with the corollary). Now we show the converse.

Theorem. *If $\mathcal{G} : \mathcal{B} \longrightarrow \mathcal{A}$ is a left adjoint of $\mathcal{F} : \mathcal{A} \longrightarrow \mathcal{B}$ and $u = (u_B) : \mathcal{I}_\mathcal{B} \longrightarrow \mathcal{F} \circ \mathcal{G}$ is a corresponding natural transformation, then for each $B \in |\mathcal{B}|$, $(u_B, \mathcal{G}(B))$ is a universal map with respect to $\mathcal{F}$.*

Proof. Let $B \in |\mathcal{B}|$. It must be shown that for each $A \in |\mathcal{A}|$ and each $f : B \longrightarrow \mathcal{F}(A)$, there is a unique $\overline{f} : \mathcal{G}(B) \longrightarrow A$ such that the diagram

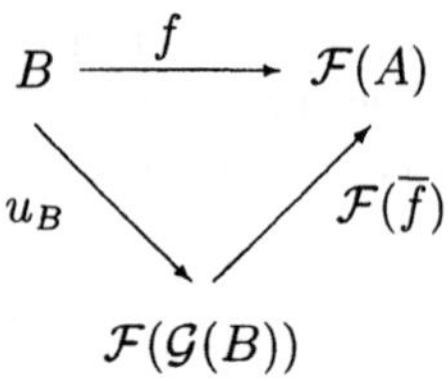

commutes. Since $\mathcal{G}$ is a left adjoint of $\mathcal{F}$, there is a natural transformation $v = (v_A) : \mathcal{G} \circ \mathcal{F} \longrightarrow \mathcal{I}_\mathcal{A}$ such that

(1) $\mathcal{F}(v_A) \circ u_{\mathcal{F}(A)} = 1_{\mathcal{F}(A)}$ for each $A \in |\mathcal{A}|$ and

(2) $v_{\mathcal{G}(B)} \circ \mathcal{G}(u_B) = 1_{\mathcal{G}(B)}$ for each $B \in |\mathcal{B}|$.

Put $\overline{f} = v_A \circ \mathcal{G}(f)$. Then by (1), $\mathcal{F}(\overline{f}) \circ u_B = \mathcal{F}(v_A) \circ \mathcal{F}(\mathcal{G}(f)) \circ u_B = \mathcal{F}(v_A) \circ u_{\mathcal{F}(A)} \circ f = 1_{\mathcal{F}(A)} \circ f = f$ because $\mathcal{F}$ is a functor and u is a natural transformation. Thus, $\overline{f}$ is the desired morphism provided that its uniqueness can be shown: If $\overline{\overline{f}} : \mathcal{G}(B) \longrightarrow A$ is a morphism such that $\mathcal{F}(\overline{\overline{f}}) \circ u_B = f$, then $\mathcal{G}(\mathcal{F}(\overline{\overline{f}})) \circ \mathcal{G}(u_B) = \mathcal{G}(f)$ and since $v : \mathcal{G} \circ \mathcal{F} \longrightarrow \mathcal{I}_\mathcal{A}$ is a natural transformation, the diagram

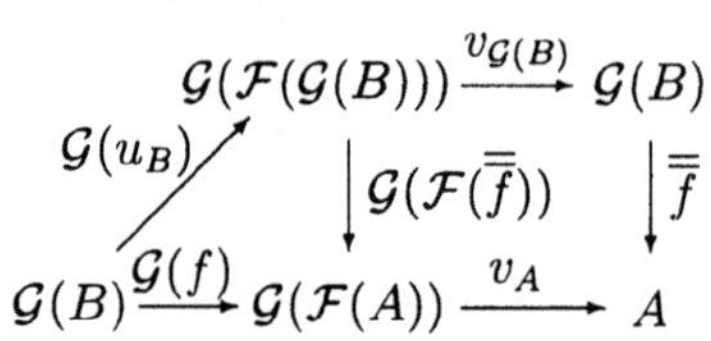

is commutative. Thus, $\overline{\overline{f}} \circ v_{\mathcal{G}(B)} \circ \mathcal{G}(u_B) = v_A \circ \mathcal{G}(f) = \overline{f}$. Consequently, by (2), $\overline{\overline{f}} = \overline{f}$.

2.1.16 Remarks. ① Remember that for every functor $\mathcal{H} : \mathcal{C} \longrightarrow \mathcal{D}$ there is a dual functor $\mathcal{H}^* : \mathcal{C}^* \longrightarrow \mathcal{D}^*$ (cf. 2.1.7.). If $(\mathcal{G} : \mathcal{B} \longrightarrow \mathcal{A},\ \mathcal{F} : \mathcal{A} \longrightarrow \mathcal{B})$ is a pair of adjoint functors by means of $((u_B) : \mathcal{I}_\mathcal{B} \longrightarrow \mathcal{F} \circ \mathcal{G},\ (v_A) : \mathcal{G} \circ \mathcal{F} \longrightarrow \mathcal{I}_\mathcal{A})$, then obviously, $(\mathcal{F}^* : \mathcal{A}^* \longrightarrow \mathcal{B}^*,\ \mathcal{G}^* : \mathcal{B}^* \longrightarrow \mathcal{A}^*)$ is a pair of adjoint functors by means of $((v_A^*) : \mathcal{I}_{\mathcal{A}^*} \longrightarrow \mathcal{G}^* \circ \mathcal{F}^*,\ (u_B^*) : \mathcal{F}^* \circ \mathcal{G}^* \longrightarrow \mathcal{I}_{\mathcal{B}^*})$. Especially, $\mathcal{G}$ is a left adjoint of $\mathcal{F}$ if and only if $\mathcal{G}^*$ is a right adjoint of $\mathcal{F}^*$ (*Duality principle for adjoint functors*).

② By the preceding theorems, *a functor $\mathcal{F} : \mathcal{A} \longrightarrow \mathcal{B}$ has a left adjoint $\mathcal{G} : \mathcal{B} \longrightarrow \mathcal{A}$ if and only if every $B \in |\mathcal{B}|$ has a universal map with respect to $\mathcal{F}$.*

③ By ② and the remark 2.1.13, for a given functor $\mathcal{F} : \mathcal{A} \longrightarrow \mathcal{B}$ a functor $\mathcal{G} : \mathcal{B} \longrightarrow \mathcal{A}$ is uniquely determined up to natural equivalence by the property of being a left adjoint of $\mathcal{F}$. Moreover, by the duality principle for adjoint functors, the functor $\mathcal{F}$ is also uniquely determined (up to natural equivalence) by the property of being a right adjoint of a given functor $\mathcal{G}$. Thus,

adjoint functors are uniquely determined by each other up to natural equivalence.

④ Let $\mathcal{F}, \mathcal{G}, \mathcal{H}$ be functors from $\mathcal{C}$ to $\mathcal{D}$ and $s = (s_X) : \mathcal{F} \longrightarrow \mathcal{G}$, $t = (t_X) : \mathcal{G} \longrightarrow \mathcal{H}$ natural transformations. Then a natural transformation $t \circ s : \mathcal{F} \longrightarrow \mathcal{H}$, called *the composition of the natural transformations s and t*, is defined by $t \circ s = (t_X \circ s_X)_{X \in |\mathcal{C}|}$. If $(\mathcal{G} : \mathcal{B} \longrightarrow \mathcal{A}, \mathcal{F} : \mathcal{A} \longrightarrow \mathcal{B})$ is a pair of adjoint functors, then there are natural transformations

$$u = (u_B) : \mathcal{I}_{\mathcal{B}} \longrightarrow \mathcal{F} \circ \mathcal{G} \text{ and}$$
$$v = (v_A) : \mathcal{G} \circ \mathcal{F} \longrightarrow \mathcal{I}_{\mathcal{A}}$$

such that

(1) $\mathcal{F}(v_A) \circ u_{\mathcal{F}(A)} = 1_{\mathcal{F}(A)}$ for each $A \in |\mathcal{A}|$ and

(2) $v_{\mathcal{G}(B)} \circ \mathcal{G}(u_B) = 1_{\mathcal{G}(B)}$ for each $B \in |\mathcal{B}|$.

If $(u' : \mathcal{I}_{\mathcal{B}} \longrightarrow \mathcal{F} \circ \mathcal{G}, v' : \mathcal{G} \circ \mathcal{F} \longrightarrow \mathcal{I}_{\mathcal{A}})$ is a second pair of natural transformations satisfying relations corresponding to (1) and (2), then by the preceding theorem, $(u_B, \mathcal{G}(B))$, $(u'_B, \mathcal{G}(B))$ are universal maps w.r.t. $\mathcal{F}$ for each $B \in |\mathcal{B}|$ and by 2.1.6 there is an isomorphism $i_B : \mathcal{G}(B) \longrightarrow \mathcal{G}(B)$ for each $B \in |\mathcal{B}|$ such that $\mathcal{F}(i_B) \circ u_B = u'_B$.

Then $j = (\mathcal{F}(i_B))_{B \in |\mathcal{B}|} : \mathcal{F} \circ \mathcal{G} \longrightarrow \mathcal{F} \circ \mathcal{G}$ is a natural equivalence satisfying $j \circ u = u'$ (cf. the big diagram in 2.1.13 and put $\mathcal{G} = \mathcal{G}'$), i.e. u and u' coincide up to a natural equivalence. Similarly, one concludes that (v_A^*) coincides with $({v'_A}^*)$ up to a natural equivalence and therefore, (v_A) coincides with (v'_A) up to a natural equivalence. Thus, the *natural transformations u, v are uniquely determined (up to natural equivalence) by* (1) *and* (2). Consequently, it makes sense to speak of *the* natural transformations u, v belonging to a pair of adjoint functors. Then $(\mathcal{G}, \mathcal{F}, u, v)$ is called an **adjoint situation**, whereas u (resp. v) is called the *unit* (resp. *counit*).

⑤ The examples under 2.1.5 read now as follows:

a) The inclusion functor $\mathcal{F}_e : \mathbf{Top_0} \longrightarrow \mathbf{Top}$ has a left adjoint assigning to each topological space a 'universal' T_0–space.

b) The inclusion functor $\mathcal{F}_e : \mathbf{CompT_2} \longrightarrow \mathbf{Tych}$ has a left adjoint $\beta : \mathbf{Tych} \longrightarrow \mathbf{CompT_2}$ assigning to each Tychonoff space X its Stone–Čech compactification $\beta(X)$. Thereby the meaning of the symbol "β" is clarified.

c) If $\mathcal{C}$ is a topological construct, then the forgetful functor $\mathcal{F}_u : \mathcal{C} \longrightarrow \mathbf{Set}$ has a left adjoint $\mathcal{D} : \mathbf{Set} \longrightarrow \mathcal{C}$ assigning to each set X the set X endowed with the discrete $\mathcal{C}$–structure ξ_d (Note: $\mathcal{F}_u : \mathcal{C} \longrightarrow \mathbf{Set}$ has also a right–adjoint $\mathcal{I} : \mathbf{Set} \longrightarrow \mathcal{C}$ assigning to each set X the set X endowed with the indiscrete $\mathcal{C}$–structure ξ_i, in other words: For each $X \in |\mathbf{Set}|$, $\mathcal{F}_u : \mathcal{C} \longrightarrow \mathbf{Set}$ has a couniversal map $((X, \xi_i), 1_X)$.)

2.2 Characterization theorems of $\mathcal{E}$–reflective and $\mathcal{M}$–coreflective subcategories

2.2.1. In the following we consider in some detail the case that an inclusion functor has a left adjoint or a right adjoint respectively.

2.2.2 Definitions. Let $\mathcal{A}$ be a subcategory of a category $\mathcal{C}$ and $\mathcal{F}_e : \mathcal{A} \longrightarrow \mathcal{C}$ the inclusion functor. Then $\mathcal{A}$ is called

a) *reflective* in $\mathcal{C}$ iff one of the two equivalent conditions is satisfied:

(1) $\mathcal{F}_e$ has a left adjoint $\mathcal{R}$.
(2) Each $X \in |\mathcal{C}|$ has a universal map with respect to $\mathcal{F}_e$, i.e. for each $X \in |\mathcal{C}|$ there exists an $\mathcal{A}$–object $X_{\mathcal{A}}$ and a $\mathcal{C}$–morphism $r_X : X \longrightarrow X_{\mathcal{A}}$ such that for each $\mathcal{A}$–object Y and each $\mathcal{C}$–morphism $f : X \longrightarrow Y$ there is a unique $\mathcal{A}$–morphism $\overline{f} : X_{\mathcal{A}} \longrightarrow Y$ such that the diagram

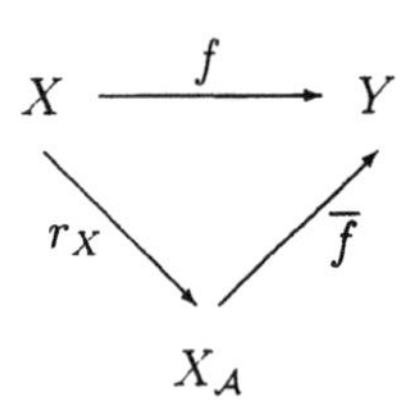

commutes.
(The functor $\mathcal{R}$ is called a *reflector.*)

b) *epireflective, extremal epireflective* or *bireflective* in $\mathcal{C}$ provided that $\mathcal{A}$ is reflective in $\mathcal{C}$ and for each $X \in |\mathcal{C}|$, the $\mathcal{C}$–morhpism $r_X : X \longrightarrow X_{\mathcal{A}}$ is an epimorphism, an extremal epimorphism or a bimorphism respectively.
[Then the functor $\mathcal{R}$ is called an *epireflector*, an *extremal epireflector* or a *bireflector* respectively.]

a′) *coreflective* in $\mathcal{C}$ iff $\mathcal{A}^*$ is reflective in $\mathcal{C}^*$, i.e. iff one of the two equivalent conditions is satisfied:
(1′) $\mathcal{F}_e$ has a right adjoint $\mathcal{R}_c$.
(2′) Each $X \in |\mathcal{C}|$ has a couniversal map with respect to $\mathcal{F}_e$, i.e. for each $X \in |\mathcal{C}|$ there exists an $\mathcal{A}$–object $X_{\mathcal{A}}$ and a $\mathcal{C}$–morphism $c_X : X_{\mathcal{A}} \longrightarrow X$ such that for each $\mathcal{A}$–object Y and each $\mathcal{C}$–morphism $f : Y \longrightarrow X$ there is a unique $\mathcal{A}$–morphism $\overline{f} : Y \longrightarrow X_{\mathcal{A}}$ such that the diagram

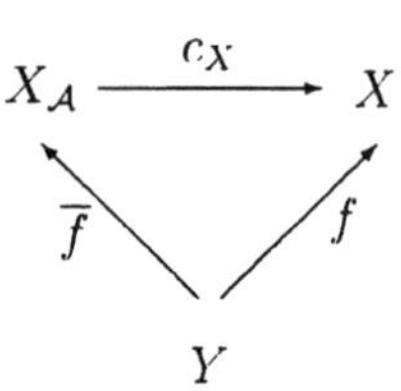

commutes.
(The functor $\mathcal{R}_c$ is called a *coreflector.*)

b′) *monocoreflective, extremal monocoreflective* or *bicoreflective* in $\mathcal{C}$ provided that $\mathcal{A}$ is coreflective in $\mathcal{C}$ and for each $X \in |\mathcal{C}|$, the $\mathcal{C}$–morphism $c_X : X_{\mathcal{A}} \longrightarrow X$ is a monomorphism, an extremal monomorphism or a bimorphism respectively.
[Then the functor $\mathcal{R}_c$ is called a *monocoreflector*, an *extremal monocoreflector* or a *bicoreflector respectively.*]

The morphisms

$r_X : X \longrightarrow X_{\mathcal{A}}$ | $c_X : X_{\mathcal{A}} \longrightarrow X$

are called

reflections of $X \in |\mathcal{C}|$ with respect to $\mathcal{A}$ in case a) and *epireflections*, *extremal epireflections* or *bireflections* (of $X \in |\mathcal{C}|$ with respect to $\mathcal{A}$) respectively in case b).

coreflections of $X \in |\mathcal{C}|$ with respect to $\mathcal{A}$ in case a′) and *monocoreflections*, *extremal monocoreflections* or *bicoreflections* (of $X \in |\mathcal{C}|$ with respect to $\mathcal{A}$) respectively in case b′).

2.2.3 Remarks. 1) Let $\mathcal{E}$ (resp. $\mathcal{M}$) be a class of epimorphisms (resp. monomorphisms) which is closed under composition with isomorphisms. Then one may introduce analogously to 2.2.2. b) (resp. 2.2.2. b′)) the concept *$\mathcal{E}$–reflective* (resp. *$\mathcal{M}$–coreflective*) *subcategory* by requiring that all reflections (resp. coreflections) belong to $\mathcal{E}$ (resp. $\mathcal{M}$).

2) In the following we will often study subcategories $\mathcal{A}$ of a category $\mathcal{C}$, which are

(1) full and
(2) *isomorphism–closed*, i.e. each $X \in |\mathcal{C}|$ which is isomorphic to an $A \in |\mathcal{A}|$ belongs to $|\mathcal{A}|$.

Obviously, we obtain

a) Each subclass $|\mathcal{A}|$ of the object class $|\mathcal{C}|$ of a category $\mathcal{C}$ can be extended to a full subcategory $\mathcal{A}$ of $\mathcal{C}$ in a natural way (Put $[A, B]_{\mathcal{A}} = [A, B]_{\mathcal{C}}$ for each $(A, B) \in |\mathcal{A}| \times |\mathcal{A}|$).

b) Full subcategories $\mathcal{A}$ of a category $\mathcal{C}$ defined by

$$|\mathcal{A}| = \{X \in |\mathcal{C}| : X \text{ has the property } P\}$$

for each property P of $\mathcal{C}$–objects which is a $\mathcal{C}$–invariant are obviously isomorphism–closed (e.g. for $\mathcal{C}$ = **Top** the property “connected” is a **Top**–invariant, i.e. a topological invariant in the usual sense).

3) Corresponding to 1.2.2.6. 2) we will use the term *weak subobject* (resp. *weak quotient object*) for the following situation: If $\mathcal{C}$ is a category and X, Y are $\mathcal{C}$–objects such that $[X, Y]_{\mathcal{C}}$ (resp. $[Y, X]_{\mathcal{C}}$) contains a monomorphism (resp. epimorphism), then X is called a weak subobject (resp. weak quotient object) of Y.

2.2.4. A full subcategory $\mathcal{A}$ of a category $\mathcal{C}$ is called

a) *closed under formation of products (resp. coproducts) in* $\mathcal{C}$ provided that whenever $(P, (p_i : P \longrightarrow A_i)_{i \in I})$ (resp. $((j_i : A_i \longrightarrow C)_{i \in I}, C)$ is a product (resp. coproduct) in $\mathcal{C}$ such that all A_i belong to $|\mathcal{A}|$, then P (resp. C) belongs to $|\mathcal{A}|$,

b) *closed under formation of [weak] subobjects (resp. [weak] quotient objects) in* $\mathcal{C}$ provided that for each $A \in |\mathcal{A}|$, each [weak] subobject (resp. [weak] quotient object) of A formed in $\mathcal{C}$ belongs to $|\mathcal{A}|$.

With this terminology the following is valid:

Theorem. *If $\mathcal{A}$ is a full and isomorphism–closed subcategory of a*

A) co–well–powered, (epi, extremal mono–factorizable category $\mathcal{C}$ that has products, then the following are equivalent:
(1) $\mathcal{A}$ is epireflective in $\mathcal{C}$.
(2) $\mathcal{A}$ is closed under formation of products and subobjects in $\mathcal{C}$.

A′) well–powered, (extremal epi, mono)– factorizable category $\mathcal{C}$ that has coproducts, then the following are equivalent:
(1) $\mathcal{A}$ is monocoreflective in $\mathcal{C}$.
(2) $\mathcal{A}$ is closed under formation of coproducts and quotient objects in $\mathcal{C}$.

B) co–well–powered, (extremal epi, mono)–factorizable category that has products, then the following are equivalent:
(1) $\mathcal{A}$ is extremal epireflective.
(2) $\mathcal{A}$ is closed under formation of products and weak subobjects.

B′) well–powered, (epi, extremal mono)–factorizable category that has coproducts, then the following are equivalent:
(1) $\mathcal{A}$ is extremal monocoreflective.
(2) $\mathcal{A}$ is closed under formation of coproducts and weak quotient objects in $\mathcal{C}$.

Proof. Because of duality it suffices to prove the left part of the above theorem.

A) (2) $\Longrightarrow$ (1). | B) (2) $\Longrightarrow$ (1).

Let $X \in |\mathcal{C}|$ and and $\{e_i : X \longrightarrow A_i\}$ be a representative set of

epimorphisms | extremal epimorphisms

such that all A_i belong to $|\mathcal{A}|$. If $(P, (p_i))$ is the product of the family (A_i), then $P \in |\mathcal{A}|$ by assumption. According to the definition of a product, there is a unique $f : X \longrightarrow P$ such that all diagrams

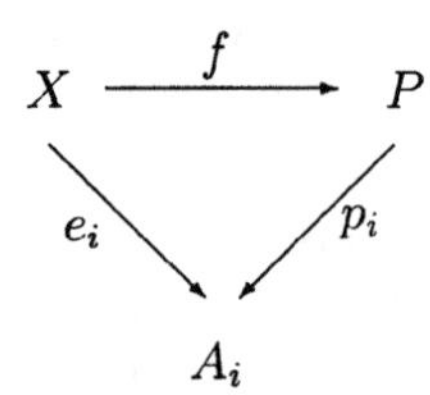

commute. Let

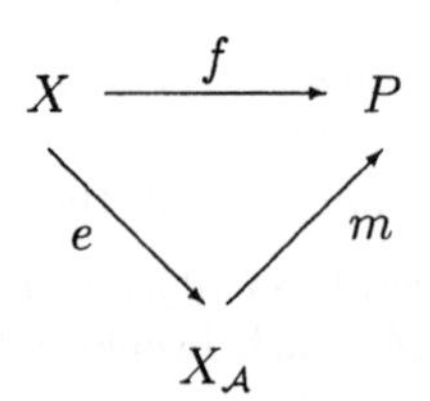

be an

(epi, extremal mono)–factorization | (extremal epi, mono)–factorization of f. By (2), $X_{\mathcal{A}} \in |\mathcal{A}|$, because $X_{\mathcal{A}}$ is a

subobject | weak subobject

of P. It remains to show that e is a reflection:
Let $g \in [X, Y]_{\mathcal{C}}$ such that $Y \in |\mathcal{A}|$. Let $X \xrightarrow{e'} A \xrightarrow{m'} Y$ be an

(epi, extremal mono)–factorization | (extremal epi, mono)–factorization

of g. Thus, $A \in |\mathcal{A}|$. Without loss of generality it may be assumed that $e' = e_i$ and $A = A_i$ for some i. Then the following diagram

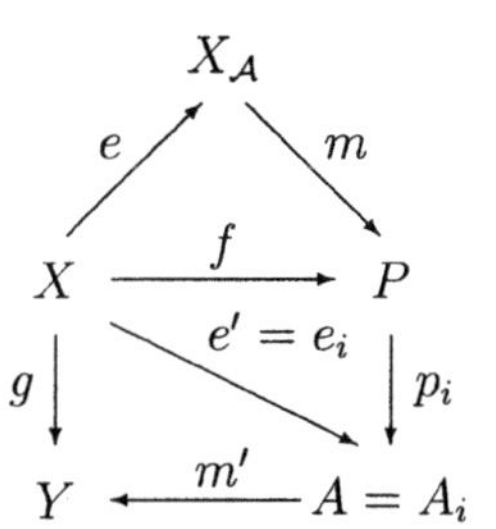

is commutative. Put $\overline{g} = m' \circ p_i \circ m$. Then $\overline{g} \circ e = g$ and since e is an epimorphism, $\overline{g}$ is the unique morphism with this property.

A) (1) $\Longrightarrow$ (2). | B) (1) $\Longrightarrow$ (2).

a) Let $(A_i)_{i \in I}$ be a family of $\mathcal{A}$–objects and $(P, (p_i))$ the product of this family in $\mathcal{C}$. If the

epireflection | extremal epireflection

of P with respect to $\mathcal{A}$ is denoted by $r_P : P \longrightarrow P_{\mathcal{A}}$, then there is a unique $\overline{p_i} : P_{\mathcal{A}} \longrightarrow A_i$ such that $\overline{p_i} \circ r_P = p_i$ for each $i \in I$. By the definition of a product, there is a unique $s_P : P_{\mathcal{A}} \longrightarrow P$ such that $\overline{p_i} = p_i \circ s_P$ for each $i \in I$. Hence, $p_i \circ 1_P = p_i = \overline{p_i} \circ r_P = pi \circ (s_P \circ r_P)$ for each $i \in I$. Consequently, since $(p_i : P \longrightarrow A_i)_{i \in I}$ is a mono–source, $s_P \circ r_P = 1_P$. Since 1_P is an extremal monomorphism and r_P is an

epimorphism, | extremal epimorphism,

r_P is an isomorphism. Consequently, $P \in |\mathcal{A}|$, since $\mathcal{A}$ is isomorphism–closed.

b) Let $f \in [X, Y]_{\mathcal{C}}$ be

an extremal monomorphism | a monomorphism

and let $Y \in |\mathcal{A}|$. If the

epireflection | extremal epireflection

of X with respect to $\mathcal{A}$ is denoted by $r_X : X \longrightarrow X_{\mathcal{A}}$, then, since $Y \in |\mathcal{A}|$, there is a unique $\overline{f} : X_{\mathcal{A}} \longrightarrow Y$ such that $\overline{f} \circ r_X = f$.

Since r_X is an epimorphism and f is an extremal monomorphism, r_X has to be an isomorphism.

Since r_X is an extremal epimorphism and additionally a monomorphism (because $\overline{f} \circ r_X$ is a monomorphism), r_X is an isomorphism.

Thus, since $\mathcal{A}$ is isomorphism–closed, $X \in |\mathcal{A}|$.

2.2.5 Remarks. 1) Obviously, the preceding theorem can be generalized as follows:

Let $\mathcal{E}$ (resp. $\mathcal{M}$) be a class of epimorphisms (resp. morphisms) which is closed under composition with isomorphisms. Then

(a) *$\mathcal{A}$ is closed under formation of products and $\mathcal{M}$–subobjects* [1] *in $\mathcal{C}$*
implies

(b) *$\mathcal{A}$ is $\mathcal{E}$–reflective in $\mathcal{C}$*

provided that $\mathcal{C}$ is $\mathcal{E}$–co–well–powered (i.e. for every $X \in |\mathcal{C}|$ there is a representative set of $\mathcal{E}$–quotient objects [1]), *$(\mathcal{E}, \mathcal{M})$–factorizable* (i.e. for every $\mathcal{C}$–morphism f there are some $e \in \mathcal{E}$ and some $m \in \mathcal{M}$ such that $f = m \circ e$) *and has products.* (a) *and* (b) *are equivalent provided that, additionally, $\mathcal{C}$ is an $(\mathcal{E}, \mathcal{M})$–category* (a category $\mathcal{C}$ is called an *$(\mathcal{E}, \mathcal{M})$–category* provided that

(1) $\mathcal{C}$ is $(\mathcal{E}, \mathcal{M})$–factorizable.
(2) For every commutative diagram

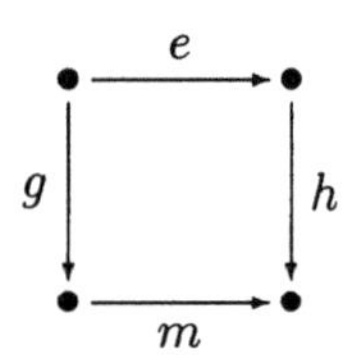

in $\mathcal{C}$ with $e \in \mathcal{E}$ and $m \in \mathcal{M}$ there exists a $\mathcal{C}$–morphism h that makes the diagram

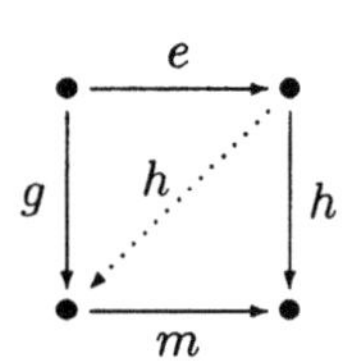

commute.
[*$(\mathcal{E}, \mathcal{M})$–diagonalization property*]).

The implication "(a) $\Longrightarrow$ (b)" is proved analogously to "(2) $\Longrightarrow$ (1)" of 2.2.4. A) (resp. B)). Conversely, let (b) be satisfied and $m : X \longrightarrow A$ be a morphism of $\mathcal{M}$ such that $A \in |\mathcal{A}|$. If $r_X : X \longrightarrow X_{\mathcal{A}}$ is the $\mathcal{E}$–reflection of X with respect to $\mathcal{A}$, then there exists an $f : X_{\mathcal{A}} \longrightarrow A$ such that $f \circ r_X = m$. Since $\mathcal{C}$ satisfies the $(\mathcal{E}, \mathcal{M})$–diagonalization property, there is an ℓ such that the diagram

[1] $X \in |\mathcal{C}|$ is called an $\mathcal{M}$–*subobject* of $Y \in |\mathcal{C}|$ provided that there is a $\mathcal{C}$–morphism $f : X \longrightarrow Y$ such that $f \in \mathcal{M}$. Similarly, an $\mathcal{E}$–*quotient object* is defined.

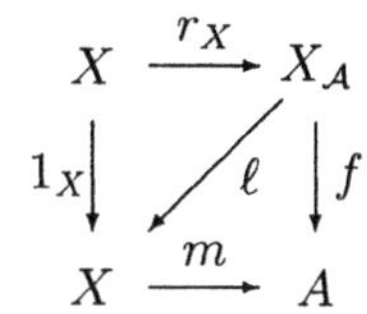

commutes. Especially $\ell \circ r_X = 1_X$. Thus, r_X is an isomorphism, because it is an epimorphism and 1_X is an extremal monomorphism. Consequently, $X \in |\mathcal{A}|$, because $\mathcal{A}$ is isomorphism–closed. (The fact that $\mathcal{A}$ is closed under formation of products in $\mathcal{C}$ is proved as above.)

2) If $\mathcal{C}$ is *a topological construct*, then $\mathcal{C}$ *is an* $(\mathcal{E},\mathcal{M})$*–category* for the following cases:

a) $\mathcal{E} = \{e \in \text{Mor}\,\mathcal{C} : e \text{ is an epimorphism}\}$,
$\mathcal{M} = \{m \in \text{Mor}\,\mathcal{C} : m \text{ is an extremal monomorphism}\}$;
shortly: $\mathcal{C}$ is an (epi, extremal mono)–category.

b) $\mathcal{E} = \{e \in \text{Mor}\,\mathcal{C} : e \text{ is an extremal epimorphism}\}$,
$\mathcal{M} = \{m \in \text{Mor}\,\mathcal{C} : m \text{ is a monomorphism}\}$;
shortly: $\mathcal{C}$ is an (extremal epi, mono)–category.

c) $\mathcal{E} = \{e \in \text{Mor}\,\mathcal{C} : e \text{ is a bimorphism}\}$,
$\mathcal{M} = \{m \in \text{Mor}\,\mathcal{C} : m \text{ is an initial morphism}\}$,
where a $\mathcal{C}$–morphism $m : (X,\xi) \longrightarrow (Y,\eta)$ is called *initial* provided that ξ is the initial $\mathcal{C}$–structure w.r.t. m.
Shortly: $\mathcal{C}$ is a (bimorphism, initial morphism)–category.

(The $(\mathcal{E},\mathcal{M})$–factorization under a) and b) have been studied before in 1.2.2.6. 3) a) and b). The $(\mathcal{E},\mathcal{M})$–diagonalization property is easily checked in these cases. Concerning c), let $f : (X,\xi) \longrightarrow (Y,\eta)$ be a $\mathcal{C}$–morphism and ξ_{in} the initial $\mathcal{C}$–structure on X w.r.t. f. Then $1_X : (X,\xi) \longrightarrow (X,\xi_{in})$ is a $\mathcal{C}$–morphism and the diagram

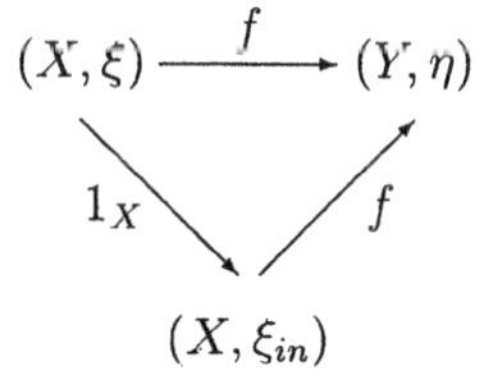

commutes. In order to prove the $(\mathcal{E},\mathcal{M})$–diagonalization property, let

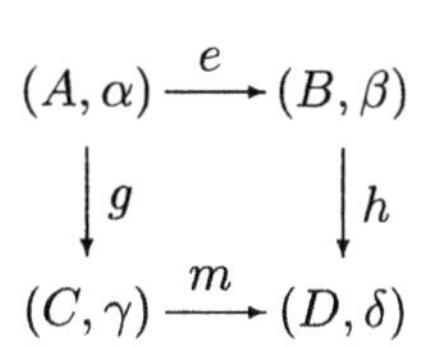

be a commutative diagram in $\mathcal{C}$ with $e \in \mathcal{E}$, i.e. e is a bimorphism, and $m \in \mathcal{M}$, i.e. m is an initial morphism. Then $g \circ e^{-1} = \ell$ is the desired diagonal morphism.

2.2.6. In a topological construct $\mathcal{C}$ an object (X, ξ) is called an *initial subobject* of a $\mathcal{C}$–object (Y, η) provided that there is an initial morphism $f : (X, \xi) \longrightarrow (Y, \eta)$. If $\mathcal{E}$ and $\mathcal{M}$ are chosen as under 2.2.5. 2) c), then $\mathcal{C}$ is an $\mathcal{E}$–co–well–powered $(\mathcal{E}, \mathcal{M})$–category with products and the result under 2.2.5. 1) yields the following

Corollary. *A full and isomorphism–closed subconstruct $\mathcal{A}$ of a topological construct $\mathcal{C}$ is bireflective in $\mathcal{C}$ iff it is closed under formation of products and initial subobjects.*

2.2.7 Definition. An object S of a category $\mathcal{C}$ is called a *separator* provided that for each pair of distinct morphisms $f, g : A \longrightarrow B$ with the same domain and the same codomain, there is a morphism $h : S \longrightarrow A$ such that $f \circ h \neq g \circ h$.

2.2.8 Example. Each object (X, ξ) in a topological construct $\mathcal{C}$ with $X \neq \emptyset$ is a separator.

2.2.9 Theorem. *Let S be a separator of a category $\mathcal{C}$. Then each coreflective subcategory $\mathcal{A}$ of $\mathcal{C}$ such that $S \in |\mathcal{A}|$ is epicoreflective, i.e. the coreflections are epimorphisms.*

Proof. Let $m : A \longrightarrow X$ be the coreflection of $X \in |\mathcal{C}|$ w.r.t. $\mathcal{A}$ and let $\alpha, \beta : X \longrightarrow Y$ be $\mathcal{C}$–morphisms such that $\alpha \circ m = \beta \circ m$. Then for each $\mathcal{C}$–morphism $h : S \longrightarrow X$ there is a unique $\mathcal{A}$–morphism $\overline{h} : S \longrightarrow A$ satisfying $m \circ \overline{h} = h$. Thus $\alpha \circ h = \alpha \circ m \circ \overline{h} = \beta \circ m \circ \overline{h} = \beta \circ h$. Since S is a separator, $\alpha = \beta$. Consequently, m is an epimorphism.

2.2.10 Theorem. *Every epicoreflective full subcategory $\mathcal{A}$ of a category $\mathcal{C}$ is bicoreflective.*
(Dually: Every monoreflective full subcategory $\mathcal{A}$ of a category $\mathcal{C}$ is bireflective.)

Corollary. *Let S be a separator. Then every coreflective full subcategory $\mathcal{A}$ of $\mathcal{C}$ such that $S \in |\mathcal{A}|$ is bicoreflective.*

Proof. Let $e_X : A_X \longrightarrow X$ be the epicoreflection of $X \in |\mathcal{C}|$ w.r.t. $\mathcal{A}$ and let $\alpha, \beta : Y \longrightarrow A_X$ be $\mathcal{C}$–morphisms such that $e_X \circ \alpha = e_X \circ \beta$. If $e_Y : A_Y \longrightarrow Y$ is the epicoreflection of Y w.r.t. $\mathcal{A}$, then

$$(*) \qquad e_X \circ \alpha \circ e_Y = e_X \circ \beta \circ e_Y \, .$$

Since $\mathcal{A}$ is full, $\alpha \circ e_Y$ and $\beta \circ e_Y$ are $\mathcal{A}$–morphisms. Thus, applying $(*)$, $\alpha \circ e_Y = \beta \circ e_Y$, because e_X is a coreflection. Since e_Y is an epimorphism, $\alpha = \beta$. Consequently, e_X is a monomorphism. Therefore the theorem is proved. The corollary is an application of this theorem together with 2.2.9.

2.2.11 Remarks. 1) By the preceding considerations *every coreflective, full and isomorphism–closed subconstruct $\mathcal{A}$ of a topological construct $\mathcal{C}$ is bicoreflective provided that $|\mathcal{A}|$ contains at least one element with non–empty underlying* set.

In this case the coreflection $c_X : (Y_A, \eta_A) \longrightarrow (X, \xi)$ of $(X, \xi) \in |\mathcal{C}|$ w.r.t. $\mathcal{A}$ is bijective. By 1.2.2.7 there is a $\mathcal{C}$-structure ξ_A on X such that $c_X : (Y_A, \eta_A) \longrightarrow (X, \xi_A)$ is an isomorphism. Obviously, *ξ_A is the coarsest one of all $\mathcal{C}$-structures ξ' which are finer than ξ and for which $(X, \xi') \in |\mathcal{A}|$.*

($(Y_A, \eta_A) \xrightarrow{c_X} (X, \xi_A) \xrightarrow{1_X} (X, \xi)$ is a $\mathcal{C}$-morphism; furthermore $c_X^{-1} : (X, \xi_A) \longrightarrow (Y_A, \eta_A)$ is a $\mathcal{C}$-morphism [even a $\mathcal{C}$-isomorphism]. Thus, $1_X = (1_X \circ c_X) \circ c_X^{-1} : (X, \xi_A) \longrightarrow (X, \xi)$ is a $\mathcal{C}$-morphism as a composition of two $\mathcal{C}$-morphisms. Consequently

$$\xi_A \leq \xi, \quad \text{i.e. } \xi_A \text{ is finer than } \xi\,.$$

If ξ' is a $\mathcal{C}$-structure on X such that $\xi' \leq \xi$ and $(X, \xi') \in |\mathcal{A}|$, then $1_X : (X, \xi') \longrightarrow (X, \xi)$ is a $\mathcal{C}$-morphism and there exists a unique $\mathcal{C}$-morphism $\overline{1_X} : (X, \xi') \longrightarrow (Y_A, \eta_A)$ such that the diagram

(∗)

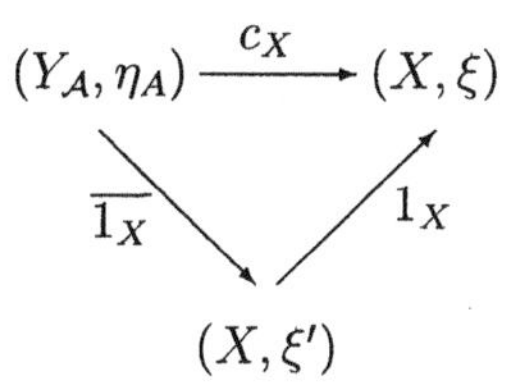

commutes.

$(X, \xi') \xrightarrow{\overline{1_X}} (Y_A, \eta_A) \underset{\cong}{\xrightarrow{c_X}} (X, \xi_A)$ is a $\mathcal{C}$-morphism whose underlying map is the identity map on X, because the diagram (∗) is commutative. Consequently, $\xi' \leq \xi_A$.)

Therefore $1_X : (X, \xi_A) \longrightarrow (X, \xi)$ is the coreflection of (X, ξ) with respect to $\mathcal{A}$, i.e. one obtains the coreflection of a $\mathcal{C}$-object (X, ξ) with respect to $\mathcal{A}$ (up to isomorphism) by a modification of the $\mathcal{C}$-structure ξ on X.
Moreover, *$\mathcal{A}$ contains obviously all discrete objects of $\mathcal{C}$* because for each discrete $\mathcal{C}$-object (X, ξ), the coreflection $1_X : (X, \xi_A) \longrightarrow (X, \xi)$ is an isomorphism.

2) By dualization of the concept "separator" one obtains the concept "*coseparator*". Obviously every indiscrete object in a topological construct whose underlying set consists at least of two elements is a coseparator.
Applying the dual assertion of 2.2.10. Cor. one obtains the following theorem.

THEOREM. *Let $\mathcal{C}$ be a topological construct and $\mathcal{A}$ a full and isomorphism-closed subconstruct of $\mathcal{C}$. Then the following are equivalent:*

(1) *$\mathcal{A}$ is bireflective in $\mathcal{C}$.*
(2) *$\mathcal{A}$ is (epi)reflective in $\mathcal{C}$ and contains all indiscrete $\mathcal{C}$-objects.*

If $r_X : (X, \xi) \longrightarrow (Y_A, \eta_A)$ is the bireflection of $(X, \xi) \in |\mathcal{C}|$ w.r.t. $\mathcal{A}$, then by 1.2.2.7. there is a unique $\mathcal{C}$-structure ξ_A on X such that $r_X^{-1} : (Y_A, \eta_A) \longrightarrow (X, \xi_A)$ is a $\mathcal{C}$-isomorphism.

Especially, $1_X : (X,\xi) \longrightarrow (X,\xi_{\mathcal{A}})$ *is (up to isomorphism) the bireflection of* (X,ξ) *with respect to* $\mathcal{A}$ *and* $\xi_{\mathcal{A}}$ *is the finest one of all* $\mathcal{C}$*–structures* ξ' *on* X *which are coarser than* ξ *and for which* $(X,\xi') \in |\mathcal{A}|$. (This is proved analogously to the corresponding assertion of 1) with respect to bicoreflections.)

2.2.12 Theorem. *Let* $\mathcal{A}$ *be a full and isomorphism–closed subconstruct of a topological construct* $\mathcal{C}$*. Then* $\mathcal{A}$ *is a topological construct, provided that* $\mathcal{A}$ *is bireflective or bicoreflective in* $\mathcal{C}$*. Particularly, if* $\mathcal{A}$ *is bireflective (resp. bicoreflective) in* $\mathcal{C}$*, then the initial structures (resp. final structures) in* $\mathcal{A}$ *are formed as in* $\mathcal{C}$*, whereas the final structures (resp. initial structures) arise from the final structures (resp. initial structures) in* $\mathcal{C}$ *by applying the bireflector (resp. bicoreflector).*

Proof. A) Let $\mathcal{A}$ be bicoreflective in $\mathcal{C}$.

(1) Let X be a set, $((X_i,\xi_i))_{i\in I}$ a family of $\mathcal{A}$–objects and $(f_i : X \longrightarrow X_i)_{i\in I}$ a family of maps. Let ξ be the initial $\mathcal{C}$–structure on X w.r.t. $(X, f_i, (X_i,\xi_i), I)$ and let $1_X : (X,\xi_{\mathcal{A}}) \longrightarrow (X,\xi)$ be the bicoreflection of (X,ξ) w.r.t. $\mathcal{A}$. Then $\xi_{\mathcal{A}}$ is the unique initial $\mathcal{A}$-structure on X w.r.t. $(X, f_i, (X_i,\xi_i), I)$:

(a) Let $g : (Y_{\mathcal{A}},\eta_{\mathcal{A}}) \longrightarrow (X,\xi_{\mathcal{A}})$ be an $\mathcal{A}$–morphism (which is also a $\mathcal{C}$–morphism). Since $f_i : (X,\xi) \longrightarrow (X_i,\xi_i)$ is a $\mathcal{C}$–morphism for each $i \in I$, $f_i = f_i \circ 1_X : (X,\xi_{\mathcal{A}}) \longrightarrow (X_i,\xi_i)$ is a $\mathcal{C}$–morphism between $\mathcal{A}$–objects, i.e. an $\mathcal{A}$–morphism, for each $i \in I$. Consequently, $f_i \circ g : (Y_{\mathcal{A}},\eta_{\mathcal{A}}) \longrightarrow (X_i,\xi_i)$ is an $\mathcal{A}$–morphism for each $i \in I$.

(b) Let $f_i \circ g : (Y_{\mathcal{A}},\eta_{\mathcal{A}}) \longrightarrow (X_i,\xi_i)$ be an $\mathcal{A}$–morphism for each $i \in I$. $1_X \circ g : (Y_{\mathcal{A}},\eta_{\mathcal{A}}) \longrightarrow (X,\xi)$ is a $\mathcal{C}$–morphism, because $f_i \circ g = f_i \circ 1_X \circ g : (Y_{\mathcal{A}},\eta_{\mathcal{A}}) \longrightarrow (X_i,\xi_i)$ is an $\mathcal{A}$–morphism ($= \mathcal{C}$–morphism) and ξ is the initial $\mathcal{C}$–structure.

Since $1_X : (X,\xi_{\mathcal{A}}) \longrightarrow (X,\xi)$ is the bicoreflection of (X,ξ) w.r.t. $\mathcal{A}$, there exists a unique $\mathcal{A}$–morphism $h : (Y_{\mathcal{A}},\eta_{\mathcal{A}}) \longrightarrow (X,\xi_{\mathcal{A}})$ such that the diagram

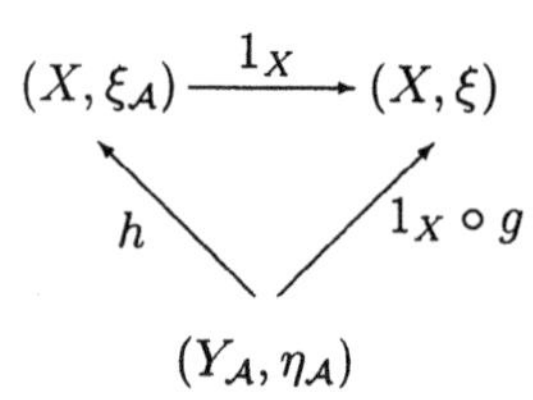

commutes, i.e. $1_X \circ h = 1_X \circ g$. Thus, h and g coincide as maps between the underlying sets. Consequently, $g : (Y_{\mathcal{A}},\eta_{\mathcal{A}}) \longrightarrow (X,\xi_{\mathcal{A}})$ is an $\mathcal{A}$–morphism.

(c) If $\xi'_{\mathcal{A}}$ is also an initial $\mathcal{A}$–structure on X w.r.t. $(X, f_i, (X_i,\xi_i), I)$, then, by the definition of initial structures, both $1_X : (X,\xi_{\mathcal{A}}) \longrightarrow (X,\xi'_{\mathcal{A}})$ and $1_X : (X,\xi'_{\mathcal{A}}) \longrightarrow (X,\xi_{\mathcal{A}})$ are $\mathcal{A}$–morphisms and therefore $\mathcal{C}$–morphisms. Thus, $\xi_{\mathcal{A}} \leq \xi'_{\mathcal{A}}$ and $\xi'_{\mathcal{A}} \leq \xi_{\mathcal{A}}$ in $\mathcal{C}$. Consequently, by 1.1.3. 2), $\xi_{\mathcal{A}} = \xi'_{\mathcal{A}}$ because $\mathcal{C}$ is topological.

(2) For each set X, $\{(Y,\eta) \in |\mathcal{A}| : Y = X\}$ is a set because $\{(Z,\zeta) \in |\mathcal{C}| : Z = X\}$ is a set.

(3) On every set X with cardinality at most one there is the indiscrete $\mathcal{A}$-structure. Since every $\mathcal{A}$-structure is a $\mathcal{C}$-structure and $\mathcal{C}$ is topological, it is the unique $\mathcal{A}$-structure on X.

(4) Let X be a set, $((X_i, \xi_i))_{i\in I}$ a family of $\mathcal{A}$-objects and $(f_i : X_i \longrightarrow X)_{i\in I}$ a family of maps. By 1.2.1.1. there exists the final $\mathcal{A}$-structure $\xi_{\mathcal{A}}$ on X with respect to $((X_i, \xi_i), f_i, X, I)$. In order to verify that $\xi_{\mathcal{A}}$ coincides with the final $\mathcal{C}$-structure $\xi_{\mathcal{C}}$ on X w.r.t. $((X_i, \xi_i), f_i, X, I)$ it suffices to prove that $(X, \xi_{\mathcal{C}}) \in |\mathcal{A}|$, in other words:
the bicoreflection $1_X : (X, (\xi_{\mathcal{C}})_{\mathcal{A}}) \longrightarrow (X, \xi_{\mathcal{C}})$ of $(X, \xi_{\mathcal{C}})$ w.r.t. $\mathcal{A}$ is an isomorphism. Obviously, $1_X : (X, \xi_{\mathcal{C}}) \longrightarrow (X, (\xi_{\mathcal{C}})_{\mathcal{A}})$ is a $\mathcal{C}$-morphism if and only if $1_X \circ f_i = f_i : (X_i, \xi_i) \longrightarrow (X, (\xi_{\mathcal{C}})_{\mathcal{A}})$ is a $\mathcal{C}$-morphism for each $i \in I$. For every $i \in I$, there exists a unique $\mathcal{C}$-morphism $g_i : (X_i, \xi_i) \longrightarrow (X, (\xi_{\mathcal{C}})_{\mathcal{A}})$ such that the diagram

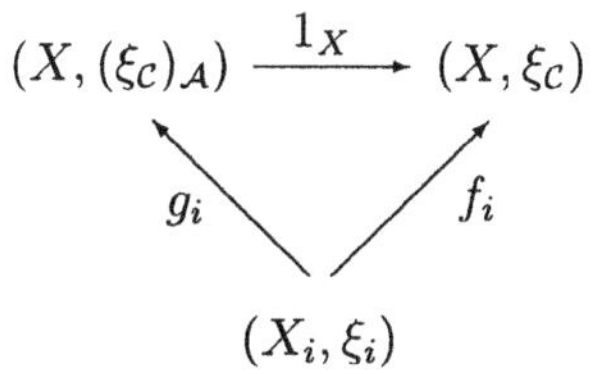

commutes, i.e. $g_i = f_i$ (as maps betwen the underlying sets).
Therefore the condition is fulfilled.

B) The case that $\mathcal{A}$ is bireflective in $\mathcal{C}$ is proved analogously to A).

2.3 Examples of bireflections and bicoreflections

2.3.1 Convergence structures

2.3.1.1 Definitions. 1) A generalized convergence space (X, q) is called

a) a *limit space* provided that the following is satisfied:

C$_4$) $(\mathcal{F} \cap \mathcal{G}, x) \in q$ whenever $(\mathcal{F}, x) \in q$ and $(\mathcal{G}, x) \in q$,

b) a *pseudotopological space* (or *Choquet space*) provided that the following is satisfied:

C$_5$) $(\mathcal{F}, x) \in q$ whenever $(\mathcal{U}, x) \in q$ for every ultrafilter $\mathcal{U} \supset \mathcal{F}$,

c) a *pretopological space* provided that the following is satisfied:

C$_6$) $(\mathcal{U}_q(x), x) \in q$ for all $x \in X$, where $\mathcal{U}_q(x) = \bigcap\{\mathcal{F} \in F(X) : (\mathcal{F}, x) \in q\}$.

2) A pretopological space (X, q) is called a *topologically pretopological space* (shortly: a *topological space*) provided that the following condition is satisfied:

C$_7$) For each $U \in \mathcal{U}_q(x)$ there is some $V \in \mathcal{U}_q(x)$ such that $U \in \mathcal{U}_q(y)$ for all $y \in V$.

3) The classes of all limit spaces, pseudotopological spaces, pretopological spaces and topologically pretopological spaces respectively define full (and isomorphism–closed) subconstructs of **GConv** which are denoted by **Lim**, **PsTop**, **PrTop** and **T–PrTop** respectively.

2.3.1.2 Remark. Each of the constructs in the following list

$$\mathbf{GConv} \supset \mathbf{KConv} \supset \mathbf{Lim} \supset \mathbf{PsTop} \supset \mathbf{PrTop} \supset \mathbf{T\text{–}PrTop}$$

is a (full and isomorphism–closed) subconstruct of the preceding ones:

a) Every pretopological space (X,q) is a pseudotopological space. (Obviously, $\mathcal{F} \xrightarrow{q} x$ iff $\mathcal{F} \supset \mathcal{U}_q(x)$ because (X,q) is pretopological. If $\mathcal{U} \xrightarrow{q} x$ for each ultrafilter $\mathcal{U} \supset \mathcal{F}$, then

$$\mathcal{U}_q(x) \subset \bigcap\{\mathcal{U} : \mathcal{U} \text{ is an ultrafilter with } \mathcal{U} \supset \mathcal{F}\} = \mathcal{F}, \text{ i .e. } \mathcal{F} \xrightarrow{q} x.)$$

b) Every pseudotopological space (X,q) is a limit space. (Namely, if C_4) were not satisfied for (X,q), then there would be filters $\mathcal{F}$, $\mathcal{G}$ on X such that $(\mathcal{F},x) \in q$ and $(\mathcal{G},x) \in q$ for some $x \in X$ but $(\mathcal{F} \cap \mathcal{G}, x) \notin q$. Thus, there would exist some ultrafilter $\mathcal{U}$ containing $\mathcal{F} \cap \mathcal{G}$ such that $(\mathcal{U},x) \notin q$. Especially, $\mathcal{U} \not\supset \mathcal{F}$ and $\mathcal{U} \not\supset \mathcal{G}$, i.e. there would exist $F \in \mathcal{F}$ and $G \in \mathcal{G}$ with $F \notin \mathcal{U}$ and $G \notin \mathcal{U}$. On the other hand we would have $F \cup G \in \mathcal{U}$. So it would follow that $F \in \mathcal{U}$ or $G \in \mathcal{U}$ since $\mathcal{U}$ is an ultrafilter. This is a contradiction.)

c) The remaining implications are obvious.

2.3.1.3 Proposition. **KConv** *is a bireflective and bicoreflective (full and isomorphism–closed) subconstruct of* **GConv**.

Proof. Let (X,q) be a generalized convergence space. Then two Kent convergence structures q_r and q_c on X are defined as follows:

$$(\mathcal{F},x) \in q_r \iff \exists\, (\mathcal{G},x) \in q \text{ such that } \mathcal{G} \cap \dot{x} \subset \mathcal{F}$$

$$(\mathcal{F},x) \in q_c \iff (\mathcal{F},x) \in q \text{ and } (\mathcal{F} \cap \dot{x}, x) \in q$$

Then $1_X : (X,q) \longrightarrow (X,q_r)$ (resp. $1_X : (X,q_c) \longrightarrow (X,q)$) is the desired bireflection (resp. bicoreflection) of (X,q) with respect to **KConv**.

2.3.1.4 Remark. The initial structures in **GConv** have been described under 1.1.6.③ b). The final structures in **GConv** are constructed as follows: Let X be a set, $((X_i,q_i))_{i\in I}$ a family of generalized convergence spaces and $(f_i : X_i \longrightarrow X)_{i \in I}$ a family of maps. Then $q = \{(\mathcal{F},x) \in F(X) \times X$: there is some $i \in I$ and some $(\mathcal{E}_i, x_i) \in q_i$ with $f_i(\mathcal{E}_i) \subset \mathcal{F}$ and $f_i(x_i) = x\} \cup \{(\dot{x},x) : x \in X\}$ is the final **GConv**–structure on X w.r.t. the given data. If $(f_i : X_i \longrightarrow X)_{i\in I}$ is an epi–sink in **Set** (i.e. $X = \bigcup_{i\in I} f_i[X_i]$), then $q = \{(\mathcal{F},x) \in F(X) \times X$: there is some $i \in I$ and some $(\mathcal{E}_i,x_i) \in q_i$ with $f_i(\mathcal{E}_i) \subset \mathcal{F}$ and $f_i(x_i) = x\}$.
It follows from 2.2.12. and 2.3.1.3. that *initial and final structures in the topological construct* **KConv** *are formed as in* **GKonv**.

2.3.1.5 Proposition. *Each of the constructs in the following list*

KConv $\supset$ **Lim** $\supset$ **PsTop** $\supset$ **PrTop** $\supset$ **T–PrTop**

is a bireflective (full and isomorphism–closed) subconstruct of the preceding ones.

Proof. Let (X,q) be a Kent convergence space. Then a **Lim**–structure $\widetilde{q}$, a **PsTop**–structure $\widehat{q}$, a **PrTop**–structure $\overline{q}$ and a **T–PrTop**–structure q^* are defined as follows:

$$(\mathcal{F},x)\in\widetilde{q} \iff \exists\,\mathcal{F}_1,\dots,\mathcal{F}_n\in F(X) \text{ such that } (\mathcal{F}_i,x)\in q \text{ for each } i\in\{1,\dots,n\} \text{ and } \mathcal{F}\supset\bigcap_{i=1}^n\mathcal{F}_i,$$

$$(\mathcal{F},x)\in\widehat{q} \iff (\mathcal{U},x)\in q \text{ for each ultrafilter } \mathcal{U}\supset\mathcal{F},$$

$$(\mathcal{F},x)\in\overline{q} \iff \mathcal{F}\supset\mathcal{U}_q(x)=\bigcap\{\mathcal{H}\in F(X):(\mathcal{H},x)\in q\},$$

$$(\mathcal{F},x)\in q^* \iff \mathcal{F}\supset\mathcal{U}_{\mathcal{X}_q}(x), \text{ where } \mathcal{U}_{\mathcal{X}_q}(x) \text{ denotes the neighborhood filter of x with respect to the topology } \mathcal{X}_q \text{ defined by}$$
$$O\in\mathcal{X}_q \iff \text{For each } x\in O \text{ and each filter } \mathcal{G} \text{ on } X \text{ with } (\mathcal{G},x)\in q,\ O\in\mathcal{G}.$$

Then $\left\{\begin{array}{l} 1_X:(X,q)\longrightarrow(X,\widetilde{q})\\ 1_X:(X,q)\longrightarrow(X,\widehat{q})\\ 1_X:(X,q)\longrightarrow(X,\overline{q})\\ 1_X:(X,q)\longrightarrow(X,q^*)\end{array}\right.$ is the bireflection of (X,q) w.r.t. $\left\{\begin{array}{l}\mathbf{Lim}\\ \mathbf{PsTop}\\ \mathbf{PrTop}\\ \mathbf{T\text{–}PrTop}\end{array}\right.$

The corresponding bireflectors are denoted by $\widetilde{\mathcal{R}}$, $\widehat{\mathcal{R}}$, $\overline{\mathcal{R}}$ and $\mathcal{R}^*$ respectively. Then the restrictions

$$\widehat{\mathcal{R}}|_{\mathbf{Lim}}:\mathbf{Lim}\longrightarrow\mathbf{PsTop},\ \overline{\mathcal{R}}|_{\mathbf{PsTop}}:\mathbf{PsTop}\longrightarrow\mathbf{PrTop} \text{ and}$$
$$\mathcal{R}^*|_{\mathbf{PrTop}}:\mathbf{PrTop}\longrightarrow\mathbf{T\text{–}PrTop}$$

are also bireflectors.

2.3.1.6 Remarks. 1) It follows from 2.3.1.5. that initial structures in **Lim**, **PsTop**, **PrTop** and **T–PrTop** respectively are formed as in **KConv** (and thus as in **GConv**) whereas the final structures in these topological constructs arise from the final structures in **KConv** (cf. 2.3.1.4.) by applying the corresponding bireflectors (cf. the proof of 2.3.1.5. for the definition of these bireflectors).

2) There is an alternative description of pretopological spaces by means of closure spaces:

a) A *closure space* is a pair (X,cl), where X is a set and $cl:\mathcal{P}(X)\longrightarrow\mathcal{P}(X)$ a map, called a closure operation on X, satisfying the following conditions:

$\mathrm{Cl}_1)$ $cl(\emptyset)=\emptyset$,
$\mathrm{Cl}_2)$ $A\subset cl(A)$ for each $A\in\mathcal{P}(X)$,
$\mathrm{Cl}_3)$ $cl(A\cup B)=cl(A)\cup cl(B)$ for each $(A,B)\in\mathcal{P}(X)\times\mathcal{P}(X)$.

b) A map $f:(X,cl)\longrightarrow(X',cl')$ between closure spaces is called *continuous* provided that $f[clA]\subset cl'(f[A])$ for each $A\in\mathcal{P}(X)$.

c) The construct of all closure spaces (and continuous maps) is denoted by **Closp**. (Closure spaces have been extensively studied in Čech's book [29]).

d) *For each generalized convergence space (X,q) a closure operation cl_q on X is* *defined by*

$$cl_q A = \{x \in X : \text{there is some } \mathcal{G} \in F(X) \text{ with } (\mathcal{G}, x) \in q \text{ and } A \in \mathcal{G}\}$$

for each $A \in \mathcal{P}(X)$.

e) If (X, q) is a generalized convergence space, then for each $x \in X$ the filter $\mathcal{U}_q(x) = \bigcap\{\mathcal{F} \in F(X) : (\mathcal{F}, x) \in q\}$ is called the *neighborhood filter* of x. Obviously,

$$\mathcal{U}_q(x) = \{U \subset X : x \notin cl_q(X \setminus U)\}$$

f) If $f : (X, q) \longrightarrow (X', q')$ is a continuous map between generalized convergence spaces, then $f : (X, cl_q) \longrightarrow (X', cl_{q'})$ is a continuous map between closure spaces.

g) If (X, cl) is a closure space, then a **PrTop**–structure q_{cl} on X is defined by

$$(\mathcal{F}, x) \in q_{cl} \Longleftrightarrow \mathcal{F} \supset \mathcal{U}_{cl}(x).$$

where $\mathcal{U}_{cl}(x)$ denotes the neighborhood filter in the closure space (X, cl), i.e. $\mathcal{U}_{cl}(x) = \{U \subset X : x \notin cl(X \setminus U)\}$.

h) It is easily checked that
α) $q_{cl_q} = q$ for each **PrTop**–structure q
β) $cl_{q_{cl}} = cl$ for each closure operation cl.

i) If $f : (X, cl) \longrightarrow (X', cl')$ is a continuous map between closure spaces, then $f : (X, q_{cl}) \longrightarrow (X', q_{cl'})$ is a continuous map between pretopological spaces.

Because of h), f) and i) *we need not distinguish between pretopological spaces and closure spaces.*
These ideas are captured by the following definition (cf. also 2.3.1.8. below).

2.3.1.7 Definition. a) Let $\mathcal{A}$ and $\mathcal{B}$ be categories. Then a functor $\mathcal{F} : \mathcal{A} \longrightarrow \mathcal{B}$ is called an *isomorphism* provided that there is a functor $\mathcal{G} : \mathcal{B} \longrightarrow \mathcal{A}$ such that $\mathcal{G} \circ \mathcal{F} = \mathcal{I}_{\mathcal{A}}$ and $\mathcal{F} \circ \mathcal{G} = \mathcal{I}_{\mathcal{B}}$.

b) Let $\mathcal{A}$ and $\mathcal{B}$ be constructs. If $\mathcal{H} : \mathcal{A} \longrightarrow$ **Set** (resp. $\mathcal{K} : \mathcal{B} \longrightarrow$ **Set**) is the forgetful functor, then a functor $\mathcal{F} : \mathcal{A} \longrightarrow \mathcal{B}$ is called *concrete* provided that $\mathcal{K} \circ \mathcal{F} = \mathcal{H}$.

c) A concrete functor $\mathcal{F} : \mathcal{A} \longrightarrow \mathcal{B}$ between constructs which is an isomorphism, is called a *concrete isomorphism.*

d) Two categories (resp. constructs) $\mathcal{A}$ and $\mathcal{B}$ are called *isomorphic* (resp. *concretely isomorphic*) provided that there is an isomorphism (resp. concrete isomorphism) $\mathcal{F} : \mathcal{A} \longrightarrow \mathcal{B}$ (notation: $\mathcal{A} \cong \mathcal{B}$).

2.3.1.8 Proposition. a) **PrTop** *and* **Closp** *are concretely isomorphic.*
b) **Top** *and* **T–PrTop** *are concretely isomorphic.*

Proof. a) Put $\mathcal{F}((X, q)) = (X, cl_q)$ for each pretopological space (X, q), and for each continuous map f between pretopological spaces, let $\mathcal{F}(f)$ be the corresponding map between closure spaces (cf. 2.3.1.6. 2) f)). Then a concrete functor

$\mathcal{F}$: **PrTop** $\longrightarrow$ **Closp** is defined. Put $\mathcal{G}((X,cl)) = (X,q_{cl})$ for each closure space (X,cl), and for each continuous map f between closure spaces, let $\mathcal{G}(f)$ be the corresponding continuous map between pretopological spaces (cf. 2.3.1.6. i)). Then a functor $\mathcal{G}$: **Closp** $\longrightarrow$ **PrTop** is defined such that $\mathcal{G} \circ \mathcal{F} = \mathcal{I}_{\mathbf{PrTop}}$ and $\mathcal{F} \circ \mathcal{G} = \mathcal{I}_{\mathbf{Closp}}$ (cf. 2.3.1.6. 2) h)). Consequently, $\mathcal{F}$: **PrTop** $\longrightarrow$ **Closp** is a concrete isomorphism.

b) α) If (X,q) is a generalized convergence space, then a topology $\mathcal{X}_q$ on X is defined by

$$O \in \mathcal{X}_q \iff \text{For each } x \in X \text{ and each filter } \mathcal{F} \text{ on } X \text{ with } (\mathcal{F},x) \in q,\ O \in \mathcal{F}.$$

[Obviously, $O \in \mathcal{X}_q$ iff $O = \mathrm{int}_q O$, where $\mathrm{int}_q O = X \setminus cl_q(X \setminus O)$ (cf. 2.3.1.6. d) for the definition of cl_q). Thus, $\{A \subset X : A = cl_q A\}$ is the set of all closed sets in $(X, \mathcal{X}_q)$].

β) If $\mathcal{X}$ is a topology on a set X, then a **T–PrTop**–structure $q_{\mathcal{X}}$ on X is defined by

$$(\mathcal{F},x) \in q_{\mathcal{X}} \iff \mathcal{F} \supset \mathcal{U}_{\mathcal{X}}(x)$$

where $\mathcal{U}_{\mathcal{X}}(x)$ denotes the usual neighborhood filter in $(X,\mathcal{X})$, i.e. $U \in \mathcal{U}_{\mathcal{X}}(x)$ iff there is some $O \in \mathcal{X}$ with $x \in O \subset U$.

γ) It is easy to check that
(1) $\mathcal{X}_{q_{\mathcal{X}}} = \mathcal{X}$ for each topology $\mathcal{X}$,
(2) $q_{\mathcal{X}_q} = q$ for each **T–PrTop**–structure q.

[Concerning (2), note that a **GConv**–structure q on a set X is a **T–PrTop**–structure iff there is a topology $\mathcal{X}$ on X such that $q = q_{\mathcal{X}}$, and use (1))].

δ) If $f : (X,q) \longrightarrow (X',q')$ is a continuous map between generalized convergence spaces, then $f : (X,\mathcal{X}_q) \longrightarrow (X',\mathcal{X}_{q'})$ is continuous (apply 2.3.1.6. 2) f) and [29; 16 A. 6.]).

ϵ) If $f : (X,\mathcal{X}) \longrightarrow (X',\mathcal{X}')$ is a continuous map between topological spaces defined by means of open sets, then $f : (X,q_{\mathcal{X}}) \longrightarrow (X',q_{\mathcal{X}'})$ is a **T–PrTop**–morphism.

It follows from γ), δ) and ϵ) that **Top** is concretely isomorphic to **T–PrTop** (cf. the corresponding procedure under a)).

2.3.1.9 Remarks. 1) Because of 2.3.1.8. b) *it is now justified to call topologically pretopological spaces also topological spaces.*

2) If (X,q) is a generalized convergence space, then $1_X : (X,q) \longrightarrow (X,q_{cl_q})$ is the bireflection of (X,q) with respect to **PrTop**. Thus, (X,q_{cl_q}) (resp. (X,cl_q)) is called the *underlying closure space of the generalized convergence space* (X,q).

3) If (X,q) is a generalized convergence space, then $1_X : (X,q) \longrightarrow (X,q_{\mathcal{X}_q})$ is the bireflection of (X,q) with respect to **T–PrTop** ($\cong$ **Top**). Thus, $(X,q_{\mathcal{X}_q})$ (resp. $(X,\mathcal{X}_q)$) is called the *underlying topological space of the generalized convergence space* (X,q).

2.3.2 Uniform convergence structures

2.3.2.1 Definitions. 1) A semiuniform convergence space $(X, \mathcal{J}_X)$ is called a *semiuniform limit space* provided that the following is satisfied:

UC$_4$) $\mathcal{F} \in \mathcal{J}_X$ and $\mathcal{G} \in \mathcal{J}_X$ imply $\mathcal{F} \cap \mathcal{G} \in \mathcal{J}_X$.

2) A semiuniform limit space $(X, \mathcal{J}_X)$ is called a *uniform limit space* provided that the following is satisfied:

UC$_5$) $\mathcal{F} \in \mathcal{J}_X$ and $\mathcal{G} \in \mathcal{J}_X$ imply $\mathcal{F} \circ \mathcal{G} \in \mathcal{J}_X$ (whenever $\mathcal{F} \circ \mathcal{G}$ exists, i.e. $F \circ G = \{(x, y) : \exists z \in X \text{ with } (x, z) \in G \text{ and } (z, y) \in F\} \neq \emptyset$ for every $F \in \mathcal{F}$, $G \in \mathcal{G}$), where $\mathcal{F} \circ \mathcal{G}$ is the filter generated by the filter base $\{F \circ G : F \in \mathcal{F}, G \in \mathcal{G}\}$.

3) A uniform limit space $(X, \mathcal{J}_X)$ is called a *principal uniform limit space* provided that there is a non–empty subset $\mathcal{F}$ of $\mathcal{P}(X \times X)$ satisfying the conditions F$_2$) and F$_3$) for a filter such that $\mathcal{J}_X = [\mathcal{F}]$, where $[\mathcal{F}] = \{\mathcal{G} \in F(X \times X) : \mathcal{G} \supset \mathcal{F}\}$.

4) The classes of all semiuniform limit spaces, uniform limit spaces and principal uniform limit spaces respectively define full (and isomorphism–closed) subconstructs of **SUConv** which are denoted by **SULim**, **ULim** and **PrULim** respectively.

2.3.2.2 Remark Obviously, **PrULim** *is concretely isomorphic to* **Unif** (note, that a non–empty subset $\mathcal{F}$ of $\mathcal{P}(X \times X)$ satisfying the conditions F$_2$) and F$_3$) for a filter is a uniformity on X if and only if $[\mathcal{F}]$ is a **PrULim**–structure, and see 1.1.6. ⑤ b)). Thus, *we need not distinguish between uniform spaces and principal uniform limit spaces.*

2.3.2.3 Proposition. *Each of the constructs in the following list*

$$\textbf{SUConv} \supset \textbf{SULim} \supset \textbf{ULim} \supset \textbf{PrULim} \ (\cong \textbf{Unif})$$

is a bireflective (full and isomorphism–closed) subconstruct of the preceding ones.

Proof. 1) Let $(X, \mathcal{J}_X)$ be a semiuniform convergence space. Then $1_X : (X, \mathcal{J}_X) \longrightarrow (X, (\mathcal{J}_X)_L)$ is the bireflection of $(X, \mathcal{J}_X)$ with respect to **SULim** where $(\mathcal{J}_X)_L = \{\mathcal{F} \in F(X \times X)$: there exist $\mathcal{F}_1, \dots, \mathcal{F}_n \in \mathcal{J}_X$ with $\mathcal{F} \supset \bigcap_{i=1}^n \mathcal{F}_i\}$.

2) Let $(X, \mathcal{J}_X)$ be a semiuniform limit space. Then $1_X : (X, \mathcal{J}_X) \longrightarrow (X, ((\mathcal{J}_X)_Q)_L)$ is the bireflection of $(X, \mathcal{J}_X)$ with respect to **ULim** provided that $(\mathcal{J}_X)_Q = \{\mathcal{F} \in F(X \times X)$: there exist $\mathcal{F}_1, \dots, \mathcal{F}_n \in \mathcal{J}_X$ with $\mathcal{F} \supset \mathcal{F}_1 \circ \dots \circ \mathcal{F}_n\}$ and L is defined as under 1) [Note: a) If $f : X \longrightarrow Y$ is a map and $\mathcal{F}$ (resp. $\mathcal{G}$) is a filter on $X \times X$, then
α) $(f \times f)(\mathcal{F}^{-1}) = ((f \times f)(\mathcal{F}))^{-1}$
β) $(f \times f)(\mathcal{F} \circ \mathcal{G}) \supset (f \times f)(\mathcal{F}) \circ (f \times f)(\mathcal{G})$ provided that $\mathcal{F} \circ \mathcal{G}$ exists (the equality is valid whenever f is injective).

b) If $\mathcal{F}, \mathcal{G} \in F(X \times X)$, then
α) $(\mathcal{F} \cap \mathcal{G})^{-1} = \mathcal{F}^{-1} \cap \mathcal{G}^{-1}$,
β) $(\mathcal{F} \circ \mathcal{G})^{-1} = \mathcal{G}^{-1} \circ \mathcal{F}^{-1}$ provided that $\mathcal{F} \circ \mathcal{G}$ exists.

c) Let $(\mathcal{F}_i)_{i\in I}$ and $(\mathcal{G}_j)_{j\in J}$ be finite families of filters on $X \times X$. Then $(\bigcap_{i\in I}\mathcal{F}_i)\circ(\bigcap_{j\in J}\mathcal{G}_j)$ exists iff there is some $(i_0, j_0) \in I \times J$ such that $\mathcal{F}_{i_0}\circ\mathcal{G}_{j_0}$ exists. If $(\bigcap_{i\in I}\mathcal{F}_i)\circ\bigcap_{j\in J}\mathcal{G}_j)$ exists, then

$(\bigcap_{i\in I}\mathcal{F}_i)\circ(\bigcap_{j\in J}\mathcal{G}_j) = \bigcap_{(i,j)\in K}\mathcal{F}_i\circ\mathcal{G}_j$, where $K = \{(i,j) \in I \times J : \mathcal{F}_i \circ \mathcal{G}_j$ exists$\}$.]

3) Let $(X, \mathcal{J}_X)$ be a uniform limit space and $\mathcal{W}$ the finest uniformity on X which is coarser than each $\mathcal{F} \in \mathcal{J}_X$ [Note: Let $\mathcal{M} = \{\mathcal{V} : \mathcal{V}$ is a uniformity on X such that $\mathcal{V} \subset \mathcal{F}$ for each $\mathcal{F} \in \mathcal{J}_X\}$. Then $\mathcal{M} \neq \emptyset$, because $\mathcal{V} = \{X \times X\} \in \mathcal{M}$. Obviously, $\mathcal{W}$ is the initial uniformity on X with respect to $(X, 1_X^{\mathcal{V}}, (X, \mathcal{V}), \mathcal{M})$, where $1_X^{\mathcal{V}} : X \longrightarrow (X, \mathcal{V})$ is the identity map for each $\mathcal{V} \in \mathcal{M}$; namely $\mathcal{V} \subset \mathcal{W}$ for each $\mathcal{V} \in \mathcal{M}$ and $\mathcal{W} \subset \mathcal{F}$ for each $\mathcal{F} \in \mathcal{J}_X$ (If $W \in \mathcal{W}$, then $W \supset \bigcap_{k=1}^n V_k$ with $V_k \in \mathcal{V}_k \in \mathcal{M}$ for each $k \in \{1, \ldots, n\}$. Since $\mathcal{V}_k \subset \mathcal{F}$ for each $\mathcal{F} \in \mathcal{J}_X$, it follows that $W \in \mathcal{F}$ for each $\mathcal{F} \in \mathcal{J}_X$.).]. Then $1_X : (X, \mathcal{J}_X) \longrightarrow (X, [\mathcal{W}])$ is the bireflection of $(X, \mathcal{J}_X)$ with respect to **PrULim**: If $(X', [\mathcal{W}']) \in |\mathbf{PrULim}|$ and $f : (X, \mathcal{J}_X) \longrightarrow (X', [\mathcal{W}'])$ is uniformly continuous, then $f \times f(\mathcal{J}_X) \subset [\mathcal{W}']$. Hence $f \times f(\mathcal{F}) \supset \mathcal{W}'$ for each $\mathcal{F} \in \mathcal{J}_X$, i.e. for each $W' \in \mathcal{W}'$, we have $(f \times f)^{-1}[W'] \in \mathcal{F}$ for each $\mathcal{F} \in \mathcal{J}_X$. Thus, the initial uniformity on X with respect to $f : X \longrightarrow (X, \mathcal{W}')$ is contained in each $\mathcal{F} \in \mathcal{J}_X$; therefore, it is contained in the finest uniformity of this kind, namely in $\mathcal{W}$. This means that $f : (X, [\mathcal{W}]) \longrightarrow (X', [\mathcal{W}'])$ is uniformly continuous.

2.3.2.4. We know already that each pseudometric on a set X induces a uniformity (cf. 1.1.6. ②). If the set $I\!R$ of real numbers is endowed with the uniformity induced by the Euclidean metric, we get a uniform space denoted by $I\!R_u$. Let $I\!R_u^{I\!R}$ be the product space in **Unif**. Then there is no pseudometric on $I\!R^{I\!R}$ inducing the uniformity on $I\!R^{I\!R}$ just described, in other words: *$I\!R_u^{I\!R}$ is not pseudometrizable* (note: 1) a) Every uniform space $(X, \mathcal{W})$ induces a topological space $(X, \mathcal{X}_{\mathcal{W}})$ described as follows: $O \in \mathcal{X}_{\mathcal{W}} \Longleftrightarrow$ For each $x \in O$ there is some $V \in \mathcal{W}$ such that $V(x) \subset O\}$.
b) If $(X, \mathcal{W})$ is pseudometrizable, then $(X, \mathcal{X}_{\mathcal{W}})$ fulfills the first axiom of countability.
2) Initial uniformities induce initial topologies (cf. 2.3.3.18).
3) The topological space $R_t^{I\!R}$ does not fulfill the first axiom of countability, where $I\!R_t$ denotes the set $I\!R$ endowed with the usual topology [2]).

On the other hand we will prove that every uniformity is 'generated' by a whole family of pseudometrics. But before doing this, we need the following

Proposition. *Let $(X, \mathcal{W})$ be a uniform space. Then the following are vaild:*
(1) *$\mathcal{B} = \{V \in \mathcal{W} : V = V^{-1}\}$ is a base for $\mathcal{W}$, i.e. the symmetric entourages form a base for $\mathcal{W}$.*

(2) *For each natural number $n \geq 1$ and each base $\mathcal{B}$ for $\mathcal{W}$, $\mathcal{B}_n = \{V^n : V \in \mathcal{B}\}$ is a base for $\mathcal{W}$, where $V^1 = V$ and $V^n = V^{n-1} \circ V$ for $n > 1$.*

---
[2] $I\!R_t$ is the topological space induced by $I\!R_u$.

Proof. In order to show that $\mathcal{M} \subset \mathcal{P}(X \times X)$ is a base for $\mathcal{W}$ it suffices to prove:

α) All $V \in \mathcal{M}$ are entourages,
β) Each entourage contains an entourage belonging to $\mathcal{M}$.

(1) α) Obvious.

β) Let $W \in \mathcal{W}$. Put $V = W \cap W^{-1}$. Then $V = V^{-1} \in \mathcal{B}$ and $W \supset V$.

(2) α) For each natural number $n \geq 1$, V^n is an entourage for each $V \in \mathcal{B} \subset \mathcal{W}$, namely $V \subset V^n$ (since $\Delta \subset V$!) and $\mathcal{W}$ is a filter.

β) For each natural number $n \geq 1$ the following holds: For each $W \in \mathcal{W}$ there is some $V \in \mathcal{B}$ such that $V^n \subset W$. This assertion is correct for $n = 1$ and because of U_3) also for $n = 2$ (since $\mathcal{B}$ is a base for $\mathcal{W}$). It is proved by induction that it is also correct for $n = 2^m$ for all natural numbers m. Thus, it is correct for each natural number $n \geq 1$ because for each $n \geq 1$ there is some natural number k such that $n + k = 2^n$, which implies $V^n \subset V^{n+k} = V^{2^n}$ (since $\Delta \subset V$!).

Theorem. *Let $(X, \mathcal{W})$ be a uniform space. Then there is a family $(d_V)_{V \in \mathcal{W}}$ of pseudometrics on X such that $\mathcal{W}$ is the coarsest uniformity on X which is finer than each $\mathcal{D}_V$ provided that for each $V \in \mathcal{W}$, $\mathcal{D}_V$ denotes the uniformity induced by d_V, in other words: $\mathcal{W} = \operatorname{glb}\{\mathcal{D}_V : V \in \mathcal{W}\}$ where the greatest lower bound is formed with respect to $\leq$ (cf. 1.1.4.).*

Proof. 1) For each $V \in \mathcal{W}$, the pseudometric d_V in the above theorem is constructed as follows: Let V_1 be a symmetric entourage such that $V_1 \subset V$ (cf. part (1) of the above proposition). By part (2) of the above proposition there is a sequence $(V_n)_{n \in I\!N}$ of symmetric entourages such that

$$V_{n+1}^3 \subset V_n \qquad (n = 1, 2, \dots)$$

(note: (V_n) is a decreasing sequence because of $V_{n+1} \subset V_{n+1}^3$).

Let $h_V : X \times X \longrightarrow [0, 1]$ be defined by

$$h_V(x,y) = \begin{cases} 0 & \text{if } (x,y) \in \bigcap_{n \in I\!N} V_n \\ 1 & \text{if } (x,y) \in (X \times X) \setminus V_1 \\ 2^{-k} & \text{if } (x,y) \in (\bigcap_{n=1}^{k} V_n) \cap ((X \times X) \setminus V_{k+1}) \end{cases}$$

for all $(x, y) \in X \times X$. Unfortunately, h_V is not a pseudometric since the triangle inequality need not be satisfied.
Put

$$M_{xy} = \Big\{\sum_{i=1}^{p} h_V(z_{i-1}, z_i) : (z_0, \dots, z_p) \text{ is a finite sequence of elements of } X \text{ such that } z_0 = x \text{ and } z_p = y,\ p \in I\!N\Big\}$$

for each $(x, y) \in X \times X$. Then for each $V \in \mathcal{W}$, a pseudometric d_V on X is defined by

$$d_V(x, y) = \operatorname{glb} M_{xy} \quad \text{for each } (x, y) \in X \times X\,.$$

Obviously, $d_V(x,y) \geq 0$ since all elements of M_{xy} are ≥ 0. Now let us check the axioms for a pseudometric:

M$_1$) $d_V(x,x) = 0$ for all $x \in X$, because $h_V(x,x) = 0 \in M_{xx}$.

M$_2$) $d_V(x,y) = d_V(y,x)$ for all $(x,y) \in X \times X$:
Obviously, $M_{xy} = M_{yx}$ for all $(x,y) \in X \times X$.

M$_3$) 'triangle inequality'. Let $(x,y,z) \in X^3$. Since $d_V(x,y)$ is the greatest lower bound of M_{xy}, for each $\varepsilon > 0$ there is a finite sequence $(z_0 = x, z_1, \dots, z_{p-1}, z_p = y)$ of elements of X such that

$$\sum_{i=1}^{p} h_V(z_{i-1}, z_i) < d_V(x,y) + \varepsilon\,.$$

Analogously, there is a finite sequence $(z_{p+1} = y, z_{p+2}, \dots, z_{p+q} = z)$ of elements of X such that

$$\sum_{i=p+1}^{p+q} h_V(z_{i-1}, z_i) < d_V(y,z) + \varepsilon\,.$$

Thus, we obtain

$$d_V(x,z) \leq \sum_{\substack{i=1 \\ z_0 = x \\ z_{p+q} = z}}^{p+q} h_V(z_{i-1}, z_i) < d_V(x,y) + d_V(y,z) + 2\varepsilon$$

for each $\varepsilon > 0$. Consequently,

$$d_V(x,z) \leq d_V(x,y) + d_V(y,z)\,.$$

B) For each $(x,y) \in X \times X$ the following is valid:

$$\frac{1}{2}\, h_V(x,y) \leq d_V(x,y) \leq h_V(x,y)\,.$$

1) $d_V(x,y) \leq h_V(x,y)$ because $h_V(x,y) \in M_{xy}$.

2) If for every finite sequence $(z_0, \dots, z_p)$ of elements of X the inequality

$$(*) \qquad \frac{1}{2}\, h_V(z_0, z_p) \leq \sum_{i=1}^{p} h_V(z_{i-1}, z_i)$$

is valid, then $\frac{1}{2}\, h_V(x,y) \leq \sum_{i=1}^{p} h_V(z_{i-1}, z_i)$ for every finite sequence $(z_0, \dots, z_p)$ of elements of X such that $z_0 = x$ and $z_p = y$. Thus, since $d_V(x,y)$ is the greatest lower bound of M_{xy},

$$\frac{1}{2}\, h_V(x,y) \leq d_V(x,y)\,.$$

Consequently, it suffices to prove $(*)$ by induction with respect to p. Obviously, $(*)$ is valid provided that $p = 1$. Let $q > 1$ be a natural number and suppose that

$(*)$ is valid for all natural numbers $p < q$. In order to prove that $(*)$ is valid for $p = q$, let $(z_0, \ldots, z_p)$ be any sequence of elements of X and put

$$a = \sum_{i=1}^{q} h_V(z_{i-1}, z_i) .$$

If $a \geq \frac{1}{2}$, $(*)$ is valid because $h_V(z_0, z_q) \leq 1$.

Hence, we can suppose that $a < \frac{1}{2}$. Since the case where $a = 0$ is trivial, we may assume that $a > 0$. Then there is a smallest integer $j \leq q$ such that

$$\sum_{i=1}^{j} h_V(z_{i-1}, z_i) > \frac{a}{2} .$$

By inductive assumption we obtain

(1) $h_V(z_0, z_{j-1}) \leq a$

and (2) $h_V(z_j, z_q) < a$

Choose a natural number k such that

$$\frac{1}{2^k} = 2^{-k} \leq a < 2^{-k+1} = \frac{2}{2^k} .$$

Since $a < \frac{1}{2}$, we have $k \geq 2$. Obviously, $h_V(z_o, z_{j-1}) \leq a < 2^{-k+1}$, $h_V(z_{j-1}, z_j) \leq a < 2^{-k+1}$ and $h_V(z_j, z_q) < a < 2^{-k+1}$. By the definition of h_V we obtain $h_V(z_0, z_{j-1}) \leq 2^{-k}$, $h_V(z_{j-1}, z_j) \leq 2^{-k}$ and $h_V(z_j, z_q) \leq 2^{-k}$ which implies $(z_0, z_{j-1}) \in V_k$, $(z_{j-1}, z_j) \in V_k$ and $(z_j, z_q) \in V_k$, i.e. $(z_0, z_q) \in V_k^3 \subset V_{k-1}$. Thus, $h_V(z_0, z_q) \leq 2^{-(k-1)} = 2 \cdot 2^{-k} \leq 2a$, i.e. $(*)$ holds for $p = q$.

C) $\mathcal{W}$ is the coarsest uniformity on X which is finer than $\mathcal{D}_V$ for each $V \in \mathcal{W}$ provided that $\mathcal{D}_V$ denotes the uniformity induced by d_V:

a) For each $V \in \mathcal{W}$, $\mathcal{D}_V \subset \mathcal{W}$:
If $W \in \mathcal{D}_V$, then there is some $\varepsilon > 0$ such that $V_\varepsilon = \{(x, y) : d_V(x, y) < \varepsilon\} \subset W$. Choose a natural number k such that $2^{-k} < \varepsilon$. Then $V_k \subset V_\varepsilon$ because $(x, y) \in V_k$ implies $h_V(x, y) \leq 2^{-k} < \varepsilon$ and thus, $d_V(x, y) \leq h_V(x, y) < \varepsilon$, i.e. $(x, y) \in V_\varepsilon$. Since $V_k \in \mathcal{W}$, we obtain $W \in \mathcal{W}$.

b) If $\mathcal{W}'$ is a uniformity on X such that $\mathcal{D}_V \subset \mathcal{W}'$ for each $V \in \mathcal{W}$, then $\mathcal{W} \subset \mathcal{W}'$:
If $V \in \mathcal{W}$, then $\{W_\varepsilon = \{(x, y) : d_V(x, y) < \varepsilon\} : \varepsilon > 0\}$ is a base for $\mathcal{D}_V$. Choose $\varepsilon = \frac{1}{2}$. Then $W_{\frac{1}{2}} \subset V_1 \subset V$, namely if $(x, y) \in W_{\frac{1}{2}}$, i.e. $d_V(x, y) < \frac{1}{2}$, then, since $\frac{1}{2} h_V(x, y) \leq d_V(x, y)$, we have $h_V(x, y) < 1$, i.e. $(x, y) \in V_1 \subset V$. Consequently, $V \in \mathcal{D}_V$, since $\mathcal{D}_V$ is a filter. Furthermore, $V \in \mathcal{W}'$, since $\mathcal{W}'$ is finer than each $\mathcal{D}_V$.

2.3.2.5 Definition. A semiuniform convergence space $(X, \mathcal{J}_X)$ is called *pseudometrizable* provided that there is a pseudometric d on X such that $\mathcal{J}_X = \{\mathcal{F} \in F(X \times X) : \mathcal{F} \supset \{V_\varepsilon : \varepsilon > 0\}\}$ where $V_\varepsilon = \{(x, y) : d(x, y) < \varepsilon\}$.

2.3.2.6 Theorem. *Let $(X, \mathcal{J}_X)$ be a semiuniform convergence space. Then* *the following are equivalent:*

(1) $(X, \mathcal{J}_X)$ is a uniform space (=principal uniform limit space).

(2) $(X, \mathcal{J}_X)$ is a subspace of a product of pseudometrizable semiuniform convergence spaces.

Proof. (1) $\Longrightarrow$ (2).

A) Let $(X, \mathcal{W})$ be a uniform space. By the theorem under 2.3.2.4 there is a family $(d_V)_{V \in \mathcal{W}}$ of pseudometrics on X such that $\mathcal{W}$ is the coarsest uniformity on X which is finer than each $\mathcal{D}_V$, where $\mathcal{D}_V$ denotes the uniformity induced by d_V for each $V \in \mathcal{W}$, in other words: $\mathcal{W}$ is the initial uniformity on X with respect to $(X, 1_X^V, (X, \mathcal{D}_V), \mathcal{W})$ provided that 1_X^V denotes the identity map for each $V \in \mathcal{W}$. Let $\prod_{V \in \mathcal{W}}(X, \mathcal{D}_V)$ be the product of the family $((X, \mathcal{D}_V))_{V \in \mathcal{W}}$ in **Unif** and let $p_V : \prod_{V \in \mathcal{W}}(X, \mathcal{D}_V) \longrightarrow (X, \mathcal{D}_V)$ be the projection for each $V \in \mathcal{W}$. By the (categorical) definition of a product, there is a unique uniformly continuous map

$$e : (X, \mathcal{W}) \longrightarrow \textstyle\prod_{V \in \mathcal{W}}(X, \mathcal{D}_V) \text{ such that } p_V \circ e = 1_X^V \text{ for each } V \in \mathcal{W}.$$

Since e is injective and $\mathcal{W}$ is the initial uniformity w.r.t. e, e is an embedding, i.e. $(X, \mathcal{W})$ is a subspace of a product of pseudometrizable spaces.

B) Since **PrULim** ($\cong$ **Unif**) is a bireflective subconstruct of **SUConv**, products and subspaces in **PrULim** (resp. **Unif**) are formed as in **SUConv**. Thus the desired implication follows from A).

(2) $\Longrightarrow$ (1). Since every pseudometrizable semiuniform convergence space is uniform, and **Unif** (resp. **PrULim**) is closed under formation of products and subspaces in **SUConv** (because it is bireflective in **SUConv**), this implication is obvious.

2.3.2.7. In order to obtain a necessary and sufficient condition for a semiuniform convergence space to be pseudometrizable we need the following

Definition. A semiuniform convergence space $(X, \mathcal{J}_X)$ fulfills the *first axiom of countability* provided that for each $\mathcal{F} \in \mathcal{J}_X$ there is a subfilter[3] $\mathcal{G} \in \mathcal{J}_X$ of $\mathcal{F}$ such that $\mathcal{G}$ has a countable base.

2.3.2.8 Theorem. *A semiuniform convergence space $(X, \mathcal{J}_X)$ is pseudometrizable iff the following conditions are satisfied:*

(1) $(X, \mathcal{J}_X)$ is a uniform space (= principal uniform limit space),

(2) $(X, \mathcal{J}_X)$ fulfills the first axiom of countability.

Proof. a) Let $(X, \mathcal{J}_X)$ be pseudometrizable: (1) Then there is a pseudometric d on X such that $\mathcal{J}_X = \{\mathcal{F} \in F(X \times X) : \mathcal{F} \supset \{V_\varepsilon : \varepsilon > 0\}\}$, i.e. $\mathcal{W}_d \subset \mathcal{F}$ for each $\mathcal{F} \in \mathcal{J}_X$, where $\mathcal{W}_d$ denotes the uniformity induced by d. Consequently, $\mathcal{J}_X = [\mathcal{W}_d]$, i.e. $(X, \mathcal{J}_X)$ is a principal uniform limit space (= uniform space).

[3]A subset $\mathcal{G}$ of a filter $\mathcal{F}$ on a set Z is called a *subfilter* of $\mathcal{F}$ provided that it is a filter on Z.

(2) $\{V_{\frac{1}{n}} : n \in I\!N\}$ is a countable base for $\mathcal{W}_d \in \mathcal{J}_X$ which is a subfilter of each $\mathcal{F} \in \mathcal{J}_X$.

b) If (1) and (2) are satisfied, we obtain $\mathcal{J}_X = [\mathcal{W}]$, where $\mathcal{W}$ is a uniformity on X such that $\mathcal{W}$ has a countable base $\{W_n : n \in I\!N\}$. Choose a symmetric entourage $V_1 \in \mathcal{W}$ such that $V_1 \subset W_1$. Furthermore, choose symmetric entourages V_n $(n = 2, 3, \dots)$ such that

$$V_n^3 \subset V_{n-1} \cap \bigcap_{i=1}^n W_i$$

(by the proposition under 2.3.2.4 this is possible because $V_{n-1} \cap \bigcap_{i=1}^n W_i \in \mathcal{W}$). Obviously, $V_n \subset W_n$ for each $n \in I\!N$. Thus, $\{V_n : n \in I\!N\}$ is also a base for $\mathcal{W}$. Put $V = W_1$. Thus, by means of the proof of the theorem under 2.3.2.4, for the sequence $(V_n)_{n \in I\!N}$ there is a pseudometric d_V on X inducing a uniformity $\mathcal{D}_V$. Put $U_n = \{(x,y) : d_V(x,y) \leq 2^{-n}\}$ for $n = 2, 3, \dots$. Then

$$(\alpha) \qquad \{U_n : n = 2, 3, \dots\} \quad \text{is a base of } \mathcal{D}_V.$$

Furthermore,

$$V_n \subset U_n \subset V_{n-1} \qquad (n = 2, 3, \dots)$$

[namely, if $(x,y) \in V_n$, then $h_V(x,y) \leq 2^{-n}$ and because of $d_V(x,y) \leq h_V(x,y)$ it follows that $d_V(x,y) \leq 2^{-n}$, i.e. $(x,y) \in U_n$, and if $(x',y') \in U_n$, i.e. $d_V(x',y') \leq 2^{-n}$, then we obtain from $\frac{1}{2} h_V(x',y') \leq d_V(x',y')$ that $h_V(x',y') \leq 2 \cdot 2^{-n} = 2^{-(n-1)}$, i.e. $(x',y') \in V_{n-1}$].

Since $V_n \subset U_n$, $U_n \in \mathcal{W}$ and since $U_n \subset V_{n-1}$,

$$(\beta) \qquad \{U_n : n = 2, 3, \dots\} \quad \text{is a base for } \mathcal{W}.$$

It follows from (α) and (β) that $\mathcal{W} = \mathcal{D}_V$, i.e. $(X, \mathcal{J}_X) = (X, [\mathcal{D}_V])$ is pseudometrizable w.r.t. d_V.

2.3.3 The linkage between convergence structures and uniform convergence structures

2.3.3.1 Definition. Let $(X, \mathcal{J}_X)$ be a semiuniform convergence space. Then a filter on X is called a *$\mathcal{J}_X$–Cauchy filter* provided that $\mathcal{F} \times \mathcal{F} \in \mathcal{J}_X$, where $\mathcal{F} \times \mathcal{F}$ denotes the filter generated by the filter base $\{F \times F : F \in \mathcal{F}\}$.

2.3.3.2 Proposition. *(1) Let $(X, \mathcal{J}_X)$ be a semiuniform convergence space. Then $(X, \gamma_{\mathcal{J}_X})$ is a filter space provided that $\gamma_{\mathcal{J}_X}$ denotes the set of all $\mathcal{J}_X$–Cauchy filters on X.*

(2) Let (X, γ) be a filter space and let $\mathcal{J}_\gamma = \{\mathcal{F} \in F(X \times X)$: there is some $\mathcal{G} \in \gamma$ with $\mathcal{F} \supset \mathcal{G} \times \mathcal{G}\}$. Then $(X, \mathcal{J}_\gamma)$ is a semiuniform convergence space such that $\gamma_{\mathcal{J}_\gamma} = \gamma$, i.e. γ is the set of all $\mathcal{J}_\gamma$–Cauchy filters on X.

Proof. (1) follows immediately from the definitions.

(2) a) $(X, \mathcal{J}_\gamma) \in |\mathbf{SUConv}|$: Evidently, UC_1) and UC_2) are valid. UC_3) follows from the fact that $(\mathcal{G} \times \mathcal{G})^{-1} = \mathcal{G} \times \mathcal{G} \subset \mathcal{F}^{-1}$ provided $\mathcal{G} \times \mathcal{G} \subset \mathcal{F}$ with $\mathcal{G} \in \gamma$.

b) $\mathcal{F} \in \gamma$ implies $\mathcal{F} \times \mathcal{F} \in \mathcal{J}_\gamma$, i.e. $\mathcal{F} \in \gamma_{\mathcal{J}_\gamma}$. Conversely, if $\mathcal{F} \in \gamma_{\mathcal{J}_\gamma}$ then $\mathcal{F} \times \mathcal{F} \in \mathcal{J}_\gamma$, i.e. there exists some $\mathcal{G} \in \gamma$ with $\mathcal{G} \times \mathcal{G} \subset \mathcal{F} \times \mathcal{F}$. Thus, $\mathcal{G} \subset \mathcal{F}$ and therefore $\mathcal{F} \in \gamma$.

2.3.3.3 Remark. Obviously, $\mathcal{J}_\gamma$ is the finest one of all semiuniform convergence structures $\mathcal{J}_X$ on X such that $\gamma_{\mathcal{J}_X} = \gamma$.

2.3.3.4 Definition. A semiuniform convergence space $(X, \mathcal{J}_X)$ is called **Fil**–*determined* provided that $\mathcal{J}_X = \mathcal{J}_{\gamma_{\mathcal{J}_X}}$ (i.e. $\mathcal{J}_X$ is 'generated' by all $\mathcal{J}_X$–Cauchy filters).

2.3.3.5 Proposition. *If* **Fil–D–SUConv** *denotes the construct of all* **Fil**–*determined semiuniform convergence spaces (and uniformly continuous maps), then* **Fil** *is concretely isomorphic to* **Fil–D–SUConv**.

Proof. 1) a) $\gamma_{\mathcal{J}_\gamma} = \gamma$ for each **Fil**–structure γ on X (cf. 2.3.3.2. 2)).
b) $\mathcal{J}_{\gamma_{\mathcal{J}_X}} = \mathcal{J}_X$ for each **Fil–D–SUConv**–structure $\mathcal{J}_X$ on X (cf. 2.3.3.4.).

2) a) If $f : (X, \gamma) \longrightarrow (X', \gamma')$ is a **Fil**–morphism, then $f : (X, \mathcal{J}_\gamma) \longrightarrow (X', \mathcal{J}_{\gamma'})$ is a **Fil–D–SUConv**–morphism (Note: For each filter $\mathcal{G}$ on X, $f \times f(\mathcal{G} \times \mathcal{G}) = f(\mathcal{G}) \times f(\mathcal{G})$).
b) If $f : (X, \mathcal{J}_X) \longrightarrow (X', \mathcal{J}_{X'})$ is a **SUConv**–morphism, then $f : (X, \gamma_{\mathcal{J}_X}) \longrightarrow (X', \gamma_{\mathcal{J}_{X'}})$ is a **Fil**–morphism (shortly: *every uniformly continuous map is Cauchy continuous).*

It follows from 1) and 2) that **Fil–D–SUConv** and **Fil** are concretely isomorphic (cf. the corresponding procedure in the proof of 2.3.1.8. a)).

2.3.3.6 Proposition. **Fil–D–SUConv** *is a bireflective and bicoreflective (full and isomorphism–closed) subconstruct of* **SUConv**.

Proof. 1) Let $(X, \mathcal{J}_X) \in |\mathbf{SUConv}|$. Then $1_X : (X, \mathcal{J}_{\gamma_{\mathcal{J}_X}}) \longrightarrow (X, \mathcal{J}_X)$ is the bicoreflection of $(X, \mathcal{J}_X)$ w.r.t. **Fil–D–SUConv**:
$\gamma' = \gamma_{\mathcal{J}_X}$ is a **Fil**–structure and therefore $\gamma' = \gamma_{\mathcal{J}_{\gamma'}}$ (cf. 2.3.3.2. 2)) which implies $\mathcal{J}_{\gamma'} = \mathcal{J}_{\gamma_{\mathcal{J}_{\gamma'}}}$, i.e. $\mathcal{J}_{\gamma'}$ is a **Fil–D–SUConv**–structure. Since $\mathcal{J}_{\gamma_{\mathcal{J}_X}} \subset \mathcal{J}_X$, $1_X : (X, \mathcal{J}_{\gamma_{\mathcal{J}_X}}) \longrightarrow (X, \mathcal{J}_X)$ is uniformly continuous.

If $f : (Y, \mathcal{J}_Y) \longrightarrow (X, \mathcal{J}_X)$ is uniformly continuous, where $(Y, \mathcal{J}_Y) \in |\mathbf{Fil\text{–}D\text{–}SUConv}|$, i.e. $\mathcal{J}_Y = \mathcal{J}_{\gamma_{\mathcal{J}_Y}}$, then $f : (Y, \gamma_{\mathcal{J}_Y}) \longrightarrow (X, \gamma_{\mathcal{J}_X})$ is Cauchy continuous (cf. 2.3.3.5. 2) b)) and $f : (Y, \mathcal{J}_{\gamma_{\mathcal{J}_Y}}) \longrightarrow (X, \mathcal{J}_{\gamma_{\mathcal{J}_X}})$ is uniformly continuous (cf. 2.3.3.5. 2) a)), i.e. $f : (Y, \mathcal{J}_Y) \longrightarrow (X, \mathcal{J}_{\gamma_{\mathcal{J}_X}})$ is uniformly continuous.

2) a) **Fil–D–SUConv** is closed under formation of products in **SUConv**: Let $((X_i, \mathcal{J}_{X_i}))_{i \in I}$ be a family of **Fil–D–SUConv**–objects. In order to show that

$\mathcal{J}_X \subset \mathcal{J}_{\gamma_{\mathcal{J}_X}}$ where $\mathcal{J}_X = \{\mathcal{F} \in F(\prod_{i\in I} X_i \times \prod_{i\in I} X_i) : (p_i \times p_i)(\mathcal{F}) \in \mathcal{J}_{X_i}$ for each projection $p_i : \prod X_i \longrightarrow X_i\}$, let $\mathcal{F} \in \mathcal{J}_X$. Since $\mathcal{J}_{X_i}$ is a **Fil–D–SUConv**–structure, there exists a filter $\mathcal{G}_i$ on X_i with $\mathcal{G}_i \times \mathcal{G}_i \in \mathcal{J}_{X_i}$ and $(p_i \times p_i)(\mathcal{F}) \supset \mathcal{G}_i \times \mathcal{G}_i$. If $j : \prod_{i\in I}(X_i \times X_i) \longrightarrow \prod_{i\in I} X_i \times \prod_{i\in I} X_i$ denotes the canonical isomorphism (in **Set**) and $p'_i : \prod_{i\in I}(X_i \times X_i) \longrightarrow X_i \times X_i$ is the i–th projection, then $p'_i(j^{-1}(\mathcal{F})) = (p_i \times p_i)(j(j^{-1}(\mathcal{F})) = (p_i \times p_i)(\mathcal{F}) \supset \mathcal{G}_i \times \mathcal{G}_i$. Hence, $j^{-1}(\mathcal{F}) \supset \prod_{i\in I}(\mathcal{G}_i \times \mathcal{G}_i)$ [4] and consequently $\mathcal{F} \supset j(\prod_{i\in I}(\mathcal{G}_i \times \mathcal{G}_i)) = \prod_{i\in I} \mathcal{G}_i \times \prod_{i\in I} \mathcal{G}_i$. Furthermore, the product filter $\prod_{i\in I} \mathcal{G}_i$ on $\prod_{i\in I} X_i$ belongs to $\gamma_{\mathcal{J}_X}$, i.e. $\prod_{i\in I} \mathcal{G}_i \times \prod_{i\in I} \mathcal{G}_i \in \mathcal{J}_X$, because $(p_i \times p_i)(\prod_{i\in I} \mathcal{G}_i \times \prod_{i\in I} \mathcal{G}_i) = p_i(\prod_{i\in I} \mathcal{G}_i) \times p_i(\prod_{i\in I} \mathcal{G}_i) = \mathcal{G}_i \times \mathcal{G}_i \in \mathcal{J}_{X_i}$ for each $i \in I$. Thus, $\mathcal{F} \in \mathcal{J}_{\gamma_{\mathcal{J}_X}}$. Since obviously, $\mathcal{J}_{\gamma_{\mathcal{J}_X}} \subset \mathcal{J}_X$, we obtain $\mathcal{J}_X = \mathcal{J}_{\gamma_{\mathcal{J}_X}}$.

b) **Fil–D–SUConv** is closed under formation of subspaces in **SUConv**: Let $(X, \mathcal{J}_X) \in |\textbf{Fil–D–SUConv}|$ and $U \subset X$. If $i : U \longrightarrow X$ denotes the inclusion map and $\mathcal{J}_U$ is the initial **SUConv**–structure on U w.r.t. i, then in order to show that $\mathcal{J}_U \subset \mathcal{J}_{\gamma_{\mathcal{J}_U}}$, let $\mathcal{F} \in \mathcal{J}_U$. Then there exists some $\mathcal{G} \in F(X)$ such that $\mathcal{G} \times \mathcal{G} \in \mathcal{J}_X$ and $(i \times i)(\mathcal{F}) \supset \mathcal{G} \times \mathcal{G}$, since $(X, \mathcal{J}_X)$ is **Fil**–determined. It follows from

$$\mathcal{F} = (i \times i)^{-1}((i \times i)(\mathcal{F})) \supset (i \times i)^{-1}(\mathcal{G} \times \mathcal{G}) = i^{-1}(\mathcal{G}) \times i^{-1}(\mathcal{G})$$

that $\mathcal{F} \in \mathcal{J}_{\gamma_{\mathcal{J}_U}}$, because $(i \times i)^{-1}(\mathcal{G} \times \mathcal{G}) \in \mathcal{J}_U$. Since obviously, $\mathcal{J}_{\gamma_{\mathcal{J}_U}} \subset \mathcal{J}_U$, we obtain $\mathcal{J}_U = \mathcal{J}_{\gamma_{\mathcal{J}_U}}$.

c) **Fil–D–SUConv** contains all indiscrete **SUConv**–objects: Let $(X, \mathcal{J}_X)$ be an indiscrete **SUConv**–object, i.e. $\mathcal{J}_X$ is equal to the set of all filters on $X \times X$. If $\mathcal{F} \in \mathcal{J}_X$ and $\mathcal{G} = \{X\}$, then $\mathcal{G} \times \mathcal{G} = \{X \times X\} \subset \mathcal{F}$, i.e. $\mathcal{F} \in \mathcal{J}_{\gamma_{\mathcal{J}_X}}$. Since $\mathcal{J}_{\gamma_{\mathcal{J}_X}} \subset \mathcal{J}_X$, we obtain $\mathcal{J}_X = \mathcal{J}_{\gamma_{\mathcal{J}_X}}$, i.e. $(X, \mathcal{J}_X) \in |\textbf{Fil–D–SUConv}|$.

d) Since **Fil–D–SUConv** is a full and isomorphism–closed subconstruct of **SUConv**, it follows from a), b) and c) that it is bireflective in **SUConv** (cf. the theorem under 2.2.11.2) as well as the theorem under 2.2.4.).

2.3.3.7 Remarks. 1) If $(X, \mathcal{J}_X)$ is a semiuniform convergence space, then its bicoreflective **Fil**–modification $(X, \mathcal{J}_{\gamma_{\mathcal{J}_X}})$ (resp. $(X, \gamma_{\mathcal{J}_X})$) is called the *underlying filter space of the semiuniform convergence space* $(X, \mathcal{J}_X)$.

2) If $(X, \mathcal{J}_X)$ is a semiuniform convergence space, then $1_X : (X, \mathcal{J}_X) \longrightarrow (X, \widehat{\mathcal{J}}_X)$ with $\widehat{\mathcal{J}}_X = \bigcap\{\overline{\mathcal{J}_X} : (X, \overline{\mathcal{J}_X}) \in |\textbf{Fil–D–SUConv}|$ and $\overline{\mathcal{J}_X} \supset \mathcal{J}_X\}$ is the bireflection of $(X, \mathcal{J}_X)$ w.r.t. **Fil–D–SUConv** (cf. 2.3.3.6. and the last part of 2.2.11. 2)). Sometimes $(X, \widehat{\mathcal{J}}_X)$ (resp. $(X, \gamma_{\widehat{\mathcal{J}}_X})$) is called *the bireflective* **Fil**–*modification of the semiuniform convergence space* $(X, \mathcal{J}_X)$.

3) Since **Fil–D–SUConv** is bireflective and bicoreflective in **SUConv**, it is a topological construct in which initial and final structures are formed as in **SUConv**. Thus, in **Fil** the following is valid:

[4] If $(X_i)_{i\in I}$ is a family of sets and $\mathcal{F}_i \in F(X_i)$ for each $i \in I$, then the **product filter** $\prod_{i\in I} \mathcal{F}_i$ is generated by the filter base $\mathcal{B} = \{\prod_{i\in I} F_i : F_i \in \mathcal{F}_i$ for each $i \in I$ and $F_i \neq X_i$ for at most finitely many $i \in I\}$. Obviously, $p_i(\prod_{i\in I} \mathcal{F}_i) = \mathcal{F}_i$ for each $i \in I$ where $p_i : \prod_{i\in I} X_i \longrightarrow X_i$ denotes the i–th projection. If $I = \{1, \ldots, n\}$, we write $\mathcal{F}_1 \times \mathcal{F}_2 \times \cdots \times \mathcal{F}_n$ instead of $\prod_{i\in I} \mathcal{F}_i$.

1) If X is a set X, $((X_i, \gamma_i))_{i\in I}$ a family of filter spaces and $(f_i : X \longrightarrow X_i)_{i\in I}$ is a family of maps, then $\gamma = \{\mathcal{F} \in F(X) : f_i(\mathcal{F}) \in \gamma_i$ for each $i \in I\}$ is the initial **Fil**–structure on X w.r.t. the given data.

2) If X is a set, $((X_i, \gamma_i))_{i\in I}$ a family of filter spaces and $(f_i : X_i \longrightarrow X)_{i\in I}$ a family of maps, then $\gamma = \{\mathcal{F} \in F(X) :$ there is some $i \in I$ and some $\mathcal{F}_i \in \gamma_i$ with $f_i(\mathcal{F}_i) \subset \mathcal{F}\} \cup \{\dot{x} : x \in X\}$ is the final **Fil**–structure on X w.r.t. the given data. If $(f_i : X_i \longrightarrow X)_{i\in I}$ is an epi–sink in **Set**, then $\gamma = \{\mathcal{F} \in F(X) :$ there is some $i \in I$ and some $\mathcal{F}_i \in \gamma_i$ with $f_i(\mathcal{F}_i) \subset \mathcal{F}\}$.

2.3.3.8 Definition. A filter space (X, γ) is called *complete* provided that for each $\mathcal{F} \in \gamma$ there is some $x \in X$ such that $\mathcal{F} \cap \dot{x} \in \gamma$.

2.3.3.9 Proposition. *The construct* **CFil** *of complete filter spaces (and Cauchy continuous maps) is a (full and isomorphism–closed) bicoreflective subconstruct of* **Fil**.

Proof. Let $(X, \gamma) \in |\mathbf{Fil}|$ and $\gamma_c = \{\mathcal{F} \in \gamma$: there is some $x \in X$ with $\mathcal{F} \cap \dot{x} \in \gamma\}$. Then (X, γ_c) is a complete filter space and $1_X : (X, \gamma_c) \longrightarrow (X, \gamma)$ is the bicoreflection of (X, γ) w.r.t. **CFil**.

2.3.3.10 Definition. A generalized convergence space (X, q) is called *symmetric* provided that the following is satisfied:

S) $\quad (\mathcal{F}, x) \in q$ and $y \in \bigcap\limits_{F\in\mathcal{F}} F$ imply $(\mathcal{F}, y) \in q$.

(note: A topological space $(X, \mathcal{X})$ is symmetric iff it is an R_0*–space*, i.e. $x \in \overline{\{y\}}$ implies $y \in \overline{\{x\}}$ for each pair $(x, y) \in X \times X$. Obviously, *every topological T_1-space and every regular topological space are symmetric.*)

2.3.3.11 Proposition. *1) a) Let (X, γ) be a filter space. Then a symmetric Kent convergence structure q_γ on X is defined by*

$$(\mathcal{F}, x) \in q_\gamma \Longleftrightarrow \mathcal{F} \cap \dot{x} \in \gamma.$$

b) If $f : (X, \gamma) \longrightarrow (X', \gamma')$ is a Cauchy continuous map between filter spaces, then $f : (X, q_\gamma) \longrightarrow (X', q_{\gamma'})$ is continuous.

2) a) Let (X, q) be a Kent convergence space. Then a complete **Fil**–*structure γ_q on X is defined by*

$$\gamma_q = \{\mathcal{F} \in F(X)\text{: there is some } x \in X \text{ with } (\mathcal{F}, x) \in q\}.$$

b) If $f : (X, q) \longrightarrow (X', q')$ is a continuous map between Kent convergence spaces, then $f : (X, \gamma_q) \longrightarrow (X', \gamma_{q'})$ is Cauchy continuous.

3) The construct **CFil** *is concretely isomorphic to the construct* $\mathbf{KConv}_S$ *of symmetric Kent convergence spaces (and continuous maps).*

Proof. 1) a) Obviously, q_γ fulfills C_1), C_2) and C_3). If $(\mathcal{F}, x) \in q_\gamma$ and

$y \in \bigcap_{F\in\mathcal{F}} F$, then $\mathcal{F} = \mathcal{F} \cap \dot{y}$ and it follows from $\mathcal{F} \cap \dot{x} \in \gamma$ that $\mathcal{F} \in \gamma$, i.e. $\mathcal{F} \xrightarrow{q_\gamma} y$.

b) Let $\mathcal{F} \xrightarrow{q_\gamma} x$, i.e. $\mathcal{F} \cap \dot{x} \in \gamma$. By assumption, $f(\mathcal{F} \cap \dot{x}) = f(\mathcal{F}) \cap f(\dot{x}) = f(\mathcal{F}) \cap f(x)^{\cdot} \in \gamma'$, i.e. $f(\mathcal{F}) \xrightarrow{q_{\gamma'}} f(x)$.

2) a) Obviously, (X, γ_q) is a filter space. Furthermore, (X, γ_q) is complete, namely if $\mathcal{F} \in \gamma_q$, then there is some $x \in X$ with $(\mathcal{F}, x) \in q$ and, since (X, q) is a Kent convergence space, $(\mathcal{F} \cap \dot{x}, x) \in q$, i.e. $\mathcal{F} \cap \dot{x} \in \gamma_q$.

b) Let $\mathcal{F} \in \gamma_q$, i.e. there is some $x \in X$ with $(\mathcal{F}, x) \in q$. By assumption $(f(\mathcal{F}), f(x)) \in q'$, i.e. $f(\mathcal{F}) \in \gamma'_q$.

3) a) $\gamma_{q_\gamma} = \gamma$ for each **CFil**–structure γ.

α) Let $\mathcal{F} \in \gamma_{q_\gamma}$. Then there is some $x \in X$ with $\mathcal{F} \cap \dot{x} \in \gamma$. Consequently, $\mathcal{F} \in \gamma$.
β) Let $\mathcal{F} \in \gamma$. Since (X, γ) is complete, there is some $x \in X$ with $\mathcal{F} \cap \dot{x} \in \gamma$. Hence, $\mathcal{F} \in \gamma_{q_\gamma}$.

b) $q_{\gamma_q} = q$ for each $\mathbf{KConv_S}$–structure q:

α) Let $(\mathcal{F}, x) \in q_{\gamma_q}$. Then $\mathcal{F} \cap \dot{x} \in \gamma_q$, i.e. $\mathcal{H} = \mathcal{F} \cap \dot{x} \xrightarrow{q} y$ for some $y \in X$. Obviously, $x \in \bigcap_{H\in\mathcal{H}} H = \bigcap_{F\in\mathcal{F}} (F \cup \{x\})$. Since (X, q) is symmetric, $\mathcal{H} \xrightarrow{q} x$. Consequently, $(\mathcal{F}, x) \in q$ because $\mathcal{F} \supset \mathcal{H}$.
β) Let $(\mathcal{F}, x) \in q$. Since (X, q) is a Kent convergence space, $(\mathcal{F} \cap \dot{x}, x) \in q$, i.e. $(\mathcal{F}, x) \in q_{\gamma_q}$.

c) The desired concrete isomorphism follows immediately from 1) b), 2) b), 3) a) and 3) b).

2.3.3.12 Remarks. 1) Because of 2.3.3.11. 3) *we need not distinguish between complete filter spaces and symmetric Kent convergence spaces.*

2) Obviously, *a filter space (X, γ) is complete iff each $\mathcal{F} \in \gamma$, i.e. each Cauchy filter, converges in (X, q_γ).*

3) a) If (X, γ) is a filter space and $1_X : (X, \gamma_c) \longrightarrow (X, \gamma)$ is the bicoreflection of (X, γ) w.r.t. **CFil**, then $\gamma_c = \gamma_{q_\gamma}$. Therefore, (X, γ_{q_γ}) (resp. (X, q_γ)) *is called the underlying symmetric Kent convergence space of the filter space* (X, γ).

b) If $(X, \mathcal{J}_X)$ is a semiuniform convergence space, then the underlying symmetric Kent convergence space of its underlying filter space $(X, \gamma_{\mathcal{J}_X})$, namely $(X, q_{\gamma_{\mathcal{J}_X}})$ is called *the underlying symmetric Kent convergence space of the semiuniform convergence space* $(X, \mathcal{J}_X)$.

2.3.3.13 Proposition. *1) Let $\mathcal{A}$ be either the topological construct of limit spaces, pseudotopological spaces, pretopological spaces or topological spaces. Then its full (and isomorphism–closed) subconstruct $\mathcal{A}_S$ of symmetric $\mathcal{A}$–objects is bireflective in $\mathcal{A}$.*

2) Each of the constructs in the following list

$$\mathbf{KConv_S} \supset \mathbf{Lim_S} \supset \mathbf{PsTop_S} \supset \mathbf{PrTop_S} \supset \mathbf{Top_S}$$

is a bireflective subconstruct of the preceding ones.

Proof. 1) We prove that $\mathcal{A}_S$ is closed under formation of initial structures in $\mathcal{A}$: Let X be a set, $((X_i, q_i))_{i\in I}$ a family of $\mathcal{A}_S$-objects and $(f_i : X \longrightarrow X_i)_{i\in I}$ a family of maps. If q denotes the initial $\mathcal{A}$-structure w.r.t. the given data, then (X, q) is symmetric; namely, if $(\mathcal{F}, x) \in q$ and $y \in \bigcap_{F\in\mathcal{F}} F$, then $f_i(y) \in f_i[F]$ for each $i \in I$ and each $F \in \mathcal{F}$, i.e. $f_i(y) \in \bigcap_{H\in f_i(\mathcal{F})} H$ for each $i \in I$, which implies that $(f_i(\mathcal{F}), f_i(y)) \in q_i$ for each $i \in I$, since each (X_i, q_i) is symmetric and $(f_i(\mathcal{F}), f_i(x)) \in q_i$ for each $i \in I$. Thus, $(\mathcal{F}, y) \in q$.

2) Let $\widetilde{\mathcal{R}} : \mathbf{KConv} \longrightarrow \mathbf{Lim}$ be the bireflector (cf. 2.3.1.5.). Then $\widetilde{\mathcal{R}}|_{\mathbf{KConv_S}}$: $\mathbf{KConv_S} \longrightarrow \mathbf{Lim}$ is a bireflector, and by 1) there is a bireflector $\mathcal{R} : \mathbf{Lim} \longrightarrow \mathbf{Lim_S}$. Consequently, $\mathcal{R}\circ\widetilde{\mathcal{R}}|_{\mathbf{KConv_S}} : \mathbf{KConv_S} \longrightarrow \mathbf{Lim_S}$ is the desired bireflector. The remaining cases are proved similarly.

2.3.3.14 Remarks. 1) $\mathbf{KConv_S}$ *is a bireflective (full and isomorphism-closed) subconstruct of* $\mathbf{KConv}$, namely if (X, q) is a Kent convergence space, then

$$1_X : (X, q) \longrightarrow (X, q_S)$$

is the bireflection of (X, q) w.r.t. $\mathbf{KConv_S}$ provided that

$$(\mathcal{F}, x) \in q_S \text{ iff there is some } z \in X \text{ such that } (\mathcal{F} \cap \dot{x}, z) \in q.$$

2) a) If $(X, \mathcal{J}_X)$ is a uniform limit space, then the underlying symmetric Kent convergence space $(X, q_{\gamma_{\mathcal{J}_X}})$ can also be described as follows:

$$(\mathcal{F}, x) \in q_{\gamma_{\mathcal{J}_X}} \iff \mathcal{F} \times \dot{x} \in \mathcal{J}_X$$

(Namely, $(\mathcal{F} \cap \dot{x}) \times (\mathcal{F} \cap \dot{x}) = (\mathcal{F} \times \mathcal{F}) \cap (\mathcal{F} \times \dot{x}) \cap (\dot{x} \times \mathcal{F}) \cap (\dot{x} \times \dot{x})$. Thus, if $(\mathcal{F}, x) \in q_{\gamma_{\mathcal{J}_X}}$, i.e. $(\mathcal{F}\times\dot{x})\cap(\mathcal{F}\times\dot{x}) \in \mathcal{J}_X$, it follows that $\mathcal{F}\times\dot{x} \supset (\mathcal{F}\cap\dot{x})\times(\mathcal{F}\cap\dot{x})$ belongs to $\mathcal{J}_X$. Conversely, if $\mathcal{F} \times \dot{x} \in \mathcal{J}_X$ then $(\mathcal{F} \times \dot{x})^{-1} = \dot{x} \times \mathcal{F} \in \mathcal{J}_X$. Furthermore, $\mathcal{F} \times \mathcal{F} = (\dot{x} \times \mathcal{F}) \circ (\mathcal{F} \times \dot{x}) \in \mathcal{J}_X$ and $\dot{x} \times \dot{x} \in \mathcal{J}_X$. Consequently, $(\mathcal{F} \cap \dot{x}) \times (\mathcal{F} \cap \dot{x}) \in \mathcal{J}_X$, i.e. $(\mathcal{F}, x) \in q_{\gamma_{\mathcal{J}_X}}$).

b) The underlying symmetric Kent convergence space $(X, q_{\gamma_{\mathcal{J}_X}})$ of a uniform limit space $(X, \mathcal{J}_X)$ is a *weakly Hausdorff limit space*, where a limit space (X, q) is called weakly Hausdorff (or T_{2W}) provided that the existence of a filter $\mathcal{F}$ on X converging to $x, y \in X$ implies $\{\mathcal{G} \in F(X) : (\mathcal{G}, x) \in q\} = \{\mathcal{H} \in F(X) : (\mathcal{H}, y) \in q\}$:

α) Let $\mathcal{F}$ be a filter on X converging to $x, y \in X$ in $(X, q_{\gamma_{\mathcal{J}_X}})$. If $\mathcal{G} \in F(X)$ such that $(\mathcal{G}, x) \in q_{\gamma_{\mathcal{J}_X}}$, then $(\mathcal{G}, y) \in q_{\gamma_{\mathcal{J}_X}}$, namely

$$\mathcal{G} \times \dot{y} = (\mathcal{F} \times \dot{y}) \circ (\dot{x} \times \mathcal{F}) \circ (\mathcal{G} \times \dot{x}) \in \mathcal{J}_X$$

since $\mathcal{G} \times \dot{x} \in \mathcal{J}_X$, $(\mathcal{F} \times \dot{x})^{-1} = \dot{x} \times \mathcal{F} \in \mathcal{J}_X$ and $\mathcal{F} \times \dot{y} \in \mathcal{J}_X$.

β) $(X, q_{\gamma_{\mathcal{J}_X}})$ fulfills the property C_4) for limit spaces because $(\mathcal{F} \cap \mathcal{G}) \times \dot{x} = (\mathcal{F} \times \dot{x}) \cap (\mathcal{G} \times \dot{x})$.

2.3.3.15 Theorem. *The underlying symmetric Kent convergence space of a uniform space (= principal uniform limit space) is a completely regular topological space.*

Proof. Let $(X, \mathcal{W})$ be a uniform space, i.e. $(X, [\mathcal{W}])$ is a principal uniform limit space. If $(X, q_{\gamma[\mathcal{W}]})$ is the underlying (symmetric) Kent convergence space, then by 2.3.3.14 2) a)

(1) $(\mathcal{F}, x) \in q_{\gamma[\mathcal{W}]} \Longleftrightarrow \mathcal{F} \times \dot{x} \in [\mathcal{W}]$.

For each $x \in X$, let $\mathcal{U}(x) = \{W(x) : W \in \mathcal{W}\}$. Then

(2) $\mathcal{F} \times \dot{x} \supset \mathcal{W} \Longleftrightarrow \mathcal{F} \supset \mathcal{U}(x)$,

namely if $W(x) \in \mathcal{U}(x)$, then there is some $V = V^{-1} \in \mathcal{W}$ such that $V \subset W$ and by $\mathcal{F} \times \dot{x} \supset \mathcal{W}$ there is some $F \in \mathcal{F}$ such that $F \times \{x\} \subset V$; consequently, $F \subset V(x) \subset W(x)$ and $W(x) \in \mathcal{F}$. Conversely, let $W \in \mathcal{W}$. Then there is some $V = V^{-1} \in \mathcal{W}$ such that $V^2 \subset W$. Obviously, for each $x \in X$,

$$V(x) \times V(x) \subset V^2 \subset W\,,$$

i.e. $W \in \mathcal{F} \times \dot{x}$, since $V(x) \in \mathcal{U}(x) \cap \dot{x}$ and $\mathcal{U}(x) \subset \mathcal{F}$.
It follows from (1) and (2) that

(3) $(\mathcal{F}, x) \in q_{\gamma[\mathcal{W}]}$ iff $\mathcal{F} \supset \mathcal{U}(x)$.

Thus, for each $x \in X$, $\bigcap\{\mathcal{F} \in F(X) : (\mathcal{F}, x) \in q_{\gamma[\mathcal{W}]}\} = \mathcal{U}(x)$ and $(\mathcal{U}(x), x) \in q_{\gamma[\mathcal{W}]}$, i.e.$(X, q_{\gamma[\mathcal{W}]})$ is a pretopological space. In order to prove that it is topological, let $W \in \mathcal{W}$. Then there is some $V \in \mathcal{W}$ such that $V^2 \subset W$. Since $V \subset V^2$, it follows that $V(x) \subset V^2(x) \subset W(x)$. If $y \in V(x)$ and $z \in V(y)$, i.e. $(x, y) \in V$ and $(y, z) \in V$, then $(x, z) \in V^2$, i.e. $z \in V^2(x) \subset W(x)$. Thus, $V(y) \subset W(x)$, i.e. $W(x) \in \mathcal{U}(y)$ for each $y \in V(x)$.

The topological space $(X, q_{\gamma[\mathcal{W}]})$ may be identified with $(X, \mathcal{X}_{\mathcal{W}})$, where the topology $\mathcal{X}_{\mathcal{W}}$ can be defined by

(4) $O \in \mathcal{X}_{\mathcal{W}} \Longleftrightarrow$ For each $x \in O$ there is some $V \in \mathcal{W}$ such that $V(x) \subset O$.

In order to prove that $(X, \mathcal{X}_{\mathcal{W}})$ is completely regular, i.e. given $x \in X$ and $U_x \in \mathcal{U}(x)$, then there is a continuous map $f : X \longrightarrow [0,1]$ such that $f(x) = 0$ and $f[X \backslash U_x] \subset \{1\}$, let $V \in \mathcal{W}$ such that $U_x = V(x)$. Since $(X, \mathcal{W})$ is a uniform space, there is a pseudometric $d_V : X \times X \longrightarrow [0,1]$ such that $\frac{1}{2} h_V(x, y) \leq d_V(x, y) \leq h_V(x, y)$, where $0 \leq h_V(x, y) \leq 1$ (cf. the proof of the theorem under 2.3.3.4.). Then $d_V^x : X \longrightarrow [0,1]$, defined by $d_V^x(y) = d_V(x, y)$ for each $y \in X$, is a continuous map w.r.t. $\mathcal{X}_{\mathcal{D}_V}$, where $\mathcal{D}_V$ denotes the uniformity induced by d_V. Since $\mathcal{D}_V \subset \mathcal{W}$, it follows that $\mathcal{X}_{\mathcal{D}_V} \subset \mathcal{X}_{\mathcal{W}}$. Thus d_V^x is also continuous w.r.t. $\mathcal{X}_{\mathcal{W}}$. Let $f : X \longrightarrow [0,1]$ be defined by $f(y) = \min\{2d_V^x(y), 1\}$ for each $y \in X$. Then f is continuous and $f(x) = 0$ since $d_V^x(x) = d_V(x, x) = 0$. Furthermore, $f[X \backslash V(x)] \subset \{1\}$, namely if $y \in X \backslash V(x)$, then $(x, y) \in (X \times X) \backslash V \subset (X \times X) \backslash V_1$, i.e. $h_V(x, y) = 1$, and, since $d_V(x, y) \geq \frac{1}{2} h_V(x, y) = \frac{1}{2}$, it follows that $2d_V^x(y) \geq 1$ which implies $f(y) = \min\{2d_V^x(y), 1\} = 1$. This completes the proof.

2.3.3.16 Definition. Let $\mathcal{C}$ be a topological construct. A source (resp. sink)

$(f_i : (X,\xi) \longrightarrow (X_i,\xi_i))_{i\in I}$ (resp. $(f_i : (X_i,\xi_i) \longrightarrow (X,\xi))_{i\in I}$) is called *initial* (resp. *final*) provided that ξ is the initial (resp. final) $\mathcal{C}$–structure on X w.r.t. the given data.

2.3.3.17 Proposition. *If $(f_i : (X,\mathcal{J}_X) \longrightarrow (X_i,\mathcal{J}_{X_i}))_{i\in I}$ is an initial source in* **SUConv**, *then*

a) $(f_i : (X,\gamma_{\mathcal{J}_X}) \longrightarrow (X_i,\gamma_{\mathcal{J}_{X_i}}))_{i\in I}$ *is an initial source in* **Fil**

and

b) $(f_i : (X,q_{\gamma_{\mathcal{J}_X}}) \longrightarrow (X_i,q_{\gamma_{\mathcal{J}_{X_i}}}))_{i\in I}$ *is an initial source in* $\mathbf{KConv_S}$.

Proof. a) Let γ be the initial **Fil**–structure w.r.t. $(X,f_i,(X_i,\gamma_{\mathcal{J}_{X_i}}),I)$. Then $\gamma = \gamma_{\mathcal{J}_X}$:

$\mathcal{F}\in\gamma \iff f_i(\mathcal{F})\in\gamma_{\mathcal{J}_{X_i}}$ for each $i\in I$

$\iff f_i(\mathcal{F})\times f_i(\mathcal{F}) = f_i\times f_i(\mathcal{F}\times\mathcal{F})\in\mathcal{J}_{X_i}$ for each $i\in I$

$\iff \mathcal{F}\times\mathcal{F}\in\mathcal{J}_X \iff \mathcal{F}\in\gamma_{\mathcal{J}_X}$.

b) Let q be the initial $\mathbf{KConv_S}$–structure w.r.t. $(X,f_i,(X_i,q_{\gamma_{\mathcal{J}_{X_i}}}),I)$. By means of 2.3.3.14. 1) this one is formed as in **KConv**. Then $q = q_{\gamma_{\mathcal{J}_X}}$:

$(\mathcal{F},x)\in q \iff (f_i(\mathcal{F}),f_i(x))\in q_{\gamma_{\mathcal{J}_{X_i}}}$ for each $i\in I \iff f_i(\mathcal{F})\cap \dot{f_i(x)}\in\gamma_{\mathcal{J}_{X_i}}$ for each $i\in I \iff (f_i(\mathcal{F})\cap\dot{f_i(x)})\times(f_i(\mathcal{F})\cap\dot{f_i(x)}) = f_i(\mathcal{F}\cap\dot{x})\times f_i(\mathcal{F}\cap\dot{x}) = f_i\times f_i((\mathcal{F}\cap\dot{x})\times(\mathcal{F}\cap\dot{x}))\in\mathcal{J}_{X_i}$ for each $i\in I \iff (\mathcal{F}\cap\dot{x})\times(\mathcal{F}\cap\dot{x})\in\mathcal{J}_X \iff (\mathcal{F},x)\in q_{\gamma_{\mathcal{J}_X}}$.

2.3.3.18 If $(X,\mathcal{W})$ is a uniform space, then by 2.3.3.15 the underlying symmetric Kent convergence space is a symmetric topological space whose topology $\mathcal{X}_\mathcal{W}$, called *the topology induced by the uniformity* $\mathcal{W}$, is given by

$$O\in\mathcal{X}_\mathcal{W} \iff \text{For each } x\in O \text{ there is some } V\in\mathcal{W} \text{ such that } V(x)\subset O$$

By means of 2.3.3.17 and 2.3.3.13 we obtain the following

Corollary. *If $(f_i : (X,\mathcal{W}) \longrightarrow (X_i,\mathcal{W}_i))_{i\in I}$ is an initial source in* **Unif**, *then $(f_i : (X,\mathcal{X}_\mathcal{W}) \longrightarrow (X_i,\mathcal{X}_{\mathcal{W}_i}))_{i\in I}$ is an initial source in* **Top** *(resp.* $\mathbf{Top_S}$*).*

2.3.3.19 Corollary. *A topological space $(X,\mathcal{X})$ is uniformizable, i .e. there is a uniformity $\mathcal{W}$ on X such that $\mathcal{X} = \mathcal{X}_\mathcal{W}$, if and only if it is completely regular.*

Proof. We know already that $(X,\mathcal{X})$ is completely regular provided that it is uniformizable (cf. 2.3.3.15). Let $(X,\mathcal{X})$ be a completely regular topological space and $\mathcal{W}$ the coarsest uniformity on X such that all continuous maps from $(X,\mathcal{X})$ into $[0,1]$ are uniformly continuous. Then by the corollary under 2.3.3.18. $\mathcal{X}_\mathcal{W}$ is the coarsest topology on X with respect to which all continuous maps from $(X,\mathcal{X})$ into $[0,1]$ are continuous. Thus, $\mathcal{X}_\mathcal{W}\subset\mathcal{X}$. In order to prove that $\mathcal{X}\subset\mathcal{X}_\mathcal{W}$, let $O\in\mathcal{X}$. Then $A = X\backslash O$ is a closed subset of X. By assumption, for each $x\in O$ there is a continuous map $f_x : (X,\mathcal{X}) \longrightarrow [0,1]$ such that $f_x(x) = 0$ and

$f_x[A] \subset \{1\}$. Hence, $x \in f_x^{-1}[[0,1)] \subset O$ for each $x \in O$, and

$$(*) \qquad O = \bigcup_{x \in O} f_x^{-1}[[0,1)]\,.$$

It follows from $(*)$ that $O \in \mathcal{X}_\mathcal{W}$ since all f_x are also continuous w.r.t. $\mathcal{X}_\mathcal{W}$.

2.3.3.20 Definition. A semiuniform convergence space $(X, \mathcal{J}_X)$ is called *complete* provided that the underlying filter space $(X, \gamma_{\mathcal{J}_X})$ is complete, i.e. each $\mathcal{J}_X$–Cauchy filter on X converges in $(X, q_{\gamma_{\mathcal{J}_X}})$.

2.3.3.21 Proposition. *Let (X,q) be a Kent convergence space. Then there is a semiuniform convergence structure $\mathcal{J}_X$ on X such that $q_{\gamma_{\mathcal{J}_X}} = q$, i.e. (X,q) is convergence semiuniformizable, if and only if (X,q) is symmetric. Furthermore, if (X,q) is symmetric, then $\mathcal{J}_{\gamma_q}$ is the finest one of all semiuniform convergence structures $\mathcal{J}_X$ on X with $q_{\gamma_{\mathcal{J}_X}} = q$, and $(X, \mathcal{J}_{\gamma_q})$ is a complete semiuniform convergence space.*

Proof. If (X,q) is symmetric, then $(X, \mathcal{J}_{\gamma_q})$ is a complete semiuniform convergence space, namely γ_q is a complete **Fil**–structure by 2.3.3.11.2) a) and $\gamma_{\mathcal{J}_{\gamma_q}} = \gamma_q$ (cf. 2.3.3.2.). Since $q_{\gamma_q} = q$ for each symmetric Kent convergence structure q (cf. 3) b) in the proof of 2.3.3.11.), it follows that $q_{\gamma_{\mathcal{J}_{\gamma_q}}} = q$. Furthermore, $\mathcal{J}_{\gamma_q} \subset \mathcal{J}_X$ for each **SUConv**–structure $\mathcal{J}_X$ on X with $q_{\gamma_{\mathcal{J}_X}} = q$, namely if $\mathcal{F} \in \mathcal{J}_{\gamma_q}$, then there are some $\mathcal{G} \in \gamma_q$ and some $x \in X$ with $\mathcal{G} \times \mathcal{G} \subset \mathcal{F}$ and $(\mathcal{G} \cap \dot{x}) \times (\mathcal{G} \cap \dot{x}) \in \mathcal{J}_X$, and consequently $\mathcal{F} \in \mathcal{J}_X$ since $\mathcal{F} \supset (\mathcal{G} \cap \dot{x}) \times (\mathcal{G} \cap \dot{x})$.

Conversely, if there is an **SUConv**–structure $\mathcal{J}_X$ on X with $q_{\gamma_{\mathcal{J}_X}} = q$, then (X,q) is symmetric (cf. 2.3.3.11. 1) a)).

2.3.3.22 Definitions. 1) A semiuniform convergence space $(X, \mathcal{J}_X)$ is called a *convergence space* provided that

$$\mathcal{J}_X = \{\mathcal{F} \in F(X \times X) : \text{ there are some } x \in X \text{ and some } \mathcal{G} \in F(X) \text{ with } (\mathcal{G}, x) \in q_{\gamma_{\mathcal{J}_X}} \text{ such that } \mathcal{G} \times \mathcal{G} \subset \mathcal{F}\}\,,$$

in other words: a convergence space is a semiuniform convergence space which is 'generated' by its convergent filters.

2) The full subconstruct of **SUConv** whose object class consists of all convergence spaces is denoted by **Conv**.

2.3.3.23 Remarks. 1) **Conv** *is concretely isomorphic to* $\mathbf{KConv_S}$: This follows immediately form the following:

a) $q_{\gamma_{\mathcal{J}_{\gamma_q}}} = q$ for each $\mathbf{KConv_S}$–structure q on a set X (cf. 2.3.3.21.)

b) $\mathcal{J}_{\gamma_{q_{\gamma_{\mathcal{J}_X}}}} = \mathcal{J}_X$ for each **Conv**–structure $\mathcal{J}_X$ on a set X (cf. the definition 2.3.3.22.)

c) If $f : (X,q) \longrightarrow (X',q')$ is a continuous map between Kent convergence spaces, then $f : (X, \mathcal{J}_{\gamma_q}) \longrightarrow (X', \mathcal{J}_{\gamma_{q'}})$ is uniformly continuous (cf. 2.3.3.11. 2) b) and 2.3.3.5. 2) a)).

d) If $f : (X, \mathcal{J}_X) \longrightarrow (X', \mathcal{J}_{X'})$ is a uniformly continuous map between semiuniform convergence spaces, then $f : (X, q_{\gamma_{\mathcal{J}_X}}) \longrightarrow (X', q_{\gamma_{\mathcal{J}_{X'}}})$ is continuous (cf. 2.3.3.5. 2) b) and 2.3.3.11. 1) b)).

Then a functor $\mathcal{F}$: **Conv** $\longrightarrow$ **KConv**$_\mathbf{S}$ (resp. $\mathcal{G}$: **KConv**$_\mathbf{S}$ $\longrightarrow$ **Conv**) can be defined by $\mathcal{F}((X, \mathcal{J}_X)) = (X, q_{\gamma_{\mathcal{J}_X}})$ and $\mathcal{F}(f) = f$ (resp. $\mathcal{G}((X, q)) = (X, \mathcal{J}_{\gamma_q})$ and $(\mathcal{G}(f) = f)$ such that $\mathcal{F} \circ \mathcal{G} = \mathcal{I}_{\mathbf{KConv_S}}$ and $\mathcal{G} \circ \mathcal{F} = \mathcal{I}_{\mathbf{Conv}}$, i.e. $\mathcal{F}$ (resp.$\mathcal{G}$) is the desired (concrete) isomorphism (note $(X, \mathcal{J}_{\gamma_q}) \in |\mathbf{Conv}|$ for each $(X, q) \in |\mathbf{KConv_S}|$ because $q_{\gamma_{\mathcal{J}_{\gamma_q}}} = q$ by 2.3.3.21.).

2) Because of 1) *we need not distinguish between convergence spaces and symmetric Kent convergence spaces.*

3) **Conv** *is bicoflective in* **SUConv**: Since **CFil** is bicoreflective in **Fil** and **Fil** ($\cong$ **Fil–D–SUConv**) is bicoreflective in **SUConv**, this follows from the fact that **CFil** $\cong$ **KConv**$_\mathbf{S}$ $\cong$ **Conv**. Thus $1_X : (X, \mathcal{J}_{\gamma_{q_{\gamma_{\mathcal{J}_X}}}}) \longrightarrow (X, \mathcal{J}_X)$ is the bicoreflection of $(X, \mathcal{J}_X) \in |\mathbf{SUConv}|$ w.r.t. **Conv**.

2.3.3.24 Corollary. *A semiuniform convergence space is a convergence space if and only if it is complete and* **Fil**–*determined.*

Proof. Obviously, for each semiuniform convergence space $(X, \mathcal{J}_X)$,

$$(*) \qquad \mathcal{J}_{\gamma_{q_{\gamma_{\mathcal{J}_X}}}} \subset \mathcal{J}_{\gamma_{\mathcal{J}_X}} \subset \mathcal{J}_X \,.$$

Furthermore, $(X, \mathcal{J}_X) \in |\mathbf{Fil–D–SUConv}|$ iff $\mathcal{J}_X = \mathcal{J}_{\gamma_{\mathcal{J}_X}}$.

1) If $(X, \mathcal{J}_X) \in |\mathbf{Fil–D–SUConv}|$ is complete, i.e. $(X, \gamma_{\mathcal{J}_X})$ is a complete filter space, then $\gamma_{q_{\gamma_{\mathcal{J}_X}}} = \gamma_{\mathcal{J}_X}$ (cf. 3) a) in the proof of 2.3.3.11.) and $\mathcal{J}_{\gamma_{q_{\gamma_{\mathcal{J}_X}}}} = \mathcal{J}_{\gamma_{\mathcal{J}_X}} = \mathcal{J}_X$, i.e. $(X, \mathcal{J}_X)$ is a convergence space.

2) If $(X, \mathcal{J}_X)$ is a convergence space, i.e. $\mathcal{J}_X = \mathcal{J}_{\gamma_{q_{\gamma_{\mathcal{J}_X}}}}$, then by $(*)$, $\mathcal{J}_X = \mathcal{J}_{\gamma_{\mathcal{J}_X}}$, i.e. $(X, \mathcal{J}_X)$ is **Fil**–determined. In order to prove that $(X, \mathcal{J}_X)$ is complete, let $\mathcal{F}$ be a $\mathcal{J}_X$–Cauchy filter. Then $\mathcal{F} \times \mathcal{F} \in \mathcal{J}_X = \mathcal{J}_{\gamma_{q_{\gamma_{\mathcal{J}_X}}}}$, i.e. there is some $\mathcal{G} \in \gamma_{q_{\gamma_{\mathcal{J}_X}}}$ such that $\mathcal{G} \times \mathcal{G} \subset \mathcal{F} \times \mathcal{F}$. Consequently, there is some $x \in X$ with $(\mathcal{G}, x) \in q_{\gamma_{\mathcal{J}_X}}$, and $(\mathcal{F}, x) \in q_{\gamma_{\mathcal{J}_X}}$ since $\mathcal{G} \subset \mathcal{F}$.

2.3.3.25 Proposition. *1) If $(X, \mathcal{J}_X)$ is a complete semiuniform convergence space, then each continuous map $f : (X, \mathcal{J}_X) \longrightarrow (Y, \mathcal{J}_Y)$ from $(X, \mathcal{J}_X)$ into a semiuniform convergence space $(Y, \mathcal{J}_Y)$ (i.e. $f : (X, q_{\gamma_{\mathcal{J}_X}}) \longrightarrow (Y, q_{\gamma_{\mathcal{J}_Y}})$ is continuous) is Cauchy continuous (i.e. $f : (X, \gamma_{\mathcal{J}_X}) \longrightarrow (Y, \gamma_{\mathcal{J}_Y})$ is Cauchy continuous).*

2) A semiuniform convergence space $(X, \mathcal{J}_X)$ is **Fil**–*determined iff each Cauchy continuous map $f : (X, \mathcal{J}_X) \longrightarrow (Y, \mathcal{J}_Y)$ from $(X, \mathcal{J}_X)$ into a semiuniform convergence space $(Y, \mathcal{J}_Y)$ is uniformly continuous.*

Proof. 1) Obvious.
2) "$\Longrightarrow$". Let $\mathcal{F} \in \mathcal{J}_X = \mathcal{J}_{\gamma_{\mathcal{J}_X}}$. Then there is some $\mathcal{G} \in \gamma_{\mathcal{J}_X}$ with $\mathcal{G} \times \mathcal{G} \subset \mathcal{F}$. By assumption, $f(\mathcal{G}) \in \gamma_{\mathcal{J}_Y}$, i.e. $f(\mathcal{G}) \times f(\mathcal{G}) = (f \times f)(\mathcal{G} \times \mathcal{G}) \in \mathcal{J}_Y$. Thus, $(f \times f)(\mathcal{F}) \in \mathcal{J}_Y$.
"$\Longleftarrow$". $1_X : (X, \mathcal{J}_{\gamma_{\mathcal{J}_X}}) \longrightarrow (X, \mathcal{J}_X)$ is the bicoreflection of $(X, \mathcal{J}_X)$ w.r.t. **Fil–D–SUConv**. Since $\gamma_{\mathcal{J}_{\gamma_{\mathcal{J}_X}}} = \gamma_{\mathcal{J}_X}$, $1_X : (X, \mathcal{J}_X) \longrightarrow (X, \mathcal{J}_{\gamma_{\mathcal{J}_X}})$ is Cauchy continuous and thus, by assumption, uniformly continuous. Consequently, $\mathcal{J}_X \subset \mathcal{J}_{\gamma_{\mathcal{J}_X}}$. Since $\mathcal{J}_{\gamma_{\mathcal{J}_X}} \subset \mathcal{J}_X$ is always valid, it follows that $\mathcal{J}_X = \mathcal{J}_{\gamma_{\mathcal{J}_X}}$, i.e. $(X, \mathcal{J}_X)$ is **Fil**–determined.

2.3.3.26 Corollary. *A semiuniform convergence space $(X, \mathcal{J}_X)$ is a convergence space iff each continuous map $f : (X, \mathcal{J}_X) \longrightarrow (Y, \mathcal{J}_Y)$ from $(X, \mathcal{J}_X)$ into a semiuniform convergence space $(Y, \mathcal{J}_Y)$ is uniformly continuous.*

Proof. "$\Longrightarrow$". Note 2.3.3.24. and apply 2.3.3.25.
"$\Longleftarrow$". $1_X : (X, \mathcal{J}_{\gamma_{q_{\gamma_{\mathcal{J}_X}}}}) \longrightarrow (X, \mathcal{J}_X)$ is the bicoreflection of $(X, \mathcal{J}_X)$ w.r.t. **Conv**. Put $\gamma_X = \gamma_{q_{\gamma_{\mathcal{J}_X}}}$. Since $\gamma_{\mathcal{J}_{\gamma_X}} = \gamma_X$ it follows that $q_{\gamma_{\mathcal{J}_{\gamma_X}}} = q_{\gamma_X} = q_{\gamma_{\mathcal{J}_X}}$ (cf. 3) b) in the proof of 2.3.3.11.). Hence $1_X : (X, \mathcal{J}_X) \longrightarrow (X, \mathcal{J}_{\gamma_X})$ is continuous and, by assumption, uniformly continuous, i.e. $\mathcal{J}_X \subset \mathcal{J}_{\gamma_X}$. Since $\mathcal{J}_{\gamma_X} \subset \mathcal{J}_X$ is always valid, $\mathcal{J}_X = \mathcal{J}_{\gamma_X}$, i.e. $(X, \mathcal{J}_X)$ is a convergence space.

2.3.3.27 Proposition. *1) Let (X, γ) be a complete filter space and $A \subset X$ a closed subset (i.e. $A = cl_{q_\gamma} A$). Then (A, γ_A) is complete, where γ_A denotes the initial* **Fil***-structure on A with respect to the inclusion map $i : A \longrightarrow X$.*

2) Let $((X_i, \gamma_i))_{i \in I}$ be a family of non-empty filter spaces. Then the product space $(\prod_{i \in I} X_i, \gamma)$ (formed in **Fil***) is complete if and only if (X_i, γ_i) is complete for each $i \in I$.*

Proof. 1) Let $\mathcal{F} \in \gamma_A$. Then $\mathcal{F} \in F(A)$ such that $i(\mathcal{F}) \in \gamma$. Since (X, γ) is complete, there is some $x \in X$ with $i(\mathcal{F}) \cap \dot{x} \in \gamma$, i.e. $(i(\mathcal{F}), x) \in q_\gamma$. Hence, it follows from $A = cl_{q_\gamma} A$ and $A \in i(\mathcal{F})$, that $x \in A$. Since $\mathcal{F} \cap \dot{x} \in \gamma_A$ (because $i(\mathcal{F} \cap \dot{x}) = i(\mathcal{F}) \cap \dot{x} \in \gamma$), i.e. $(\mathcal{F}, x) \in q_{\gamma_A}$, (A, γ_A) is complete.

2) a) "$\Longleftarrow$". Let $\mathcal{F} \in \gamma$. Then $p_i(\mathcal{F}) \in \gamma_i$ for each $i \in I$. Since (X_i, γ_i) is complete, for each $i \in I$, there is some $x_i \in X_i$ with $(p_i(\mathcal{F}), x_i) \in q_{\gamma_i}$. Thus, $(\mathcal{F}, x) \in q_\gamma$ with $x = (x_i)_{i \in I}$.

b) "$\Longrightarrow$". For each $i \in I$, let $\mathcal{F}_i \in \gamma_i$. Then the product filter $\prod_{i \in I} \mathcal{F}_i$ belongs to γ. Since $(\prod_{i \in I} X_i, \gamma)$ is complete, there is some $x = (x_i)_{i \in I} \in \prod_{i \in I} X_i$ with $(\prod_{i \in I} \mathcal{F}_i, x) \in q_\gamma$. Then $p_i(\prod_{i \in I} \mathcal{F}_i) = \mathcal{F}_i$ converges to $p_i(x) = x_i$ in (X, q_{γ_i}). Thus, all (X_i, γ_i) are complete.

2.3.3.28 Corollary. *1) Let $(X, \mathcal{J}_X)$ be a complete semiuniform convergence space and $A \subset X$ a closed subset (i.e. $A = cl_{q_{\gamma_{\mathcal{J}_X}}} A$). Then $(A, \mathcal{J}_A)$ is complete, where $\mathcal{J}_A$ denotes the initial* **SUConv***-structure on A with respect to the inclusion map $i : A \longrightarrow X$.*

2) Let $((X_i, \mathcal{J}_{X_i}))_{i\in I}$ be a family of non-empty semiuniform convergence spaces. Then the product space $(\prod_{i\in I} X_i, \mathcal{J}_X)$ (formed in **SUConv***) is complete if and only $(X_i, \mathcal{J}_{X_i})$ is complete for each $i \in I$.*

Proof. Use 2.3.3.27. and apply 2.3.3.17.

2.3.3.29 Remarks. 1) 2.3.3.28 is also valid for completeness in uniform spaces because products and subspaces in **Unif** are formed as in **SUConv** (**Unif** is bireflective in **SUConv**!). Note that the closed subsets in a uniform space $(X, \mathcal{W})$ are exactly the closed subsets in the underlying topological space $(X, \mathcal{X}_{\mathcal{W}})$ (cf. 2.3.3.15. and 2.3.3.18.). Furthermore, the $[\mathcal{W}]$–Cauchy filters are exactly the Cauchy filters on $(X, \mathcal{W})$ (cf. 1.1.6. ④).

2) A subspace of a complete semiuniform convergence space need not be complete, e.g. if $\mathbb{Q}_u$ denotes the uniform subspace of rational numbers of the usual uniform space $\mathbb{R}_u$ of real numbers, then $\mathbb{Q}_u$ is not complete, but $\mathbb{R}_u$ is complete.

2.3.3.30 Corollary. *1) Let $(X, \mathcal{J}_X)$ be a convergence space and $(A, \mathcal{J}_A)$ a closed subspace (formed in* **SUConv***). Then $(A, \mathcal{J}_A)$ is a convergence space.*

2) Let $((X_i, \mathcal{J}_{X_i}))_{i\in I}$ be a family of non-empty convergence spaces. Then the product space $(\prod_{i\in I} X_i, \mathcal{J}_X)$ (formed in **SUConv***) is a convergence space if and only if $(X_i, \mathcal{J}_{X_i})$ is a convergence space for each $i \in I$.*

Proof. Use 2.3.3.24., 2.3.3.28. and the fact that **Fil–D–SUConv** is closed under formation of products and subspaces (both formed in **SUConv**) [because **Fil–D–SUConv** is bireflective in **SUConv** (cf. 2.3.3.6.)]. The remaining part of 2) follows from 2.3.3.23.3) and exercise 22)b)α).

2.3.3.31. We have proved in this section that the *filter spaces are related to uniform convergence structures as well as to convergence structures.* Let us summarize the relationships by means of the following diagram, where r (resp. c) stands for embedding as a bireflective (resp. bicoreflective) full and isomorphism–closed subconstruct:

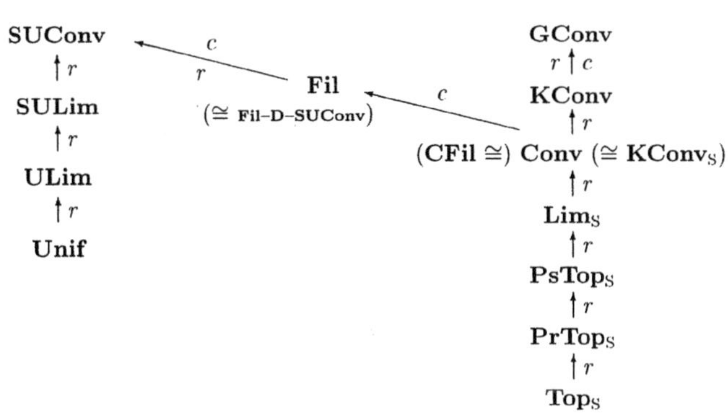

Chapter 3

Topological Universes

Several convenient properties for topological constructs are studied in this chapter, namely

1) cartesian closedness, i.e. the existence of natural function spaces,
2) extensionality, i.e. the existence of one–point extensions, and
3) the fact that products of quotient maps are quotient maps.

According to the introduction, none of them is satisfied for the construct **Top**, and even **Unif** satisfies neither the first nor the second one. Natural function spaces are useful e.g. in homotopy theory [fundamental groups] (cf. the appendix) or for constructing completions (cf. chapter 4). Furthermore, cartesian closedness may also be defined by means of a pair $(\mathcal{G}, \mathcal{F})$ of adjoint functors, where neither $\mathcal{F}$ nor $\mathcal{G}$ is an inclusion functor as in the preceding chapter dealing with reflections and coreflections. The second of the above convenient properties implies that quotients are hereditary. The theory of connection and disconnection (chapter 5) profits from this fact, e.g. the statement that the quotient space of a uniform space **X**, obtained by the decomposition of **X** into its uniform components is uniformly disconnected is true whenever quotients are formed in **SUConv**, where they are hereditary, but it is false when quotients are formed in **Unif**. Topological constructs are cartesian closed and extensional iff they form a quasitopos in the sense of J. Penon [106]. Following L.D. Nel [104], we call a topological construct which is a quasitopos, a topological universe. The notion of one–point extensions goes back to H. Herrlich [70]. Since cartesian closedness of topological constructs implies that quotient maps are finitely productive, but not necessarily productive (i.e. not closed under formation of arbitrary products), the convenient property 3) makes sense, and a topological universe fulfilling it is called strong. Thus, strong topological universes are extremely useful, and among them the construct **SUConv** of semiuniform convergence spaces plays an essential role. By the way, the structure of continuous convergence for (symmetric) limit spaces can be derived from the natural function space structure in **SUConv**. At the end of this chapter, we are able to explain the aim of Convenient Topology.

3.1 Cartesian closed topological constructs

3.1.1 Definition. A category $\mathcal{C}$ is called *cartesian closed* provided that the following conditions are satisfied:

(1) For each pair (A, B) of $\mathcal{C}$–objects there exists a product $A \times B$ in $\mathcal{C}$.

(2) For each $\mathcal{C}$–object A holds: For each $\mathcal{C}$–object B, there exists some $\mathcal{C}$–object B^A (called *power object*) and some $\mathcal{C}$–morphism $e_{A,B} : A \times B^A \longrightarrow B$ (called *evaluation morphism*) such that for each $\mathcal{C}$–object C and each $\mathcal{C}$–morphism $f : A \times C \longrightarrow B$, there exists a unique $\mathcal{C}$–morphism $\overline{f} : C \longrightarrow B^A$ such that the diagram

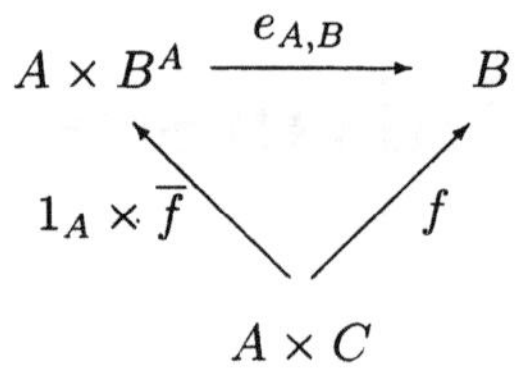

commutes.

3.1.2 Remarks. 1) Whenever a category $\mathcal{C}$ has finite products, each $\mathcal{C}$–object A defines a functor $\mathcal{F}_A : \mathcal{C} \longrightarrow \mathcal{C}$ in the following way:

$$\begin{aligned} \mathcal{F}_A(B) &= A \times B \quad \text{for each } B \in |\mathcal{C}|, \\ \mathcal{F}_A(f) &= 1_A \times f \quad \text{for each } f \in \operatorname{Mor} \mathcal{C}. \end{aligned}$$

Instead of $\mathcal{F}_A$ one writes $A \times -$. Then the assertion (2) in 3.1.1. means that the functor $A \times - : \mathcal{C} \longrightarrow \mathcal{C}$ has a right adjoint for each $A \in |\mathcal{C}|$ or equivalently, for each $A \in |\mathcal{C}|$, $A \times -$ has a couniversal map $(B^A, e_{A,B})$ for each $B \in |\mathcal{C}|$.

2) In topological constructs the condition (1) in 3.1.1. is automatically fulfilled.

3.1.3 Theorem. *Let $\mathcal{C}$ be a topological construct. Then the following are equivalent:*

(1) *$\mathcal{C}$ is cartesian closed.*
(2) *For any $A \in |\mathcal{C}|$ and any set–indexed family $(B_i)_{i \in I}$ of $\mathcal{C}$–objects, the following are satisfied:*
(*a*) *$A \times \coprod_{i \in I} B_i \cong \coprod_{i \in I}(A \times B_i)$ (more exactly: $A \times -$ preserves coproducts).*
(*b*) *If f is a quotient map, then so is $1_A \times f$, i.e. $A \times -$ preserves quotient maps.*

(3) (*a*) *For any $A \in |\mathcal{C}|$ and any set–indexed family $(B_i)_{i \in I}$ of $\mathcal{C}$–objects the following is satisfied:*

$A \times \coprod_{i \in I} B_i \cong \coprod_{i \in I} A \times B_i$ *(more exactly: $A \times -$ preserves coproducts).*

(*b*) *In $\mathcal{C}$ the product $f \times g$ of any two quotient maps f and g is a quotient map.*

(4) *For each $\mathcal{C}$–object A holds: For any final epi-sink $(f_i : B_i \longrightarrow B)_{i \in I}$ in $\mathcal{C}$, $(1_A \times f_i : A \times B_i \longrightarrow A \times B)_{i \in I}$ is a final epi-sink, i.e. $A \times -$ preserves final epi–sinks.*

(5) *For any pair $(A, B) \in |\mathcal{C}| \times |\mathcal{C}|$, the set $[A, B]_{\mathcal{C}}$ can be endowed with the structure of a $\mathcal{C}$-object denoted by B^A such that the following are satisfied:*

(*a*) *The evaluation map $e_{A,B} : A \times B^A \longrightarrow B$ defined by $e_{A,B}(a, g) = g(a)$ for each $(a, g) \in A \times B^A$ is a $\mathcal{C}$-morphism.*
(*b*) *For each $\mathcal{C}$-object C and each $\mathcal{C}$-morphism $f : A \times C \longrightarrow B$ the map $\overline{f} : C \longrightarrow B^A$ defined by $\overline{f}(c)(a) = f(a, c)$ is a $\mathcal{C}$-morphism.*

Proof. "(5) $\Longrightarrow$ (1)" is trivial.
(1) $\Longrightarrow$ (2). Dualizing 2.1.10., for each $A \in |\mathcal{C}|$, $A \times -$ preserves coproducts and coequalizers. By 1.2.2.6. 1)a), in $\mathcal{C}$ coequalizers coincide with quotient maps.

(2) $\Longrightarrow$ (3). Let $f : A \longrightarrow B$ and $g : C \longrightarrow D$ be quotient maps in $\mathcal{C}$. Then $f \times g = (1_B \times g) \circ (f \times 1_C)$ is a quotient map as a composition of two quotient maps (note that if $1_C \times f$ is a quotient map, then so is $f \times 1_C$, because the following diagram

$$\begin{array}{ccc} C \times A & \xrightarrow{1_C \times f} & C \times B \\ {\scriptstyle i_1}\downarrow & & \downarrow{\scriptstyle i_2} \\ A \times C & \xrightarrow{f \times 1_C} & B \times C \end{array}$$

in which i_1, i_2 are the canonical isomorphisms [i.e. $i_1((c, a)) = (a, c)$ for each $(c, a) \in C \times A$ and $i_2((c, b)) = (b, c)$ for each $(c, b) \in C \times B$] is commutative and hence, $f \times 1_C = i_2 \circ (1_C \times f) \circ i_1^{-1}$).

(3) $\Longrightarrow$ (4). Let $A \in |\mathcal{C}|$ and let $(f_i : B_i \longrightarrow B)_{i \in I}$ be a final epi-sink in $\mathcal{C}$.
(a) Let I be a set. If $(\coprod_{i \in I} B_i, (j_i)_{i \in I})$ is the coproduct of the family $(B_i)_{i \in I}$ in $\mathcal{C}$, then the $\mathcal{C}$-morphism $f : \coprod_{i \in I} B_i \longrightarrow B$ uniquely determined by $f \circ j_i = f_i$ for each $i \in I$ is a quotient map, i.e. a final epimorphism (Obviously, f is an epimorphism, and if $h : B \longrightarrow C$ is a map for which $h \circ f$ is a $\mathcal{C}$-morphism, then $(h \circ f) \circ j_i = h \circ f_i$ is also a $\mathcal{C}$-morphism for each $i \in I$. Thus, by assumption, h is a $\mathcal{C}$-morphism.).

Then applying (3) (b), $1_A \times f : A \times \coprod B_i \longrightarrow A \times B$ is a quotient map. Furthermore, $(A \times \coprod B_i, (1_A \times j_i)_{i \in I})$ is the coproduct of the family $(A \times B_i)_{i \in I}$. Since $A \times -$ is a functor and $f \circ j_i = f_i$ for each $i \in I$, the diagram

$$\begin{array}{ccc} A \times \coprod B_i & \xrightarrow{1_A \times f} & A \times B \\ {\scriptstyle 1_A \times j_i}\nwarrow & & \nearrow{\scriptstyle 1_A \times f_i} \\ & A \times B_i & \end{array}$$

is commutative for each $i \in I$. Since additionally the $\mathcal{C}$-structure of $A \times \coprod_{i \in I} B_i$ is final with respect to $(1_A \times j_i)_{i \in I}$, it results that the $\mathcal{C}$-structure of $A \times B$ is final with respect to $(1_A \times f_i)_{i \in I}$. Since $(f_i : B_i \longrightarrow B)_{i \in I}$ is an epi-sink in $\mathcal{C}$,

$\bigcup_{i\in I} f_i[B_i]$=B [1]
(1. If $|B| < 2$, then the assertion is trivial.
2. Let $|B| \geq 2$. If $B \neq \bigcup_{i\in I} f_i[B_i]$, then there were $b_0 \in \bigcup_{i\in I} f_i[B_i]$ and $b_1 \in B\setminus\bigcup_{i\in I} f_i[B_i]$. Endowing $\{b_0, b_1\}$ with the indiscrete $\mathcal{C}$–structure one obtains a $\mathcal{C}$–object Z. Hence $\alpha : B \longrightarrow Z$ defined by $\alpha(b) = b_0$ for each $b \in B$ and $\beta : B \longrightarrow Z$ defined by

$$\beta(b) = \begin{cases} b_0 & \text{if } b \in \bigcup_{i\in I} f_i[B_i] \\ b_1 & \text{otherwise} \end{cases}$$

are $\mathcal{C}$–morphisms such that $\alpha \circ f_i = \beta \circ f_i$ for each $i \in I$. Obviously, $\alpha \neq \beta$ in contradiction to the fact that $(f_i)_{i\in I}$ is an epi–sink.). Then it is easily verified that $(1_A \times f_i : A \times B_i \longrightarrow A \times B)_{i\in I}$ is an epi–sink (namely if $\alpha, \beta : A \times B \longrightarrow D$ are $\mathcal{C}$–morphisms such that $\alpha \circ (1_A \times f_i) = \beta \circ (1_A \times f_i)$ for each $i \in I$ and if $(a, b) \in A \times B$, then since $B = \bigcup_{i\in I} f_i[B_i]$, there is some $i \in I$ and some $b_i \in B_i$ with $f_i(b_i) = b$; hence $\alpha((a, b)) = \alpha((a, f_i(b_i))) = \alpha((1_A \times f_i)((a, b_i))) = \beta((1_A \times f_i)((a, b_i))) = \beta((a, f_i(b_i))) = \beta((a, b))$, i.e. $\alpha = \beta$).

(b) Let I be a proper class. Then there is a set $K \subset I$ such that $(f_i : B_i \longrightarrow B)_{i\in K}$ is a final epi–sink in $\mathcal{C}$ too (Put $K = I' \cup I''$, where I' and I'' are determined as follows:
1. Since $B = \bigcup_{i\in I} f_i[B_i]$, there is a set $I' \subset I$ such that $B = \bigcup_{j\in I'} f_j[B_j]$.
2. Since the final structure ζ of B w.r.t. (f_i) is the supremum [with respect to $\leq$] of the final structures ζ_i on the underlying set of B generated by each f_i, it follows from 1.1.2. (2) that there is a set $I'' \subset I$ such that $\{\zeta_i : i \in I\} = \{\zeta_j : j \in I''\}$.).
As under (a) one concludes that the $\mathcal{C}$–structure of $A \times B$ is final w.r.t. $(1_A \times f_i)_{i\in K}$ and therefore it is final w.r.t. $(1_A \times f_i)_{i\in I}$. Additionally, $(1_A \times f_i)_{i\in K}$ is an epi–sink, which implies that $(1_A \times f_i)_{i\in I}$ is an epi–sink too.

(4) $\Longrightarrow$ (5). Let $(A, B) \in |\mathcal{C}| \times |\mathcal{C}|$. Then $[A, B]_{\mathcal{C}}$ is endowed with the final $\mathcal{C}$–structure with respect to the class of all maps $f_i : C_i \longrightarrow [A, B]_{\mathcal{C}}$ from $\mathcal{C}$–objects C_i into $[A, B]_{\mathcal{C}}$ for which $e_{A,B} \circ (1_A \times f_i) : A \times C_i \longrightarrow B$ is a $\mathcal{C}$–morphism. The resulting object is denoted by B^A. The resulting sink is a (final) epi–sink:

Let $\alpha \circ f_i = \beta \circ f_i$ for each i, where α and β are $\mathcal{C}$–morphisms with domain B^A. If $e \in [A, B]_{\mathcal{C}}$, then $P = \{e\}$ is endowed with the uniquely determined $\mathcal{C}$–structure. A map $g : P \longrightarrow [A, B]_{\mathcal{C}}$ is defined by $g(e) = e$. Then $e_{A,B} \circ (1_A \times g) = e \circ p_A$, where $p_A : A \times P \longrightarrow A$ denotes the projection, is a $\mathcal{C}$–morphism. Thus, $P = C_i$ and $g = f_i$ for a suitable i and consequently $\alpha(e) = (\alpha \circ f_i)(e) = (\beta \circ f_i)(e) = \beta(e)$. Hence $\alpha = \beta$.

By assumption the $\mathcal{C}$–structure of $A \times B^A$ is final w.r.t. the family of all maps $1_A \times h : A \times C \longrightarrow A \times B^A$ for which $e_{A,B} \circ (1_A \times h)$ is a $\mathcal{C}$–morphism. Therefore $e_{A,B} : A \times B^A \longrightarrow B$ is a $\mathcal{C}$–morphism. Now let $f : A \times C \longrightarrow B$ be a $\mathcal{C}$–morphism. For each $c \in C$, the map $f_c : A \longrightarrow B$ defined by $f_c(a) = f((a, c))$ for each $a \in A$ is a $\mathcal{C}$–morphism. For the map $\overline{f} : C \longrightarrow [A, B]_{\mathcal{C}}$ defined by $\overline{f}(c) = f_c$

[1]Sometimes, we do not make a notational distinction between an object in a topological construct and its underlying set.

for each $c \in C$, $e_{A,B} \circ (1_A \times \overline{f}) = f$ is a $\mathcal{C}$–morphism. Hence $\overline{f} : C \longrightarrow B^A$ is a $\mathcal{C}$–morphism.

3.1.4 Remark. Because of 3.1.3. (5) the power objects in a cartesian closed topological construct are also called *natural function spaces.*

3.1.5 Corollary. *Let $\mathcal{C}$ be a cartesian closed topological construct. Then the following are satisfied:*

(1) *First exponential law:* $A^{B\times C} \cong (A^B)^C$

(2) *Second exponential law:* $(\prod_{i\in I} A_i)^B \cong \prod_{i\in I} A_i^B$

(3) *Third exponential law:* $A^{\coprod_{i\in I} B_i} \cong \prod_{i\in I} A^{B_i}$

(4) *Distributive law:* $A \times \coprod_{i\in I} B_i \cong \coprod_{i\in I} A \times B_i$

Proof. Concerning (4), see 3.1.3. (2) (a).
(2) follows from the fact that the right adjoint of $B\times -$ denoted by $(-)^B$ preserves products (cf. 2.1.10. and 2.1.16. ②).

(1) If $(\mathcal{F}_1, \mathcal{G}_1)$ and $(\mathcal{F}_2, \mathcal{G}_2)$ are pairs of adjoint functors and if the composition $\mathcal{F}_2 \circ \mathcal{F}_1$ is defined, then $(\mathcal{F}_2 \circ \mathcal{F}_1,\ \mathcal{G}_1 \circ \mathcal{G}_2)$ is again a pair of adjoint functors (exercise!). Especially, $(B\times -, (-)^B)$ as well as $(C\times -, (-)^C)$ are pairs of adjoint functors and consequently, $(B \times - \circ C \times -,\ (-)^C \circ (-)^B)$ is a pair of adjoint functors. Since $B \times - \circ C \times - \approx (B \times C) \times -$ and $(-)^{B\times C}$ is the corresponding right adjoint, the functors $(-)^C \circ (-)^B$ and $(-)^{B\times C}$ are naturally equivalent so that

$$(-)^{B\times C}(A) = A^{B\times C} \cong ((-)^C \circ (-)^B)(A) = (A^B)^C$$

(3) For each $A \in |\mathcal{C}|$ let us consider the contravariant hom–functor $H_A : \mathcal{C} \longrightarrow$ **Set** defined as follows:

$H_A(B) = [B, A]_{\mathcal{C}}$ for each $B \in |\mathcal{C}|$,
$H_A(f)(g) = g \circ f$ for each $f \in [C, B]_{\mathcal{C}}$ and each $g \in [B, A]_{\mathcal{C}}$
(i.e. $H_A(f) : H_A(B) \longrightarrow H_A(C)$).

It is easily verified that $H_A : \mathcal{C} \longrightarrow$ **Set** converts coproducts into products, i.e.

$$H_A(\coprod_{i\in I} B_i) = [\coprod_{i\in I} B_i, A]_{\mathcal{C}} \cong \prod_{k\in I} H_A(B_k) = \prod_{k\in I}[B_k, A]_{\mathcal{C}}\,.$$

The one–one correspondence $[r_i]_{i\in I} \longleftrightarrow (r_k)_{k\in I}$ describes this isomorphism (i.e. this bijective map), where the following diagram

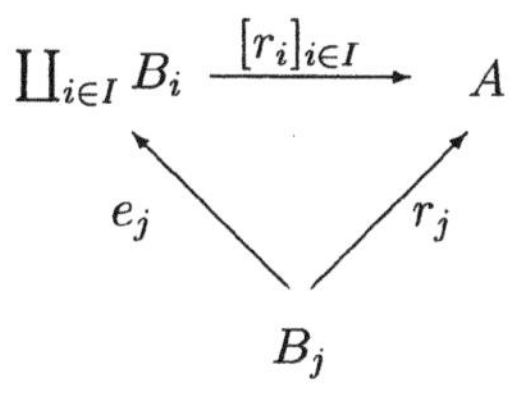

(e_j: j–the injection)

commutes for each $j \in I$. Now the above bijection is "lifted" to an isomorphism in $\mathcal{C}$. Obviously,

$$m : A^{\coprod_{i\in I} B_i} \longrightarrow \prod_{k\in I} A^{B_k}$$

defined by $m(f) = (f \circ e_k)_{k\in I}$ is a $\mathcal{C}$–morphism such that $m([r_i]_{i\in I}) = (r_k)_{k\in I}$ (note: for each $i \in I$, $p_i \circ m = \hom\,(e_i, 1_A)$ is a $\mathcal{C}$–morphism [cf. 3.1.6], where $p_i : \prod_{k\in I} A^{B_k} \longrightarrow A^{B_i}$ denotes the i–th projection). Since $\mathcal{C}$ is cartesian closed, for any $g : \coprod_{i\in I} B_i \times \prod_{k\in I} A^{B_k} \longrightarrow A$, there is precisely one $\mathcal{C}$–morphism $g^* : \prod_{k\in I} A^{B_k} \longrightarrow A^{\amalg_{i\in I} B_i}$ such that the diagram

$$\begin{array}{ccc}
\coprod_{i\in I} B_i \times A^{\amalg_{i\in I} B_i} & \xrightarrow{\;e_{\amalg B_i,A}\;} & A \\
{\scriptstyle 1_{\amalg B_i} \times g^*}\nwarrow & & \nearrow{\scriptstyle g} \\
& \coprod_{i\in I} B_i \times \prod_{k\in I} A^{B_k} &
\end{array}$$

commutes. In the following we write $[r_i]$ instead of $[r_i]_{i\in I}$ (resp. (r_k) instead of $(r_k)_{k\in I}$). In order to get an inverse of m we need g such that

$$(*) \qquad g((b,(r_k))) = r_j(b_j) \text{ with } b = (b_j, j)$$

(note: $\coprod_{i\in I} B_i = \bigcup_{i\in I} B_i \times \{i\}$). Then $g^*((r_k)) = [r_i]$; namely if $b \in \coprod_{i\in I} B_i$, then $b = (b_j, j) = e_j(b_j)$ and $g^*((r_k))(b) = e_{\amalg B_i,A}(b, g^*((r_k))) = (e_{\amalg B_i,A} \circ 1_{\amalg B_i} \times g^*)((b,(r_k))) = g((b,(r_k))) = r_j(b_j) = [r_i] \circ e_j(b_j) = [r_i](b)$; i.e. g^* is an inverse of m, that means m is an isomorphism.

Thus, it remains to construct $g : \coprod_{i\in I} B_i \times \prod_{k\in I} A^{B_k} \longrightarrow A$ such that $(*)$ is satisfied. For each $C \in |\mathcal{C}|$, $C\times - \approx - \times C$, hence $-\times C$ also preserves coproducts. Thus for $C = \prod_{k\in I} A^{B_k}$ there exists an isomorphism $h : \coprod_{i\in I} B_i \times \prod_{k\in I} A^{B_k} \longrightarrow \coprod_{i\in I}(B_i \times \prod_{k\in I} A^{B_k})$ such that for each $i \in I$, the diagram

$$\begin{array}{ccccccc}
& & B_i \times \prod_{k\in I} A^{B_k} & \xrightarrow{\;1_{B_i} \times p_i\;} & B_i \times A^{B_i} & & \\
& \swarrow{\scriptstyle e_i \times 1_{\prod_{k\in I} A^{B_k}}} & \downarrow{\scriptstyle e_i'} & & \downarrow{\scriptstyle e_i''} & \searrow{\scriptstyle e_{B_i,A}} & \\
\coprod_{i\in I} B_i \times \prod_{k\in I} A^{B_k} & \xrightarrow{\;h\;} & \coprod_{i\in I}(B_i \times \prod A^{B_k}) & \xrightarrow{\;\amalg_{i\in I}(1_{B_i}\times p_i)\;} & \coprod_{i\in I}(B_i \times A^{B_i}) & \xrightarrow{\;[e_{B_i,A}]\;} & A
\end{array}$$

commutes, where e_i, e_i' and e_i'' are the natural injections and p_i the projections. If $b = (b_j, j) \in \coprod_{i\in I} B_i$, then $h((b,(r_k))) = e_j'((b_j,(r_k))) = ((b_j,(r_k)), j)$.

Then $g = [e_{B_i,A}] \circ (\coprod_{i\in I}(1_{B_i} \times p_i)) \circ h$ is a $\mathcal{C}$–morphism such that $g((b,(r_k))) = [e_{B_i,A}](\coprod_{i\in I} 1_{B_i} \times p_i(e_j'((b_j,(r_k))))) = [e_{B_i,A}](e_j'' \circ 1_{B_j} \times p_j((b_j,(r_k)))) = [e_{B_i,A}](e_j''((b_j, r_j))) = e_{B_j,A}((b_j, r_j)) = r_j(b_j)$, i.e. $(*)$ is fulfilled.

3.1.6 Remark. For every cartesian closed topological construct $\mathcal{C}$, there exists an *internal hom–functor* $\hom : \mathcal{C}^* \times \mathcal{C} \longrightarrow \mathcal{C}$ which is defined as follows:

1) $\hom\,(A,B) = B^A$ for each $(A,B) \in |\mathcal{C}^* \times \mathcal{C}| = |\mathcal{C}^*| \times |\mathcal{C}|$,

2) $\hom\,(f,g) : \hom\,(A',B) \longrightarrow \hom\,(A,B')$ for each $(f,g) \in [A,A']_{\mathcal{C}} \times [B,B']_{\mathcal{C}}$ is defined by $\hom\,(f,g)(u) = g \circ u \circ f$ for each $u \in [A',B]_{\mathcal{C}}$.

(It has been shown that $\hom\,(A,B) = B^A$ is a $\mathcal{C}$–object. Hence it remains to prove that $\hom\,(f,g)$ is a $\mathcal{C}$–morphism:

a) Put $\widehat{g} = e_{A',B} \circ (e'_{A,A'} \times 1_{B^{A'}})$. Then there is a unique $\mathcal{C}$–morphism $\widehat{g}^* : (A')^A \times B^{A'} \longrightarrow B^A$ such that the diagram

$$\begin{array}{ccc} A \times B^A & \xrightarrow{e_{A,B}} & B \\ & {\scriptstyle 1_A \times \widehat{g}^*} \nwarrow \quad \nearrow {\scriptstyle \widehat{g}} & \\ & A \times (A')^A \times B^{A'} & \end{array}$$

commutes. Define $- \circ f : B^{A'} \longrightarrow B^A$ by $- \circ f = \widehat{g}^* \circ \alpha$, where $\alpha : B^{A'} \longrightarrow (A')^A \times B^{A'}$ is given by $\alpha(h) = (f, h)$. Then $- \circ f$ is a $\mathcal{C}$–morphism such that $(- \circ f)(u) = u \circ f$ for each $u \in B^{A'}$ which can be easily checked.

b) Put $\widetilde{g} = e_{B,B'} \circ (e_{A,B} \times 1_{(B')^B})$. Then there is a unique $\mathcal{C}$–morphism $\widetilde{g}^* : B^A \times (B')^B \longrightarrow (B')^A$ such that the diagram

$$\begin{array}{ccc} A \times (B')^A & \xrightarrow{e_{A,B'}} & B' \\ & {\scriptstyle 1_A \times \widetilde{g}^*} \nwarrow \quad \nearrow {\scriptstyle \widehat{g}} & \\ & A \times B^A \times (B')^B & \end{array}$$

commutes. Define $g \circ - : B^A \longrightarrow (B')^A$ by $g \circ - = \widetilde{g}^* \circ \beta$, where $\beta : B^A \longrightarrow B^A \times (B')^B$ is given by $\beta(k) = (k, g)$. Then $g \circ -$ is a $\mathcal{C}$–morphism such that $(g \circ -)(v) = g \circ v$ for each $v \in B^A$ which can be readily checked.
Thus hom $(f, g) = (g \circ -) \circ (- \circ f)$ is a $\mathcal{C}$–morphism.)

3.1.7 Corollary. *Let $\mathcal{C}$ be a cartesian closed topological construct and $\mathcal{A}$ a (full and isomorphism–closed) bicoreflective subconstruct which is closed under formation of finite products in $\mathcal{C}$. Then $\mathcal{A}$ is cartesian closed, and the power objects in $\mathcal{A}$ arise from the corresponding power objects in $\mathcal{C}$ by applying the bicoreflector.*

Proof. Let A, B be $\mathcal{A}$–objects and $([A, B], \xi)$ the power–object in $\mathcal{C}$, where $[A, B] = [A, B]_{\mathcal{A}} = [A, B]_{\mathcal{C}}$. If $e_{A,B} : A \times ([A, B], \xi) \longrightarrow B$ is an evaluation morphism in $\mathcal{C}$ and $1_{[A,B]} : ([A, B], \xi_{\mathcal{A}}) \longrightarrow ([A, B], \xi)$ the bicoreflection of $([A, B], \xi)$ w.r.t. $\mathcal{A}$, then $e_{A,B} \circ (1_A \times 1_{[A,B]}) : A \times ([A, B], \xi_{\mathcal{A}}) \longrightarrow B$ is easily seen to be an evaluation morphism in $\mathcal{A}$.

3.1.8 Remark. It is easily checked that a (full and isomorphism–closed) subconstruct $\mathcal{A}$ of a cartesian closed topological construct $\mathcal{C}$ which is closed under formation of finite products and power objects in $\mathcal{C}$ is cartesian closed. In particular, *if $\mathcal{A}$ is a bireflective (full and isomorphism–closed) subconstruct of $\mathcal{C}$ which is closed under formation of power objects in $\mathcal{C}$, then $\mathcal{A}$ is a cartesian closed topological construct, and the power–objects in $\mathcal{A}$ are formed as in $\mathcal{C}$.*

3.1.9 Examples. ① **Top** is not cartesian closed because in **Top** quotient maps are not finitely productive (cf. ② of the introduction).

② **Unif** is not cartesian closed: Let $\mathcal{W}$ be the uniformity on $\mathbb{R}$ induced by the Euclidean metric and put $\mathbb{R}_u = (\mathbb{R}, \mathcal{W})$. Further, let $\mathcal{D}$ be the discrete uniformity on $[\mathbb{R}_u, \mathbb{R}_u] = [\mathbb{R}_u, \mathbb{R}_u]_{\mathbf{Unif}}$, i.e. $\mathcal{D}$ is generated by the diagonal $\Delta_{[I\!R_u,I\!R_u]}$ of $[I\!R_u, I\!R_u] \times [I\!R_u, I\!R_u]$. It suffices to prove that the evaluation map $ev : I\!R_u \times ([I\!R_u, I\!R_u], \mathcal{D}) \longrightarrow I\!R_u$ is not uniformly continuous (namely if there were a uniformity $\mathcal{V}$ on $[I\!R_u, I\!R_u]$ such that the evaluation map $ev : I\!R_u \times ([I\!R_u, I\!R_u], \mathcal{V}) \longrightarrow I\!R_u$ were uniformly continous, then $1_{I\!R} \times 1_{[I\!R_u,I\!R_u]} : I\!R_u \times ([I\!R_u, I\!R_u], \mathcal{D}) \longrightarrow I\!R_u \times ([I\!R_u, I\!R_u], \mathcal{V})$ were uniformly continuous and consequently

$$ev \circ (1_{I\!R} \times 1_{[I\!R_u,I\!R_u]}) = ev : I\!R_u \times ([I\!R_u, I\!R_u], \mathcal{D}) \longrightarrow I\!R_u$$

were also uniformly continuous). If the evaluation map were uniformly continuous, then for each $V \in \mathcal{W}$ there were some $W_V \in \mathcal{W}$ such that $\Delta_{[I\!R_u,I\!R_u]}[W_V] = \{(f(a), f(b)) : f \in [I\!R_u, I\!R_u] \text{ and } (a,b) \in W_V\} \subset V$ (note: the uniform continuity of the evaluation map means that for each $V \in \mathcal{W}$ there exists some M belonging to the product uniformity on $I\!R \times [I\!R_u, I\!R_u]$ such that $((x,f),(y,g)) \in M$ implies $(f(x), g(y)) \in V$; obviously, for such an M there is some $W_V \in \mathcal{W}$ with $M \supset (p_{I\!R} \times p_{I\!R})^{-1}[W_V] \cap (p_{[I\!R_u,I\!R_u]} \times p_{[I\!R_u,I\!R_u]})^{-1}[\Delta_{[I\!R_u,I\!R_u]}]$, where $p_{I\!R} : I\!R_u \times ([I\!R_u, I\!R_u], \mathcal{D}) \longrightarrow I\!R_u$ [resp. $p_{[I\!R_u,I\!R_u]} : I\!R_u \times ([I\!R_u, I\!R_u], \mathcal{D}) \longrightarrow ([I\!R_u, I\!R_u], \mathcal{D})$] denotes the projection; thus, $(a,b) \in W_V$ and $f \in [I\!R_u, I\!R_u]$ imply $((a,f),(b,f)) \in M$ and consequently $(f(a), f(b)) \in V$). In particular, we would have $\Delta_{[I\!R_u,I\!R_u]}[W_{V_1}] \subset V_1 = \{(x,y) : |x-y| < 1\}$. Since for each real number $\alpha > 0$, $\alpha \cdot 1_{I\!R} \in [I\!R_u, I\!R_u]$ we would obtain $(\alpha x, \alpha y) \in V_1$ for all $(x,y) \in W_{V_1}$ and all α. On the other hand W_{V_1} belongs to $\mathcal{W}$, i.e. there is some $\varepsilon > 0$ such that $V_\varepsilon = \{(x,y) : |x-y| < \varepsilon\} \subset W_{V_1}$. Thus $r = \frac{\varepsilon}{2} > 0$ and $(0,r) \in W_{V_1}$. Furthermore, if $\alpha = \frac{1}{r}$, $(\alpha 0, \alpha r) = (0,1) \notin V_1$ -- a contradiction.

③ **SUConv** is cartesian closed: Let $\mathbf{X} = (X, \mathcal{J}_X)$ and $\mathbf{Y} = (Y, \mathcal{J}_Y)$ be semiuniform convergence spaces. Then the power object $\mathbf{Y}^{\mathbf{X}}$ is the set $[\mathbf{X}, \mathbf{Y}]_{\mathbf{SUConv}}$ of all uniformly continuous maps from $\mathbf{X}$ into $\mathbf{Y}$ endowed with the **SUConv**–structure $\mathcal{J}_{X,Y} = \{\Phi \in F([\mathbf{X}, \mathbf{Y}]_{\mathbf{SUConv}} \times [\mathbf{X}, \mathbf{Y}]_{\mathbf{SUConv}}) : \Phi(\mathcal{F}) \in \mathcal{J}_Y \text{ for each } \mathcal{F} \in \mathcal{J}_X\}$, where $\Phi(\mathcal{F}) = (\{A(F) : A \in \Phi,\ F \in \mathcal{F}\})$ with $A(F) = \{(f(a), g(b)) : (f,g) \in A,$ $(a,b) \in F\}$.
$\mathcal{J}_{X,Y}$ is called *the uniformly continuous* **SUConv**–*structure.*

(1. $\mathcal{J}_{X,Y}$ is a **SUConv**–structure:
UC$_1$) $\dot{f} \times \dot{f} \in \mathcal{J}_{X,Y}$ for each $f \in [\mathbf{X}, \mathbf{Y}]_{\mathbf{SUConv}}$ because $\dot{f} \times \dot{f}(\mathcal{F}) = f \times f(\mathcal{F}) \in \mathcal{J}_Y$ for each $\mathcal{F} \in \mathcal{J}_X$.

UC$_2$) follows immediately from $\Phi(\mathcal{F}) \subset \Theta(\mathcal{F})$ for all $\Phi, \Theta \in F([\mathbf{X}, \mathbf{Y}]_{\mathbf{SUConv}} \times [\mathbf{X}, \mathbf{Y}]_{\mathbf{SUConv}})$ with $\Phi \subset \Theta$ and all $\mathcal{F} \in F(X \times X)$.

UC$_3$) Let $\Phi \in \mathcal{J}_{X,Y}$. Then $\Phi^{-1}(\mathcal{F}) = (\Phi(\mathcal{F}^{-1}))^{-1}$ for each $\mathcal{F} \in \mathcal{J}_X$ [obviously, $A^{-1}[F] = (A[F^{-1}])^{-1}$ for all $A \in \Phi$ and all $F \in \mathcal{F}$]. Since $\mathcal{F}^{-1} \in \mathcal{J}_X$, $\Phi(\mathcal{F}^{-1}) \in \mathcal{J}_Y$. Thus, $(\Phi(\mathcal{F}^{-1}))^{-1} \in \mathcal{J}_Y$ because $\mathbf{Y}$ fulfills UC$_3$). Consequently, $\Phi^{-1} \in \mathcal{J}_{X,Y}$.

2. a) The evaluation map $e_{X,Y} : \mathbf{X} \times ([\mathbf{X}, \mathbf{Y}]_{\mathbf{SUConv}}, \mathcal{J}_{X,Y}) \longrightarrow \mathbf{Y}$ is uniformly continuous because $e_{X,Y} \times e_{X,Y}(\mathcal{F} \times \Phi) = \Phi(\mathcal{F})$.

b) Let $\mathbf{Z} = (Z, \mathcal{J}_Z) \in |\mathbf{SUConv}|$ and let $f : \mathbf{X} \times \mathbf{Z} \longrightarrow \mathbf{Y}$ be a uniformly continuous map. Then

$$\overline{f} : \mathbf{Z} \longrightarrow ([\mathbf{X}, \mathbf{Y}]_{\mathbf{SUConv}}, \mathcal{J}_{X,Y})$$

defined by $\overline{f}(z)(x) = f(x,z)$ for all $x \in X$ and all $z \in Z$ is uniformly continuous too, because by assumption $(\overline{f} \times \overline{f})(\mathcal{G})(\mathcal{F}) = (f \times f)(\mathcal{F} \times \mathcal{G}) \in \mathcal{J}_Y$ for all $\mathcal{G} \in \mathcal{J}_Z$ and all $\mathcal{F} \in \mathcal{J}_X$, i.e. $\overline{f} \times \overline{f}(\mathcal{G}) \in \mathcal{J}_{X,Y}$ for all $\mathcal{G} \in \mathcal{J}_Z$.)

④ **Fil** is cartesian closed: Since **Fil** can be embedded into **SUConv** as a bireflective and bicoreflective subconstuct, it follows from ③ and 3.1.7. that **Fil** is cartesian closed and that *the power objects in* **Fil** *arise from the corresponding power objects in* **SUConv** by applying the bicoreflector, i.e. *by forming the underlying filter space.* Thus, we obtain the power objects in **Fil** as follows: Let $\mathbf{X} = (X, \gamma)$ and $\mathbf{X}' = (X', \gamma')$ be filter spaces; then the power object $\mathbf{X}'^{\mathbf{X}}$ is the set $[\mathbf{X}, \mathbf{X}']_{\mathbf{Fil}}$ of all Cauchy continuous maps from $\mathbf{X}$ into $\mathbf{X}'$ endowed with the **Fil**–structure $\widehat{\gamma} = \{\theta \in F([\mathbf{X}, \mathbf{X}']_{\mathbf{Fil}}) : \theta(\mathcal{F}) \in \gamma' \text{ for each } \mathcal{F} \in \gamma\}$ where $\theta(\mathcal{F})$ is the filter generated by $\{A(F) : A \in \theta, F \in \mathcal{F}\}$ with $A(F) = \{f(x) : f \in A, x \in F\}$.
$\widehat{\gamma}$ is called the *Cauchy continuous* **Fil**–*structure.*

(Let $(X, \mathcal{J}_\gamma)$ [resp. $(X', \mathcal{J}_{\gamma'})$] be the semiuniform convergence space corresponding to (X, γ) [resp. (X', γ')] and $\mathcal{J}_{X,X'}$ the uniformly continuous **SUConv**–structure on $[(X, \mathcal{J}_\gamma), (X', \mathcal{J}_{\gamma'})]_{\mathbf{SUConv}} = [(X, \gamma), (X', \gamma')]_{\mathbf{Fil}}$. The bicoreflective modification of $\mathcal{J}_{X,X'}$ is $\mathcal{J}_{\gamma_{\mathcal{J}_{X,X'}}}$, or $\gamma_{\mathcal{J}_{X,X'}}$, provided that **Fil–D–SUConv** and **Fil** are identified. Then $\widehat{\gamma} = \gamma_{\mathcal{J}_{X,X'}}$:

a) Let $\mathcal{F} \in \widehat{\gamma}$ (i.e. $\mathcal{F}(\mathcal{K}) \in \gamma'$ for each $\mathcal{K} \in \gamma$). In order to prove $\mathcal{F} \times \mathcal{F}(\mathcal{H}) \in \mathcal{J}_{\gamma'}$ for each $\mathcal{H} \in \mathcal{J}_\gamma$ (i.e. $\mathcal{F} \times \mathcal{F} \in \mathcal{J}_{X,X'}$ or equivalently $\mathcal{F} \in \gamma_{\mathcal{J}_{X,X'}}$), choose some $\mathcal{H} \in \mathcal{J}_\gamma$. Then there is some $\mathcal{K} \in \gamma$ such that $\mathcal{K} \times \mathcal{K} \subset \mathcal{H}$. Thus, $\mathcal{F} \times \mathcal{F}(\mathcal{H}) \supset \mathcal{F} \times \mathcal{F}(\mathcal{K} \times \mathcal{K}) = \mathcal{F}(\mathcal{K}) \times \mathcal{F}(\mathcal{K})$ which implies $\mathcal{F} \times \mathcal{F}(\mathcal{H}) \in \mathcal{J}_{\gamma'}$, because $\mathcal{F}(\mathcal{K}) \in \gamma'$.

b) Let $\mathcal{F} \in \gamma_{\mathcal{J}_{X,X'}}$, i.e. $\mathcal{F} \times \mathcal{F} \in \mathcal{J}_{X,X'}$ or equivalently $\mathcal{F} \times \mathcal{F}(\mathcal{H}) \in \mathcal{J}_{\gamma'}$ for each $\mathcal{H} \in \mathcal{J}_\gamma$. In order to prove $\mathcal{F}(\mathcal{G}) \in \gamma'$ for each $\mathcal{G} \in \gamma$ (i.e. $\mathcal{F} \in \widehat{\gamma}$), choose some $\mathcal{G} \in \gamma$. Then $\mathcal{G} \times \mathcal{G} \in \mathcal{J}_\gamma$ and consequently, $\mathcal{F} \times \mathcal{F}(\mathcal{G} \times \mathcal{G}) = \mathcal{F}(\mathcal{G}) \times \mathcal{F}(\mathcal{G}) \in \mathcal{J}_{\gamma'}$, which implies $\mathcal{F}(\mathcal{G}) \in \gamma'$.)

⑤ **Conv** is cartesian closed: Since **Conv** is a bicoreflective subconstruct of **SUConv** which is closed under formation of products in **SUConv** (cf. 2.3.3.23.3) and 2.3.3.30.2)), it follows from 3.1.7. that **Conv** is cartesian closed and that the power objects in **Conv** arise from the power objects in **SUConv** by applying the bicoreflector.

⑥ a) **GConv** is cartesian closed: Let $\mathbf{X} = (X, q)$ and $\mathbf{X}' = (X', q')$ be generalized convergence spaces; then the power object $\mathbf{X}'^{\mathbf{X}}$ is the set $[\mathbf{X}, \mathbf{X}']_{\mathbf{GConv}}$ of all continuous maps from $\mathbf{X}$ into $\mathbf{X}'$ endowed with the **GConv**–structure $\widehat{q}$ defined by

$$(\psi, f) \in \widehat{q} \Longleftrightarrow (\psi(\mathcal{F}), f(x)) \in q' \text{ for each } (\mathcal{F}, x) \in q,$$

where $\psi(\mathcal{F}) = e_{X,X'}(\mathcal{F} \times \psi)$, i.e. $\psi(\mathcal{F}) = (\{A(F) : A \in \psi, F \in \mathcal{F}\})$ where $A(F) = \{f(z) : f \in A, z \in F\}$.
$\widehat{q}$ is called the **GConv**–*structure of continuous convergence.*

(1. $\widehat{q}$ is a **GConv**–structure.
C_1) $(\dot{f}, f) \in \widehat{q}$: Let $(\mathcal{F}, x) \in q$. Since $e_{X,X'}(\mathcal{F} \times \dot{f}) = f(\mathcal{F})$, we obtain from the continuity of f that $(e_{X,X'}(\mathcal{F} \times \dot{f}), f(x)) \in q'$.

C$_2$) Let $(\psi, f) \in \widehat{q}$ and $\psi' \supset \psi$. Then for any $(\mathcal{F}, x) \in q$, $e_{X,X'}(\mathcal{F} \times \psi') \supset e_{X,X'}(\mathcal{F} \times \psi)$. Since $(e_{X,X'}(\mathcal{F} \times \psi), f(x)) \in q'$, we obtain $(e_{X,X'}(\mathcal{F} \times \psi'), f(x)) \in q'$.

2. a) It follows immediately from the definition of $\widehat{q}$ that the evaluation map $e_{X,X'} : \mathbf{X} \times ([\mathbf{X}, \mathbf{X}']_{\mathbf{GConv}}, \widehat{q}) \longrightarrow \mathbf{X}'$ is continuous.

b) Let $\mathbf{X}'' = (X'', q'')$ be a generalized convergence space and $f : \mathbf{X} \times \mathbf{X}'' \longrightarrow \mathbf{X}'$ a continuous map. Then $\overline{f} : \mathbf{X}'' \longrightarrow ([\mathbf{X}, \mathbf{X}'], \widehat{q})$ defined by $\overline{f}(x'')(x) = f(x, x'')$ is continuous: Let $(\mathcal{H}, x'') \in q''$. In order to prove that $(\overline{f}(\mathcal{H}), \overline{f}(x'')) \in \widehat{q}$, we have to show that $(e_{X,X'}(\mathcal{F} \times \overline{f}(\mathcal{H})), \overline{f}(x'')(x)) \in q'$ whenever $(\mathcal{F}, x) \in q$. Since $e_{X,X'}(\mathcal{F} \times \overline{f}(\mathcal{H})) = f(\mathcal{F} \times \mathcal{H})$ and $\overline{f}(x'')(x) = f(x, x'')$, this follows immediately from the continuity of f.)

b) **Lim** is cartesian closed: If $\mathbf{X} = (X, q) \in |\mathbf{GConv}|$ and $\mathbf{X}' = (X', q') \in |\mathbf{Lim}|$, then the **GConv**–structure $\widehat{q}$ of continuous convergence on $[\mathbf{X}, \mathbf{X}']$ is a **Lim**–structure (namely, if $(\psi, f) \in \widehat{q}$ and $(\psi', f) \in \widehat{q}$, i.e. $(\psi(\mathcal{F}), f(x)) \in q'$ and $(\psi'(\mathcal{F}), f(x)) \in q'$ for each $(\mathcal{F}, x) \in q$, it follows from

$$\psi(\mathcal{F}) \cap \psi'(\mathcal{F}) = \psi \cap \psi'(\mathcal{F})$$

[note: $(A \cup B)(F) = A(F) \cup B(F)$ for all $A \in \psi$, $B \in \psi'$ and $F \in \mathcal{F}$] that $(\psi \cap \psi'(\mathcal{F}),\ f(x)) \in q'$ for each $(\mathcal{F}, x) \in q$ because q' fulfills C$_4$), i.e. $(\psi \cap \psi', f) \in \widehat{q}$). Thus, **Lim** is closed under formation of power objects in **GConv**. Since **GConv** is a cartesian closed topological construct (cf. a)) and **Lim** is a bireflective subconstruct, it follows from 3.1.8. that **Lim** is cartesian closed and that *the natural function space structure in* **Lim** *is the structure of continuous convergence*.

c) $\mathbf{Lim_S}$ is cartesian closed: If $\mathbf{X} = (X, q)$ is a limit space and $\mathbf{X}' = (X', q')$ a symmetric limit space, then the **Lim**–structure $\widehat{q}$ of continuous convergence on $[\mathbf{X}, \mathbf{X}']$ is a $\mathbf{Lim_S}$–structure (namely, let $(\psi, f) \in \widehat{q}$, i.e. $(\psi(\mathcal{F}), f(x)) \in q'$ for each $(\mathcal{F}, x) \in q$, and $g \in \bigcap_{A \in \psi} A$: If $(\mathcal{F}, x) \in q$ and $\mathcal{F}' = \mathcal{F} \cap \dot{x}$, then $(\mathcal{F}', x) \in q$ and $(\psi(\mathcal{F}'), f(x)) \in q'$; since $g(x) \in \bigcap\{H : H \in \psi(\mathcal{F}')\}$ and (X', q') is symmetric, $(\psi(\mathcal{F}'), g(x)) \in q'$, which implies $(\psi(\mathcal{F}), g(x)) \in q'$ because $\psi(\mathcal{F}') \subset \psi(\mathcal{F})$ and (X', q') fulfills C$_2$); thus, $(\psi, g) \in \widehat{q}$.). Hence, $\mathbf{Lim_S}$ is closed under formation of power objects in **Lim**, and since **Lim** is cartesian closed and contains $\mathbf{Lim_S}$ as a bireflective subconstruct, it follows from 3.1.8. that $\mathbf{Lim_S}$ is cartesian closed and that *the natural function space structure in* $\mathbf{Lim_S}$ *is the structure of continuous convergence*.

3.1.10 Theorem. *Let (X, q) (resp. (X', q')) be a symmetric limit space and (X, γ_q) (resp. $(X', \gamma_{q'})$) the corresponding (complete) filter space (i.e. γ_q [resp. $\gamma_{q'}$] consists of all convergent filters in (X, q) [resp. (X', q')]). Then the* $\mathbf{Lim_S}$*–structure $\widehat{q}$ of continuous convergence is the bicoreflective* $\mathbf{KConv_S}$*–modification of the Cauchy continuous* **Fil***–structure $\widehat{\gamma}$ on* $[(X, \gamma_q), (X', \gamma_{q'})]_{\mathbf{Fil}} = [(X, q), (X', q')]_{\mathbf{KConv}_S} = [(X, q), (X', q')]_{\mathbf{Lim}_S}$.

Proof. 1) $e_{X,X'}(\mathcal{F} \times (\psi \cap \dot{f}))$ converges in(X', q') for each convergent filter $\mathcal{F}$ in (X, q) iff $e_{X,X'}(\mathcal{F} \times \psi) \longrightarrow f(x)$ in (X', q') for each filter $\mathcal{F}$ in (X, q) with $\mathcal{F} \longrightarrow x$:

a) "$\Longrightarrow$". Let $\mathcal{F} \xrightarrow{q} x$. Then $\mathcal{F}' = \mathcal{F} \cap \dot{x} \xrightarrow{q} x$. By assumption, there is some $x' \in X'$ such that $e_{X,X'}(\mathcal{F}' \times \psi') \xrightarrow{q'} x'$ where $\psi' = \psi \cap \dot{f}$. Since (X', q') is symmetric and $f(x) \in \bigcap\{H : H \in e_{X,X'}(\mathcal{F}' \times \psi')\}$ it follows that $e_{X,X'}(\mathcal{F}' \times \psi') \xrightarrow{q'} f(x)$. Because of $e_{X,X'}(\mathcal{F}' \times \psi') \subset e_{X,X'}(\mathcal{F} \times \psi)$ we obtain $e_{X,X'}(\mathcal{F} \times \psi) \xrightarrow{q'} f(x)$.

b) "$\Longleftarrow$". α) $(e_{X,X'}(\mathcal{F} \times \psi)) \cap f(\mathcal{F}) = e_{X,X'}(\mathcal{F} \times \psi')$ which is easily verified.

β) Let $\mathcal{F} \xrightarrow{q} x$. Since f is continuous, it follows that $f(\mathcal{F}) \xrightarrow{q'} f(x)$. By assumption, $e_{X,X'}(\mathcal{F} \times \psi) \xrightarrow{q'} f(x)$. Since (X', q') is a limit space, we obtain $(e_{X,X'}(\mathcal{F} \times \psi)) \cap f(\mathcal{F}) \xrightarrow{q'} f(x)$. Thus, by α), $e_{X,X'}(\mathcal{F} \times \psi') \xrightarrow{q'} f(x)$.

2) In order to prove the above theorem we have to show that $q_{\widehat{\gamma}} = \widehat{q}$ (note: if $1_X : (X, \gamma_c) \longrightarrow (X, \gamma)$ denotes the bicoreflection of $(X, \gamma) \in |\mathbf{Fil}|$ w.r.t. **CFil** [$\cong$ **KConv**$_\mathbf{S}$] (cf. 2.3.3.9.), then $q_{\gamma_c} = q_\gamma$):

a) $q_{\widehat{\gamma}} \subset \widehat{q}$: Let $(\Theta, f) \in q_{\widehat{\gamma}}$, i.e. $\Theta \cap \dot{f} \in \widehat{\gamma}$. Then $(\Theta \cap \dot{f})(\mathcal{F}) = e_{X,X'}(\mathcal{F} \times (\Theta \cap \dot{f})) \in \gamma_{q'}$, i.e. $e_{X,X'}(\mathcal{F} \times (\Theta \cap \dot{f}))$ converges in (X', q'), for each $\mathcal{F} \in \gamma_q$, i.e. for all filters $\mathcal{F}$ converging in (X, q). Thus, by 1) for each filter $\mathcal{F}$ on X with $\mathcal{F} \xrightarrow{q} x$, $e_{X,X'}(\mathcal{F} \times \Theta) \xrightarrow{q'} f(x)$, i.e. $(\Theta, f) \in \widehat{q}$.

b) $\widehat{q} \subset q_{\widehat{\gamma}}$: Let $(\psi, f) \in \widehat{q}$, i.e. $e_{X,X'}(\mathcal{F} \times \psi) \xrightarrow{q'} f(x)$ for each $(\mathcal{F}, x) \in q$. Then, by 1), $e_{X,X'}(\mathcal{F} \times (\psi \cap \dot{f}))$ converges in (X', q'), i.e. $e_{X,X'}(\mathcal{F} \times (\psi \cap \dot{f})) \in \gamma_{q'}$, for all filters $\mathcal{F}$ converging in (X, q), i.e. for all $\mathcal{F} \in \gamma_q$. Thus, $\psi \cap \dot{f} \in \widehat{\gamma}$, i.e. $(\psi, f) \in q_{\widehat{\gamma}}$.

3.1.11 Remarks. 1) It follows from 3.1.10. together with 3.1.9. ④ that *the* **Lim**$_\mathbf{S}$*-structure of continuous convergence can be derived from the natural function space structure in* **SUConv**, *namely from the uniformly continuous* **SUConv***-structure.*

2) The **Lim**$_\mathbf{S}$–structure of continuous convergence is the coarsest **Lim**$_\mathbf{S}$–structure for which the evaluation map is continuous. For each cartesian closed topological construct $\mathcal{C}$ the natural function space structure is the coarsest $\mathcal{C}$–structure for which the evaluation map is a $\mathcal{C}$–morphism as the following proposition shows.

3.1.12 Proposition. *Let $\mathcal{C}$ be a cartesian closed topological construct. If A and B are $\mathcal{C}$-objects and $([A, B]_\mathcal{C}, \xi)$ is the natural function space, then ξ is the coarsest $\mathcal{C}$-structure for which the evaluation map $e_{A,B} : A \times ([A, B]_\mathcal{C}, \xi) \longrightarrow B$ is a $\mathcal{C}$-morphism.*

Proof. Let ξ' be a $\mathcal{C}$–structure on $[A, B]_\mathcal{C}$ such that the evaluation map $e_{A,B} : A \times ([A, B]_\mathcal{C}, \xi') \longrightarrow B$ is a $\mathcal{C}$–morphism. Since $\mathcal{C}$ is cartesian closed, it follows from 3.1.3. (5) (b) that $\overline{e_{A,B}} : ([A, B], \xi') \longrightarrow ([A, B], \xi)$ is a $\mathcal{C}$–morphism such that the diagram

$$\begin{array}{ccc} A\times([A,B],\xi) & \xrightarrow{\quad e_{A,B}\quad} & B \\ & {\scriptstyle 1_A\times\overline{e_{A,B}}}\nwarrow \qquad \nearrow{\scriptstyle e_{A,B}} & \\ & A\times([A,B],\xi') & \end{array}$$

commutes. Obviously, $\overline{e_{A,B}} = 1_{[A,B]}$, i.e. $\xi' \leq \xi$.

3.2 Extensional topological constructs

3.2.1 Definitions. 1) In a topological construct $\mathcal{C}$, a *partial morphism* from A to B is a $\mathcal{C}$–morphism $f : C \longrightarrow B$ whose domain is a subobject of A.

2) A topological construct $\mathcal{C}$ is called *extensional* (or *hereditary*) provided that every $\mathcal{C}$–object B has a *one-point extension* $B^* \in |\mathcal{C}|$, i.e. every $B \in |\mathcal{C}|$ can be embedded via the addition of a single point ∞_B into a $\mathcal{C}$–object B^* such that, for every partial morphism $f : C \longrightarrow B$ from A to B, the map $f^* : A \longrightarrow B^*$, defined by

$$f^*(a) = \begin{cases} f(a), & \text{if } a \in C \\ \infty_B, & \text{if } a \notin C \end{cases},$$

is a $\mathcal{C}$–morphism.

3) Let $\mathcal{C}$ be a topological construct. Then a final sink $(f_i : A_i \longrightarrow A)_{i\in I}$ in $\mathcal{C}$ is called *hereditary* provided that the following is satisfied: If B is a subspace [2] of A, B_i a subspace of A_i with underlying set $f_i^{-1}[B]$ and $g_i : B_i \longrightarrow B$ is the corresponding restriction of f_i, then $(g_i : B_i \longrightarrow B)_{i\in I}$ is a final sink in $\mathcal{C}$, too.

3.2.2 Proposition. *In a topological construct $\mathcal{C}$ the following are equivalent:*

(1) *Final sinks in $\mathcal{C}$ are hereditary.*
(2) *Final epi–sinks in $\mathcal{C}$ are hereditary.*
(3) *Quotients and coproducts in $\mathcal{C}$ (regarded as final epi–sinks) are hereditary.* [3]

Proof. (3) $\Longrightarrow$ (2). Let $(f_i : A_i \longrightarrow A)_{i\in I}$ be a final epi–sink. Then there is a set $J \subset I$ such that $(f_j : A_j \longrightarrow A)_{j\in J}$ is a final epi–sink. Hence $f : \amalg_{j\in J} A_j \longrightarrow A$ defined by $f \circ e_j = f_j$ for each $j \in J$ is a quotient map provided that $e_j : A_j \longrightarrow \amalg A_j$ denotes the j–th injection. By assumption, the final epi–sinks $(e_j : A_j \longrightarrow \amalg A_j)_{j\in J}$ and $(f : \amalg A_j \longrightarrow A)$ are hereditary. Hence, $(f_j : A_j \longrightarrow A)_{j\in J}$ is hereditary. Consequently, $(f_i : A_i \longrightarrow A)_{i\in I}$ is hereditary.

(2) $\Longrightarrow$ (1). Let $(f_i : A_i \longrightarrow A)_{i\in I}$ be a final sink in $\mathcal{C}$. Blow this one up to a final epi–sink by adding sufficiently many constant $\mathcal{C}$–morphisms. If B is a subspace

[2] By 1.2.2.5., subspaces may be assumed to be subsets endowed with the initial structure w.r.t. the inclusion map.

[3] *Quotients are hereditary* means that for each quotient map $f : A \longrightarrow B$, the final epi-sink $(f : A \longrightarrow B)$ is hereditary.

of A, then the corresponding restrictions $(g_j : B_j \longrightarrow B)_{j\in J}$ form a final epi–sink by assumption. Since for each $j \in J\backslash I$, the map g_j is constant (observe that the empty map is considered to be constant) and hence does not contribute anything to the finality, the sink $(g_i : B_i \longrightarrow B)_{i\in I}$ must be final.

(1) $\Longrightarrow$ (3). This implication is obvious.

3.2.3 Theorem. *For a topological construct $\mathcal{C}$ the following are equivalent:*

(1) *$\mathcal{C}$ is extensional.*
(2) *In $\mathcal{C}$ final sinks are hereditary.*

Proof. (1) $\Longrightarrow$ (2). Let $(f_i : A_i \longrightarrow A)_{i\in I}$ be a final sink in $\mathcal{C}$, let B be a subspace of A, and let $g_i : B_i \longrightarrow B$ for each $i \in I$ be the corresponding restriction of f_i. In order to prove that $(g_i : B_i \longrightarrow B)_{i\in I}$ is final, let $f : B \longrightarrow C$ be a map such that $f \circ g_i$ is a $\mathcal{C}$–morphism for each $i \in I$. Let C^* be the one–point extension of C:

$$\begin{array}{ccc} B_i & \hookrightarrow & A_i \\ {\scriptstyle g_i}\downarrow & & \downarrow{\scriptstyle f_i} \\ B & \hookrightarrow & A \\ {\scriptstyle f}\downarrow & & \downarrow{\scriptstyle f^*} \\ C & \hookrightarrow & C^* \end{array}$$

Then $f^* \circ f_i = (f \circ g_i)^*$ is a $\mathcal{C}$–morphism for each $i \in I$. Since $(f_i : A_i \longrightarrow A)_{i\in I}$ is final, $f^* : A \longrightarrow C^*$ is a $\mathcal{C}$–morphism. Hence, by initiality of $C \hookrightarrow C^*$, $f : B \longrightarrow C$ is a $\mathcal{C}$–morphism, i.e. $(g_i : B_i \longrightarrow B)_{i\in I}$ is final.

(2) $\Longrightarrow$ (1). Let $B \in |\mathcal{C}|$. Consider the collection $(f_i : C_i \longrightarrow B)_{i\in I}$ of all partial morphisms from A_i into B. Enlarge I to $J = I \cup \{\infty\}$ by adding a single element ∞, denote the underlying set of B by $|B|$ and define:

$$\begin{array}{rcl} g_i & = & f_i^* : A_i \longrightarrow |B| \cup \{\infty_B\} \text{ for } i \in I\,, \\ g_\infty & : & A_\infty = B \hookrightarrow |B| \cup \{\infty_B\} \text{ the inclusion map.} \end{array}$$

$$\begin{array}{ccc} C_i & \hookrightarrow & A_i \\ {\scriptstyle f_i}\downarrow & & \downarrow{\scriptstyle f_i^* = g_i} \\ A_\infty = B & \underset{g_\infty}{\hookrightarrow} & |B| \cup \{\infty_B\} \end{array}$$

Endowing $|B| \cup \{\infty_B\}$ with the final $\mathcal{C}$–structure ξ^* w.r.t. $(g_j)_{j\in J}$ one obtains a $\mathcal{C}$–object B^*. It remains to be shown that $g_\infty : B \longrightarrow B^*$ is an embedding. Hence, if $g_\infty : \widehat{B} \longrightarrow B^*$ is an embedding (i.e. $\widehat{B} = (|B|, \widehat{\xi})$ where $\widehat{\xi}$ denotes the initial $\mathcal{C}$–structure on $|B|$ w.r.t. g_∞), it must be proved that $\widehat{\xi} = \xi$, where ξ denotes the $\mathcal{C}$–structure of B (i.e. $B = (|B|, \xi)$). Obviously, $1_{|B|} : (|B|, \xi) \longrightarrow (|B|, \widehat{\xi})$

is a $\mathcal{C}$–morphism, since $\widehat{\xi}$ is initial and $g_\infty : (|B|, \xi) \longrightarrow (|B| \cup \{\infty_B\}, \xi^*)$ is a $\mathcal{C}$–morphism. Thus, $\xi \leq \widehat{\xi}$. In order to prove the converse, observe that the diagrams in $\mathcal{C}$

$$\begin{array}{ccc} C_i = g_i^{-1}[\widehat{B}] & \hookrightarrow & A_i \\ (g_i|g_i^{-1}[\widehat{B}])' \big\downarrow = f_i & & \big\downarrow g_i \\ \widehat{B} & \underset{g_\infty}{\hookrightarrow} & B^* \end{array} \quad (i \in I) \quad \text{and} \quad \begin{array}{ccc} C_\infty = g_i^{-1}[\widehat{B}] = B & \xrightarrow{1_B} & B \\ f_\infty = (g_\infty|C)' \big\downarrow = 1_{|B|} & & \big\downarrow g_\infty \\ \widehat{B} & \underset{g_\infty}{\hookrightarrow} & B^* \end{array}$$

commute. By assumption, $(f_j : C_j \longrightarrow \widehat{B})_{j \in J}$ is a final sink. Thus, $\widehat{\xi}$ is the finest $\mathcal{C}$–structure such that $f_j : C_j \longrightarrow (|B|, \widehat{\xi})$ is a $\mathcal{C}$–morphism for each $j \in J$. On the other hand $f_j : C_j \longrightarrow (|B|, \xi)$ is a $\mathcal{C}$–morphism for each $j \in J$. Consequently, $\widehat{\xi} \leq \xi$.

3.2.4 Proposition. *Let $\mathcal{C}$ be an extensional topological construct and let $(Y, \eta) \in |\mathcal{C}|$. If (Y^*, η^*) denotes the one-point extension of (Y, η), then η^* is the coarsest $\mathcal{C}$–structure on $Y^* = Y \cup \{\infty_Y\}$ such that (Y, η) is a subspace of (Y^*, η^*).*

Proof. If (Y, η) is a subspace of $(Y^*, \eta^{**}) \in |\mathcal{C}|$, then $1_Y : (Y, \eta) \longrightarrow (Y, \eta)$ is a partial morphism from (Y^*, η^{**}) to (Y, η). Hence, $(1_Y)^* = 1_{Y^*} : (Y^*, \eta^{**}) \longrightarrow (Y^*, \eta^*)$ is a $\mathcal{C}$–morphism, i.e. $\eta^{**} \leq \eta^*$.

3.2.5 Proposition. *Let $\mathcal{C}$ be an extensional topological construct and $\mathcal{A}$ a (full and isomorphism–closed) bicoreflective subconstruct which is closed under formation of subspaces in $\mathcal{C}$. Then $\mathcal{A}$ is extensional, and the one-point extensions in $\mathcal{A}$ arise from the corresponding one-point extensions in $\mathcal{C}$ by applying the bicoreflector.*

Proof. Let $(Y, \eta) \in |\mathcal{A}|$ and let (Y^*, η^*) be its one–point extension in $\mathcal{C}$. If $1_{Y^*} : (Y^*, (\eta^*)_{\mathcal{A}}) \longrightarrow (Y^*, \eta^*)$ denotes the bicoreflection of (Y^*, η^*) w.r.t. $\mathcal{A}$, then (Y, η) is a subspace of $(Y^*, (\eta^*)_{\mathcal{A}})$, i.e. η is the initial $\mathcal{A}$–structure w.r.t. the inclusion map $i : Y \longrightarrow (Y^*, (\eta^*)_{\mathcal{A}})$ (namely, if $(Z, \zeta) \in |\mathcal{A}|$ and $h : (Z, \zeta) \longrightarrow (Y, \eta)$ is a map such that $i \circ h : (Z, \zeta) \longrightarrow (Y^*, (\eta^*)_{\mathcal{A}})$ is a $\mathcal{C}$–morphism, then $1_Y^* \circ i \circ h = i \circ h : (Z, \zeta) \longrightarrow (Y^*, \eta^*)$ is a $\mathcal{C}$–morphism which implies that $h : (Z, \zeta) \longrightarrow (Y, \eta)$ is a $\mathcal{C}$–morphism [= $\mathcal{A}$–morphism], since $i : (Y, \eta) \longrightarrow (Y^*, \eta^*)$ is initial; furthermore, it follows from the universal property of the bicoreflection that $i : (Y, \eta) \longrightarrow (Y^*, (\eta^*)_{\mathcal{A}})$ is a $\mathcal{C}$–morphism, since $i : (Y, \eta) \longrightarrow (Y^*, \eta^*)$ is a $\mathcal{C}$–morphism.) In order to prove that $(Y^*, (\eta^*)_{\mathcal{A}})$ is the one–point extension of (Y, η) in $\mathcal{A}$, let $f : (Z, \zeta) \longrightarrow (Y, \eta)$ be a partial morphism from (X, ξ) to (Y, η) in $\mathcal{A}$. This one is also a partial $\mathcal{C}$–morphism, since subspaces in $\mathcal{A}$ are formed as in $\mathcal{C}$. Thus, $f^* : (X, \xi) \longrightarrow (Y^*, \eta^*)$ is a $\mathcal{C}$–morphism and, by the universal property of the bicoreflection, $f^* : (X, \xi) \longrightarrow (Y^*, (\eta^*)_{\mathcal{A}})$ is an $\mathcal{A}$–morphism.

3.2.6 Remark. Since a bireflective (full and isomorphism–closed) subconstruct of a topological construct is closed under formation of subspaces, we obtain: *If a bireflective (full and isomorphism–sloced) subconstruct $\mathcal{A}$ of an extensional topological construct $\mathcal{C}$ is closed under formation of one–point extensions in $\mathcal{C}$, then it is an extensional topological construct, and the one–point extensions in $\mathcal{A}$ are formed as in $\mathcal{C}$.*

3.2.7 Examples. ① **Top** is not extensional, since in **Top** quotients are not hereditary (cf. ③ of the introduction).

② **Unif** and **ULim** are not extensional: Let $\mathcal{A}$ be the construct **Unif** or the construct **ULim** respectively. If D_2 denotes the discrete uniform space with underlying set $\{0,1\}$, then there is no one–point extension of D_2 in $\mathcal{A}$. If there were a one–point extension of D_2, its $\mathcal{A}$–structure ξ would be the coarsest $\mathcal{A}$–structure on $\{0,1,2\}$ such that D_2 is a subspace of $(\{0,1,2,\},\xi)$. On the other hand, there are two uniform spaces (i.e. these spaces belong to $|\mathcal{A}|$) containing D_2 as a subspace, namely $(\{0,1,2\},[\mathcal{V}])$ and $(\{0,1,2\},[\mathcal{W}])$ where $\mathcal{V}$ [resp. $\mathcal{W}$] denotes the filter generated by the sets $V = \{(0,0),(1,1),(1,2),(2,1),(2,2)\}$ [resp. $W = \{(0,0),(1,1),(0,2),(2,0),(2,2)\}$]. Thus, $\mathcal{V}$ and $\mathcal{W}$ would belong to ξ. Since ξ fulfills UC_5), $\mathcal{V}\circ\mathcal{W}\circ\mathcal{V} = (\{V\circ W\circ V\}) = \{\{0,1,2\}^2\} \in \xi$, i.e. ξ is the indiscrete $\mathcal{A}$–structure. Consequently, D_2 is not a subspace of $(\{0,1,2\},\xi)$ – a contradiction.

③ **SUConv** is extensional: Let $(X,\mathcal{J}_X)$ be a semiuniform convergence space. Put $X^* = X\cup\{\infty_X\}$ with $\infty_X \notin X$. For each $M\subset X^*\times X^*$, let $M^* = M\cup(X^*\times\{\infty_X\})\cup(\{\infty_X\}\times X^*)$. For each $\mathcal{F}\in\mathcal{J}_X$, the filter $\mathcal{F}^* = \{F^* : F\in\mathcal{F}\}$ belongs to

$$\begin{aligned}\mathcal{J}_{X^*} &= \{\mathcal{H}\in F(X^*\times X^*) : \text{the trace of } \mathcal{H} \text{ on } X\times X \text{ exists and belongs to}\\ &\quad \mathcal{J}_X \text{ or } \{(\infty_X,\infty_X)\}^*\in\mathcal{H}\}\end{aligned}$$

which is an **SUConv**–structure for X^*:

UC_1) If $i : X \longrightarrow X^*$ denotes the inclusion map, then for each $x\in X$, $(i\times i)^{-1}(\dot{x}\times\dot{x}) = i^{-1}(\dot{x})\times i^{-1}(\dot{x}) = \dot{x}\times\dot{x}\in\mathcal{J}_X$, i.e. $\dot{x}\times\dot{x}\in\mathcal{J}_{X^*}$ for each $x\in X$. Furthermore, $\dot{\infty}_X\times\dot{\infty}_X\in\mathcal{J}_{X^*}$ since $\{(\infty_X,\infty_X)\}^*\in\dot{\infty}_X\times\dot{\infty}_X$.

UC_2) Let $\mathcal{H}\in\mathcal{J}_{X^*}$ and $\mathcal{G}\in F(X^*\times X^*)$ with $\mathcal{G}\supset\mathcal{H}$. If $\{(\infty_X,\infty_X)\}^*\in\mathcal{G}$, $\mathcal{G}\in\mathcal{J}_{X^*}$ by definition. If $\{(\infty_X,\infty_X)\}^*\notin\mathcal{G}$, $(i\times i)^{-1}(\mathcal{G})$ exists and belongs to $\mathcal{J}_X$, since $(i\times i)^{-1}(\mathcal{H})\in\mathcal{J}_X$ is coarser than $(i\times i)^{-1}(\mathcal{G})$. Thus, $\mathcal{G}\in\mathcal{J}_{X^*}$.

UC_3) Let $\mathcal{H}\in\mathcal{J}_{X^*}$. If $\{(\infty_X,\infty_X)\}^*\in\mathcal{H}^{-1}$, $\mathcal{H}^{-1}\in\mathcal{J}_{X^*}$. If $\{(\infty_X,\infty_X)\}^*\notin\mathcal{H}^{-1}$, $\{(\infty_X,\infty_X)\}^*\notin\mathcal{H}$, i.e. $(i\times i)^{-1}(\mathcal{H})\in\mathcal{J}_X$. Furthermore, $(i\times i)^{-1}(\mathcal{H}^{-1}) = ((i\times i)^{-1}(\mathcal{H}))^{-1}\in\mathcal{J}_X$, i.e. $\mathcal{H}^{-1}\in\mathcal{J}_{X^*}$.
It remains to prove that $(X^*,\mathcal{J}_{X^*})$ is the one–point extension of $(X,\mathcal{J}_X)$:

a) $(X,\mathcal{J}_X)$ is a subspace of $(X^*,\mathcal{J}_{X^*})$: Let $\widehat{\mathcal{J}}_X = \{\mathcal{F}\in F(X\times X) : (i\times i)(\mathcal{F})\in\mathcal{J}_{X^*}\}$. In order to prove $\mathcal{J}_X = \widehat{\mathcal{J}}_X$, let $\mathcal{F}\in\widehat{\mathcal{J}}_X$, i.e. $(i\times i)(\mathcal{F})\in\mathcal{J}_{X^*}$. Since $\{(\infty_X,\infty_X)\}^*\notin(i\times i)(\mathcal{F})$, $(i\times i)^{-1}((i\times i)(\mathcal{F})) = \mathcal{F}\in\mathcal{J}_X$. Conversely, let $\mathcal{F}\in\mathcal{J}_X$. Then $(i\times i)^{-1}(\mathcal{F}^*) = \mathcal{F}$ and consequently, $(i\times i)(\mathcal{F})\supset\mathcal{F}^*\in\mathcal{J}_{X^*}$, i.e. $(i \times i)(\mathcal{F}) \in \mathcal{J}_{X^*}$. Hence $\mathcal{F} \in \widehat{\mathcal{J}}_X$.

b) Let $(Y, \mathcal{J}_Y)$ be a semiuniform convergence space, $(Z, \mathcal{J}_Z)$ a subspace of $(Y, \mathcal{J}_Y)$ and $f : (Z, \mathcal{J}_Z) \longrightarrow (X, \mathcal{J}_X)$ a uniformly continuous map. In order to prove that $f^* : (Y, \mathcal{J}_Y) \longrightarrow (X^*, \mathcal{J}_X^*)$ is uniformly continuous, let $\mathcal{G} \in \mathcal{J}_Y$ and let $j : Z \longrightarrow Y$ be the inclusion map. Obviously, $(j \times j)^{-1}(\mathcal{G})$ exists iff $(Y \times Y)\backslash(Z \times Z) \notin \mathcal{G}$. If $(j \times j)^{-1}(\mathcal{G})$ exists, then $(j \times j)^{-1}(\mathcal{G}) \in \mathcal{J}_Z$. Since f is uniformly continuous, $(f \times f)((j \times j)^{-1}(\mathcal{G})) \in \mathcal{J}_X$. Thus, $((f \times f)((j \times j)^{-1}(\mathcal{G})))^* \in \mathcal{J}_X^*$. Since $(f \times f)((j \times j)^{-1}(\mathcal{G})))^* \subset (f^* \times f^*)(\mathcal{G})$, $(f^* \times f^*)(\mathcal{G}) \in \mathcal{J}_X^*$ (note: $f^* \times f^*[G] = (f^* \times f^*)[G \cap (Z \times Z)] \cup (f^* \times f^*)[G \cap (Y \times Y \backslash Z \times Z)] \subset (f \times f)[G \cap (Z \times Z)] \cup (\{\infty_X\} \times X^*) \cup (X^* \times \{\infty_X\}) = ((f \times f)[G \cap (Z \times Z)])^*$). If $(j \times j)^{-1}(\mathcal{G})$ does not exist, $(Y \times Y)\backslash(Z \times Z) \in \mathcal{G}$. Then $f^* \times f^*[Y \times Y \backslash Z \times Z] = f^* \times f^*[((Y\backslash Z) \times (Y\backslash Z)) \cup ((Z \times (Y\backslash Z)) \cup ((Y\backslash Z) \times Z)] = (f^*[Y\backslash Z] \times f^*[Y\backslash Z]) \cup (f^*[Z] \times f^*[Y\backslash Z]) \cup (f^*[Y\backslash Z] \times f^*[Z]) = (\{\infty_X\} \times \{\infty_X\}) \cup (f[Z] \times \{\infty_X\}) \cup (\{\infty_X\} \times f[Z]) \subset \{(\infty_X, \infty_X)\}^*$, i.e. $\{(\infty_X, \infty_X)\}^* \in (f^* \times f^*)(\mathcal{G})$. Hence, $(f^* \times f^*)(\mathcal{G}) \in \mathcal{J}_X^*$.

④ **Fil** is extensional: Since **Fil** can be embedded into **SUConv** as a bireflective and bicoreflective subconstruct, it follows from ③ and 3.2.5. that **Fil** is extensional and that the one–point extensions arise from the corresponding one–point extensions in **SUConv** by applying the bicoreflectior, i.e. by forming the underlying filter space. Thus, we obtain the one–point extensions in **Fil** as follows: Let (X, γ) be a filter space and $(X, \mathcal{J}_\gamma)$ its corresponding semiuniform convergence space. If $(X^*, \mathcal{J}_\gamma^*)$ is the one–point extension of $(X, \mathcal{J}_\gamma)$ in **SUConv**, then (X^*, γ^*) is the one–point extension of (X, γ) in **Fil** provided that $\gamma^* = \gamma_{\mathcal{J}_\gamma^*} = \{\mathcal{F} \in F(X^*) : \mathcal{F} \neq \dot{\infty}_X \text{ and } i^{-1}(\mathcal{F}) \in \gamma\} \cup \{\dot{\infty}_X\}$.

⑤ a) **GConv** is extensional: Let (X, q) be a generalized convergence space. Put $X^* = X \cup \{\infty_X\}$ with $\infty_X \notin X$. Let $i : X \longrightarrow X^*$ be the inclusion map. A **GConv**–structure q^* on X^* is defined by

$$(\mathcal{F}, x) \in q^* \iff x = \infty_X \text{ or } \mathcal{F} = \dot{\infty}_X \text{ or } (i^{-1}(\mathcal{F}), x) \in q\,,$$

i.e. all filters on X^* converge to ∞_X, and $\dot{\infty}_X$ converges to all elements of X^*, while in any other case the convergence behaviour of a filter $\mathcal{F}$ on X^* is determined by the convergence behaviour of the trace $i^{-1}(\mathcal{F})$ on X (note: $i^{-1}(\mathcal{F})$ exists iff $\mathcal{F} \neq \dot{\infty}_X$). Then (X^*, q^*) is the one–point extension of (X, q):

a$_1$) (X, q) is a subspace of (X^*, q^*): Let $\widehat{q}$ be the initial **GConv**–structure on X w.r.t. $i : X \longrightarrow (X^*, q^*)$. In order to prove $q = \widehat{q}$, let $(\mathcal{F}, x) \in \widehat{q}$. Then $(i(\mathcal{F}), i(x)) \in q^*$, i.e. $(i^{-1}(i(\mathcal{F})), i(x)) = (\mathcal{F}, x) \in q$ since $x \neq \infty_X$ and $i(\mathcal{F}) \neq \dot{\infty}_X$. Thus, $\widehat{q} \subset q$. Conversely, let $(\mathcal{F}, x) = (i^{-1}(i(\mathcal{F})), i(x)) \in q$. Then $(i(\mathcal{F}), i(x)) \in q^*$, i.e. $(\mathcal{F}, x) \in \widehat{q}$.

a$_2$) Let $f : (X'', q'') \longrightarrow (X, q)$ be a partial continuous map from (X', q') to (X, q). In order to prove that $f^* : (X', q') \longrightarrow (X^*, q^*)$ is continuous, let $(\mathcal{F}', x') \in q'$. Then

α) $X'\backslash X'' \in \mathcal{F}'$ or
β) $X'\backslash X'' \notin \mathcal{F}'$.

In case α) $\{\infty_X\} = f^*[X'\backslash X''] \in f^*(\mathcal{F}')$, i.e. $f^*(\mathcal{F}') = \dot{\infty}_X$. Consequently, $(f^*(\mathcal{F}'), f^*(x')) \in q^*$. In case β), $j^{-1}(\mathcal{F}')$ exists provided that $j : X'' \longrightarrow X'$

denotes the inclusion map. Let us consider the following cases:

1. $f^*(x') = \infty_X$, i.e. $x' \in X'\backslash X''$ or
2. $f^*(x') \neq \infty_X$, i.e. $x' \in X''$.

In the first case, we obtain $(f^*(\mathcal{F}), f^*(x')) \in q^*$. In the second case $(j^{-1}(\mathcal{F}'), x') \in q''$, since $j(j^{-1}(\mathcal{F}')) \supset \mathcal{F}'$, $(\mathcal{F}', x') \in q'$ and (X'', q'') is a subspace of (X', q'). It follows from the continuity of f, that $(f(j^{-1}(\mathcal{F}'), f(x')) = (i^{-1}(f^*(\mathcal{F}')), f(x')) \in q$, i.e. $(f^*(\mathcal{F}'), f(x')) \in q^*$.

b) **KConv** is extensional: **KConv** is closed under formation of one–point extensions in **GConv**, i.e. if (X, q) is a Kent convergence space, then (X^*, q^*) is a Kent convergence space where (X^*, q^*) denotes the one–point extension of (X, q) in **GConv**, namely if $(\mathcal{F}, x) \in q^*$, then $x = \infty_X$ (and consequently $(\mathcal{F} \cap \dot{x}, x) \in q^*$) or $x \in X$:

1. $\mathcal{F} = \dot{\infty}_X$, i.e. $\mathcal{F}' = \mathcal{F} \cap \dot{x} = (\{x, \infty_X\})$, and consequently $i^{-1}(\mathcal{F}') = \dot{x} \xrightarrow{q} x$, i.e. $(\mathcal{F}', x) \in q^*$.

or

2. $\mathcal{F} \neq \dot{\infty}_X$, i.e. $i^{-1}(\mathcal{F})$ exists, and consequently $(i^{-1}(\mathcal{F}), x) \in q$; thus, $i^{-1}(\mathcal{F} \cap \dot{x}) = i^{-1}(\mathcal{F}) \cap i^{-1}(\dot{x}) = i^{-1}(\mathcal{F}) \cap \dot{x}$ converges to x, since (X, q) fulfills C_3); hence $(\mathcal{F} \cap \dot{x}, x) \in q^*$.

Since **KConv** is a bireflective subconstruct of **GConv**, it follows from 3.2.6. that **KConv** is extensional and that the one–point extensions in **KConv** are formed as in **GConv**.

c) $\mathbf{KConv_S}$ ($\cong$ **Conv**) is not extensional: Let (Y, q_Y) be a symmetric Kent convergence space. Put $Y^* = Y \cup \{\infty_Y\}$ with $\infty_Y \notin Y$. If $j : Y \longrightarrow Y^*$ denotes the inclusion map, then a symmetric Kent convergence structure q_Y^* on Y^* is defined by

$$(\mathcal{F}, x) \in q_Y^* \iff \begin{array}{l}(x \in Y \text{ and } \mathcal{F} \supset j(\mathcal{G}) \cap \dot{\infty}_Y \text{ for some } (\mathcal{G}, x) \in q_Y) \\ \text{or } (x = \infty_Y \text{ and } \mathcal{F} = \mathcal{H} \cap \dot{z} \text{ for some } z \in Y \\ \text{implies } (\mathcal{H}, z) \in q_Y^*).\end{array}$$

It is easily checked that q_Y^* is the coarsest $\mathbf{KConv_S}$–structure on Y^* such that (Y, q_Y) is a subspace of (Y^*, q_Y^*). Consider the following objects of $\mathbf{KConv_S}$:

1. (X, q_X) defined by $X = \mathbb{Z}$ and $(\mathcal{F}, x) \in q_X$ iff $(x \neq 0$ and $\mathcal{F} = \dot{x})$ or $(x = 0$ and $\mathcal{F} \supset \mathcal{F}_c \cap \dot{0})$, where $\mathcal{F}_c = \{U \in \mathcal{P}(X) : X\backslash U \text{ is finite}\}$.

and

2. (Y, q_Y) and (Z, q_Z) are the subspaces of (X, q_X) with $Y = \{x \in \mathbb{Z} : x > 0\}$ and $Z = \mathbb{Z}\backslash\{0\}$.

Obviously, (Y, q_Y) and (Z, q_Z) are discrete. The map $f : Z \longrightarrow Y$ defined by $f(x) = 1$ if $x < 0$ and $f(x) = x$ if $x > 0$, is continuous, hence a partial morphism from (X, q_X) to (Y, q_Y). However, $f^* : X \longrightarrow Y^*$ is not continuous: since $\mathcal{H}_c \cap \dot{\infty}_Y$ does not converge to 1 in Y^* (where $\mathcal{H}_c = \{V \in \mathcal{P}(Y^*)\colon Y^*\backslash V \text{ is finite}\}$), $f^*(\mathcal{F}_c \cap \dot{0}) = \mathcal{H}_c \cap \dot{\infty}_Y \cap \dot{1}$ does not converge to $\infty_Y = f^*(0)$. Therefore, $\mathbf{KConv_S}$ is not extensional (cf. 3.2.4.).

3.3 Strong topological universes

3.3.1 Definitions. Let $\mathcal{C}$ be a topological construct. Consider the following *convenient properties:*

CP$_1$) $\mathcal{C}$ is cartesian closed.
CP$_2$) $\mathcal{C}$ is extensional.
CP$_3$) In $\mathcal{C}$ products of quotient maps are quotient maps.

Then $\mathcal{C}$ is called

1) *strongly cartesian closed* provided that $\mathcal{C}$ fulfills CP$_1$) and CP$_3$),
2) a *topological universe* provided that $\mathcal{C}$ fulfills CP$_1$) and CP$_2$), and
3) a *strong topological universe* provided that $\mathcal{C}$ fulfills CP$_1$), CP$_2$) and CP$_3$).

3.3.2 Proposition. *Let $\mathcal{A}$ be a topological construct.*

1) *If $\mathcal{B}$ is a bicoreflective (full and isomorphism–closed) subconstruct of $\mathcal{A}$ which is closed under formation of products in $\mathcal{A}$, then $\mathcal{B}$ fulfills CP$_3$) whenever $\mathcal{A}$ fulfills CP$_3$).*

2) *If $\mathcal{B}$ is a bireflective (full and isomorphism–closed) subconstruct of $\mathcal{A}$ which is closed under formation of quotient objects in $\mathcal{A}$, then $\mathcal{B}$ fulfills CP$_3$) whenever $\mathcal{A}$ fulfills CP$_3$).*

Proof. In both cases products and quotients in $\mathcal{B}$ are formed as in $\mathcal{A}$. Thus, $\mathcal{B}$ fulfills CP$_3$) whenever $\mathcal{A}$ fulfills CP$_3$).

3.3.3 Examples. ① **Top** does not fulfill any of the above convenient properties (cf. ② of the introduction, 3.1.9. ① and 3.2.6. ①).

② **Unif** fulfills CP$_3$) which has been proved by Hušek and Rice in 1978 (cf. [78]), but it does not fulfill CP$_1$) and CP$_2$) (cf. 3.1.9. ② and 3.2.7. ②). The question whether **ULim** fulfills CP$_3$) is unsolved.

③ **SUConv** is a strong topological universe: Since **SUConv** is a topological universe (cf. 3.1.9. ③ and 3.2.7. ③), it remains to prove that **SUConv** fulfills CP$_3$). At first let us consider the final structures in **SUConv**: Let X be a set, $((X_i, \mathcal{J}_{X_i}))_{i\in I}$ a family of semiuniform convergence spaces and $(f_i : X_i \longrightarrow X)_{i\in I}$ a family of maps. Then $\mathcal{J}_X = \{\mathcal{F} \in F(X \times X)$: there exists some $i \in I$ and some $\mathcal{F}_i \in \mathcal{J}_{X_i}$ with $(f_i \times f_i)(\mathcal{F}_i) \subset \mathcal{F}\} \cup \{\dot{x} \times \dot{x} : x \in X\}$ is the final **SUConv**–structure on X w.r.t. the given data. If $(f_i : X_i \longrightarrow X)_{i\in I}$ is an epi–sink in **Set** (i.e. $X = \bigcup_{i\in I} f_i[X_i]$), then $\mathcal{J}_X = \{\mathcal{F} \in F(X \times X)$: there exists some $i \in I$ and some $\mathcal{F}_i \in \mathcal{J}_{X_i}$ with $(f_i \times f_i)(\mathcal{F}_i) \subset \mathcal{F}\}$. In order to prove that **SUConv** fulfills CP$_3$), let $(f_i : (X_i, \mathcal{J}_{X_i}) \longrightarrow (Y_i, \mathcal{J}_{Y_i}))_{i\in I}$ be a non–empty family of quotient maps in **SUConv** and let

$$\begin{array}{ccc} (X, \mathcal{J}_X) & \xrightarrow{\ \prod f_i\ } & (Y, \mathcal{J}_Y) \\ {\scriptstyle p_i}\downarrow & & \downarrow{\scriptstyle p_i'} \\ (X_i, \mathcal{J}_{X_i}) & \xrightarrow{\ f_i\ } & (Y_i, \mathcal{J}_{Y_i}) \end{array}$$

be the corresponding product diagram in **SUConv**, where $\prod_{i\in I}(X_i, \mathcal{J}_{X_i}) = (X, \mathcal{J}_X)$ and $\prod_{i\in I}(Y_i, \mathcal{J}_{Y_i}) = (Y, \mathcal{J}_Y)$. Since all f_i are surjective, $\prod f_i$ is surjective. For each $i \in I$, $\mathcal{J}_{Y_i} = \{\mathcal{F} \in F(Y_i \times Y_i) :$ there exists $\mathcal{G}_i \in \mathcal{J}_{X_i}$ with $(f_i \times f_i)(\mathcal{G}_i) \subset \mathcal{F}\}$, because f_i is a quotient map. Let $\mathcal{J}'_Y = \{\mathcal{K} \in F(Y \times Y) :$ there exists $\mathcal{G} \in \mathcal{J}_X$ with $(\prod f_i \times \prod f_i)(\mathcal{G}) \subset \mathcal{K}\}$. Then $\mathcal{J}_Y = \{\mathcal{H} \in F(Y \times Y) : (p'_i \times p'_i)(\mathcal{H}) \in \mathcal{J}_{Y_i}$ for each $i \in I\}$ is equal to $\mathcal{J}'_Y$, i.e. $\prod f_i$ is a quotient map:

1) If $\mathcal{K} \in \mathcal{J}'_Y$, then there exists $\mathcal{G} \in \mathcal{J}_X$ with $(\prod f_i \times \prod f_i)(\mathcal{G}) \subset \mathcal{K}$. Furthermore, $(f_i \times f_i)((p_i \times p_i)(\mathcal{G})) = (p'_i \times p'_i)((\prod f_i \times \prod f_i)(\mathcal{G})) \subset (p'_i \times p'_i)(\mathcal{K})$ for each $i \in I$, implies $(p'_i \times p'_i)(\mathcal{K}) \in \mathcal{J}_{Y_i}$ for each $i \in I$, i.e. $\mathcal{K} \in \mathcal{J}_Y$.

2) If $\mathcal{H} \in \mathcal{J}_Y$, then $(p'_i \times p'_i)(\mathcal{H}) \in \mathcal{J}_{Y_i}$ for each $i \in I$. Thus, for each $i \in I$, there is some $\mathcal{G}_i \in \mathcal{J}_{X_i}$ with $(f_i \times f_i)(\mathcal{G}_i) \subset (p'_i \times p'_i)(\mathcal{H})$. If $j : \prod_{i\in I}(X_i \times X_i) \longrightarrow \prod_{i\in I} X_i \times \prod_{i\in I} X_i$ denotes the canonical isomorphism (i.e. $j((x_i, y_i)) = ((x_i), (y_i))$) and $\prod_{i\in I} \mathcal{G}_i$ the product filter on $\prod_{i\in I}(X_i \times X_i)$, then $j(\prod \mathcal{G}_i)$ is a filter on $\prod X_i \times \prod X_i$ with $(p_i \times p_i)(j(\prod \mathcal{G}_i)) = \mathcal{G}_i$ for each $i \in I$. Thus $j(\prod \mathcal{G}_i) \in \mathcal{J}_X$. If $\tilde{j} : \prod_{i\in I} Y_i \times Y_i \longrightarrow \prod_{i\in I} Y_i \times \prod_{i\in I} Y_i$ denotes the canonical isomorphism, then

$$\tilde{j}^{-1}(\prod f_i \times \prod f_i(j(\prod \mathcal{G}_i)) \subset \prod(f_i \times f_i)(\mathcal{G}_i) \subset \prod(p'_i \times p'_i)(\mathcal{H}) \subset \tilde{j}^{-1}(\mathcal{H}).$$

Thus, $(\prod f_i \times \prod f_i)(j(\prod \mathcal{G}_i)) \subset \mathcal{H}$, i.e. $\mathcal{H} \in \mathcal{J}'_Y$.

④ **Fil** is a strong topological universe: Since **Fil** is a topological universe (cf. 3.1.9. ④ and 3.2.7. ④), it remains to prove that **Fil** fulfills CP_3). But this is obvious, because **Fil** is a bireflective and bicoreflective subconstuct of **SUConv** and ③ as well as 3.3.2. are valid.

⑤ **Conv** is strongly cartesian closed, but it is not a strong topological universe: Since **Conv** is cartesian closed (cf. 3.1.9. ⑤) and not extensional (cf. 3.2.7. ⑤c)), it remains to verify that **Conv** fulfills CP_3). But this is obvious, because **Conv** is bicoreflective in **SUConv** and closed under formation of products in **SUConv** (cf. 2.3.3.30.2)) [note: ③ and 3.3.2.].

3.3.4 Remarks. 1) There are other strong topological universes, e.g. **SULim**, **GConv**, **KConv**, **Lim** and **PsTop** (exercise!).

2) a) The category **Chy** of Cauchy spaces (and Cauchy continuous maps) is a cartesian closed topological construct which fulfills neither CP_2) nor CP_3) (cf. [12]). Note that a filter space (X, γ) is called a *Cauchy space* provided that the following is satisfied:

(C) If $\mathcal{F}$ and $\mathcal{G}$ belong to γ such that every member of $\mathcal{F}$ meets every member of $\mathcal{G}$ (i.e. $\sup\{\mathcal{F}, \mathcal{G}\}$ exists), then $\mathcal{F} \cap \mathcal{G}$ belongs to γ.

The underlying filter space $(X, \gamma_{\mathcal{J}_X})$ of a uniform limit space $(X, \mathcal{J}_X)$ is a Cauchy space, in particular the underlying filter space of a uniform space is a Cauchy space: Let $\mathcal{F}, \mathcal{G} \in \gamma_{\mathcal{J}_X}$ such that $\sup\{\mathcal{F}, \mathcal{G}\}$ exists. Consequently, $\mathcal{F} \times \mathcal{G} = (\mathcal{G} \times \mathcal{G}) \circ (\mathcal{F} \times \mathcal{F}) \in \mathcal{J}_X$ and $(\mathcal{F} \times \mathcal{G})^{-1} = \mathcal{G} \times \mathcal{F} \in \mathcal{J}_X$. Thus

$$(\mathcal{F} \cap \mathcal{G}) \times (\mathcal{F} \cap \mathcal{G}) = (\mathcal{F} \times \mathcal{F}) \cap (\mathcal{F} \times \mathcal{G}) \cap (\mathcal{G} \times \mathcal{F}) \cap (\mathcal{G} \times \mathcal{G}) \in \mathcal{J}_X,$$

i.e. $\mathcal{F} \cap \mathcal{G} \in \gamma_{\mathcal{J}_X}$.

b) The category **SChy** of semi–Cauchy spaces (and Cauchy continuous maps) is a topological universe which is not strong (cf. [113]). Note that a filter space (X, γ) is called a *semi–Cauchy space* provided that the following is satisfied:

(SC) A filter $\mathcal{F}$ on X belongs to γ whenever there exist finitely many $\mathcal{F}_1 \dots, \mathcal{F}_n \in \gamma$ such that $\mathcal{F} \times \mathcal{F} \supset \bigcap_{i=1}^{n} \mathcal{F}_i \times \mathcal{F}_i$.

The underlying filter space $(X, \gamma_{\mathcal{J}_X})$ of a semiuniform limit space $(X, \mathcal{J}_X)$ is a semi–Cauchy space: Let $\mathcal{F} \in F(X)$ such that there exist finitely many $\mathcal{F}_1, \dots, \mathcal{F}_n \in \gamma_{\mathcal{J}_X}$ with $\mathcal{F} \times \mathcal{F} \supset \bigcap_{i=1}^{n} \mathcal{F}_i \times \mathcal{F}_i$. Since $\mathcal{F}_i \times \mathcal{F}_i \in \mathcal{J}_X$ for each $i \in \{1, \dots, n\}$ and $(X, \mathcal{J}_X)$ fulfills UC_4) and UC_2), we obtain $\mathcal{F} \times \mathcal{F} \in \mathcal{J}_X$, i.e. $\mathcal{F} \in \gamma_{\mathcal{J}_X}$.

3) **Convenient Topology** *consists in the study of strong topological universes in which convergence structures and uniform convergence structures are available. Furthermore, such a strong topological universe should be easily described by means of suitable axioms and should not be too big. The strong topological universe* **SUConv** *of semiuniform convergence spaces fulfills these criteria. Thus, in the realm of Convenient Topology we are mainly concerned with semiuniform convergence spaces or, more exactly, with the study of* **SUConv***–invariants, i.e. properties of semiuniform convergence spaces which are preserved by isomorphisms in* **SUConv**. *This includes the study of full and isomorphism–closed subconstructs of* **SUConv** (such as **Unif** and $\mathbf{Top}_\mathbf{S}$ as well as all the other subconstructs of **SUConv** according to 2.3.3.31.).

Chapter 4

Completions of Semiuniform Convergence Spaces

The best known completion is the construction of the reals from the rationals. The procedure of G. Cantor [23] and Ch. Méray ([97] and [98]) has been generalized by F. Hausdorff [58] to metric spaces and later on by A. Weil [147] to (separated) uniform spaces. In this chapter the so–called Hausdorff completion of uniform spaces is introduced at first by means of categorical methods in order to emphasize its universal character. The next step is a more concrete construction of the Hausdorff completion of (separated) uniform spaces via natural function spaces in **SUConv**. More exactly, for a certain class of semiuniform convergence spaces, so–called u–embedded semiuniform convergence spaces, including separated uniform spaces, a completion is constructed, which is due to R.J. Gazik, D.C. Kent and G.D. Richardson [54]. It follows from the universal property of this Gazik-Kent–Richardson completion that it coincides with the Hausdorff completion provided that it is applied to separated uniform spaces. By the way, the theory of regular semiuniform convergence spaces, including uniform spaces and regular topological spaces (the latter were introduced by L. Vietoris [145]), is developed, where a useful extension theorem for uniformly continuous maps is proved. After an alternative description of uniform spaces by means of uniform covers due to W. Tukey [141], precompactness (= total boundedness) and compactness are introduced in the realm of semiuniform convergence spaces. In particular, a semiuniform convergence space is compact, iff it is precompact and weakly complete, where the concepts 'weakly complete' and 'complete' are equivalent for uniform limit spaces. Precompact uniform spaces are identified here with proximity spaces, since their study is equivalent to the study of proximity spaces in the sense of Efremovič ([39], [40], [41] and [42]), who axiomatized the concept of nearness between two sets. This equivalence has been proved in the papers of Ju.M. Smirnov ([133] and [134]), E.M. Alfsen and A.J. Fenstad [3], and I.S. Gál [52]. The first one who introduced 'proximity structures' was F. Riesz [126], but he did not develop a complete theory. Furthermore, Hausdorff compactifications (and extensions) of topological spaces are constructed by means of the Hausdorff completion, e.g. the Stone–Čech compactification, the Hewitt realcompactification, the Banascheswki compactification and the Alexandroff compactification. The decisive step in this direction goes back to Ju.M. Smirnov ([133] and [134]), who proved that there is a one–one correspondence between the Hausdorff compactifications and the separated proximity structures on a completely regular Hausdorff space. Concerning uniform limit spaces as a generalization of uniform spaces, O. Wyler [150] has constructed a completion characterized by a universal property. It will be considered in the last part of the present chapter. Furthermore, a completion for arbitrary semiuniform convergence spaces, namely the simple completion, is introduced. It can be described easily, and it turns out that the underlying uniform limit space of the simple completion of a separated uniform limit space is its Wyler completion, and that the underlying uniform space of the simple completion of a separated uniform space is its Hausdorff completion. The simple completion applied to filter spaces is a filter space completion, which has been introduced by A. Császár [34]. Correspondingly, the Wyler completion applied to Cauchy spaces (= underlying filter spaces of uniform limit spaces) is a Cauchy space completion characterized by means of a universal property. Cauchy spaces form a common generalization of proximity spaces and limit spaces satisfying a weakening of the axiom T_2, called T_{2W}. They have been introduced by H.H. Keller [81], who has characterized them axiomatically.

4.1 Completion of uniform spaces

4.1.1 Definition. A uniform space $(X, \mathcal{W})$ is called *separated* provided that $\bigcap_{W \in \mathcal{W}} W = \Delta$.

4.1.2 Proposition. *Let $(X, \mathcal{W})$ be a uniform space and $(X, \mathcal{X}_{\mathcal{W}})$ the underlying topological space* (cf. 2.3.3.18.). *Then the following are equivalent:*

(1) *$(X, \mathcal{W})$ is separated.*
(2) *$(X, \mathcal{X}_{\mathcal{W}})$ is a T_0–space.*
(3) *$(X, \mathcal{X}_{\mathcal{W}})$ is a T_1–space.*
(4) *$(X, \mathcal{X}_{\mathcal{W}})$ is a T_2–space.*
(5) *$(X, \mathcal{X}_{\mathcal{W}})$ is a regular Hausdorff space.*
(6) *$(X, \mathcal{X}_{\mathcal{W}})$ is a Tychonoff space.*

Proof. Since $(X, \mathcal{X}_{\mathcal{W}})$ is completely regular (cf. 2.3.3.15. and 2.3.3.18.), it suffices to prove that (1) and (2) are equivalent:

(1) $\Longrightarrow$ (2). Let $x, y \in X$ with $x \neq y$. Then $(x, y) \notin \Delta = \bigcap_{W \in \mathcal{W}} W$, i.e. there is some $W \in \mathcal{W}$ with $(x, y) \notin W$. Thus, $y \notin W(x) \in \mathcal{U}(x)$ (cf. also the proof of 2.3.3.15), i.e. $(X, \mathcal{X}_{\mathcal{W}})$ is a T_0–space.

(2) $\Longrightarrow$ (1). If (1) is not satisfied, then there are $x, y \in X$ such that $x \neq y$ and $(x, y) \in \bigcap_{W \in \mathcal{W}} W$. Consequently, y belongs to each neighborhood of x, namely to

each $W(x)$ with $W \in \mathcal{W}$. Thus, $(X, \mathcal{X}_{\mathcal{W}})$ is not a T_1–space and, since $(X, \mathcal{X}_{\mathcal{W}})$ is completely regular, it is also not a T_0–space.

4.1.3 Proposition. *Let* **SepUnif** *denote the construct of separated uniform spaces and uniformly continuous maps. Then a uniformly continuous map $f : (X, \mathcal{W}) \longrightarrow (Y, \mathcal{R})$ between separated uniform spaces is an epimorphism in* **SepUnif** *if and only if $f[X]$ is dense in $(Y, \mathcal{Y}_{\mathcal{R}})$ provided that $(Y, \mathcal{Y}_{\mathcal{R}})$ denotes the underlying topological space of $(Y, \mathcal{R})$.*

Proof. 1) "$\Longleftarrow$". If $\overline{f[X]} = Y$ and $\alpha, \beta : (Y, \mathcal{R}) \longrightarrow (Z, \mathcal{S})$ are uniformly continuous maps into a separated uniform space $(Z, \mathcal{S})$ such that $\alpha \circ f = \beta \circ f$, i.e. α and β coincide on the dense subset $f[X]$ of Y, then $\alpha = \beta$, since $(Y, \mathcal{Y}_{\mathcal{R}})$ is a T_2–space (cf. 4.1.2.). Consequently, f is an epimorphism in **SepUnif**.

2) "$\Longrightarrow$" (indirectly). If $f[X]$ is not dense in Y, then there are $y_0 \in Y$ and $V = V^{-1} \in \mathcal{R}$ such that $V(y_0) \cap f[X] = \emptyset$. For V, there is a pseudometric d_V on Y (cf. the theorem under 2.3.2.4.). Put $A = f[X]$. Then a uniformly continuous map $g_V : (Y, d_V) \longrightarrow ([0,1], d)$ is defined by $g_V(y) = d_V(y, A) = \inf\{d_V(y,z) : z \in A\}$ for each $y \in Y$ provided that d denotes the metric induced by the Euclidean metric of the real numbers (note: $|d_V(y', A) - d_V(y'', A)| \leq d_V(y', y'')$ for each $(y', y'') \in Y \times Y$). Obviously, g_V fulfills $g_V[A] = \{0\}$. If $\mathcal{D}_V$ denotes the uniformity induced by d_V, then $\mathcal{D}_V \subset \mathcal{R}$ and consequently $1_Y : (Y, \mathcal{R}) \longrightarrow (Y, \mathcal{D}_V)$ is uniformly continuous. Thus, $\alpha = g_V \circ 1_Y : (Y, \mathcal{R}) \longrightarrow ([0,1], \mathcal{D})$ is also uniformly continuous provided that $\mathcal{D}$ denotes the uniformity induced by d.
$\beta : (Y, \mathcal{R}) \longrightarrow ([0,1], \mathcal{D})$ defined by $\beta(y) = 0$ for each $y \in Y$ is a uniformly continuous map differing from α : since $\frac{1}{2} \leq d_V(y_0, z) \leq 1$ for each $z \in A$ (note that $\frac{1}{2} h_V(y_0, z) \leq d_V(y_0, z) \leq h_V(y_0, z)$ and since $A \cap V(y_0) = \emptyset$, $(y_0, z) \notin V$, i.e. $z \notin V(y_0)$, so that $h_V(y_0, z) = 1$), $\alpha(y_0) = g_V(y_0) = \inf\{d_V(y_0, z) : z \in A\} \neq 0$, i.e. $\alpha(y_0) \neq \beta(y_0) = 0$.

Since the uniformity $\mathcal{D}$ is separated, α and β are **SepUnif**–morphisms such that $\alpha \circ f = \beta \circ f$ (because $\alpha[A] = \beta[A] = \{0\}$). Consequently, f is not an epimorphism in **SepUnif**.

4.1.4 Proposition. 1) *The monomorphisms in* **SepUnif** *are the injective uniformly continuous maps.*

2) *The extremal monomorphisms in* **SepUnif** *are the closed embeddings, i.e. the subobjects in* **SepUnif** *are the closed subspaces.*

3) **SepUnif** *is (epi, extremal mono)–factorizable.*

4) **SepUnif** *is co–well–powered.*

5) **SepUnif** *has products.*

Proof. 1) is proved analogously to the corresponding fact in topological constructs.

2) a) Let $f : X \longrightarrow Y$ be an extremal monomorphism in **SepUnif**. Consider the factorization $f = h \circ g$, where $g : X \longrightarrow \overline{f[X]}$ is defined by $g(x) = f(x)$ for

each $x \in X$ and $h : \overline{f[X]} \longrightarrow Y$ is the inclusion map. Then the epimorphism g must be an isomorphism and consequently, $f[X] = g[X] = \overline{f[X]}$.

b) Let $f : X \longrightarrow Y$ be a closed embedding between the separated uniform spaces X and Y, i.e. $f' : X \longrightarrow f[X]$ defined by $f'(x) = f(x)$ for each $x \in X$ is an isomorphism and $f[X] = \overline{f[X]}$. If $i : f[X] \longrightarrow Y$ denotes the inclusion map, then $f = i \circ f'$ is a monomorphism as a composition of two monomorphisms. Let $f = h \circ g$ be a factorization, where $g : X \longrightarrow Z$ is an epimorphism. Since $h[Z] = h[\overline{g[X]}] \subset \overline{h[g[X]]} = \overline{f[X]} = f[X]$ [continuity of h!], a morphism $h' : Z \longrightarrow f[X]$ is defined by $i \circ h' = h$. Obviously, $f' = h' \circ g$. Thus, since f' is an extremal monomorphism, g has to be an isomorphism. Consequently, f is an extremal monomorphism.

3) Let $f : X \longrightarrow Y$ be a **SepUnif**–morphism. Let $g : X \longrightarrow \overline{f[X]}$ be defined by $g(x) = f(x)$ for each $x \in X$ and let $h : \overline{f[X]} \longrightarrow Y$ be the inclusion map. Then $f = h \circ g$ is the desired (epi, extremal mono)–factorization of f.

4) A) If k is a cardinal number, then there is a set $\mathcal{Q}$ of uniform spaces such that every uniform space $(Y, \mathcal{R})$ with $|Y| \leq k$ is isomorphic to a space of $\mathcal{Q}$ (cf. part A) of the proof of 1.2.2.9.).
B) If $f : X \longrightarrow Y$ is an epimorphism in **SepUnif**, then by 4.1.3. $\overline{f[X]} = Y$. Thus, for each $y \in Y$, the trace of the neighborhood filter $\mathcal{U}(y)$ on $f[X]$, i.e. $i^{-1}(\mathcal{U}(y))$ [$i : f[X] \longrightarrow Y$ inclusion map], exists. Hence, a map $g : Y \longrightarrow \mathcal{P}(\mathcal{P}(f[X]))$ is defined by $g(y) = i^{-1}(\mathcal{U}(y))$ for each $y \in Y$. Obviously, g is injective (let $g(y_1) = g(y_2)$ for $y_1, y_2 \in Y$, i.e. $i^{-1}(\mathcal{U}(y_1)) = i^{-1}(\mathcal{U}(y_2))$; then $\mathcal{U}(y_1) \subset i(i^{-1}(\mathcal{U}(y_1))) = i(i^{-1}(\mathcal{U}(y_2))) \supset \mathcal{U}(y_2)$ and consequently $y_1 = y_2$, since each filter on Y converges to at most one point).
Since there is an injective map from $f[X]$ to X (which assigns to each $y \in f[X]$ a unique $x \in f^{-1}(y)$), there is an injective map $h : \mathcal{P}(\mathcal{P}(f[X])) \longrightarrow \mathcal{P}(\mathcal{P}(X))$ (note: for every injective map j from a set M to a set N, an injective map $j^* : \mathcal{P}(M) \longrightarrow \mathcal{P}(N)$ can be defined by $j^*(A) = j[A]$ for each $A \in \mathcal{P}(M)$). Hence the map $h \circ g : Y \longrightarrow \mathcal{P}(\mathcal{P}(X))$ is injective, i.e. $|Y| \leq k = |\mathcal{P}(\mathcal{P}(X))| = 2^{2^{|X|}}$. Consequently, by A), there is a representative set of epimorphisms with domain X.

5) It suffices to verify that products of separated uniform spaces formed in **Unif** are separated. But this is obvious, since initial uniformities induce initial topologies and in **Top** products of Hausdorff spaces are Hausdorff spaces.

4.1.5 Theorem. *The construct* **CSepUnif** *of complete separated uniform spaces (and uniformly continuous maps) is epireflective in* **SepUnif** *and reflective in* **Unif**.

Proof. 1) **SepUnif** is epireflective in **Unif** : Since **Unif** is a topological construct, it fulfills the assumptions of the characterization theorem for epireflective subconstructs (cf. 2.2.4.). Obviously, **SepUnif** is closed under formation of products and subspaces in **Unif** (since initial uniformities induce initial topologies and in **Top** products and subspaces of Hausdorff spaces are Hausdorff spaces).

Thus, **SepUnif** is epireflective in **Unif**.

2) **CSepUnif** is epireflective in **SepUnif** : By 4.1.4. 3), 4) and 5) **SepUnif** fulfills the assumptions of the characterization theorem for epireflective subcategories. Products and closed subspaces of complete uniform spaces are complete uniform spaces (cf. 2.3.3.28 and note that initial structures in **Unif** are formed as in **SUConv**). Thus, together with 1), it follows that **CSepUnif** is closed under formation of products and subobjects (= closed subspaces) in **SepUnif**. Consequently, **CSepUnif** is epireflective in **SepUnif**.

3) It follows from 1) and 2) that **CSepUnif** is reflective in **Unif**.

4.1.6 Definitions. 1) If $(X, \mathcal{V})$ is a separated uniform space and $s_X : (X, \mathcal{V}) \longrightarrow (X', \mathcal{V}')$ the epireflection of $(X, \mathcal{V})$ w.r.t. **CSepUnif**, then $(X', \mathcal{V}')$ is called the *complete hull of* $(X, \mathcal{V})$.

2) If $(X, \mathcal{V})$ is a uniform space and $r_X : (X, \mathcal{V}) \longrightarrow (X^*, \mathcal{V}^*)$ the reflection of $(X, \mathcal{V})$ w.r.t. **CSepUnif**, then $(X^*, \mathcal{V}^*)$ is called *the Hausdorff completion of* $(X, \mathcal{V})$.

4.2 Regular completion of semiuniform convergence spaces

4.2.1 Regular spaces

4.2.1.1 Definitions. 1) A semiuniform convergence space $(X, \mathcal{J}_X)$ is called *regular* provided that for each $\mathcal{F} \in \mathcal{J}_X$ the subfilter $\overline{\mathcal{F}}$ generated by the filter base $\{\overline{F} : F \in \mathcal{F}\}$ belongs to $\mathcal{J}_X$, where $\overline{F}$ denotes the closure of F in the underlying Kent convergence space of the product space $(X, \mathcal{J}_X) \times (X, \mathcal{J}_X)$.

2) A generalized convergence space (X, q) is called *regular* provided that $(\overline{\mathcal{F}}, x) \in q$ for each $(\mathcal{F}, x) \in q$, where $\overline{\mathcal{F}} = (\{\overline{F} : F \in \mathcal{F}\})$ and $\overline{F}$ denotes the closure of F, i.e. $\overline{F} = cl_q F$ (cf. 2.3.1.6. 1) d) for the definition of cl_q).

4.2.1.2 Proposition. *Let $(X, \mathcal{J}_X)$ be a regular semiuniform convergence space. Then the underlying (symmetric) Kent convergence space $(X, q_{\gamma_{\mathcal{J}_X}})$ is regular.*

Proof. Let $(\mathcal{F}, x) \in q_{\gamma_{\mathcal{J}_X}}$, i.e. $(\mathcal{F} \cap \dot{x}) \times (\mathcal{F} \cap \dot{x}) \in \mathcal{J}_X$. Then $\overline{(\mathcal{F} \cap \dot{x}) \times (\mathcal{F} \cap \dot{x})} = \overline{(\mathcal{F} \cap \dot{x})} \times \overline{(\mathcal{F} \cap \dot{x})} \subset (\overline{\mathcal{F}} \cap \dot{x}) \times (\overline{\mathcal{F}} \cap \dot{x})$ and consequently, $(\overline{\mathcal{F}} \cap \dot{x}) \times (\overline{\mathcal{F}} \cap \dot{x}) \in \mathcal{J}_X$, i.e. $(\overline{\mathcal{F}}, x) \in q_{\gamma_{\mathcal{J}_X}}$, since $\overline{(\mathcal{F} \cap \dot{x}) \times (\mathcal{F} \cap \dot{x})} \in \mathcal{J}_X$ [note: $H \in \overline{\mathcal{F} \cap \dot{x}} \Longrightarrow H \supset \overline{F \cup \{x\}} = \overline{F} \cup \overline{\{x\}} \supset \overline{F} \cup \{x\}$, i.e. $H \in \overline{\mathcal{F}} \cap \dot{x}$].

4.2.1.3 Corollaries. 1) *Let (X, q) be a symmetric Kent convergence space and $(X, \mathcal{J}_{\gamma_q})$ the corresponding semiuniform convergence space. Then (X, q) is regular (in the sense of* 4.2.1.1. 2)) *if and only if $(X, \mathcal{J}_{\gamma_q})$ is regular (in the sense* *of* 4.2.1.1. 1)).

2) *Let* $(X, \mathcal{X})$ *be a symmetric topological space and* $(X, \mathcal{J}_{\gamma_{q_\mathcal{X}}})$ *the corresponding semiuniform convergence space. Then* $(X, \mathcal{X})$ *is regular (in the usual sense) if and only if* $(X, \mathcal{J}_{\gamma_{q_\mathcal{X}}})$ *is regular (in the sense of* 4.2.1.1. 1)).

Proof. 1) a) "$\Longrightarrow$". Let $\mathcal{F} \in \mathcal{J}_{\gamma_q}$. Then there is some $(\mathcal{G}, x) \in q$ such that $\mathcal{G} \times \mathcal{G} \subset \mathcal{F}$. Since (X, q) is regular, $(\overline{\mathcal{G}}, x) \in q$. Consequently $\overline{\mathcal{G}} \times \overline{\mathcal{G}} = \overline{\mathcal{G} \times \mathcal{G}} \subset \overline{\mathcal{F}}$, i.e. $\overline{\mathcal{F}} \in \mathcal{J}_{\gamma_q}$.

b) "$\Longleftarrow$". If $(X, \mathcal{J}_{\gamma_q})$ is regular, then (X, q) is regular by 4.2.1.2., since $q_{\gamma_{\mathcal{J}_{\gamma_q}}} = q$ (cf. 2.3.3.21.).

2) Obviously, $(X, \mathcal{X})$ is regular (in the usual sense) iff $(X, q_\mathcal{X})$ is regular. Thus, the desired results follows from 1).

4.2.1.4. In order to answer the question whether each uniform space is regular, we need the following

Proposition. *Let* $(X, \mathcal{W})$ *be a uniform space. Then the closed*[1] *(resp. open*[1]*), symmetric entourages form a base for* $\mathcal{W}$.

Proof. 1) Let $W \in \mathcal{W}$ and $V = V^{-1} \in \mathcal{W}$ such that $V^3 \subset W$. Then

$$V \subset \overline{V} \subset V^3,$$

namely $(x, y) \in \overline{V}$ implies $(V(x) \times V(y)) \cap V \neq \emptyset$ (since $V(x) \times V(y)$ is a neighborhood of (x, y) in $X \times X$); thus, there is $(x', y') \in V$ such that $x' \in V(x)$, i.e. $(x, x') \in V$, and $y' \in V(y)$, i.e. $(y, y') \in V$ and consequently, $(x, y) \in V^3$, since V is symmetric. Furthermore, $\overline{V}$ is symmetric:

a) $\overline{V}^{-1} \subset \overline{V}$: If $(x, y) \in \overline{V}^{-1}$, then $(y, x) \in \overline{V}$. Thus, for each open neighborhood O_x of x and each open neighborhood O_y of y, $(O_y \times O_x) \cap V \neq \emptyset$ and consequently, $((O_y \times O_x) \cap V)^{-1} = (O_x \times O_y) \cap V^{-1} = (O_x \times O_y) \cap V \neq \emptyset$, i.e. $(x, y) \in \overline{V}$.

b) $\overline{V} \subset \overline{V}^{-1}$ is proved analogously to a).
Obviously, $\overline{V} \in \mathcal{W}$. Hence, $\overline{V}$ is the desired closed symmetric entourage which is contained in W.

2) Since the intersection of an open subset of $X \times X$ with its inverse is an open symmetric subset of $X \times X$, it suffices to prove that the open entourages form a base for $\mathcal{W}$: Let $W \in \mathcal{W}$ and choose some $V = V^{-1} \in \mathcal{W}$ such that $V^3 \subset W$. Then $V \subset W^0$; namely if $(x, y) \in V$, then $V(x) \times V(y) \subset V^3 \subset W$. Hence, $W^0 \in \mathcal{W}$ and consequently, the assertion is proved.

4.2.1.5 Proposition. *Every uniform space is regular.*

Proof. Let $(X, \mathcal{J}_X)$ be a principal uniform limit space (= uniform space), i.e. $\mathcal{J}_X = [\mathcal{W}]$, where $\mathcal{W}$ is a uniformity on X. If $\mathcal{F} \in \mathcal{J}_X$, i.e. $\mathcal{F} \supset \mathcal{W}$, then

[1] with respect to the underlying topological space of the product space $(X, \mathcal{W}) \times (X, \mathcal{W})$

$\overline{\mathcal{W}} = \mathcal{W} \subset \overline{\mathcal{F}}$, i.e. $\overline{\mathcal{F}} \in \mathcal{J}_X$ (note that in $(X, \mathcal{W})$ the entourages being closed in $(X, \mathcal{X}_\mathcal{W}) \times (X, \mathcal{X}_\mathcal{W})$ form a base for $\mathcal{W}$ [cf. 4.2.1.4.]).

4.2.1.6 Proposition. *Let $((X_i, \mathcal{J}_{X_i}))_{i\in I}$ be a family of regular semiuniform convergence spaces, X a set and $(f_i : X \longrightarrow X_i)_{i\in I}$ a family of maps. If $\mathcal{J}_X$ denotes the initial* **SUConv**-*structure on X w.r.t. $(f_i)_{i\in I}$, then $(X, \mathcal{J}_X)$ is regular.*

Proof. Let $\mathcal{F} \in \mathcal{J}_X$. Then $(f_i \times f_i)(\mathcal{F}) \in \mathcal{J}_{X_i}$ for each $i \in I$. By assumption, $\overline{(f_i \times f_i(\mathcal{F}))} \in \mathcal{J}_{X_i}$ for each $i \in I$. Furthermore, for each $i \in I$, $\overline{(f_i \times f_i(\mathcal{F}))} \subset (f_i \times f_i)(\overline{\mathcal{F}})$ since $f_i \times f_i$ is a continuous map w.r.t. the induced closure spaces (apply first the bicoreflector **SUConv** $\longrightarrow$ **KConv**$_\mathbf{S}$ and then the restriction of the bireflector **KConv** $\longrightarrow$ **PrTop** to **KConv**$_\mathbf{S}$ to the uniformly continuous map $f_i \times f_i : (X, \mathcal{J}_X) \times (X, \mathcal{J}_X) \longrightarrow (X_i, \mathcal{J}_{X_i}) \times (X_i, \mathcal{J}_{X_i})$). Thus, $(f_i \times f_i)(\overline{\mathcal{F}}) \in \mathcal{J}_{X_i}$ for each $i \in I$. Consequently, $\overline{\mathcal{F}} \in \mathcal{J}_X$.

4.2.1.7 Corollary. *The construct* **Reg** *of regular semiuniform convergence spaces (and uniformly continuous maps) is a bireflective (full and isomorphism-closed) subconstruct of* **SUConv** *and thus a topological construct.*

4.2.1.8 Regularity plays an essential role in the realm of extensions of (uniformly) continuous maps as we will see later on. In order to prove the uniqueness of such extensions we need the following

Definitions. 1) A generalized convergence space (X, q) is called a T_2*-space* (or *separated*) provided that each filter $\mathcal{F}$ on X converges to at most one element of X.

2) A semiuniform convergence space $(X, \mathcal{J}_X)$ is called a T_2*-space* (or *separated*) provided that $(X, q_{\gamma_{\mathcal{J}_X}})$ is separated.

3) If (X, q) is a generalized convergence space, then a subset A of X is called *dense* (resp. *topologically dense*) provided that $X = cl_q A$ (resp. $X = \overline{A}^t$, where $\overline{A}^t = \bigcap\{B : B = cl_q B \supset A\}$).

4.2.1.9 Proposition. *Let (X, q), (X', q') be generalized convergence spaces and A a topologically dense subset of X. If (X', q') is a T_2-space and $f, g : (X, q) \longrightarrow (X', q')$ are continuous maps such that $f|A = g|A$, then $f = g$.*

Proof. Let $K = \{x : f(x) = g(x)\}$. Then $K = cl_q K$: If $x \in cl_q K$, then there is some $\mathcal{F} \in F(X)$ such that $(\mathcal{F}, x) \in q$ and $K \in \mathcal{F}$. Since f, g are continuous maps, $(f(\mathcal{F}), f(x)) \in q'$ and $(g(\mathcal{F}), g(x)) \in q'$. For each $F \in \mathcal{F}$, $F \cap K \in \mathcal{F}$ and $g[F \cap K] = f[F \cap K]$. Thus, $f(\mathcal{F}) = g(\mathcal{F})$. Since (X', q') is a T_2-space, we obtain $f(x) = g(x)$, i.e. $x \in K$. Consequently, $cl_q K \subset K$ and since $K \subset cl_q K$ is always valid, $K = cl_q K$.
Since $K = cl_q K$ is equivalent to $K = \overline{K}^t$, it follows immediately from $A \subset K \subset X$ that $X = \overline{A}^t \subset \overline{K}^t = K \subset X$, i.e. $K = X$. Thus, $f = g$.

4.2.1.10 Corollary. *Let $(X,q),(X',q')$ be generalized convergence spaces and A a $\mathcal{F} \in F(X)$ with $(\mathcal{F},x) \in q$ and dense subset of X. If (X',q') is a T_2-space and $f,g:(X,q) \longrightarrow (X',q')$ are continuous maps such that $f|A = g|A$, then $f=g$.*

Proof. Since $X = cl_q A$ implies $X = \overline{A}^t$, the corollary follows from 4.2.1.9. ($x \in cl_q A$ implies the existence of some $\mathcal{G} \in F(X)$ with $(\mathcal{G},x) \in q$ and $A \in \mathcal{G}$. If x would not belong to $\overline{A}^t$, there were some $B = cl_q B \supset A$ with $x \notin B$. Thus, $x \in X \setminus B$ and $X \setminus B \in \mathcal{X}_q$. Consequently, $X \setminus B \in \mathcal{G}$ and therefore $X \setminus A \in \mathcal{G}$ – a contradiction. Hence, $cl_q A = X \subset \overline{A}^t \subset X$, i.e. $X = \overline{A}^t$).

4.2.1.11 Theorem. *Let (X,q) be a topological space (= topologically pretopological space), (Y,r) a regular generalized convergence space and A a dense subset of X. If $f: A \longrightarrow Y$ is a continuous map, then there is a continuous extension of f to X, i.e. a continuous map $g: X \longrightarrow Y$ such that $g|A = f$, if and only if for each $x \in X$, there is some $y \in Y$ such that for each $\mathcal{F} \in F(X)$ with $(\mathcal{F},x) \in q$ and $A \in \mathcal{F}$, $(f(i^{-1}(\mathcal{F})),y) \in r$, where $i: A \longrightarrow X$ denotes the inclusion map. If (Y,r) is a T_2-space, then g is uniquely determined.*

Proof. 1) For each $x \in X \setminus A$ choose exactly one $y_x \in \{y \in Y : (f(i^{-1}(\mathcal{F})),y) \in r$ for each $\mathcal{F} \in F(X)$ with $(\mathcal{F},x) \in q$ and $A \in \mathcal{F}\}$. Define $g: X \longrightarrow Y$ by

$$g(x) = \begin{cases} f(x) & \text{if } x \in A \\ y_x & \text{if } x \in X \setminus A\,. \end{cases}$$

In order to prove that q is continuous, let $(\mathcal{F},x) \in q$, i.e. $\mathcal{F} \supset \mathcal{U}_q(x)$. Since (X,q) is a topological space,

$$\mathcal{U}_q(x) = \sup\{\bigcap_{y \in U_x} \mathcal{U}_q(y) : U_x \in \mathcal{U}_q(x)\}\,.$$

Since A is dense in X, $i^{-1}(\mathcal{U}_q(x))$ exists. Since $i^{-1}(\mathcal{U}_q(x)) = \sup\{\bigcap_{y \in U_x} i^{-1}(\mathcal{U}_q(y)) : U_x \in \mathcal{U}_q(x)\}$, it is easily checked that

$$(*) \qquad f(i^{-1}(\mathcal{U}_q(x)) = \sup\{\bigcap_{y \in U_x} f(i^{-1}(\mathcal{U}_q(y))) : U_x \in \mathcal{U}_q(x)\}\,.$$

If $x \in A$, $(f(i^{-1}(\mathcal{U}_q(x))), f(x) = g(x)) \in r$ because f is continuous and $(i^{-1}(\mathcal{U}_q(x)),x) \in q_A$, where $(\mathcal{H},z) \in q_A$ iff $(i(\mathcal{H}),i(z)) \in q$. If $x \notin A$, i.e. $x \in X \setminus A$, $(f(i^{-1}(\mathcal{U}_q(x)),g(x)) \in r$ by definition of g (note: $\mathcal{F} = i(i^{-1}(\mathcal{U}_q(x))) \supset \mathcal{U}_q(x)$, i.e. $(\mathcal{F},x) \in q$, $A \in \mathcal{F}$ and $i^{-1}(\mathcal{F}) = i^{-1}(\mathcal{U}_q(x))$). Thus, for each $x \in X$, $(f(i^{-1}(\mathcal{U}_q(x))),g(x)) \in r$. For each $H \in f(i^{-1}(\mathcal{U}_q(x)))$, there is some $U_x \in \mathcal{U}_q(x)$ such that $H \in \bigcap_{y \in U_x} f(i^{-1}(\mathcal{U}_q(y)))$ [note $(*)$ and remember that $\mathcal{U}_q(x)$ is a filter]. Consequently,

$$g[U_x] \subset \overline{H}$$

[$y \in U_x$ implies $g(y) \in \overline{H}$ since $(f(i^{-1}(\mathcal{U}_q(y))),g(y)) \in r$ and $H \in f(i^{-1}(\mathcal{U}_q(y)))$].

Thus, $\overline{f(i^{-1}(\mathcal{U}_q(x)))} \subset g(\mathcal{U}_q(x))$. Since (Y,r) is regular, $(\overline{f(i^{-1}(\mathcal{U}_q(x)))}, g(x)) \in r$ Thus, $(g(\mathcal{U}_q(x)), g(x)) \in r$ and consequently, $(g(\mathcal{F}), g(x)) \in r$.

2) Let f have a continuous extension g, i.e. $g \circ i = f$. Furthermore, let $x \in X$ and $\mathcal{F} \in F(X)$ such that $(\mathcal{F}, x) \in q$ and $A \in \mathcal{F}$. Then $i^{-1}(\mathcal{F})$ exists and $f(i^{-1}(\mathcal{F})) = g(i(i^{-1}(\mathcal{F}))) \supset g(\mathcal{F})$. Since g is continuous, $(g(\mathcal{F}), g(x)) \in r$ and consequently, $(f(i^{-1}(\mathcal{F})), g(x) = y) \in r$.

3) If (Y,r) is a T_2–space, then the uniqueness of g follows from 4.2.1.9.

4.2.1.12 Remark. In the above theorem the topological space (X,q) may be substituted by a generalized convergence space (X,q) fulfilling the following *diagonal property*:

(DP) $(\mathcal{F}, x) \in q$ for some $x \in X$ and $(\mathcal{F}_y, y) \in q$ for each $y \in X$ imply $(\sup\{\bigcap_{y\in F} \mathcal{F}_y : F \in \mathcal{F}\}, x) \in q$

(Note that the proof of this assertion is similar to the proof of 4.2.1.11. provided that the filter $\mathcal{U}_q(x)$ belonging to $(\mathcal{F}, x) \in q$ is substituted by the following filter $\mathcal{G}$:
Since $\overline{A} = X$, i.e. $cl_q A = X$, there is $(\mathcal{F}_y, y) \in q$ with $A \in \mathcal{F}_y$ for each $y \in X$. If $\mathcal{G} = \sup\{\bigcap_{y\in F} \mathcal{F}_y : F \in \mathcal{F}\}$, then by (DP), $(\mathcal{G}, x) \in q$. Since $A \in \mathcal{G}$, $i^{-1}(\mathcal{G})$ exists and so on).

4.2.1.13 Theorem. *Let $(X, \mathcal{J}_X)$ be a principal uniform limit space (= uniform space) and $(Y, \mathcal{J}_Y)$ a complete, separated and regular semiuniform convergence space. If A is a dense subset of X and $f : A \longrightarrow Y$ a uniformly continuous map, then there is a unique uniformly continuous extension $g : X \longrightarrow Y$ of f.*

Proof. For each $x \in X$, there is some $\mathcal{F} \in F(X)$ such that $(\mathcal{F}, x) \in q_{\gamma\mathcal{J}_X}$ and $A \in \mathcal{F}$, since A is dense in X. If $i : A \longrightarrow X$ denotes the inclusion map, then $i^{-1}(\mathcal{F})$ exists and it is a Cauchy filter. Consequently, $f(i^{-1}(\mathcal{F}))$ is a Cauchy filter. Since $(Y, \mathcal{J}_Y)$ is complete, there is some $y \in Y$ such that $f(i^{-1}(\mathcal{F})) \longrightarrow y$. If $\mathcal{G} \in F(X)$ with $(\mathcal{G}, x) \in q_{\gamma\mathcal{J}_X}$ and $A \in \mathcal{G}$, then $i^{-1}(\mathcal{G})$ exists, and it is a Cauchy filter. Thus $f(i^{-1}(\mathcal{G})) \longrightarrow y'$ for some $y' \in Y$. Since $(\mathcal{F} \cap \mathcal{G}, x) \in q_{\gamma\mathcal{J}_X}$, $\mathcal{F} \cap \mathcal{G}$ is a Cauchy filter. Because of $A \in \mathcal{F} \cap \mathcal{G}$, $i^{-1}(\mathcal{F} \cap \mathcal{G}) = i^{-1}(\mathcal{F}) \cap i^{-1}(\mathcal{G})$ exists, and it is a Cauchy filter. Consequently, $f(i^{-1}(\mathcal{F})) \cap f(i^{-1}(\mathcal{G})) = f(i^{-1}(\mathcal{F}) \cap i^{-1}(\mathcal{G}))$ converges to some $y'' \in Y$, i.e. $f(i^{-1}(\mathcal{F})) \longrightarrow y''$ and $f(i^{-1}(\mathcal{G})) \longrightarrow y''$. Since $(Y, \mathcal{J}_Y)$ is separated, $y'' = y$ and $y'' = y'$, i.e. y is uniquely determined. Furthermore, $(Y, q_{\gamma\mathcal{J}_Y})$ is regular (cf. 4.2.1.2.) and separated. Thus, the assumptions of 4.2.1.11. are fulfilled and the (uniformly) continuous map $f : A \longrightarrow Y$ has a unique continuous extension $g : X \longrightarrow Y$. It remains to prove that g is uniformly continuous: Let $\mathcal{J}_X = [\mathcal{W}]$, where $\mathcal{W}$ is a uniformity on X. It suffices to verify that $g \times g(\mathcal{W}) \in \mathcal{J}_Y$. Let $U \in \mathcal{W}$. By 4.2.1.4. there is some open symmetric $V \in \mathcal{W}$ such that $V \subset U$. Then

$$(*) \quad V = V \cap (X \times X) = V \cap \overline{A \times A} \subset \overline{U \cap (A \times A)};$$

namely if $z \in V \subset \overline{A \times A} = X \times X$, there is some $\mathcal{F} \in F(X \times X)$ such that

$\mathcal{F} \longrightarrow z$ and $A \times A \in \mathcal{F}$; since U is a neighborhood of z, $U \in \mathcal{F}$, and consequently, $U \cap (A \times A) \in \mathcal{F}$, i.e. $z \in \overline{U \cap (A \times A)}$.

It follows from (*) that

$$\overline{\mathcal{W}_A} \subset \mathcal{W}$$

where $\mathcal{W}_A$ denotes the initial uniformity on A w.r.t. the inclusion map $i : A \longrightarrow (X, \mathcal{W})$. Since g is continuous, $g \times g$ is also continuous, and thus, for each $W \in \mathcal{W}$,

$$g \times g\left[\overline{W \cap (A \times A)}\right] \subset \overline{g \times g\left[W \cap (A \times A)\right]} = \overline{f \times f\left[W \cap (A \times A)\right]},$$

i.e. $\overline{f \times f(\mathcal{W}_A)} \subset g \times g(\overline{\mathcal{W}_A}) \subset g \times g(\mathcal{W})$. Since f is uniformly continuous, $f \times f(\mathcal{W}_A) \in \mathcal{J}_Y$, and since $(Y, \mathcal{J}_Y)$ is regular, $\overline{f \times f(\mathcal{W}_A)} \in \mathcal{J}_Y$, which implies $g \times g(\mathcal{W}) \in \mathcal{J}_Y$.

4.2.1.14 Remark. Let (X, q) be a generalized convergence space and let A, U be subsets of X. Then U is called a *neighborhood* of A provided that $U \in \mathcal{U}_q(x)$ for each $x \in A$. For each $\mathcal{F} \in F(X)$, $\mathcal{U}_q(\mathcal{F}) = \{U \subset X : U$ is a neighborhood of some $F \in \mathcal{F}\}$ is called the *neighborhood filter* of $\mathcal{F}$. A semiuniform convergence space $(X, \mathcal{J}_X)$ fulfills the *neighborhood condition* provided that the following is satisfied:

(NC) For each $\mathcal{F} \in \mathcal{J}_X$, $\mathcal{U}_q(\mathcal{F}) \in \mathcal{J}_X$, where $(X \times X, q)$ is the underlying Kent convergence space of the product space $(X, \mathcal{J}_X) \times (X, \mathcal{J}_X)$.

If $(X, \mathcal{J}_X) \in |\mathbf{SUConv}|$ fulfills the neighborhood condition, then $(X, q_{\gamma_{\mathcal{J}_X}})$ has the diagonal property (cf. 4.2.1.12.). In the above theorem the uniform space $(X, \mathcal{J}_X)$ may be substituted by a semiuniform convergence space $(X, \mathcal{J}_X)$ fulfilling the neighborhood condition (Note, that each uniform space fulfills the neighborhood condition since 4.2.1.4. is valid).
The proof parallels the corresponding proof for uniform limit spaces in [132].

4.2.2 Regular completion via function spaces

4.2.2.1 Proposition. *Let* $\mathbf{X} = (X, \mathcal{J}_X)$ *be a semiuniform convergence space and* $\mathbf{Y} = (Y, \mathcal{J}_Y)$ *a regular semiuniform limit space. Then the natural function space* $([\mathbf{X}, \mathbf{Y}], \mathcal{J}_{X,Y})$ *in* **SUConv** *is a complete semiuniform limit space provided that* $\mathbf{Y}$ *is complete.*

Proof. Obviously, $([\mathbf{X}, \mathbf{Y}], \mathcal{J}_{X,Y})$ is a semiuniform limit space, since $\mathbf{Y}$ is a semiuniform limit space (note: $\phi \cap \psi(\mathcal{F}) = \phi(\mathcal{F}) \cap \psi(\mathcal{F})$ provided that $\phi, \psi \in \mathcal{J}_{X,Y}$ and $\mathcal{F} \in \mathcal{J}_X$). Furthermore, if $(Z, \mathcal{J}_Z)$ is a semiuniform limit space, then a $\mathcal{J}_Z$–Cauchy filter $\mathcal{F}$ on Z converges to $z \in Z$ in $(Z, q_{\gamma_{\mathcal{J}_Z}})$ iff $\mathcal{F} \times \dot{z} \in \mathcal{J}_Z$ (note: $(\mathcal{F} \cap \dot{z}) \times (\mathcal{F} \cap \dot{z}) = (\mathcal{F} \times \mathcal{F}) \cap (\mathcal{F} \times \dot{z}) \cap (\dot{z} \times \mathcal{F}) \cap (\dot{z} \times \dot{z})$).
Let $\mathcal{F}$ be a $\mathcal{J}_{X,Y}$–Cauchy filter. Since $\dot{x} \times \dot{x} \in \mathcal{J}_X$ for each $x \in X$, $\mathcal{F} \times \mathcal{F}(\dot{x} \times \dot{x}) = \mathcal{F}(\dot{x}) \times \mathcal{F}(\dot{x}) \in \mathcal{J}_Y$, i.e. $\mathcal{F}(\dot{x})$ is a $\mathcal{J}_Y$–Cauchy filter. $(Y, \mathcal{J}_Y)$ being complete, $\mathcal{F}(\dot{x})$ converges in $(Y, q_{\gamma_{\mathcal{J}_Y}})$. Let $f(x)$ be a limit of $\mathcal{F}(\dot{x})$. First we show that f is uniformly continuous. Let $\mathcal{G} \in \mathcal{J}_X$ and denote $(\mathcal{F} \times \mathcal{F})(\mathcal{G})$ by $\mathcal{H}$. Since

$\mathcal{F} \times \mathcal{F} \in \mathcal{J}_{X,Y}$, $\mathcal{H} \in \mathcal{J}_Y$. For each $F \in \mathcal{F}$, $f(x) \in \overline{F(x)}$, because $\mathcal{F}(\dot{x}) = (\{F(x) : F \in \mathcal{F}\}) \longrightarrow f(x)$. Thus, for every $(x, y) \in A \in \mathcal{G}$,

$$(f(x), f(y)) \in \overline{F(x)} \times \overline{F(y)} = \overline{F(x) \times F(y)} = \overline{(F \times F)(x, y)} \subset \overline{(F \times F)(A)} \in \overline{\mathcal{H}},$$

i.e. $(f \times f)[A] \subset \overline{(F \times F)(A)}$, which implies $\overline{\mathcal{H}} \subset (f \times f)(\mathcal{G})$. Since $(Y, \mathcal{J}_Y)$ is regular, $\overline{\mathcal{H}} \in \mathcal{J}_Y$ and consequently, $(f \times f)(\mathcal{G}) \in \mathcal{J}_Y$, i.e. f is uniformly continuous. It remains to be shown that $\mathcal{F} \times \dot{f} \in \mathcal{J}_{X,Y}$. Obviously, for every $(x, y) \in A \in \mathcal{G}$,

$$F(x) \times \{f(y)\} \subset F(x) \times \overline{F(y)} \subset \overline{F(x)} \times \overline{F(y)} \subset \overline{(F \times F)(A)}.$$

Thus, $(F \times \{f\})(A) \subset \overline{F \times F(A)}$ and consequently, $\overline{\mathcal{H}} \subset (\mathcal{F} \times \dot{f})(\mathcal{G})$. Hence $(\mathcal{F} \times \dot{f})(\mathcal{G}) \in \mathcal{J}_Y$ for each $\mathcal{G} \in \mathcal{J}_X$, i.e. $\mathcal{F} \times \dot{f} \in \mathcal{J}_{X,Y}$. Therefore $([\mathbf{X}, \mathbf{Y}], \mathcal{J}_{X,Y})$ is complete.

4.2.2.2 Proposition. *Let* $\mathbf{X} = (X, \mathcal{J}_X)$ *and* $\mathbf{Y} = (Y, \mathcal{J}_Y)$ *be semiuniform convergence spaces and let* $\mathcal{J}^s_{X,Y}$ *be the subspace structure of the product structure of* $\mathbf{Y}^X$ *on* $[\mathbf{X}, \mathbf{Y}]_{\mathbf{SUConv}}$, *i.e. the so-called structure of simple convergence. Then the uniformly continuous* **SUConv**-*structure* $\mathcal{J}_{X,Y}$ *on* $[\mathbf{X}, \mathbf{Y}]_{\mathbf{SUConv}}$ *is finer than* $\mathcal{J}^s_{X,Y}$.

Proof. Let $i : [\mathbf{X}, \mathbf{Y}] \longrightarrow Y^X$ be the inclusion map and let $p_x : Y^X \longrightarrow Y$ be the x-th projection for each $x \in X$. If $\Phi \in \mathcal{J}_{X,Y}$, then $\Phi(\dot{x} \times \dot{x}) \subset (p_x \circ i \times p_x \circ i)(\Phi) = p_x \times p_x(i \times i(\Phi))$ for each $x \in X$. Since $\dot{x} \times \dot{x} \in \mathcal{J}_X$, $\Phi(\dot{x} \times \dot{x}) \in \mathcal{J}_Y$ and consequently, $p_x \times p_x(i \times i(\Phi)) \in \mathcal{J}_Y$, i.e. $\Phi \in \mathcal{J}^s_{X,Y}$.

4.2.2.3 Corollary. *The natural function space* $([\mathbf{X}, \mathbf{Y}], \mathcal{J}_{X,Y})$ *in* **SUConv** *is a* T_2-*space provided that* $\mathbf{Y}$ *is a* T_2-*space.*

Proof. It is easily checked that in the realm of generalized convergence spaces subspaces and products of T_2-spaces are T_2-spaces. Since initial semiuniform convergence structures induce initial $\mathbf{KConv_S}$-structures and $\mathbf{KConv_S}$ is bireflective in **GConv**, products and subspaces of T_2-spaces in **SUConv** are T_2-spaces. Thus $([\mathbf{X}, \mathbf{Y}], \mathcal{J}^s_{X,Y})$ is a T_2-space, because $\mathbf{Y}$ is a T_2-space. By 4.2.2.2. $1_{[\mathbf{X},\mathbf{Y}]} : ([\mathbf{X}, \mathbf{Y}], \mathcal{J}_{X,Y}) \longrightarrow ([\mathbf{X}, \mathbf{Y}], \mathcal{J}^s_{X,Y})$ is uniformly continuous and consequently $1_{[\mathbf{X},\mathbf{Y}]} : ([\mathbf{X}, \mathbf{Y}], q_{\gamma_{\mathcal{J}_{X,Y}}}) \longrightarrow ([\mathbf{X}, \mathbf{Y}], q_{\gamma_{\mathcal{J}^s_{X,Y}}})$ is continuous. Hence $([\mathbf{X}, \mathbf{Y}], \mathcal{J}_{X,Y})$ is a T_2-space.

4.2.2.4. Let $\mathbf{X} = (X, \mathcal{J}_X)$ be a semiuniform convergence space and $U(\mathbf{X}) = ([\mathbf{X}, \mathbb{R}_u], \mathcal{J}_{X,\mathbb{R}})$ the natural function space of all uniformly continuous maps from $\mathbf{X}$ into the (usual) uniform space $\mathbb{R}_u$ of real numbers. Put $U(U(\mathbf{X})) = U^2(\mathbf{X})$. Then the map

$$i_X : \mathbf{X} \longrightarrow U^2(\mathbf{X})$$

defined by $i_X(x)(f) = f(x)$ for each $x \in X$ and each $f \in [X, \mathbb{R}_u]$ is uniformly continuous (If $\mathcal{W}_d$ denotes the uniformity induced by the Euclidean metric d on $\mathbb{R}$, then we have to prove that $((i_X \times i_X)(\mathcal{F}))(\Phi) \supset \mathcal{W}_d$ for each $\mathcal{F} \in \mathcal{J}_X$ and each $\Phi \in \mathcal{J}_{X,\mathbb{R}}$. But this is obvious, since $((i_X \times i_X)(\mathcal{F}))(\Phi) = \Phi(\mathcal{F})$ [note: $A(F) = ((i_X \times i_X)[F])(A)$ for each $F \in \mathcal{F}$ and each $A \in \Phi$]).

Definition. A semiuniform convergence space $\mathbf{X} = (X, \mathcal{J}_X)$ is called *u–embedded* provided that the map

$$i_X : \mathbf{X} \longrightarrow U^2(\mathbf{X})$$

is an embedding in **SUConv**.

4.2.2.5 Proposition. *Every separated uniform space* $(X, \mathcal{J}_X)$ *is u–embedded.*

Proof. Let $\mathcal{W}$ be a uniformity on X and $\mathcal{J}_X = [\mathcal{W}]$. Furthermore, let $\mathcal{D}_V$ be the uniformity on X induced by one of the pseudometrics d_V generating $\mathcal{W}$ (cf. 2.3.2.4 theorem). Then $d_V^x : X \longrightarrow \mathbb{R}$ defined by $d_V^x(y) = d_V(x,y)$ is uniformly continuous. Put $H_V = \{(d_V^x, d_V^x) : x \in X\}$. If $A \subset X \times X$, it is easily verified that

(*) $H_V(A) \subset \{(x,y) \in \mathbb{R}^2 : |x-y| < \varepsilon\} \Longleftrightarrow A \subset \{(x,y) \in X \times X : d_V(x,y) < \varepsilon\}$.

It follows from (*) that

$$\mathcal{W}_d \subset \dot{H}_V(\mathcal{D}_V) \subset \dot{H}_V(\mathcal{W})$$

where $\mathcal{W}_d$ denotes the uniformity on $\mathbb{R}$ induced by the Euclidean metric d, and $\dot{H}_V = (\{H_V\})$. Thus, $\dot{H}_V(\mathcal{W}) \in [\mathcal{W}_d]$, i.e. $\dot{H}_V \in \mathcal{J}_{X,\mathbb{R}}$. Furthermore, for each $\mathcal{F} \in F(X \times X)$,

(**) $\dot{H}_V(\mathcal{F}) \supset \mathcal{W}_d \Longleftrightarrow \mathcal{F} \supset \mathcal{D}_V$.

In order to prove that $[\mathcal{W}]$ is the initial structure w.r.t. to i_X, it suffices to show that $\mathcal{J}'_X = \{\mathcal{F} \in F(X \times X) : (i_X \times i_X)(\mathcal{F})$ belongs to the structure of $U^2(\mathbf{X})\} \subset [\mathcal{W}] = \mathcal{J}_X$. Let $\mathcal{F} \in \mathcal{J}'_X$, i.e. $((i_X \times i_X)(\mathcal{F}))(\Phi) \supset \mathcal{W}_d$ for each $\Phi \in \mathcal{J}_{X,\mathbb{R}}$. Consequently, for $\Phi = \dot{H}_V$, $((i_X \times i_X)(\mathcal{F}))(\dot{H}_V) = \dot{H}_V(\mathcal{F}) \supset \mathcal{W}_d$. By (**), we get $\mathcal{F} \supset \mathcal{D}_V$ for each $V \in \mathcal{W}$. Thus, since $V \in \mathcal{D}_V$ for each $V \in \mathcal{W}$, $\mathcal{F} \supset \mathcal{W}$, i.e. $\mathcal{F} \in \mathcal{J}_X$.
It remains to show that i_X is injective provided that $(X, \mathcal{J}_X)$ is separated, i.e. $\bigcap_{V \in \mathcal{W}} V = \Delta$: If $x, y \in X$ such that $x \neq y$, then there is some $V \in \mathcal{W}$ with $(x,y) \notin V$. Consequently, $\frac{1}{2} \leq d_V(x,y) \leq 1$ (note: $h_V(x,y) = 1$ and $\frac{1}{2}h_V(x,y) \leq d_V(x,y) \leq h_V(x,y)$), i.e. $d_V(x,y) \neq 0$. Thus, $i_X(x)(d_V^x) = d_V^x(x) = 0 \neq i_X(y)(d_V^x) = d_V(x,y)$, i.e. i_X is injective.

4.2.2.6 Proposition. *Every u–embedded semiuniform convergence space* $\mathbf{X}$ *is regular and separated.*

Proof. 1) Since $\mathbb{R}_u$ is a T_2–space, $U(\mathbf{X})$ as well as $U^2(\mathbf{X})$ are T_2–spaces (cf. 4.2.2.3). Consequently, $\mathbf{X}$ is a T_2–space (as a subspace of a T_2–space).

2) Since $\mathbf{X} = (X, \mathcal{J}_X)$ is u–embedded, we obtain

(∗) $\mathcal{J}_X = \{\mathcal{F} \in F(X \times X) : \Phi(\mathcal{F}) \supset \mathcal{W}_d \text{ for each } \Phi \in \mathcal{J}_{X,\mathbb{R}}\}$

where $\mathbb{R}_u = (\mathbb{R}, [\mathcal{W}_d])$, and $\mathcal{J}_{X,\mathbb{R}}$ denotes the uniformly continuous **SUConv**–structure on $[\mathbf{X}, \mathbb{R}_u]$: Put $\mathcal{J}_{X,\mathbb{R}} = \mathcal{J}$, and let $\mathcal{J}^2$ be the **SUConv**–structure of $U^2(\mathbf{X})$, i.e.

$$\mathcal{J}^2 = \{\Psi \in F([[\mathbf{X}, \mathbb{R}_u], \mathbb{R}_u] \times [[\mathbf{X}, \mathbb{R}_u], \mathbb{R}_u]) : \Psi(\Phi) \supset \mathcal{W}_d \quad \text{for each} \quad \Phi \in \mathcal{J}\}.$$

Since $i_X : \mathbf{X} \longrightarrow U^2(\mathbf{X})$ is initial,

$$\mathcal{J}_X = \{\mathcal{F} \in F(X \times X) : (i_X \times i_X)(\mathcal{F}) \in \mathcal{J}^2\}.$$

Obviously, $(i_X \times i_X)(\mathcal{F}) \in \mathcal{J}^2 \iff (i_X \times i_X)(\mathcal{F})(\Phi) = \Phi(\mathcal{F}) \supset \mathcal{W}_d$ for each $\Phi \in \mathcal{J}$. Thus, (∗) is valid.

Let $\mathcal{F} \in \mathcal{J}_X$, i.e. $\Phi(\mathcal{F}) \supset \mathcal{W}_d$ for each $\Phi \in \mathcal{J}$. For each closed entourage $U \in \mathcal{W}_d$ there are some $F \in \mathcal{F}$ and some $A \in \Phi$ such that $A(F) \subset U$. Then

(∗∗) $\quad A(\overline{F}) \subset U$:

If $(f, g) \in A$, then $f \times g : \mathbf{X} \times \mathbf{X} \longrightarrow \mathbb{R}_u \times \mathbb{R}_u$ is uniformly continuous and consequently, it is continuous. Thus $(f \times g)[\overline{F}] \subset \overline{(f \times g)[F]} \subset \overline{A(F)}$, and $\bigcup_{(f,g) \in A}(f \times g)[\overline{F}] = A[\overline{F}] \subset \overline{A(F)} \subset \overline{U} = U$.

It follows from (∗∗), $\Phi(\overline{\mathcal{F}}) \supset \mathcal{W}_d$ for each $\Phi \in \mathcal{J}$, i.e. $\overline{\mathcal{F}} \in \mathcal{J}_X$.

4.2.2.7 Proposition. *$U(\mathbf{X})$ is u–embedded for each semiuniform convergence space $\mathbf{X} = (X, \mathcal{J}_X)$.*

Proof. Obviously, $U : \mathbf{SUConv} \longrightarrow \mathbf{SUConv}$ is a contravariant functor defined by $U(\mathbf{X}) = ([\mathbf{X}, \mathbb{R}_u], \mathcal{J}_{X,\mathbb{R}})$ for each $\mathbf{X} \in |\mathbf{SUConv}|$, and $U(f)(g) = g \circ f$ for all **SUConv**–morphisms $f : \mathbf{X} \longrightarrow \mathbf{Y}$ and $g : \mathbf{Y} \longrightarrow \mathbb{R}_u$, i.e. $U(f) : U(\mathbf{Y}) \longrightarrow U(\mathbf{X})$), namely a contravariant hom–functor.

The following diagram

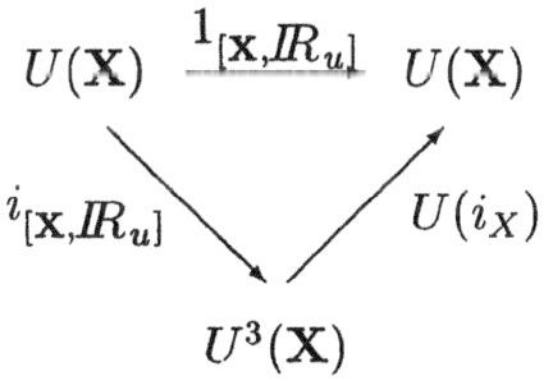

commutes [let $f \in U(\mathbf{X})$; then $U(i_X)(i_{[\mathbf{X},\mathbb{R}_u]}(f))(x) = (i_{[\mathbf{X},\mathbb{R}_u]}(f) \circ i_X)(x) = i_{[\mathbf{X},\mathbb{R}_u]}(f)(i_X(x)) = i_X(x)(f) = f(x)$ for each $x \in X$, i.e. $U(i_X)(i_{[\mathbf{X},\mathbb{R}_u]})(f) = f$]. Let $i'_{[\mathbf{X},\mathbb{R}_u]} : U(\mathbf{X}) \longrightarrow i_{[\mathbf{X},\mathbb{R}_u]}[U(\mathbf{X})]$ be defined by $i'_{[\mathbf{X},\mathbb{R}_u]}(f) = i_{[\mathbf{X},\mathbb{R}_u]}(f)$ for each $f \in U(\mathbf{X})$. Then $U(i_X)|Z \circ i'_{[\mathbf{X},\mathbb{R}_u]} = 1_{[\mathbf{X},\mathbb{R}_u]}$, where $Z = i_{[\mathbf{X},\mathbb{R}_u]}[U(\mathbf{X})]$. Since $1_{[\mathbf{X},\mathbb{R}_u]}$ is an isomorphism (and consequently an extremal monomorphism) and $i'_{[\mathbf{X},\mathbb{R}_u]}$ is an epimorphism, $i'_{[\mathbf{X},\mathbb{R}_u]}$ must be an isomorphism, i.e. $i_{[\mathbf{X},\mathbb{R}_u]} : U(\mathbf{X}) \longrightarrow U^3(\mathbf{X})$ is an embedding.

4.2.2.8 Remark. *A u–embedded semiuniform convergence space need not be a* *uniform space* as the following example shows: $U(I\!R_u)$ is u–embedded (cf. 4.2.2.7), but not uniform (let $\mathcal{W}_d$ be the usual uniformity on $I\!R$ and $I\!R_u = (I\!R, [\mathcal{W}_d])$; if $U(I\!R_u) = ([I\!R_u, I\!R_u], \mathcal{J}_{I\!R,I\!R})$ were uniform, then the filter $\dot{\Delta}_{[I\!R_u,I\!R_u]}$ generated by the diagonal $\Delta_{[I\!R_u,I\!R_u]}$ of $[I\!R_u, I\!R_u] \times [I\!R_u, I\!R_u]$ would belong to $\mathcal{J}_{I\!R,I\!R}$, but this is not true: let $f \in U(I\!R_u)$ be the identity map; then for each real number $k > 0$, $(kf, kf) \in \Delta_{[I\!R_u,I\!R_u]}$; consequently, $\dot{\Delta}_{[I\!R_u,I\!R_u]}(\mathcal{W}_d) \not\supset \mathcal{W}_d$, i.e. $\dot{\Delta}_{[I\!R_u,I\!R_u]} \notin \mathcal{J}_{I\!R,I\!R}$).

4.2.2.9 Proposition. *Every subspace of a u–embedded space is u-embedded.*

Proof. Let $f : \mathbf{X} \longrightarrow \mathbf{Y}$ be an embedding in **SUConv** and let $\mathbf{Y}$ be u–embedded. Then the diagram

$$\begin{array}{ccc} \mathbf{X} & \xrightarrow{\;f\;} & \mathbf{Y} \\ {\scriptstyle i_X}\downarrow & & \downarrow{\scriptstyle i_Y} \\ U^2(\mathbf{X}) & \xrightarrow{\;U^2(f)\;} & U^2(\mathbf{Y}) \end{array}$$

commutes. Consequently, $U^2(f) \circ i_X$ is an embedding. Thus, $i_X : \mathbf{X} \longrightarrow U^2(\mathbf{X})$ is an embedding, i.e. injective and initial.

4.2.2.10 Remarks. 1) *Let (X, d) be a metric space and $\mathcal{W}_d$ the uniformity on X induced by d. Then $(X, [\mathcal{W}_d])$ is a complete semiuniform convergence space iff (X, d) is complete in the usual sense (i.e. each Cauchy sequence in (X, d) converges to some $x \in X$).* A Cauchy filter $\mathcal{F}$ in $(X, [\mathcal{W}_d])$ is a filter on X such that $\mathcal{F} \times \mathcal{F} \in [\mathcal{W}_d]$ or equivalently, a filter $\mathcal{F}$ on X containing arbitrarily small sets (i.e. for each $W \in \mathcal{W}_d$ there is some $F \in \mathcal{F}$ such that $F \times F \subset W$). If $(X, [\mathcal{W}_d])$ is complete, then obviously, (X, d) is complete. If (X, d) is complete and $\mathcal{F}$ is a Cauchy filter on $(X, [\mathcal{W}_d])$, then for each $n \in I\!N$ there is some $F_n \in \mathcal{F}$ such that $F_n \times F_n \subset V_{\frac{1}{n}} = \{(x, y) : d(x, y) < \frac{1}{n}\}$, since $\{V_{\frac{1}{n}} : n \in I\!N\}$ is a base for $\mathcal{W}_d$. Choose a unique $x_n \in F_n$ for each $n \in I\!N$. It is easily checked that $(x_n)_{n \in I\!N}$ is a Cauchy sequence in (X, d). Thus, $(x_n)_{n \in I\!N}$ converges to some $x \in X$. In order to prove that $\mathcal{F}$ converges to x, it suffices to show that $\mathcal{F} \supset \{V_{\frac{1}{n}}(x) : n \in I\!N\} = \mathcal{B}(x)$, since $\mathcal{B}(x)$ is a neighborhood base at x: Let $j \in I\!N$. Since (x_n) converges to x, there is some $n_0 \in I\!N$ such that for all natural numbers $n \geq n_0$, $d(x, x_n) < \frac{1}{2j}$. Put $i = \max\{n_0, 2j\}$. Then $F_i \subset V_{\frac{1}{j}}(x)$, since it follows from $y \in F_i$, $d(x, y) \leq d(x, x_i) + d(x_i, y) < \frac{1}{2j} + \frac{1}{i} \leq \frac{1}{j}$, i.e. $y \in V_{\frac{1}{j}}(x)$. Hence $V_{\frac{1}{j}}(x) \in \mathcal{F}$ for each $j \in I\!N$.

2) Obviously, *every u–embedded semiuniform convergence space* $\mathbf{X}$ *is a semiuniform limit space* (Since $U^2(\mathbf{X})$ is a semiuniform limit space and **SULim** is bireflective in **SUConv**, $\mathbf{X}$ is a semiuniform limit space as a subspace of $U^2(\mathbf{X})$.).

4.2.2.11 Proposition. *Every complete subspace $(Y, \mathcal{J}_Y)$ of a separated semiuniform convergence space $(X, \mathcal{J}_X)$ is closed.*

Proof. Let $i : Y \longrightarrow X$ denote the inclusion map and $y \in \overline{Y} = cl_{q_{\gamma\mathcal{J}_X}} Y$. Then there is some $(\mathcal{F}, y) \in q_{\gamma\mathcal{J}_X}$ such that $Y \in \mathcal{F}$. Consequently, $i^{-1}(\mathcal{F})$ exists, and it is a Cauchy filter. Since $(Y, \mathcal{J}_Y)$ is complete, there is some $z \in Y$ such that $(i^{-1}(\mathcal{F}), z) \in q_{\gamma\mathcal{J}_Y}$. It follows from the continuity of i that $i(i^{-1}(\mathcal{F})) \overset{q_{\gamma\mathcal{J}_X}}{\longrightarrow} i(z) = z$. Since $i(i^{-1}(\mathcal{F})) \supset \mathcal{F}$, $(i(i^{-1}(\mathcal{F}), y) \in q_{\gamma\mathcal{J}_X}$. Thus, $z = y$, because $(Y, \mathcal{J}_Y)$ is separated. Consequently, $\overline{Y} \subset Y$.

4.2.2.12 Theorem. *Let $\mathbf{X} = (X, \mathcal{J}_X)$ be a u-embedded semiuniform convergence space and $i_X : \mathbf{X} \longrightarrow U^2(\mathbf{X})$ the corresponding embedding, defined by $i_X(f) = f(x)$ for each $f \in U(\mathbf{X})$ and each $x \in X$. Put $\widehat{X} = \overline{i_X[X]}^t$ and let $\mathcal{J}_{\widehat{X}}$ be the initial* **SUConv***-structure on $\widehat{X}$ w.r.t. the inclusion map $j_X : \widehat{X} \longrightarrow [[\mathbf{X}, I\!R_u], I\!R_u]$. Then $\widehat{\mathbf{X}} = (\widehat{X}, \mathcal{J}_{\widehat{X}})$ is a complete, u-embedded semiuniform limit space containing $\mathbf{X}$ as a topologically dense subspace. If $\mathbf{Y} = (Y, \mathcal{J}_Y)$ is any complete u-embedded semiuniform limit space and $g : \mathbf{X} \longrightarrow \mathbf{Y}$ a uniformly continuous map, then there is a unique uniformly continuous map $h : \widehat{\mathbf{X}} \longrightarrow \mathbf{Y}$ such that the diagram*

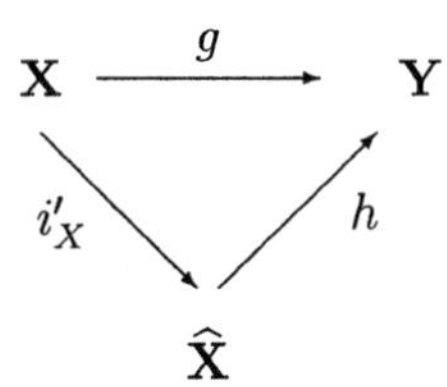

commutes, where $i'_X(x) = i_X(x)$ for each $x \in X$.

Proof. By 4.2.2.7. and 4.2.2.9., $\widehat{\mathbf{X}}$ is u–embedded. Since the uniform space $I\!R_u$ is complete (cf. 4.2.2.10. 1)) and regular (cf. 4.2.1.5.), it follows from 4.2.2.1. that $U(\mathbf{X})$ as well as $U^2(\mathbf{X})$ is complete. Thus, $\widehat{\mathbf{X}}$ is complete as a closed subspace of $U^2(\mathbf{X})$ (cf. 2.3.3.28.). Furthermore, $\widehat{\mathbf{X}}$ is a semiuniform limit space (cf. 4.2.2.10. 2)).

By assumption, $i_Y[Y]$ is complete (note that completeness is a **SUConv**–invariant). Since $(Y, \mathcal{J}_Y)$ is separated, it follows from 4.2.2.11. that $i_Y[Y] = \overline{i_Y[Y]}^t = \widehat{Y}$. Thus, $i'_Y : \mathbf{Y} \longrightarrow \widehat{\mathbf{Y}}$ is an isomorphism. If $g : \mathbf{X} \longrightarrow \mathbf{Y}$ is uniformly continuous, then $U^2(g) : U^2(\mathbf{X}) \longrightarrow U^2(\mathbf{Y})$ is a uniformly continuous map, defined by $U^2(g)(k) = k \circ U(g)$ for each $k \in U^2(\mathbf{X})$. Since $U^2(g)$ is continuous w.r.t. the induced topological spaces, $U^2(g)\,[\overline{i_X[X]}^t] \subset \overline{U^2(g)\,[i_X[X]]}^t$. On the other hand $U^2(g)\,[i_X[X]] \subset i_Y[Y]$; namely if $f \in i_X[X]$, i.e. there is some $x \in X$ such that $f = i_X(x)$, then $U^2(g)(f) = U^2(g)(i_X(x)) = i_X(x) \circ U(g)$ coincides with $i_Y(y)$ provided that $y = g(x)$, i.e. $U^2(g)(f) \in i_Y[Y]$. Consequently, $\widehat{g} : \widehat{\mathbf{X}} \longrightarrow \widehat{\mathbf{Y}}$ defined by $\widehat{g}(\widehat{x}) = U^2(g)(\widehat{x})$ for each $\widehat{x} \in \widehat{X}$ is uniformly continuous, and the diagram

$$\begin{array}{ccc} \mathbf{X} & \xrightarrow{g} & \mathbf{Y} \\ {\scriptstyle i'_X}\downarrow & & \downarrow{\scriptstyle i'_Y} \\ \widehat{\mathbf{X}} & \xrightarrow{\widehat{g}} & \widehat{\mathbf{Y}} \end{array}$$

commutes. $h = (i'_Y)^{-1} \circ \widehat{g}$ is the desired uniformly continuous map from $\widehat{\mathbf{X}}$ into $\mathbf{Y}$. The uniqueness of h follows from 4.2.1.9.

4.2.2.13 Definition. Let $\mathbf{X}$ be a u–embedded semiuniform convergence space. Then $i'_X : \mathbf{X} \longrightarrow \widehat{\mathbf{X}}$ (cf. 4.2.2.12.) is called the *Gazik–Kent–Richardson completion* of $\mathbf{X}$. Occasionally, $\widehat{\mathbf{X}}$ is already called the Gazik–Kent–Richardson completion of $\mathbf{X}$.

4.2.2.14. A semiuniform convergence space $(X, \mathcal{J}_X)$ is called *metrizable* provided that there is a metric d on X such that $\mathcal{J}_X = [\mathcal{W}_d]$, where $\mathcal{W}_d$ denotes the uniformity induced by d. Equivalently, a semiuniform space $(X, \mathcal{J}_X)$ is metrizable iff it is a separated uniform space satisfying the first axiom of countability (cf. 2.3.2.8. and note that a pseudometric d on X is a metric iff $(X, \mathcal{W}_d)$ is separated).

4.2.2.15 Theorem. *A semiuniform convergence space $(X, \mathcal{J}_X)$ is a separated uniform space (= separated principal uniform limit space) iff it is a subspace of a product of metrizable semiuniform convergence spaces.*

Proof. "$\Longleftarrow$". This implication is obvious, since **PrULim** ($\cong$ **Unif**) is a bireflective subconstruct of **SUConv** and products as well as subspaces of T_2–space are T_2–spaces.

"$\Longrightarrow$". 1) If (X, d) is a pseudometric space, then the following are valid:
a) An equivalence relation R on X is defined by $x \; R \; y \Longleftrightarrow d(x, y) = 0$.
b) There is a unique metric d_R on X/R such that $d_R(\omega_R(x), \omega_R(y)) = d(x, y)$ for each $(x, y) \in X \times X$, where $\omega_R : X \longrightarrow X/R$ denotes the natural map.
c) $\omega_R : (X, \mathcal{W}_d) \longrightarrow (X/R, \mathcal{W}_{d_R})$ is initial, i.e. $\mathcal{W}_d$ is the initial uniformity w.r.t. ω_R.

2) Let $(X, \mathcal{W})$ be a separated uniform space. By the theorem under 2.3.2.4. there is a family $(d_V)_{V \in \mathcal{W}}$ of pseudometrics on X such that $\mathcal{W}$ is the coarsest uniformity on X which is finer than each $\mathcal{D}_V$, where $\mathcal{D}_V$ denotes the uniformity induced by d_V, in other words: $(1_X^V : (X, \mathcal{W}) \longrightarrow (X, \mathcal{D}_V))_{V \in \mathcal{W}}$ is an initial source where $1_X^V : X \longrightarrow X$ denotes the identity map for each $V \in \mathcal{W}$. For each $V \in \mathcal{W}$, let d'_V be the metric induced by d_V according to 1) and $\mathcal{D}'_V$ the uniformity induced by d'_V. Since the canonical map $\omega_V : (X, \mathcal{D}_V) \longrightarrow (Y_V, \mathcal{D}'_V)$ is initial (where Y_V is the set of all equivalence classes w.r.t. the equivalence relation on X induced by d_V according to 1)), $(\omega_V \circ 1_X^V : (X, \mathcal{W}) \longrightarrow (Y_V, \mathcal{D}'_V))_{V \in \mathcal{W}}$ is an initial source. Put $f_V = \omega_V \circ 1_X^V$ for each $V \in \mathcal{W}$. Let $\prod_{V \in \mathcal{W}} (Y_V, \mathcal{D}'_V)$ be the product of the family $((Y_V, \mathcal{D}'_V))_{V \in \mathcal{W}}$ in **Unif** and let $p_V : \prod_{V \in \mathcal{W}} (Y_V, \mathcal{D}'_V) \longrightarrow (Y_V, \mathcal{D}'_V)$ be the projection for each $V \in \mathcal{W}$. By the (categorical) definition of a product, there

is a unique uniformly continuous map $f : (X, \mathcal{W}) \longrightarrow \prod_{V \in \mathcal{W}} (Y_V, \mathcal{D}'_V)$ such that $p_V \circ f = f_V$ for each $V \in \mathcal{W}$.
Obviously, f is initial. Furthermore, f is injective; namely, since $(X, \mathcal{W})$ is separated, for any two distinct elements $x, y \in X$ there is some $V \in \mathcal{W}$ such that $(x, y) \notin V$, and consequently, $d_V(x, y) \neq 0$; thus $f_V(x) \neq f_V(y)$ and therefore $f(x) \neq f(y)$. Hence f is an embedding, i.e. $(X, \mathcal{W})$ is a subspace of a product of metrizable uniform spaces.

3) Since in **PrULim** ($\cong$ **Unif**) products and subspaces are formed as in **SUConv**, the desired implication follows from 2).

4.2.2.16 Corollary. *Every separated uniform space $(X, \mathcal{W})$ is a dense subspace of a complete separated uniform space $(X', \mathcal{W}')$, namely of its complete hull, which coincides with its Hausdorff completion $(X^*, \mathcal{W}^*)$ (up to isomorphism).*

Proof. As is well–known every metric space can be embedded (isometrically) into a complete metric space. By 4.2.2.15. a separated uniform space $(X, \mathcal{W})$ can be embedded into a product of metrizable uniform spaces and thus, by our previous remark, into a product of complete metrizable uniform spaces. The closure of X in the latter product (which is complete!) is denoted by X' [resp. $(X', \mathcal{W}')$, when it is regarded as a subspace]. Thus, $(X', \mathcal{W}')$ is a complete uniform space containing $(X, \mathcal{W})$ as a dense subspace. Furthermore, $(X', \mathcal{W}')$ is separated (as a subspace of a product of separated uniform spaces). Without loss of generality let us consider X to be a subset of X' and let $i_X : X \longrightarrow X'$ be the inclusion map. The universal property of $i_X : (X, \mathcal{W}) \longrightarrow (X', \mathcal{W}')$ follows immediately from 4.2.1.13., i.e. $i_X : (X, \mathcal{W}) \longrightarrow (X', \mathcal{W}')$ is the epireflection of $(X, \mathcal{W})$ w.r.t. **CSepUnif**. Thus, $(X', \mathcal{W}')$ is the complete hull of $(X, \mathcal{W})$, in other words: the Hausdorff completion of $(X, \mathcal{W})$.

4.2.2.17 Corollary. *If $\mathbf{X}$ is a separated uniform space, then its Gazik–Kent–Richardson completion $\widehat{\mathbf{X}}$ is a (separated) uniform space, which is isomorphic to the Hausdorff completion $\mathbf{X}^*$ of $\mathbf{X}$.*

Proof. Let $i'_X : \mathbf{X} \longrightarrow \widehat{\mathbf{X}}$ ($r_X : \mathbf{X} \longrightarrow \mathbf{X}^*$) be the Gazik–Kent–Richardson completion (the Hausdorff completion) of the separated uniform space $\mathbf{X}$. Without loss of generality let us assume that i'_X (resp. r_X) is an inclusion map. Because of the universal property of i'_X, there is a unique uniformly continuous map $f : \widehat{\mathbf{X}} \longrightarrow \mathbf{X}^*$ such that the diagram

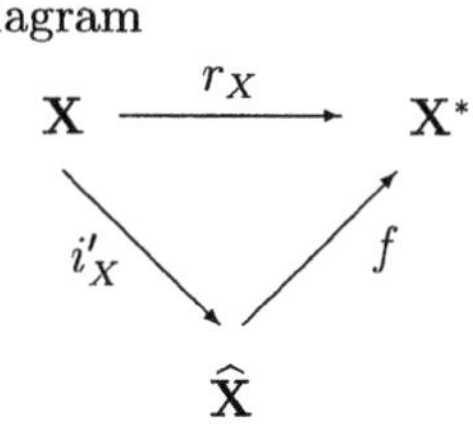

commutes. By 4.2.1.13. there is a unique uniformly continuous map $g : \mathbf{X}^* \longrightarrow \widehat{\mathbf{X}}$ such that the diagram

$$\begin{array}{ccc} \mathbf{X} & \xrightarrow{i'_X} & \widehat{\mathbf{X}} \\ & {\scriptstyle r_X}\searrow \quad \nearrow{\scriptstyle g} & \\ & \mathbf{X}^* & \end{array}$$

commutes. Thus, $f \circ i'_X = r_X$ and $g \circ r_X = i'_X$, and consequently, $f \circ g \circ r_X = r_X$ and $g \circ f \circ i'_X = i'_X$, i.e. $(f \circ g)|X = 1_{X^*}|X$ and $(g \circ f)|X = 1_{\widehat{X}}|X$. Since $\widehat{\mathbf{X}}$ and $\mathbf{X}^*$ are T_2–spaces and X is topologically dense in $\widehat{\mathbf{X}}$ (resp. $\mathbf{X}^*$) it follows from 4.2.1.9. that $g \circ f = 1_{\widehat{X}}$ and $g \circ f = 1_{X^*}$, i.e. f (resp. g) is an isomorphism.

4.2.2.18 Proposition. *The construct* **UE–SUConv** *of u–embedded semiuniform convergence spaces (and uniformly continuous maps) is an epireflective (full and isomorphism–closed) subconstruct of* **SUConv**.

Proof. Let $\mathbf{X} = (X, \mathcal{J}_X) \in |\mathbf{SUConv}|$ and let $\widetilde{\mathbf{X}}$ be the subspace of $U^2(\mathbf{X})$ with underlying set $i_X[X]$. If $j_X : X \longrightarrow i_X[X]$ is defined by $j_X(x) = i_X(x)$ for each $x \in X$, then $j_X : \mathbf{X} \longrightarrow \widetilde{\mathbf{X}}$ is the desired epireflection: Obviously, $\widetilde{\mathbf{X}}$ is u–embedded (cf. 4.2.2.9.). Furthermore, if $f : \mathbf{X} \longrightarrow \mathbf{Y}$ is a uniformly continuous map from $\mathbf{X}$ into a u–embedded semiuniform convergence space $\mathbf{Y}$, then the diagram

$$\begin{array}{ccc} \mathbf{X} & \xrightarrow{f} & \mathbf{Y} \\ {\scriptstyle j_X}\big\downarrow & & \big\downarrow{\scriptstyle j_Y}\ \cong \\ \widetilde{\mathbf{X}} & \xrightarrow{\widetilde{f}} & \widetilde{\mathbf{Y}} \end{array}$$

commutes, where $\widetilde{f}$ is defined by $\widetilde{f}(\widetilde{x}) = U^2(f)(\widetilde{x})$ for each $\widetilde{x} \in i_X[X]$. Thus, $g = j_Y^{-1} \circ \widetilde{f}$ is the unique uniformly continuous map such that $g \circ j_X = f$, since j_X is an epimorphism.

4.2.2.19 Remarks. 1) If **CUE–SUConv** denotes the construct of all complete u–embedded semiuniform convergence spaces (and uniformly continuous maps), then **CUE–SUConv** *is an epireflective (full and isomorphism–closed) subconstruct of* **UE–SUConv** *where the epireflection of $X \in |\mathbf{UE–SUConv}|$ with respect to* **CUE–SUConv** *is its Gazik-Kent-Richardson completion* (cf. 4.2.2.12. and note that $i'_X : \mathbf{X} \longrightarrow \widehat{\mathbf{X}}$ is an epimorphism [use 4.2.1.9.]). Together with 4.2.2.18. we obtain that **CUE–SUConv** *is a reflective subconstruct of* **SUConv**, i.e. for each semiuniform convergence space $\mathbf{X}$, $r_X = i'_{i_X[X]} \circ j_X : \mathbf{X} \longrightarrow \widehat{(\widetilde{\mathbf{X}})}$ is the reflection of $\mathbf{X}$ w.r.t. **CUE–SUConv**.

2) It is easily verified that *the Gazik-Kent-Richardson completion of the (usual) uniform space $\mathbb{Q}_u$ of rational numbers is (up to isomorphism) the uniform space $\mathbb{R}_u$ of real numbers.* (If $i : \mathbb{Q}_u \longrightarrow \mathbb{R}_u$ denotes the inclusion map, then for each uniformly continuous map $f : \mathbb{Q}_u \longrightarrow \mathbf{Y}$ from $\mathbb{Q}_u$ into a complete u–embedded semiuniform convergence space $\mathbf{Y}$, there is a unique uniformly continuous extension $\overline{f} : \mathbb{R}_u \longrightarrow \mathbf{Y}$ according to 4.2.1.13. [remember, that a u–embedded space

is regular and separated]. By 4.2.1.9., $i : \mathbb{Q}_u \longrightarrow \mathbb{R}_u$ is an epimorphism in **UE–SUConv**, and $\mathbb{R}_u$ is complete. Thus, by 1), $\mathbb{R}_u \cong \widehat{\mathbb{Q}_u})$.

4.3 Application to compactifications

4.3.1 An alternative description of uniform spaces by means of uniform covers

4.3.1.1 Definitions. 1) A *merotopic space* is a pair (X,μ), where X is a set and μ is a non-empty set of non-empty covers [2] of X such that the following are satisfied:

Mer$_1$) $\mathcal{A} \prec \mathcal{B}$ [3] and $\mathcal{A} \in \mu$ imply $\mathcal{B} \in \mu$.
Mer$_2$) $\mathcal{A} \in \mu$ and $\mathcal{B} \in \mu$ imply $\mathcal{A} \wedge \mathcal{B} = \{A \cap B : A \in \mathcal{A} \text{ and } B \in \mathcal{B}\} \in \mu$.
The members of μ are called *uniform covers.*

2) a) Let X be a set, $\mathcal{U}$ a cover of X and $A \subset X$. Then

$$St(A,\mathcal{U}) = \bigcup\{U \in \mathcal{U} : A \cap U \neq \emptyset\}$$

is called the *star* of A *with respect to* $\mathcal{U}$.
b) If $\mathcal{U}$ and $\mathcal{V}$ are covers of the set X, then $\mathcal{U}$ is called a *star–refinement* of $\mathcal{V}$ (denoted by $\mathcal{U} \ {}^{*}\!\!\prec \mathcal{V}$) provided that for each $U \in \mathcal{U}$ there is some $V \in \mathcal{V}$ such that $St(U,\mathcal{U}) \subset V$.
c) A *uniformly merotopic space* (shortly: a *uniform space*) is a merotopic space (X,μ) such that the following is satisfied:
U) For each $\mathcal{A} \in \mu$ there is some $\mathcal{B} \in \mu$ such that $\mathcal{B} \ {}^{*}\!\!\prec \mathcal{A}$.

3) A map $f : (X,\mu) \longrightarrow (Y,\eta)$ between merotopic spaces is called *uniformly continuous* provided that $f^{-1}\mathcal{A} = \{f^{-1}[A] : A \in \mathcal{A}\} \in \mu$ for each $\mathcal{A} \in \eta$.

4.3.1.2 Theorem. *The construct* **U–Mer** *of uniformly merotopic spaces (and uniformly continuous maps) is (concretely) isomorphic to* **Unif.**

Proof. (a) *Let (X,μ) be a uniformly merotopic space. Then $\mathcal{B}_\mu = \{\bigcup_{A\in\mathcal{A}} A \times A : \mathcal{A} \in \mu\}$ is a base for a uniformity $\mathcal{W}_\mu$ on X:*
Since μ is a non-empty set of non-empty covers, $\mathcal{B}_\mu \subset \mathcal{P}(X \times X)$ is non-empty.

BU$_1$) If $B_\mu \in \mathcal{B}_\mu$, then there exists some $\mathcal{A} \in \mu$ such that $B_\mu = \bigcup_{A\in\mathcal{A}} A \times A$. Since $\mathcal{A}$ is a cover of X, $(x,x) \in \Delta$ implies the existence of some $A \in \mathcal{A}$ with $x \in A$, so that $(x,x) \in A \times A \subset B_\mu$. Thus, $\Delta \subset B_\mu$.

BU$_2$) Let $B_\mu \in \mathcal{B}_\mu$, i.e. $B_\mu = \bigcup_{A\in\mathcal{A}} A{\times}A$ for some $\mathcal{A} \in \mu$. Obviously, $(\bigcup_{A\in\mathcal{A}} A{\times}A)^{-1} = \bigcup_{A\in\mathcal{A}} A \times A$.

[2]If $X = \emptyset$, then $\{\emptyset\}$ is a non–empty cover of X whereas $\emptyset$ is an empty cover of X. Thus, in order to fulfill 1.1.2. (3) we restrict ourselves to non–empty covers.

[3]$\mathcal{A} \prec \mathcal{B} \Longleftrightarrow \forall\, A \in \mathcal{A}\ \exists\, B \in \mathcal{B}\ \ A \subset B \Longleftrightarrow \mathcal{A}$ *refines* $\mathcal{B}$.

BU$_3$) Let $B_\mu \in \mathcal{B}_\mu$. Then there exists some $\mathcal{A} \in \mu$ such that $B_\mu = \bigcup_{A\in\mathcal{A}} A \times A$. Since U) is satisfied, there is some $\mathcal{B} \in \mu$ such that $\mathcal{B}\ ^* \!\prec \mathcal{A}$. Then $B'_\mu = \bigcup_{B\in\mathcal{B}} B \times B$ belongs to $\mathcal{B}_\mu$ and $B'_\mu \circ B'_\mu \subset B_\mu$ (If $f(x,y) \in B'_\mu \circ B'_\mu$, then there exists some $z \in X$ such that $(x,z) \in B'_\mu$ and $(z,y) \in B'_\mu$. Hence, there are $B_1, B_2 \in \mathcal{B}$ such that $(x,z) \in B_1 \times B_1$ and $(z,y) \in B_2 \times B_2$, and thus $\{x,y\} \subset St(B_1,\mathcal{B})$. Further, there exists some $A \in \mathcal{A}$ satisfying $St(B_1,\mathcal{B}) \subset A$. Therefore $(x,y) \in A \times A \subset \bigcup_{A'\in\mathcal{A}} A' \times A' = B_\mu$.).

BU$_4$) Let $B_\mu, B'_\mu \in \mathcal{B}_\mu$. Then there exist $\mathcal{A}, \mathcal{B} \in \mu$ such that $B_\mu = \bigcup_{A\in\mathcal{A}} A \times A$ and $B'_\mu = \bigcup_{B\in\mathcal{B}} B \times B$. Thus by Mer$_2$), $\mathcal{A} \wedge \mathcal{B} \in \mu$. Consequently, $B''_\mu = \bigcup_{C\in\mathcal{A}\wedge\mathcal{B}} C \times C$ belongs to $\mathcal{B}_\mu$ and $B''_\mu \subset B_\mu \cap B'_\mu$.

(b) *Let $(X, \mathcal{W})$ be a uniform space. Then there is a unique* **U–Mer**-*structure* $\mu_\mathcal{W}$ *on* X *such that* $\mathcal{W}_{\mu_\mathcal{W}} = \mathcal{W}$:
α) For every $V \in \mathcal{W}$, let $\mathcal{A}_V = \{V(x) : x \in X\}$. Then the set $\mu_\mathcal{W}$ of all covers of X for which there exists some $V \in \mathcal{W}$ such that $\mathcal{A}_V \prec \mathcal{A}$ is a **U–Mer**-structure on X:
Obviously, $\mu_\mathcal{W}$ is a non-empty set of non-empty covers of X which satisfies the following:

Mer$_1$) by definition.
Mer$_2$) Let $\mathcal{A}_1, \mathcal{A}_2 \in \mu_\mathcal{W}$. Then there exist $V, W \in \mathcal{W}$ such that $\mathcal{A}_V \prec \mathcal{A}_1$ and $\mathcal{A}_W \prec \mathcal{A}_2$. Hence

$$\mathcal{A}_{V\cap W} = \{(V\cap W)(x) : x \in X\} = \{V(x)\cap W(x) : x \in X\} \prec \mathcal{A}_V \wedge \mathcal{A}_W \prec \mathcal{A}_1 \wedge \mathcal{A}_2 .$$

Since $V \cap W \in \mathcal{W}$, $\mathcal{A}_1 \wedge \mathcal{A}_2 \in \mu_\mathcal{W}$.

U) Let $\mathcal{A} \in \mu_\mathcal{W}$. Then there exists some $V \in \mathcal{W}$ such that $\mathcal{A}_V \prec \mathcal{A}$. For each $V \in \mathcal{W}$, there is a symmetric $V' \in \mathcal{W}$ such that $V'^2 \subset V$. Then for each $x \in X$, $St(\{x\}, \mathcal{A}_{V'}) \subset V(x)$ (If $y \in St(\{x\}, \mathcal{A}_{V'}) = \bigcup\{V'(z) : z \in X \text{ and } x \in V'(z)\}$, then there is some $z \in X$ with $x, y \in V'(z)$, i.e. $(z,x) \in V'$ and $(z,y) \in V'$. Thus by the symmetry of V', $(x,y) \in V^2 \subset V$, i.e. $y \in V(x)$.). Hence $\mathcal{A}_{V'}$ is a barycentric refinement[4] of $\mathcal{A}_V$. Thereby everything has already been shown because it can be easily checked that the following holds for a set X and covers $\mathcal{U}, \mathcal{V}, \mathcal{W}$ of X:

$$\mathcal{U} \,\Delta\, \mathcal{V} \text{ and } \mathcal{V} \,\Delta\, \mathcal{W} \text{ imply } \mathcal{U}\ ^*\!\!\prec \mathcal{W} .$$

β) $\mathcal{W}_{\mu_\mathcal{W}} = \mathcal{W}$:

(1) If $W \in \mathcal{W}_{\mu_\mathcal{W}}$, then there exists $\mathcal{A} \in \mu_\mathcal{W}$ and $V \in \mathcal{W}$ such that $\{V(x) : x \in X\} = \mathcal{A}_V \prec \mathcal{A}$ and $\bigcup_{A\in\mathcal{A}} A \times A \subset W$. Thus $V \subset W$ ($(x,y) \in V$ implies $y \in V(x) \subset A$ for a suitable $A \in \mathcal{A}$, i.e. $(x,y) \in A \times A \subset W$ and consequently $W \in \mathcal{W}$.

(2) If $W \in \mathcal{W}$, then there exists a symmetric $V \in \mathcal{W}$ such that $V^2 \subset W$. Thus $\bigcup_{A\in\mathcal{A}_V} A \times A \subset W$ (If $(x,y) \in \bigcup_{A\in\mathcal{A}_V} A \times A$, then there is some $z \in X$ with $(x,y) \in V(z) \times V(z)$. Hence by the symmetry of V, $(x,y) \in V^2 \subset W$), i.e. $W \in \mathcal{W}_{\mu_\mathcal{W}}$.

[4] If $\mathcal{U}$ and $\mathcal{V}$ are covers of a set X, then $\mathcal{U}$ is called a *barycentric refinement* of $\mathcal{V}$ (denoted by $\mathcal{U} \,\Delta\, \mathcal{V}$) provided that $\{St(\{x\},\mathcal{U}) : x \in X\} \prec \mathcal{V}$.

γ) Let μ be a **U–Mer** structure on X such that $\mathcal{W}_\mu = \mathcal{W}$. Then $\mu = \mu_\mathcal{W}$:

(1) If $\mathcal{A} \in \mu_\mathcal{W}$, then there exists some $V \in \mathcal{W}$ such that $\{V(x) : x \in X\} = \mathcal{A}_V \prec \mathcal{A}$. Since $V \in \mathcal{W}_\mu = \mathcal{W}$, there is some $\mathcal{B} \in \mu$ such that $\bigcup_{B\in\mathcal{B}} B \times B \subset V$. Thus, $\mathcal{B} \prec \mathcal{A}_V$ (If $B \in \mathcal{B}$ and $x \in B$, then for every $y \in B$, the pair (x, y) belongs to $B \times B$ and hence to V, i.e. $B \subset V(x)$) and consequently (by Mer_1)), $\mathcal{A} \in \mu$.

(2) Let $\mathcal{A} \in \mu$. Then there is some $\mathcal{B} \in \mu$ such that $\mathcal{B} \,{}^*\!\!\prec \mathcal{A}$. Hence $V = \bigcup_{B\in\mathcal{B}} B \times B \in \mathcal{B}_\mu \subset \mathcal{W}_\mu = \mathcal{W}$ and $\mathcal{A}_V \prec \mathcal{A}$, i.e. $\mathcal{A} \in \mu_\mathcal{W}$ (If $x \in X$, then $V(x) = (\bigcup_{B\in\mathcal{B}} B \times B)(x) = \bigcup_{B\in\mathcal{B}}(B \times B)(x) = St(\{x\}, \mathcal{B})$ [1. $y \in \bigcup_{B\in\mathcal{B}}(B \times B)(x)$ implies the existence of some $B \in \mathcal{B}$ with $(x, y) \in B \times B$. Thus, $y \in B \subset St(\{x\}, \mathcal{B})$.
2. $y \in St(\{x\}, \mathcal{B})$ implies the existence of some $B \in \mathcal{B}$ with $x, y \in B$, hence $(x, y) \in B \times B$, i.e. $y \in (B \times B)(x) \subset V(x)$] and since $\mathcal{B}$ is a cover of X, there exists some $B \in \mathcal{B}$ with $x \in B$ and by $\mathcal{B} \,{}^*\!\!\prec \mathcal{A}$, there is some $A \in \mathcal{A}$ such that $V(x) = St(\{x\}, \mathcal{B}) \subset St(B, \mathcal{B}) \subset A$).

(c) *Let $(X, \mu), (Y, \nu)$ be uniformly merotopic spaces and $f : X \longrightarrow Y$ a map. Then the following are equivalent:*
(1) *$f : (X, \mu) \longrightarrow (Y, \nu)$ is uniformly contiuous.*
(2) *$f : (X, \mathcal{W}_\mu) \longrightarrow (Y, \mathcal{W}_\nu)$ is uniformly continous.*

Proof. (1) $\Longrightarrow$ (2): Let $B_\nu \in \mathcal{B}_\nu$, i.e. there exists some $\mathcal{A} \in \nu$ with $B_\nu = \bigcup_{A\in\mathcal{A}} A \times A$. Then by assumption, $f^{-1}\mathcal{A} \in \mu$ and hence $B_\mu = \bigcup_{A\in\mathcal{A}}(f^{-1}[A] \times f^{-1}[A]) \in \mathcal{B}_\mu$. Thus $(x, y) \in B_\mu$ implies the existence of some $A \in \mathcal{A}$ such that $(x, y) \in f^{-1}[A] \times f^{-1}[A]$, i.e. $(f(x), f(y)) \in A \times A \subset B_\nu$. Therefore (2) has been proved.

(2) $\Longrightarrow$ (1): Let $\mathcal{A} \in \nu$. Then there exists some $\mathcal{B} \in \nu$ such that $\mathcal{B} \,{}^*\!\!\prec \mathcal{A}$. Since $B_\nu = \bigcup_{B\in\mathcal{B}} B \times B \in \mathcal{B}_\nu$, $(f \times f)^{-1}[B_\nu] \in \mathcal{W}_\mu$ by (2), i.e. there exists some $\mathcal{C} \in \mu$ such that $\bigcup_{C\in\mathcal{C}} C \times C \subset (f \times f)^{-1}[B_\nu]$. Then $\mathcal{C} \prec f^{-1}\mathcal{A}$ and thus $f^{-1}\mathcal{A} \in \mu$ (Let $C \in \mathcal{C}$ and $x \in C$. Then there is some $B \in \mathcal{B}$ with $f(x) \in B$ and some $A \in \mathcal{A}$ such that $St(\{f(x)\}, \mathcal{B}) \subset St(B, \mathcal{B}) \subset A$. Hence $C \subset f^{-1}[A]$ [$y \in C$ implies $(x, y) \in C \times C \subset (f \times f)^{-1}[B_\nu]$. Thus $(f(x), f(y)) \in B_\nu$, so that there is some $B' \in \mathcal{B}$ with $f(x), f(y) \in B'$. Consequently, $f(y) \in St(\{f(x)\}, \mathcal{B}) \subset A$, i.e. $y \in f^{-1}[A]$]).

(d) Put $\mathcal{F}((X, \mu)) = (X, \mathcal{W}_\mu)$ for every uniformly merotopic space (X, μ), and for each uniformly continuous map $f : (X, \mu) \longrightarrow (Y, \nu)$, let $\mathcal{F}(f)$ be the corresponding uniformly continuous map $f : (X, \mathcal{W}_\mu) \longrightarrow (Y, \mathcal{W}_\nu)$ (cf. (c)). Thereby a (concrete) functor $\mathcal{F}$: **U–Mer** $\longrightarrow$ **Unif** has been defined. Put $\mathcal{G}((X, \mathcal{W})) = (X, \mu_\mathcal{W})$ for each uniform space $(X, \mathcal{W})$, and for each uniformly continuous map $f : (X, \mathcal{W}) \longrightarrow (Y, \mathcal{R})$, let $\mathcal{G}(f)$ be the corresponding uniformly continuous map $f : (X, \mu_\mathcal{W}) \longrightarrow (Y, \mu_\mathcal{R})$ (note (c) and $\mathcal{W}_{\mu_\mathcal{W}} = \mathcal{W}$ [resp. $\mathcal{R}_{\mu_\mathcal{R}} = \mathcal{R}$]). Thus a functor $\mathcal{G}$: **Unif** $\longrightarrow$ **U–Mer** has been defined and the following holds:

(1) $\mathcal{G} \circ \mathcal{F} = \mathcal{I}_{\mathbf{U\text{–}Mer}}$ (note $\mu_{\mathcal{W}_\mu} = \mu$ for every **U–Mer**–structure μ) and
(2) $\mathcal{F} \circ \mathcal{G} = \mathcal{I}_{\mathbf{Unif}}$ (note $\mathcal{W}_{\mu_\mathcal{W}} = \mathcal{W}$ for every uniformity $\mathcal{W}$).

Consequently, **U–Mer** and **Unif** are (concretely) isomorphic.

4.3.1.3 Remarks. 1) Because of 4.3.1.2. we need not distinguish between uniform spaces and uniformly merotopic spaces. In other words: *The definition of uniform spaces by means of uniform covers* (in the sense of Tukey [141]) *is equivalent to the definition of uniform spaces by means of entourages* (in the sense of A. Weil [147]).

2) The category **Mer** of merotopic spaces (and uniformly continuous maps) is a topological construct in which initial and final structures are formed as follows:
a) If X is a set, $((X_i, \mu_i))_{i\in I}$ a family of merotopic spaces and $(f : X \longrightarrow X_i)_{i\in I}$ a family of maps, then

$$\mu = \{\mathcal{A} \subset \mathcal{P}(X) : \text{ there exist finitely many elements } \mathcal{A}_1, \dots, \mathcal{A}_n \text{ of } \mu' \text{ such that } \mathcal{A}_1 \wedge \dots \wedge \mathcal{A}_n \prec \mathcal{A}\}$$

is the *initial merotopic structure* on X with respect to $(X, f_i, (X_i, \mu_i), I)$ where $\mu' = \{f_i^{-1}\mathcal{A}_i : \mathcal{A}_i \in \mu_i \text{ and } i \in I\}$.

b) If X is a set, $((X_i, \mu_i))_{i\in I}$ a family of merotopic spaces and $(f_i : X_i \longrightarrow X)_{i\in I}$ a family of maps, then

$$\mu = \{\mathcal{A} : \mathcal{A} \text{ is a cover of } X \text{ and } f_i^{-1}\mathcal{A} \in \mu_i \text{ for each } i \in I\}$$

is the *final merotopic structure* on X with respect to $((X_i, \mu_i), f_i, X, I)$.

3) a) **U–Mer** (≅ **Unif**) *is a bireflective* (full and isomorphism–closed) *subconstruct of* **Mer**, namely if (X, μ) is a merotopic space and μ_u is the set of all $\mathcal{A} \in \mu$ for which there exists a sequence $(\mathcal{A}_n)_{n\in I\!N}$ in μ such that $\mathcal{A}_1 = \mathcal{A}$ and $\mathcal{A}_{n+1} \,{}^*\!\!\prec \mathcal{A}_n$ for each $n \in I\!N$, then $1_X : (X, \mu) \longrightarrow (X, \mu_u)$ is the bireflection of (X, μ) w.r.t. **U–Mer** (note: Since $\{X\} \in \mu$, $\{X\} \,{}^*\!\!\prec \{X\}$ and thus $\{X\} \in \mu_u$, i.e. μ_u is a non–empty set of non–empty covers):

Mer$_1$) Let $\mathcal{A} \in \mu_u$ and $\mathcal{A} \prec \mathcal{B}$. Put $\mathcal{B}_1 = \mathcal{B}$ and $\mathcal{B}_n = \mathcal{A}_n$ for $n > 1$. Then a sequence $(\mathcal{B}_n)_{n\in I\!N}$ in μ with the desired property has been found, i.e. $\mathcal{B} \in \mu_u$.

Mer$_2$) Let $\mathcal{A}, \mathcal{B} \in \mu_u$. Then there exist sequences $(\mathcal{A}_n)_{n\in I\!N}$ and $(\mathcal{B}_n)_{n\in I\!N}$ in μ such that $\mathcal{A}_1 = \mathcal{A}$, $\mathcal{B}_1 = \mathcal{B}$, $\mathcal{A}_{n+1} \,{}^*\!\!\prec \mathcal{A}_n$ and $\mathcal{B}_{n+1} \,{}^*\!\!\prec \mathcal{B}_n$ for each $n \in N$. Thus $(\mathcal{A}_n \wedge \mathcal{B}_n)_{n\in I\!N}$ is a sequence in μ such that $\mathcal{A}_1 \wedge \mathcal{B}_1 = \mathcal{A} \wedge \mathcal{B}$ and $\mathcal{A}_{n+1} \wedge \mathcal{B}_{n+1} \,{}^*\!\!\prec \mathcal{A}_n \wedge \mathcal{B}_n$ for each $n \in I\!N$. Consequently, $\mathcal{A} \wedge \mathcal{B} \in \mu_u$.

U) is fulfilled by the definition of μ_u.

By the construction of μ_u, $\mu_u \subset \mu$. Hence $1_X : (X, \mu) \longrightarrow (X, \mu_u)$ is a uniformly continuous map. If (Y, ν) is a uniformly merotopic space and $f : (X, \mu) \longrightarrow (Y, \nu)$ is a uniformly continuous map, then $f : (X, \mu_u) \longrightarrow (Y, \nu)$ is a uniformly continuous map; for if $\mathcal{A} \in \nu$, then, since ν is a **U–Mer**–structure, there exists a sequence $(\mathcal{A}_n)_{n\in I\!N}$ in ν such that $\mathcal{A}_1 = \mathcal{A}$ and $\mathcal{A}_{n+1} \,{}^*\!\!\prec \mathcal{A}_n$ for each $n \in I\!N$. Hence $(f^{-1}\mathcal{A}_n)_{n\in I\!N}$ is a sequence in μ such that $f^{-1}\mathcal{A}_1 = f^{-1}\mathcal{A}$ and $f^{-1}\mathcal{A}_{n+1} \,{}^*\!\!\prec f^{-1}\mathcal{A}_n$ for each $n \in I\!N$, i.e. $f^{-1}\mathcal{A} \in \mu_u$.).

b) It follows from a) that **U–Mer** is a topological construct and that the initial structure in **U–Mer** is formed as in **Mer**, whereas the final structures in **U–Mer** arise from the final structures in **Mer** by applying the bireflector $\mathcal{R}$: **Mer** $\longrightarrow$ **U–Mer**.

4.3.2 Precompactness and compactness

4.3.2.1 Definition. 1) a) A filter space (X, γ) is called *precompact* (or *totally bounded*) provided that each ultrafilter $\mathcal{U}$ on X belongs to γ.

b) A semiuniform convergence space $(X, \mathcal{J}_X)$ is called *precompact* (or *totally bounded*) provided that the underlying filter space $(X, \gamma_{\mathcal{J}_X})$ is precompact, i.e. each ultrafilter $\mathcal{U}$ on X is a $\mathcal{J}_X$–Cauchy filter.

2) a) A generalized convergence space (X, q) is called *compact* provided that each ultrafilter $\mathcal{U}$ on X converges in (X, q).
b) A semiuniform convergence space $(X, \mathcal{J}_X)$ is called *compact* provided that the underlying (symmetric) Kent convergence space $(X, q_{\gamma_{\mathcal{J}_X}})$ is compact.

4.3.2.2 Proposition. *The construct* **PC–SUConv** *of precompact semiuniform convergence spaces (and uniformly continuous maps) is a bireflective (full and isomorphism–closed) subconstruct of* **SUConv**. *Thus,* **PC–SUConv** *is closed under formation of subspaces and products in* **SUConv** *and contains all indiscrete* **SUConv**–*objects.*

Proof. If $(X, \mathcal{J}_X) \in |\mathbf{SUConv}|$, then $1_X : (X, \mathcal{J}_X) \longrightarrow (X, \mathcal{J}'_X)$ is the desired bireflection of $(X, \mathcal{J}_X)$ w.r.t. **PC–SUConv**, where $\mathcal{J}'_X = \mathcal{J}_X \cup \{\mathcal{G} \in F(X \times X)$: there is some ultrafilter $\mathcal{U}$ on X with $\mathcal{G} \supset \mathcal{U} \times \mathcal{U}\}$.

4.3.2.3 Corollary. *The construct* **PC–Fil** *of precompact filter spaces (and Cauchy continuous maps) is a bireflective (full and isomorphism–closed) subconstruct of* **Fil**.

Proof. The restriction of the bireflector $\mathcal{R}$: **SUConv** $\longrightarrow$ **PC–SUConv** to **Fil–D–SUConv** ($\cong$ **Fil**) leads to the desired bireflection.

4.3.2.4 Proposition. 1) *Every compact semiuniform convergence space is precompact.*
2) *A complete precompact semiuniform convergence space is compact.*

Proof. 1) is obvious (note: each filter converging in $(X, q_{\gamma_{\mathcal{J}_X}})$ is a $\mathcal{J}_X$–Cauchy filter).
2) Since each ultrafilter $\mathcal{U}$ on X is a $\mathcal{J}_X$–Cauchy filter and $(X, \mathcal{J}_X)$ is complete, it follows that $\mathcal{U}$ converges in $(X, q_{\gamma_{\mathcal{J}_X}})$.

4.3.2.5 Corollary. *A convergence space $(X, \mathcal{J}_X)$ is compact iff it is precompact.*

Proof. Use 4.3.2.4. and 2.3.3.24.

4.3.2.6 Definition. Let (X, q) be a generalized convergence space and $\mathcal{F}$ a filter on X. Then $\mathcal{F}$ is said to have an *adherence point* $x \in X$ provided that there is some filter $\mathcal{G}$ on X such that $\mathcal{G} \supset \mathcal{F}$ and $(\mathcal{G}, x) \in q$.

4.3.2.7 Proposition. *Let $(X, \mathcal{J}_X)$ be a uniform limit space and $\mathcal{F}$ a $\mathcal{J}_X$-Cauchy filter on X with an adherence point x (w.r.t. $(X, q_{\gamma_{\mathcal{J}_X}})$). Then $\mathcal{F}$ converges to x (in $(X, q_{\gamma_{\mathcal{J}_X}})$).*

Proof. By 3.3.4. 2) a), $(X, \gamma_{\mathcal{J}_X})$ is a Cauchy space. By assumption there is some $\mathcal{G} \in F(X)$ such that $\mathcal{G} \supset \mathcal{F}$ and $\mathcal{G} \cap \dot{x} \in \gamma_{\mathcal{J}_X}$. Thus, since $\mathcal{F} \in \gamma_{\mathcal{J}_X}$, $\mathcal{F} \cap (\mathcal{G} \cap \dot{x}) = \mathcal{F} \cap \dot{x} \in \gamma_{\mathcal{J}_X}$, i.e. $(\mathcal{F}, x) \in q_{\gamma_{\mathcal{J}_X}}$.

4.3.2.8 Proposition. *A uniform limit space $(X, \mathcal{J}_X)$ is compact iff it is complete and precompact.*

Proof. "$\Longleftarrow$": cf. 4.3.2.4. 2).
"$\Longrightarrow$": 1) By 4.3.2.4. 1), $(X, \mathcal{J}_X)$ is precompact.
2) Let $\mathcal{F} \in \gamma_{\mathcal{J}_X}$. Then there is some ultrafilter $\mathcal{U}$ on X with $\mathcal{U} \supset \mathcal{F}$. Since $(X, \mathcal{J}_X)$ is compact, $\mathcal{U}$ converges to some $x \in X$. Thus, x is an adherence point of $\mathcal{F}$. By 4.3.2.7., $\mathcal{F}$ converges to x. Hence, $(X, \mathcal{J}_X)$ is complete.

4.3.2.9 Proposition. 1) *A semiuniform convergence space $(X, \mathcal{J}_X)$ is precompact iff for each filter $\mathcal{F}$ on X there is a $\mathcal{J}_X$-Cauchy filter $\mathcal{G}$ on X such that $\mathcal{F} \subset \mathcal{G}$.*
2) *A semiuniform convergence space $(X, \mathcal{J}_X)$ is compact iff each filter $\mathcal{F}$ on X has an adherence point (w.r.t. $(X, q_{\gamma_{\mathcal{J}_X}})$).*

Proof. Obvious.

4.3.2.10 Proposition. *Let $(X, \mathcal{J}_X)$, $(X', \mathcal{J}_{X'})$ be semiuniform convergence spaces and $f : X \longrightarrow X'$ a surjective map such that $f : (X, \gamma_{\mathcal{J}_X}) \longrightarrow (X', \gamma_{\mathcal{J}_{X'}})$ (resp. $f : (X, q_{\gamma_{\mathcal{J}_X}}) \longrightarrow (X', q_{\gamma_{\mathcal{J}_{X'}}})$ is Cauchy continuous (resp. continuous). If $(X, \mathcal{J}_X)$ is precompact (resp. compact), then $(X', \mathcal{J}_{X'})$ is also precompact (resp. compact).*

Proof. Let $\mathcal{F} \in F(X')$. Then $f^{-1}(\mathcal{F}) \in F(X)$ and by assumption there is some $\mathcal{G} \in \gamma_{\mathcal{J}_X}$ (resp. $(\mathcal{G}, x) \in q_{\gamma_{\mathcal{J}_X}}$) such that $\mathcal{G} \supset f^{-1}(\mathcal{F})$. Since f is Cauchy continuous (resp. continuous), $f(\mathcal{G}) \supset f(f^{-1}(\mathcal{F})) = \mathcal{F}$ belongs to $\gamma_{\mathcal{J}_{X'}}$ (resp. converges to $f(x)$).

4.3.2.11 Theorem (Tychonoff). *Let $((X_i, \mathcal{J}_{X_i}))_{i \in I}$ be a family of non-empty semiuniform convergence spaces. Then the product space $(\prod_{i \in I} X_i, \mathcal{J}_X)$ of this family is compact iff $(X_i, \mathcal{J}_{X_i})$ is compact for each $i \in I$.*

Proof. "$\Longrightarrow$". For each $i \in I$, the i-th projection $p_i : (\prod_{i \in I} X_i, \mathcal{J}_X) \longrightarrow (X_i, \mathcal{J}_{X_i})$ is surjective and uniformly continuous. Thus, by 4.3.2.10., $(X_i, \mathcal{J}_{X_i})$ is compact.

"$\Longleftarrow$". For each $i \in I$, $(X_i, \mathcal{J}_{X_i})$ is compact. Thus, $(X_i, q_{\gamma_{\mathcal{J}_{X_i}}})$ is a compact (symmetric) Kent convergence space. Hence, the product space $\prod_{i\in I}(X_i, q_{\gamma_{\mathcal{J}_{X_i}}})$ in **KConv** (resp. **KConv**$_S$) is compact, namely if $\mathcal{U}$ is an ultrafilter on $\prod_{i\in I}(X_i, q_{\gamma_{\mathcal{J}_{X_i}}})$, then $p_i(\mathcal{U})$ is an ultrafilter on $\prod_{i\in I} X_i$ for each $i \in I$ (where $p_i : \prod X_i \longrightarrow X_i$ denotes the i–th projection) which converges to some $x_i \in X_i$ by assumption, and hence $\mathcal{U}$ converges to $(x_i) = x \in \prod_{i\in I} X_i$ in $\prod_{i\in I}(X_i, q_{\gamma_{\mathcal{J}_{X_i}}})$. Thus, by 2.3.3.17. b), $\prod_{i\in I}(X_i, \mathcal{J}_{X_i})$ is compact.

4.3.2.12 Remarks. 1) *In Tychonoff's theorem, compactness can be replaced by precompactness* (cf. 4.3.2.2. and 4.3.2.10.).

2) a) *Tychonoff's theorem is also valid in* **Conv** ($\cong$ **KConv**$_S$) because products in **Conv** are formed as in **SUConv** (cf. 2.3.3.30. 2)).
b) *Tychonoff's theorem is also valid in* **Top**$_S$, because it is valid in **KConv**$_S$ (cf. a)) and products in **Top**$_S$ are formed as in **KConv**$_S$ (cf. 2.3.3.13. 2.)).

3) a) *In the realm of semiuniform convergence spaces (or filter spaces) compactness does not imply completeness and thus a Cauchy filter with an adherence point x need not converge to x (in constrast to the situation for uniform limit spaces [or Cauchy spaces] acccording to* 4.3.2.7.) as the following *example* shows: Let X be a set with $|X| \geq \aleph_0$. Put $\gamma = \{\mathcal{F} \in F(X)$: there is some ultrafilter $\mathcal{U}$ on X and some $x \in X$ with $\mathcal{F} \supset \mathcal{U} \cap \dot{x}\} \cup \{\mathcal{F} \in F(X) : \mathcal{F} \supset \mathcal{G}\}$, where $\mathcal{G} = \{U \subset X : X \setminus U \text{ is finite}\}$.
Then (X, γ) is a compact filter space, but the Cauchy filter $\mathcal{G}$ does not converge to any $z \in X$; namely 1. $\mathcal{G} \cap \dot{z} \not\supset \mathcal{G}$, because $X \setminus \{z\} \in \mathcal{G}$ does not belong to any element of $\mathcal{G} \cap \dot{z}$, and 2. $\mathcal{G} \cap \dot{z} \not\supset \mathcal{U} \cap \dot{x}$ for any ultrafilter $\mathcal{U}$ on X and any $x \in X$, because either a) $\{x, z\} \in \mathcal{U}$ and thus $\{x, z\} \in \mathcal{U} \cap \dot{x}$, but $\{x, z\} \notin \mathcal{G} \cap \dot{z}$ or b) $X \setminus \{x, z\} \in \mathcal{U}$ and thus $X \setminus \{z\} \in \mathcal{U} \cap \dot{x}$, but $X \setminus \{z\} \notin \mathcal{G} \cap \dot{z}$.

b) A semiuniform convergence space $(X, \mathcal{J}_X)$ is called *weakly complete* provided that each $\mathcal{J}_X$–Cauchy filter has an adherence point (or equivalently: each $\mathcal{J}_X$–Cauchy ultrafilter converges in $(X, q_{\gamma_{\mathcal{J}_X}})$). Obviously, in uniform limit spaces weak completeness coincides with completeness. Furthermore, *a semiuniform convergence space $(X, \mathcal{J}_X)$ is compact iff it is precompact and weakly complete.* It is easy to check that *products and closed subspaces of weakly complete semiuniform convergence space are weakly complete* (cf. the corresponding proofs for completeness).

4.3.2.13 Proposition. *Let $(X, \mathcal{J}_X)$ be a compact semiuniform convergence space and $A \subset X$ a closed subset (i.e. $A = cl_{\gamma_{\mathcal{J}_X}} A$). Then $(A, \mathcal{J}_A)$ is compact, where $\mathcal{J}_A$ denotes the initial* **SUConv**-*structure* on A w.r.t. the inclusion map $i : A \longrightarrow X$.

Proof. Let $\mathcal{U}$ be an ultrafilter on A. Then $i(\mathcal{A})$ is an ultrafilter on X with $A \in i(\mathcal{U})$. Since $(X, \mathcal{J}_X)$ is compact, there is some $x \in X$ with $(i(\mathcal{U}), x) \in q_{\gamma_{\mathcal{J}_X}}$. Since A is closed, $x \in A$. Obviously, $(\mathcal{U}, x) \in q_{\gamma_{\mathcal{J}_A}}$ (note 2.3.3.17. b)). Thus $(A, \mathcal{J}_A)$ is compact.

4.3.2.14 Remark. *Every weakly complete subspace $(Y, \mathcal{J}_Y)$ of a separated semiuniform convergence space $(X, \mathcal{J}_X)$ is closed* (this is proved analogously to 4.2.2.11.). *Thus, every compact subspace $(Y, \mathcal{J}_Y)$ of a separated semiuniform convergence space $(X, \mathcal{J}_X)$ is closed,* because every compact semiuniform convergence space is weakly complete.

4.3.2.15 Proposition. *Let $(X, \mathcal{W})$ be a uniform space (in the sense of Weil) and $\mu_\mathcal{W}$ the corresponding set of uniform covers. Then the following are equivalent:*

(1) *$(X, \mathcal{W})$ is precompact, i.e. $(X, [\mathcal{W}])$ is precompact.*
(2) *For each $W \in \mathcal{W}$ there is some finite cover $\mathcal{C}$ of X such that $C \times C \subset W$ for each $C \in \mathcal{C}$.*
(3) *For each $W \in \mathcal{W}$ there is some finite subset F of X such that $W[F] = X$.*
(4) *For each $\mathcal{U} \in \mu_\mathcal{W}$ there is some finite subset $\mathcal{B} \subset \mathcal{U}$ such that $\mathcal{B} \in \mu_\mathcal{W}$, i.e. every uniform cover has a finite uniform subcover.*

Proof. "(1) $\Longrightarrow$ (3)" (indirectly). If (3) is not satisfied, then there is some $W \in \mathcal{W}$ such that for all finite subsets F of X, $W[F] = \bigcup_{x\in F} W(x) \neq X$, i.e. $X \setminus (W[F]) \neq \emptyset$.
$\mathcal{B} = \{X \setminus (W[F]) : F \subset X \text{ is finite}\}$ is a filter base. Then there is an ultrafilter $\mathcal{F}' \supset \mathcal{F} = (\mathcal{B})$ which is not a Cauchy filter; namely, if there were some $F' \in \mathcal{F}'$ such that $F' \times F' \subset W$, then for some $x \in F'$, $F' \subset W(x)$, i.e. $W(x) \in \mathcal{F}'$, whereas $X \setminus (W(x)) \in \mathcal{F} \subset \mathcal{F}'$ and consequently $W(x) \cap (X \setminus (W(x)) = \emptyset \in \mathcal{F}'$ – a contradiction. Thus, (1) is not fulfilled.

(3) $\Longrightarrow$ (4). Let $\mathcal{A} \in \mu_\mathcal{W}$. Then there is some $V \in \mathcal{W}$ such that $\mathcal{A}_V \prec \mathcal{A}$. Furthermore, there exists some symmetric $V' \in \mathcal{W}$ with $V'^2 \subset V$. By assumption, there is some finite $F \subset X$ such that $V'[F] = \bigcup_{x\in F} V'(x) = X$. Then

$$\mathcal{A}_{V'} \prec \{V(x) : x \in F\} \prec \mathcal{A}_V \prec \mathcal{A}$$

($z \in X$ implies the existence of some $x \in F$ with $z \in V'(x)$; hence $V'(z) \subset V'^2(x) \subset V(x)$). Thus, $\{V(x) : x \in F\} \in \mu_\mathcal{W}$ is finite. Consequently, there is some finite $\mathcal{B} \in \mu_\mathcal{W}$ with $\mathcal{B} \subset \mathcal{A}$.

(4) $\Longrightarrow$ (2). If $W \in \mathcal{W}$, then there exists some $\mathcal{B} \in \mu_\mathcal{W}$ such that $\bigcup_{B\in\mathcal{B}} B \times B \subset W$ (note: $\mathcal{W} = \mathcal{W}_{\mu_\mathcal{W}}$). Furthermore, there exists some finite $\mathcal{C} \in \mu_\mathcal{W}$ with $\mathcal{C} \subset \mathcal{B}$ and for each $C \in \mathcal{C}$,

$$C \times C \subset \bigcup_{B\in\mathcal{B}} B \times B \subset W .$$

Hence, $\mathcal{C}$ is a finite cover of X with the desired property.

(2) $\Longrightarrow$ (1). Let $\mathcal{U}$ be an ultrafilter on X and $W \in \mathcal{W}$. Then there is some finite cover $\mathcal{C}$ of X such that $\bigcup_{C\in\mathcal{C}} C \times C \subset W$. Since $X = \bigcup_{C\in\mathcal{C}} C \in \mathcal{U}$, it follows that there is at least one $C_0 \in \mathcal{C}$ with $C_0 \in \mathcal{U}$. Thus, $C_0 \times C_0 \subset W$, and $\mathcal{U}$ is a Cauchy filter.

4.3.2.16 Remark. Let (X, d) be a metric space and A a non–empty subset of

X. Then the *diameter* $d(A)$ of A is defined by $d(A) = \text{lub}\,\{d(x,y) : x, y \in A\}$, and (X, d) is precompact (w.r.t. the uniformity $\mathcal{W}_d$ induced by d) iff for each $\varepsilon > 0$, there is a finite cover $(A_i)_{i \in I}$ of X such that $d(A_i) < \varepsilon$ for each $i \in I$, i.e. iff (X, d) is totally bounded in the usual sense (note: $\mathcal{B} = \{V_\varepsilon : \varepsilon < 0\}$ is a base for $\mathcal{W}_d$, where $V_\varepsilon = \{(x,y) : d(x,y) < \varepsilon)$.

4.3.2.17 Proposition. *The Hausdorff completion $(X^*, \mathcal{W}^*)$ of a uniform space $(X, \mathcal{W})$ preserves precompactness.*

Proof. Let $(X, \mathcal{W})$ be a precompact uniform space and $\mathcal{U}$ an ultrafilter on X^*. In order to prove that $\mathcal{U}$ is a Cauchy filter it suffices to prove that for each closed symmetric entourage $W^* \in \mathcal{W}^*$ there is some $U \in \mathcal{U}$ such that $U \times U \subset W^*$ (cf. 4.2.1.4.). If $r_X : (X, \mathcal{W}) \longrightarrow (X^*, \mathcal{W}^*)$ is the reflection of $(X, \mathcal{W})$ w.r.t. **CSepUnif**, then $W = (r_X \times r_X)^{-1}[W^*] \in \mathcal{W}$ and, since $(X, \mathcal{W})$ is precompact, there is a finite cover $(A_i)_{i \in I}$ of X such that $\bigcup_{i \in I} A_i \times A_i \subset W$. Put $B_i = r_X[A_i]$ for each $i \in I$. Then $(B_i)_{i \in I}$ is a finite cover of $r_X[X]$ such that $B_i \times B_i \subset W^*$ for each $i \in I$ (namely if $x', x'' \in r_X[A_i]$, i.e. $x' = r_X(a'_i)$ and $x'' = r_X(a''_i)$ with $a'_i, a''_i \in A_i$, then it follows from $a'_i, a''_i \in W$ immediately $(x', x'') \in W^*$). Since $r_X[X]$ is dense in X^* (cf. part 1) and part 2) of the proof under 4.1.5. and see 4.1.3.), $X^* = \bigcup_{i \in I} \overline{B_i}$. Furthermore; $\overline{B_i} \times \overline{B_i} = \overline{B_i \times B_i} \subset \overline{W^*} = W^*$ for each $i \in I$. At least one $\overline{B_i}$ belongs to $\mathcal{U}$, since $X^* \in \mathcal{U}$ and $\mathcal{U}$ is an ultrafilter. Thus, $\mathcal{U}$ is a Cauchy filter. Consequently, $(X^*, \mathcal{W}^*)$ is precompact.

4.3.2.18 Definition. A precompact uniform space is called a *proximity space.*

4.3.2.19. Since **Unif** is bireflective in **SUConv** and 4.3.2.2. is valid, the construct **Prox** of proximity spaces (and uniformly continuous maps) is closed under formation of initial structures in **Unif**, i.e. **Prox** is bireflective in **Unif**. In the following the bireflective **Prox**–modification of a uniform space will be described by means of uniform covers.

Proposition. *If (X, μ) is a uniform space (in the sense of Tukey) and μ_p is the set of all $\mathcal{A} \in \mu$ for which there is some finite $\mathcal{B} \subset \mathcal{A}$ such that $\mathcal{B} \in \mu$, then $1_X : (X, \mu) \longrightarrow (X, \mu_p)$ is the bireflection of (X, μ) w.r.t.* **Prox**.

Proof. 1) (X, μ_p) is a proximity space: a) (X, μ_p) is a uniform space; namely μ_p is a non–empty set of non–empty covers of X such that the following are satisfied:

Mer$_1$) If $\mathcal{A} \in \mu_p$ and $\mathcal{A} \prec \mathcal{B}$, then there exists some finite $\mathcal{C} \in \mu$ such that $\mathcal{C} \subset \mathcal{A} \prec \mathcal{B}$. Hence $\mathcal{B} \in \mu_p$.

Mer$_2$) If $\mathcal{A}_1, \mathcal{A}_2 \in \mu_p$, then there exist finite covers $\mathcal{B}_1, \mathcal{B}_2 \in \mu$ with $\mathcal{B}_i \subset \mathcal{A}_i$ $(i \in \{1,2\})$. Thus, $\mathcal{B}_1 \wedge \mathcal{B}_2 \in \mu$ and $\mathcal{B}_1 \wedge \mathcal{B}_2 \subset \mathcal{A}_1 \wedge \mathcal{A}_2$. Consequently, $\mathcal{A}_1 \wedge \mathcal{A}_2 \in \mu_p$.

U) Let $\mathcal{A} \in \mu_p$. Then there exists some finite $\mathcal{B} \in \mu$ with $\mathcal{B} \subset \mathcal{A}$. Furthermore, there exists some $\mathcal{C} \in \mu$ with $\mathcal{C} \, {}^*\!\!\prec \mathcal{B}$. An equivalence relation R on $\mathcal{C}$ is defined by

$$C_1 R\, C_2 \Longleftrightarrow \forall\, B \in \mathcal{B}\,[(C_1 \subset B \Longleftrightarrow C_2 \subset B) \text{ and } (St\,(C_1, \mathcal{C}) \subset B \Longleftrightarrow St\,(C_2, \mathcal{C}) \subset B)]\,.$$

If $\mathcal{B}$ has at most n elements, then there are at most 4^n equivalence classes with respect to R. Put $\widetilde{C} = \bigcup\{C' \in \mathcal{C} : C' R\, C\}$ for every $C \in \mathcal{C}$. Then $\widetilde{\mathcal{C}} = \{\widetilde{C} : C \in \mathcal{C}\}$ is finite. Since obviously $\mathcal{C} \prec \widetilde{\mathcal{C}}$, $\widetilde{\mathcal{C}} \in \mu$ and thus, $\widetilde{\mathcal{C}} \in \mu_p$. Furthermore, $\widetilde{\mathcal{C}} \overset{*}{\prec} \mathcal{A}$ [For each $C \in \mathcal{C}$, there exists some $B \in \mathcal{B}$ such that $St\,(C, \mathcal{C}) \subset B$. Additionally $St\,(\widetilde{C}, \widetilde{\mathcal{C}}) \subset B$ (if $\widetilde{C}_1 \in \widetilde{\mathcal{C}}$ such that $\widetilde{C}_1 \cap \widetilde{C} \neq \emptyset$, then there exists some $x \in \widetilde{C}_1 \cap \widetilde{C}$, i.e. there exist $C_2 \in \mathcal{C}$ and $C_3 \in \mathcal{C}$ with $x \in C_2 \cap C_3$ and $C_2\ R\ C_1$ as well as $C_3\ R\ C$. Hence $C_3 \subset St\,(C_2, \mathcal{C}) \subset B$ and thus $C_4 \subset B$ for each C_4 such that $C_4\ R\ C_3$. Finally, $\widetilde{C}_1 = \widetilde{C}_2 \subset B$ and thus, $St\ (\widetilde{C}, \widetilde{\mathcal{C}}) \subset B$)].
b) (X, μ_p) is precompact by construction of μ_p.

2) If (X', μ') is a precompact uniform space and $f : (X, \mu) \longrightarrow (X', \mu')$ is a uniformly continuous map, then $f : (X, \mu_p) \longrightarrow (X', \mu')$ is also a uniformly continuous map; namely if $\mathcal{A} \in \mu'$, then there exists some finite $\mathcal{B} \in \mu'$ with $\mathcal{B} \subset \mathcal{A}$ and $f^{-1}\mathcal{B}$ is a finite uniform cover such that $f^{-1}\mathcal{B} \subset f^{-1}\mathcal{A}$. Consequently, $f^{-1}\mathcal{A} \in \mu_p$.

4.3.2.20 Proposition. *Let (X, μ) be a uniform space and $1_X : (X, \mu) \longrightarrow (X, \mu_p)$ the bireflection of (X, μ) w.r.t.* **Prox**. *Then the underlying topological space of (X, μ_p) coincides with the underlying topological space of (X, μ).*

Proof. 1) For each uniform space (Y, η) (described by means of uniform covers) the topology $\mathcal{Y}_\eta$ of the underlying topological space is given by

$$\mathcal{Y}_\eta = \{O \subset Y : \text{ For each } y \in O,\ \ \{O, Y \setminus \{y\}\} \in \eta\} :$$

a) For each $y \in O$, $\{O, Y \setminus \{y\}\} \in \eta$ iff there is some $\mathcal{A} \in \eta$ such that $St\ (\{y\}, \mathcal{A}) \subset O$.
(1) "$\Longleftarrow$": $St\ (\{y\}, \mathcal{A}) \subset O$ for some $\mathcal{A} \in \eta$ and some $y \in Y$ implies $\mathcal{A} \prec \{O, Y \setminus \{y\}\}$ and thus $\{O, Y \setminus \{y\}\} \in \eta$.
2) "$\Longrightarrow$": Since $O = St\ (\{y\}, \{O, Y \setminus \{y\}\})$, this implication is obvious.)

b) $\mathcal{B}_\eta = \{\bigcup_{A \in \mathcal{A}} A \times A : \mathcal{A} \in \eta\}$ is a base for $\mathcal{W}_\eta$, where $\mathcal{W}_\eta$ is the uniformity corresponding to η. Then for each $y \in Y$, $\mathcal{B}_\eta(y) = \{B(y) : B \in \mathcal{B}_\eta\}$ is a neighborhood base of y with respect to the topology $\mathcal{Y}_{\mathcal{W}_\eta}$ induced by $\mathcal{W}_\eta$. For each $B \in \mathcal{B}_\eta$, there is some $\mathcal{A} \in \eta$ such that $B = \bigcup_{A \in \mathcal{A}} A \times A$. Then $B(y) = \{z : (y, z) \in B\} = \{z : (y, z) \in A \times A \text{ for some } A \in \mathcal{A}\} = St\ (\{y\}, \mathcal{A})$. Consequently, $\mathcal{Y}_{\mathcal{W}_\eta} = \{O \subset Y\!: \text{ for each } y \in O, \text{ there is some } \mathcal{A} \in \eta \text{ with } St\ (\{y\}, \mathcal{A}\} \subset O\}$. Thus, by a) $\mathcal{Y}_{\mathcal{W}_\eta} = \mathcal{Y}_\eta$.

2) A finite cover of X belongs to μ_p iff it belongs to μ. Thus, according to 1), the underlying topological space of (X, μ_p) coincides with the underlying topological space of (X, μ).

4.3.2.21. In the last part of this section we will describe the relationship between compact Hausdorff (topological) spaces and uniform spaces. Since every compact Hausdorff space is a Tychonoff space, we start with completely regular topological spaces.

Proposition. *Let $(X, \mathcal{X})$ be a completely regular topological space. Then there is a finest uniformity on X inducing $\mathcal{X}$.*

Proof. According to 2.3.3.19. there is a uniformity on X inducing $\mathcal{X}$, i.e. there is a **U–Mer**–structure μ on X such that $\mathcal{X} = \mathcal{X}_\mu = \{O \subset X\colon$ For each $x \in O$, $\{O, X \setminus \{x\}\} \in \mu\}$ (cf. part 1) of the proof under 4.3.2.20). Let $\{\mu_i : i \in I\}$ be the set of all **U–Mer**–structures on X inducing $\mathcal{X}$. Put $\mu' = \bigcup\{\mu_i : i \in I\}$. Then $\mu_F = \{\mathcal{A} \subset \mathcal{P}(X)\colon$ there are finitely many $\mathcal{A}_1, \dots, \mathcal{A}_n \in \mu'$ with $\mathcal{A}_1 \wedge \dots \wedge \mathcal{A}_n \prec \mathcal{A}\}$ is a **U–Mer**–structure on X such that $\mu_i \subset \mu_F$ for each $i \in I$, i.e. μ_F is finer than each μ_i. Thus, $\mathcal{X} = \mathcal{X}_{\mu_i} \subset \mathcal{X}_{\mu_F}$ for each $i \in I$. Since for each $x \in X$, all finite intersections of elements of $\{ St(\{x\}, \mathcal{A}) : \mathcal{A} \in \mu'\}$ form a neighborhood base at x w.r.t. $\mathcal{X}_{\mu_F}$, we obtain $\mathcal{X}_{\mu_F} \subset \mathcal{X}$.

4.3.2.22 Definitions. 1) Let $(X, \mathcal{X})$ be a completely regular topological space. Then the finest uniform structure μ_F (resp. $\mathcal{W}_{\mu_F}$) inducing $\mathcal{X}$ is called the *fine uniform structure*, and (X, μ_F) (resp. $(X, \mathcal{W}_{\mu_F})$) is called the *fine uniform space.*

2) An open cover $\mathcal{U}$ of a topological space $(X, \mathcal{X})$ is called *normally open* provided that there is a sequence $(\mathcal{U}_n)_{n \in I\!N}$ of open covers of X such that $\mathcal{U}_{n+1} \mathbin{{}^*\!\!\prec} \mathcal{U}_n$ for each $n \in I\!N$ and $\mathcal{U}_1 = \mathcal{U}$.

4.3.2.23 Proposition. *Let $(X, \mathcal{X})$ be a completely regular topological space and (X, μ_F) the fine uniform space. Then*

$$\mu_F = \{\mathcal{A} \subset \mathcal{P}(X) : \textit{ there is a normally open cover } \mathcal{B} \textit{ of } X \textit{ with } \mathcal{B} \prec \mathcal{A}\}.$$

Proof. 1) In a uniform space (X', μ'), each uniform cover is refined by some open uniform cover:
According to 4.2.1.4. the open elements of $\mathcal{W}_{\mu'}$ form a base for $\mathcal{W}_{\mu'}$. Then each $\mathcal{A} \in \mu' = \mu_{\mathcal{W}_{\mu'}}$ is refined by some $\mathcal{A}_V$, for V open in $\mathcal{W}_{\mu'}$. But $\mathcal{A}_V = \{V(z) : z \in X'\}$ and, since V is open in $X' \times X'$, $V(z)$ is open in X' for each $z \in X'$.

2) a) According to 1), each $\mathcal{A} \in \mu_F$ is refined to some open element of μ_F. Since μ_F is a uniform structure and 1) is valid, each open element of μ_F is normally open.
b) Let $\mathcal{A}$ be a cover of X which is refined by some normally open cover $\mathcal{B}$ of X, i.e. there is a sequence $(\mathcal{B}_n)_{n \in I\!N}$ of open covers of X such that $\mathcal{B}_{n+1} \mathbin{{}^*\!\!\prec} \mathcal{B}_n$ for each $n \in I\!N$ and $\mathcal{B}_1 = \mathcal{B}$. Let μ be any uniform structure on X inducing $\mathcal{X}$. Put $\mu' = \mu \cup \{\mathcal{B}_1, \mathcal{B}_2, \dots\}$. Then $\mu^* = \{\mathcal{C} \subset \mathcal{P}(X)\colon$ there exist finitely many $\mathcal{C}_1, \dots, \mathcal{C}_n \in \mu'$ with $\mathcal{C}_1 \wedge \dots \wedge \mathcal{C}_n < \mathcal{C}\}$ is a uniform structure on X inducing $\mathcal{X}$. Thus, $\mathcal{A} \in \mu^* \subset \mu_F$.

4.3.2.24 Proposition. *Every continuous map from a fine uniform space to* *some uniform space is uniformly continuous.*

Proof. Let (X, μ_F) be a fine uniform space and $f : (X, \mu_F) \longrightarrow (Y, \eta)$ a continuous map from (X, μ_F) into a uniform space (Y, η). If $\mathcal{U} \in \eta$, then there is an open cover $\mathcal{V} \in \eta$ such that $\mathcal{V} \prec \mathcal{U}$ (cf. part 1) of the proof of 4.3.2.23.). Since f is continuous, $f^{-1}\mathcal{V}$ is open, and since η is a uniform structure, $f^{-1}\mathcal{V}$ is even a normally open refinement of $f^{-1}\mathcal{U}$, i.e. $f^{-1}\mathcal{U} \in \mu_F$.

4.3.2.25 Proposition. *A compact Hausdorff space $(X, \mathcal{X})$ is uniquely uniformizable. Its unique uniformity inducing $\mathcal{X}$ consists of all neighborhoods (in $X \times X$) of the diagonal Δ and coincides with the fine uniform structure $\mathcal{W}_{\mu_F}$.*

Proof. Since $(X, \mathcal{X})$ is a compact Hausdorff space, $(X, \mathcal{X})$ is completely regular. Let $\mathcal{W}$ be a uniformity on X inducing $\mathcal{X}$, i.e. $\mathcal{X}_{\mathcal{W}} = \mathcal{X}$. By 4.2.1.4., for each $V \in \mathcal{W}$, $\Delta \subset V^0 \subset V$, i.e. V is a neighborhood of the diagonal Δ. Conversely, all neighborhoods of the diagonal Δ belong to $\mathcal{W}$: If there were a neighborhood U of Δ such that $U \notin \mathcal{W}$, then $U^0 \subset U$ would also not belong to $\mathcal{W}$. Hence $i^{-1}(\mathcal{W})$ would exist, where $i : (X \times X) \setminus U^0 \longrightarrow X \times X$ denotes the inclusion map (obviously, $V \cap ((X \times X) \setminus U^0) \neq \emptyset$ for each $V \in \mathcal{W}$, because otherwise $V \subset U^0$ for some $V \in \mathcal{W}$ and thus $U^0 \in \mathcal{W}$). Since $(X \times X) \setminus U^0$ is compact (as a closed subspace of the compact space $X \times X$), $i^{-1}(\mathcal{V})$ has an adherence point $(x, y) \in (X \times X) \setminus U^0$, i.e.

$$
\begin{aligned}
(x, y) \in \bigcap_{V \in \mathcal{W}} \overline{i^{-1}[V]} &= \bigcap_{V \in \mathcal{W}} \overline{((X \times X) \setminus U^0) \cap V} \subset \bigcap_{V \in \mathcal{W}} \overline{((X \times X) \setminus U^0)} \cap \overline{V} \\
&= ((X \times X) \setminus U^0) \cap (\bigcap_{V \in \mathcal{W}} \overline{V}) = ((X \times X) \setminus U^0) \cap \Delta = \emptyset
\end{aligned}
$$

(note: The closed entourages form a base for $\mathcal{W}$ [cf. 4.2.1.4.] and $(X, \mathcal{W})$ is separated, i.e. $\bigcap_{V \in \mathcal{W}} V = \bigcap_{V \in \mathcal{W}} \overline{V} = \Delta$). This is a contradiction.

4.3.2.26 Remarks. 1) If (X, d) is a metric space, then two subsets A, B of X are called *near* iff $d(A, B) = 0$, where

$$
d(A, B) = \begin{cases} \infty, & \text{if } A = \emptyset \text{ or } B = \emptyset \\ \text{glb}\,\{d(a, b) : a \in A,\ b \in B\}, & \text{if } A \neq \emptyset \neq B \end{cases}
$$

is called the *distance between A and B*. By $A\delta B$ iff $d(A, B) = 0$ a relation δ on $\mathcal{P}(X)$ is defined such that the following are satisfied (instead of $(A, B) \notin \delta$ we write $A \not\delta B$):

Prox$_1$) $A\delta B \Longrightarrow B\delta A$.
Prox$_2$) $(A \cup B)\,\delta C \Longleftrightarrow A\delta C$ or $B\delta C$.
Prox$_3$) $\{x\}\delta\{x\}$ for each $x \in X$.
Prox$_4$) $\emptyset \not\delta X$.
Prox$_5$) $A \not\delta B \Longrightarrow \exists\, C, D \in \mathcal{P}(X)$ with $A \subset C$, $B \subset D$ and $C \cap D = \emptyset$ such that $A \not\delta (X \setminus C)$ and $B \not\delta (X \setminus D)$.

An *Efremovič proximity space* is a pair (X, δ) where X is a set and δ a relation on $\mathcal{P}(X)$ such that Prox$_1$) – Prox$_5$) are satisfied. *Thus, the concept of near-*

ness between two sets is axiomatized. A map $f : (X, \delta) \longrightarrow (X', \delta')$ between Efremovič proximity spaces is called *proximally continuous* provided that $A\delta B$ implies $f[A]\delta' f[B]$ for each $(A, B) \in \mathcal{P}(X) \times \mathcal{P}(X)$.
The construct **E–Prox** of Efremovič proximity spaces (and proximally continuous maps) is (concretely) isomorphic to **Prox** (cf. [108; 10.4.5.]).

2) A topological space $(X, \mathcal{X})$ is called *fully normal* provided that every open cover of X has an open star–refinement. A fully normal T_1–space $(X, \mathcal{X})$ is called *paracompact*. Equivalently, a topological space $(X, \mathcal{X})$ is paracompact iff it is a Hausdorff space and each open cover $\mathcal{U}$ of X has an open locally finite refinement $\mathcal{V}$ (i.e. for every $x \in X$, there is a neighborhood U_x of x which intersects only finitely many $V \in \mathcal{V}$) (cf. [149; 20.15.]). Thus, every *compact Hausdorff space is paracompact*. But also every *metrizable topological space is paracompact* (This has been proved by A.H. Stone [138] in 1948 (cf. also [43; 4.4.1.])). Since every paracompact topological space $(X, \mathcal{X})$ is normal (cf. [149; 20.10.]), it is completely regular and its corresponding fine uniform structure μ_F can be described as follows:

$$\mu_F = \{\mathcal{A} \subset \mathcal{P}(X) : \text{ there is some open cover } \mathcal{O} \text{ of } X \text{ with } \mathcal{O} \prec \mathcal{A}\}.$$

3) a) Let $(X, \mathcal{J}_X)$ be a semiuniform convergence space and $1_X : (X, \mathcal{J}_X) \longrightarrow (X, [\mathcal{W}])$ the bireflection of $(X, \mathcal{J}_X)$ w.r.t. **PrULim** (cf. 2.3.2.3.). Then $(X, \mathcal{W})$ is called the *underlying uniform space* of $(X, \mathcal{J}_X)$.
b) *If $(X, \mathcal{X})$ is a paracompact topological space and $(X, \mathcal{J}_{\gamma_{q\mathcal{X}}})$ the corresponding semiuniform convergence space, the underlying uniform space of $(X, \mathcal{J}_{\gamma_{q\mathcal{X}}})$ is the fine uniform space.*

(Let $(X, \mathcal{W})$ be the underlying uniform space of $(X, \mathcal{J}_{\gamma_{q\mathcal{X}}})$, where $(X, \mathcal{X})$ is a paracompact topological space. In order to prove $\mathcal{W} = \mathcal{W}_{\mu_F}$, let $W \in \mathcal{W}_{\mu_F}$. Then there is some $\mathcal{A} \in \mu_F$ (i.e. $\mathcal{A}$ is refined by some open cover $\mathcal{O}$ of X) such that $W \supset \bigcup_{A \in \mathcal{A}} A \times A$. For each $x \in X$, there is some $O \in \mathcal{O}$ with $x \in O$. Furthermore, there is some $A \in \mathcal{A}$ with $O \subset A$, and consequently, $W \supset O \times O$ with $x \in O$. Hence $W \in \mathcal{U}_{\mathcal{X}}(x) \times \mathcal{U}_{\mathcal{X}}(x)$ and thus,

$$(*) \qquad \mathcal{W}_{\mu_F} \subset \mathcal{U}_{\mathcal{X}}(x) \times \mathcal{U}_{\mathcal{X}}(x) \quad \text{for each} \quad x \in X$$

provided that $\mathcal{U}_{\mathcal{X}}(x)$ denotes the neighborhood filter of x w.r.t. $\mathcal{X}$. Put $\mathcal{J}_X = \mathcal{J}_{\gamma_{q\mathcal{X}}}$ and let $\mathcal{J}_X^*$ be the bireflective modification of $\mathcal{J}_X$ w.r.t. **ULim**. Then it follows from $(*)$

$$(**) \qquad \mathcal{W}_{\mu_F} \subset \mathcal{F} \quad \text{for each} \quad \mathcal{F} \in \mathcal{J}_X^* .$$

Because of $(**)$, $\mathcal{W}_{\mu_F} \subset \mathcal{W}$, since $\mathcal{W}$ is the finest uniformity on X which is coarser than each $\mathcal{F} \in \mathcal{J}_X^*$. Conversely, let $V \in \mathcal{W}$. Then $V \in \mathcal{F}$ for each $\mathcal{F} \in \mathcal{J}_X^*$. Since $\mathcal{J}_X \subset \mathcal{J}_X^*$, we obtain $V \in \mathcal{U}_{\mathcal{X}}(x) \times \mathcal{U}_{\mathcal{X}}(x)$ for each $x \in X$, i.e. for each $x \in X$, there is an open neighborhood O_x of x such that $O_x \times O_x \subset V$. Hence, $\bigcup_{x \in X} O_x \times O_x \subset V$ and $\mathcal{O} = \{O_x : x \in X\} \in \mu_F$ since $\mathcal{O}$ is an open cover of X, i.e. $V \in \mathcal{W}_{\mu_F}$.)

Examples. I) Let $\mathbb{R}_t$ be the usual topological space of real numbers (which may be regarded as a semiuniform convergence space). Then the underlying uniform space of $I\!R_t$ is the fine uniform space $I\!R_F$ of real numbers. It does not coincide with the usual uniform space $I\!R_u$ of real numbers (cf. 4.3.2.24.)

II) Let $[0,1]_t$ be the usual topological space of the unit interval $[0,1] \subset I\!R$ (which may be regarded as a semiuniform convergence space). Then the underlying uniform space of $[0,1]_t$ is the usual uniform space $[0,1]_u$ (induced by the Euclidean metric) [cf. 4.3.2.25.].

4) If (X,q) is a symmetric Kent convergence space and $(X,\mathcal{J}_{\gamma_q})$ its corresponding semiuniform convergence space, then $\mathcal{J}_{\gamma_q}$ is the finest semiuniform convergence structure inducing q (cf. 2.3.3.21.) and every continuous map $f:(X,\mathcal{J}_{\gamma_q}) \longrightarrow (Y,\mathcal{J}_Y)$ from $(X,\mathcal{J}_{\gamma_q})$ into a semiuniform convergence space $(Y,\mathcal{J}_Y)$ is uniformly continuous (cf. 2.3.3.26. and note that $(X,\mathcal{J}_{\gamma_q})$ is a convergence space. This parallels the result for fine uniform spaces (cf. 4.3.2.24.) in the realm of uniform spaces.

5) a) If $(X,\mathcal{W})$ is a uniform space and $1_X:(X,\mathcal{W}) \longrightarrow (X,\mathcal{W}_p)$ the bireflection of $(X,\mathcal{W})$ w.r.t. **Prox** (cf. 4.3.2.19.) then the Hausdorff completion $(X^*,(\mathcal{W}_p)^*)$ of $(X,\mathcal{W}_p)$ is compact (cf. 4.3.2.17. and 4.3.2.8.). It is called the *Samuel compactification* of the uniform space $(X,\mathcal{W})$.

b) If $(X,\mathcal{W})$ is a proximity space, then the Samuel compactification of $(X,\mathcal{W})$ is also called the *Smirnov compactification* of $(X,\mathcal{W})$.

c) Obviously, every separated proximity space is densely embedded in its Smirnov compactification.

4.3.3 Hausdorff compactifications of topological spaces

4.3.3.1 Definition. A dense embedding $f:(X,\mathcal{X}) \longrightarrow (Y,\mathcal{Y})$ from a topological space $(X,\mathcal{X})$ into a compact (topological) Hausdorff space $(Y,\mathcal{Y})$ is called a *Hausdorff compactification* of $(X,\mathcal{X})$.

4.3.3.2 Theorem. *If $f:(X,\mathcal{X}) \longrightarrow (Y,\mathcal{Y})$ is a Hausdorff compactification of a topological space $(X,\mathcal{X})$, then there is a unique separated* **Prox***-structure $\mathcal{W}$ on X such that $\mathcal{X}_\mathcal{W} = \mathcal{X}$ and there exists a homeomorphism $h:(X^*,\mathcal{X}_{\mathcal{W}^*}) \longrightarrow (Y,\mathcal{Y})$ with $h \circ r_X = f$, where $r_X:(X,\mathcal{W}) \longrightarrow (X^*,\mathcal{W}^*)$ denotes the Hausdorff completion of $(X,\mathcal{W})$.*

Proof. 1) Let $\mathcal{V}$ be the unique uniformity on Y such that $\mathcal{Y}_\mathcal{V} = \mathcal{Y}$ (cf. 4.3.2.25.) and let $\mathcal{W}$ be the initial uniformity on X w.r.t. $(X,f,(Y,\mathcal{V}))$. Then $f:(X,\mathcal{W}) \longrightarrow (Y,\mathcal{V})$ is an embedding. Since subspaces of precompact separated uniform spaces are precompact and separated, it follows that $\mathcal{W}$ is a separated **Prox**–structure. Obviously, $\mathcal{X}_\mathcal{W} = \mathcal{X}$ (cf. the corollary under 2.3.3.18.). By 4.3.2.8., $(Y,\mathcal{V})$ is complete. Hence $f:(X,\mathcal{W}) \longrightarrow (Y,\mathcal{V})$ is a dense embedding into a complete separated uniform space and thus an epireflection of $(X,\mathcal{W}) \in |\mathbf{SepUnif}|$ w.r.t. **CSepUnif**. Consequently, there is an isomorphism $h:(X^*,\mathcal{W}^*) \longrightarrow (Y,\mathcal{V})$ with $h \circ r_X = f$ (cf. 2.1.6.). Then the homeomorphism induced by h fulfills the desired property.

2) Let $(X, \widetilde{\mathcal{W}})$ be a separated proximity space with $\mathcal{X}_{\widetilde{\mathcal{W}}} = \mathcal{X}$ such that there exists a homeomorphism $\widetilde{h} : (\widetilde{X}^*, \mathcal{X}_{\widetilde{\mathcal{W}}^*}) \longrightarrow (Y, \mathcal{Y})$ with $\widetilde{h} \circ \widetilde{r}_X = f$, where $\widetilde{r}_X : (X, \widetilde{\mathcal{W}}) \longrightarrow (\widetilde{X}^*, \widetilde{\mathcal{W}}^*)$ denotes the Hausdorff completion of $(X, \widetilde{\mathcal{W}})$. Then $\widehat{h} = \widetilde{h}^{-1} \circ h : (X^*, \mathcal{X}_{\mathcal{W}^*}) \longrightarrow (\widetilde{X}^*, \mathcal{X}_{\widetilde{\mathcal{W}}^*})$ is a homeomorphism with $\widetilde{r}_X = \widehat{h} \circ r_X$. Since $(X^*, \mathcal{W}^*)$ and $(\widetilde{X}^*, \widetilde{\mathcal{W}}^*)$ are fine uniform spaces, namely compact separated uniform spaces, $\widehat{h} : (X^*, \mathcal{W}^*) \longrightarrow (\widetilde{X}^*, \widetilde{\mathcal{W}}^*)$ is an isomorphism.

Consider the following diagrams

$$\begin{array}{ccc} (X, \mathcal{W}) & \xrightarrow{1_X} & (X, \widetilde{\mathcal{W}}) \\ {\scriptstyle r_X}\downarrow & & \downarrow{\scriptstyle \widetilde{r}_X} \\ (X^*, \mathcal{W}^*) & \xrightarrow{\widehat{h}} & (\widetilde{X}^*, \widetilde{\mathcal{W}}^*) \end{array} \qquad\qquad \begin{array}{ccc} (X, \widetilde{\mathcal{W}}) & \xrightarrow{1_X} & (X, \mathcal{W}) \\ {\scriptstyle \widetilde{r}_X}\downarrow & & \downarrow{\scriptstyle r_X} \\ (\widetilde{X}^*, \widetilde{\mathcal{W}}^*) & \xrightarrow{\widehat{h}^{-1}} & (X^*, \mathcal{W}^*) \end{array}$$

These diagrams commute w.r.t. to the underlying sets and maps. Since $\mathcal{W}$ (resp. $\widetilde{\mathcal{W}}$) is initial w.r.t. r_X (resp. $\widetilde{r}_X$), we get that $1_X : (X, \mathcal{W}) \longrightarrow (X, \widetilde{\mathcal{W}})$ and $1_X : (X, \widetilde{\mathcal{W}}) \longrightarrow (X, \mathcal{W})$ are uniformly continuous so that $\widetilde{\mathcal{W}} \subset \mathcal{W}$ and $\mathcal{W} \subset \widetilde{\mathcal{W}}$, i.e. $\mathcal{W} = \widetilde{\mathcal{W}}$.

4.3.3.3 Remark. The above result on Hausdorff compactifications of a topological space $(X, \mathcal{X})$ can be partially enlarged to *completely uniformizable T_0–extensions* of $(X, \mathcal{X})$, i.e. to dense embeddings of $(X, \mathcal{X})$ into completely uniformizable T_0–spaces (a topological space $(Y, \mathcal{Y})$ is called *completely uniformizable* provided that there is a complete uniformity $\mathcal{W}$ on X inducing $\mathcal{Y}$) as the following theorem shows:

Theorem. If $f : (X, \mathcal{X}) \longrightarrow (Y, \mathcal{Y})$ is a completely uniformizable T_0–extension of $(X, \mathcal{X})$, then there is a separated uniformity $\mathcal{W}$ on X such that $\mathcal{X}_{\mathcal{W}} = \mathcal{X}$ and there exists a homeomorphism $h : (X^*, \mathcal{X}_{\mathcal{W}^*}) \longrightarrow (Y, \mathcal{Y})$ with $h \circ r_X = f$ provided that $r_X : (X, \mathcal{W}) \longrightarrow (X^*, \mathcal{W}^*)$ denotes the Hausdorff completion of $(X, \mathcal{W})$.

(The *proof* parallels part 1) of the proof under 4.3.3.2.).

Obviously, a compact Hausdorff space $(Y, \mathcal{Y})$ is a completely uniformizable T_0–space; but there are completely uniformizable T_0–spaces which are non–compact, e.g. all non–compact realcompact topological spaces (a topological space $(Y, \mathcal{Y})$ is called *realcompact* provided that there is an embedding from $(Y, \mathcal{Y})$ into a closed subspace of $(\mathbb{R}_t)^I$ for some set I); namely, if $(Y, \mathcal{Y})$ is realcompact, it may be considered to be a closed subspace of $(\mathbb{R}_t)^I$ for some I and consequently, the uniform subspace $(Y, \mathcal{W})$ of $(\mathbb{R}_u)^I$ is closed, because $\mathcal{Y}_{\mathcal{W}} = \mathcal{Y}$ (initial uniformities induce initial topologies!), and thus, $(Y, \mathcal{W})$ is complete and separated (cf. also 4.3.3.4. ②).

4.3.3.4 Examples. ① Let $(X, \mathcal{X})$ be a Tychonoff space and $\mathcal{W}$ the fine uniform structure on X. If $1_X : (X, \mathcal{W}) \longrightarrow (X, \mathcal{W}_p)$ denotes the bireflection of $(X, \mathcal{W})$ w.r.t. **Prox**, then the Hausdorff completion $(X^*, (\mathcal{W}_p)^*)$ of $(X, \mathcal{W}_p)$ is compact and its underlying topological space is the *Stone–Čech compactification* $\beta(X)$ of $(X,\mathcal{X})$: Let $(Y,\mathcal{Y})$ be a compact Hausdorff space and $f:(X,\mathcal{X})\longrightarrow(Y,\mathcal{Y})$ a continuous map. Furthermore, let $\mathcal{R}$ be the unique uniformity with $\mathcal{Y}_{\mathcal{R}}=\mathcal{Y}$. Then $f:(X,\mathcal{W})\longrightarrow(Y,\mathcal{R})$ is uniformly continuous, because $(X,\mathcal{W})$ is a fine uniform space. Since $1_X:(X,\mathcal{W})\longrightarrow(X,\mathcal{W}_p)$ is a reflection, $f:(X,\mathcal{W}_p)\longrightarrow(Y,\mathcal{R})$ is also uniformly continuous and $\mathcal{X}_{\mathcal{W}_p}=\mathcal{X}$ (cf. 4.3.2.20.). It follows from the reflection property of $r_X:(X,\mathcal{W}_p)\longrightarrow(X^*,(\mathcal{W}_p)^*)$ that there is a unique uniformly continuous map $f^*:(X^*,(\mathcal{W}_p)^*)\longrightarrow(Y,\mathcal{R})$ such that $f^*\circ r_X=f$; since $(X^*,(\mathcal{W}_p)^*)$ is a fine uniform space, f^* is the unique continuous map with $f^*\circ r_X=f$. The continuous map $r_X:(X,\mathcal{X})\longrightarrow(X^*,\mathcal{X}_{(\mathcal{W}_p)^*})=\beta(X)$ is also denoted by β_X; it is a dense embedding (cf. also 2.1.5. ②).

② Let $(X,\mathcal{X})$ be a Tychonoff space and $(f_i:(X,\mathcal{X})\longrightarrow I\!R_t)_{i\in I}$ the family of all continuous maps from $(X,\mathcal{X})$ into the usual topological space $I\!R_t$ of real numbers. If $\mathcal{W}$ denotes the initial uniformity on X w.r.t. $(X,f_i,I\!R_u,I)$, then the underlying topological space of the Hausdorff completion $(X^*,\mathcal{W}^*)$ of $(X,\mathcal{W})$ is denoted by $v(X)$ and called the *Hewitt realcompactification* of $(X,\mathcal{X})$. It is characterized by the following universal property: Let $f:(X,\mathcal{X})\longrightarrow(Y,\mathcal{Y})$ be a continuous map from $(X,\mathcal{X})$ into a realcompact space $(Y,\mathcal{Y})$. Then there is a unique continuous map $\overline{f}:v(X)\longrightarrow(Y,\mathcal{Y})$ such that $\overline{f}\circ v_X=f$ provided that $v_X:(X,\mathcal{X})\longrightarrow v(X)$ denotes the canonical dense embedding, i.e. the underlying continuous map of $r_X:(X,\mathcal{W})\longrightarrow(X^*,\mathcal{W}^*)$.

[1) $(X,\mathcal{X}_{\mathcal{W}})=(X,\mathcal{X})$ (cf. the proof of 2.3.3.19.).
2) $g:(X,\mathcal{W})\longrightarrow\prod_{i\in I}I\!R^i_u$ is an embedding provided that $I\!R^i_u=I\!R_u$ for each $i\in I$ and $(g(x))_i=f_i(x)$ for each $i\in I$ and each $x\in X$:

a) Let $h:(Y,\mathcal{R})\longrightarrow(X,\mathcal{W})$ be a map from a uniform space $(Y,\mathcal{R})$ into $(X,\mathcal{W})$ such that $g\circ h$ is uniformly continuous. Since $p_i\circ g=f_i$ for each $i\in I$ ($p_i:\prod_{i\in I}I\!R^i_u\longrightarrow I\!R_u$ denotes the i-th projection!), it follows that $f_i\circ h=p_i\circ(g\circ h)$ is uniformly continuous for each $i\in I$. Consequently, h is uniformly continuous, because $\mathcal{W}$ is initial w.r.t. $(X,f_i,I\!R_u,I)$. Thus, g is initial.

b) g is injective, because $(X,\mathcal{X})$ is a Tychonoff space.

3) $v(X)=(X^*,\mathcal{X}_{\mathcal{W}^*})$ is realcompact: $\overline{g[X]}\subset\prod_{i\in I}I\!R^i_u$ is complete and separated (as a closed subspace of a complete space) and contains $g[X]\cong X$ as a dense subspace. Consequently, $(X^*,\mathcal{W}^*)\cong\overline{g[X]}$ and $(X^*,\mathcal{X}_{\mathcal{W}^*})$ is realcompact.

4) Let $f:(X,\mathcal{X})\longrightarrow(Y,\mathcal{Y})$ be a continuous map from $(X,\mathcal{X})$ into a realcompact space $(Y,\mathcal{Y})$. Then $(Y,\mathcal{Y})$ may be considered to be a closed subspace of $I\!R^J_t$ for some J. Hence $p_j\circ i_Y\circ f:(X,\mathcal{X})\longrightarrow I\!R_t$ is a continuous map provided that $p_j:I\!R^J_t\longrightarrow I\!R_t$ denotes the j–th projection for each $j\in J$ and $i_Y:Y\longrightarrow I\!R^J_t$ is the inclusion map. By the definition of $\mathcal{W}$, $f_j=p_j\circ i_Y\circ f:(X,\mathcal{W})\longrightarrow I\!R_u$ is uniformly continuous for each $j\in J$. Then for each $j\in J$, there is a unique uniformly continuous map $\overline{f_j}:(X^*,\mathcal{W}^*)\longrightarrow I\!R_u$ with $\overline{f_j}\circ r_X=f_j$. Thus, $\overline{f_j}:v(X)\longrightarrow I\!R_t$ is a continuous map with $\overline{f_j}\circ v_X=f_j$ for each $j\in J$. Consequently, $\overline{f}:v(X)\longrightarrow I\!R^J_t$ defined by $p_j(\overline{f}(z))=\overline{f_j}(z)$ for each point z of $v(X)$ is a continuous extension of $i_Y\circ f$ (without loss of generality X may be assumed to be a subset of $v(X)$!). Since X is dense in $v(X)$, $\overline{f}[v(X)]=\overline{f}\,[\overline{X}]\subset$

$\overline{\overline{f}[X]} = \overline{f[X]} \subset \overline{Y} = Y$, i.e. $f^* : v(X) \longrightarrow (Y, \mathcal{Y})$ defined by $f^*(z) = \overline{f}(z)$ for each point z of $v(X)$ is a continuous extension of f. Since X is dense in $v(X)$ and $(Y, \mathcal{Y})$ is a Hausdorff space, f^* is the desired unique continuous extension of f.]

③ A topological space $(X, \mathcal{X})$ is called *zerodimensional* provided that it is (up to isomorphism) a subspace of D_2^I for some set I, where D_2 denotes the two–point discrete topological space. Since D_2 may be regarded as a subspace of $\mathbb{R}_t$, every zerodimensional topological space is a Tychonoff space. Furthermore, since products and subspaces of totally disconnected topological spaces are totally disconnected, every zerodimensional space is totally disconnected. The *Cantor set* $D_2^{\mathbb{N}}$ is an example of a compact zerodimensional space. Now let $(X, \mathcal{X})$ be a zerodimensional space and $(f_i : (X, \mathcal{X}) \longrightarrow D_2)_{i \in I}$ the set of all continuous maps from $(X, \mathcal{X})$ into D_2. If D_2^{Δ} denotes the two–point discrete uniform space, i.e. $D_2^{\Delta} = (\{0, 1\}, (\{\Delta\}))$ with $\Delta = \{(0, 0), (1, 1)\}$, and $\mathcal{W}$ is the initial uniformity on X w.r.t. $(X, f_i, D_2^{\Delta}, I)$, then the underlying topological space of the Hausdorff completion $(X^*, \mathcal{W}^*)$ of $(X, \mathcal{W})$ is denoted by $\xi(X)$ and called the *Banaschewski compactification* of $(X, \mathcal{X})$. It is characterized by the following universal property: Let $f : (X, \mathcal{X}) \longrightarrow (Y, \mathcal{Y})$ be a continuous map from $(X, \mathcal{X})$ into a compact zerodimensional space $(Y, \mathcal{Y})$. Then there is a unique continuous map $\overline{f} : \xi(X) \longrightarrow (Y, \mathcal{Y})$ such that $\overline{f} \circ \xi_X = f$ provided that $\xi_X : (X, \mathcal{X}) \longrightarrow \xi(X)$ denotes the canonical dense embedding, i.e. the underlying continuous map of $r_X : (X, \mathcal{W}) \longrightarrow (X^*, \mathcal{W}^*)$ (The proof follows the pattern of the proof of the corresponding property for $v(X)$ provided that $\mathbb{R}_u$ is replaced by D_2^{Δ}. Note $\mathcal{X}_{\mathcal{W}} = \mathcal{X}$, since $(f_i : (X, \mathcal{X}) \longrightarrow D_2)_{i \in I}$ is initial.).

④ Let $(X, \mathcal{X})$ be a locally compact, non–compact Hausdorff space. Then a **U–Mer**–structure μ on X is defined by

$\mu = \{\mathcal{A} \subset \mathcal{P}(X) :$ there is an open cover $\mathcal{O}$ of X with $\mathcal{O} \prec \mathcal{A}$ and there is some $U \in \mathcal{O}$ such that $X \setminus U$ is compact in $(X, \mathcal{X})\}$.

The underlying topological space of the Hausdorff completion $(X^*, (\mathcal{W}_\mu)^*)$ of $(X, \mathcal{W}_\mu)$ is called the *Alexandroff compactification* of $(X, \mathcal{X})$. It can be obtained by the addition of a single point: Let $Y = X \cup \{\infty\}$ with $\infty \notin X$ and $\mathcal{Y} = \mathcal{X} \cup \{\mathrm{O} \subset X : \infty \in \mathrm{O}$ and $Y \setminus \mathrm{O}$ is a compact subset of $(X, \mathcal{X})\}$. Then $(Y, \mathcal{Y})$ is a compact Hausdorff space containing $(X, \mathcal{X})$ as a dense subspace (exercise!). According to 4.3.3.2. $(Y, \mathcal{Y})$ is homeomorphic to $(X^*, \mathcal{X}_{(\mathcal{W}_\mu)^*})$, i.e. to the Alexandroff compactification of $(X, \mathcal{X})$, because μ is the subspace structure of the fine uniform structure μ_F on Y.

Therefore, the Alexandroff compactification of $(X, \mathcal{X})$ is also called the *one–point compactification* of $(X, \mathcal{X})$ and is denoted by $\omega(X)$. Note that in Function Theory the one–point compactification of the complex plane is considered; it is homeomorphic to the two–dimensional sphere S^2.

4.4 The simple completion and the Wyler completion

In the last section of this chapter the question is solved, whether every semiuniform convergence space $(X, \mathcal{J}_X)$ can be densely embedded into a complete semiuniform convergence space, i.e. whether every semiuniform convergence space $(X, \mathcal{J}_X)$ has a completion.

4.4.1 Definitions. 1) A filter $\mathcal{F}$ on a set X is called *fixed* (resp. *free*) provided that $\bigcap \mathcal{F} \neq \emptyset$ (resp. $\bigcap \mathcal{F} = \emptyset$).

2) Let (X, γ) be a filter space. Then $\mathcal{F} \in \gamma$ is called *γ-free* provided that there is no fixed filter $\mathcal{G} \in \gamma$ with $\mathcal{G} \subset \mathcal{F}$.

4.4.2 Proposition. *If (X, γ) is a filter space, then there is an equivalence relation $\sim$ on γ defined by*

$$\mathcal{F} \sim \mathcal{G} \iff \begin{array}{l} \text{There exist finitely many } \mathcal{F}_0, \cdots, \mathcal{F}_n \in \gamma \text{ with } \mathcal{F}_0 = \mathcal{F} \\ \mathcal{F}_n = \mathcal{G} \text{ and such that } \sup\{\mathcal{F}_{i-1}, \mathcal{F}_i\} \text{ exists} \\ \text{for each } i \in \{1, \dots, n\}. \end{array}$$

(The *proof* is obvious.)

4.4.3 Remark. If (X, γ) is a filter space and $\mathcal{F} \in \gamma$, then the equivalence class of $\mathcal{F}$ with respect to $\sim$ is denoted by $[\mathcal{F}]$. If (X, γ) is a Cauchy space, then

$$\mathcal{F} \sim \mathcal{G} \iff \mathcal{F} \cap \mathcal{G} \in \gamma .$$

4.4.4 Proposition. *Let (X, γ) be a filter space and $\mathcal{F} \in \gamma$. Then the following are equivalent:*

(1) *$\mathcal{F}$ is γ-free.*
(2) *$\mathcal{F}$ does not converge in the underlying (symmetric) Kent convergence space.*

If (X, γ) is a Cauchy space, then each of the above conditions is equivalent to

(3) *$[\mathcal{F}] \neq [\dot{x}]$ for each $x \in X$.*

Proof. "(1) $\Longrightarrow$ (2)" (indirectly). If $\mathcal{F} \in \gamma$ converges to $x \in X$ in (X, q_γ), then $\mathcal{F} \cap \dot{x} \in \gamma$. Since $\mathcal{F} \cap \dot{x} \subset \mathcal{F}$ is fixed, $\mathcal{F}$ is not γ-free.

"(2) $\Longrightarrow$ (1)" (indirectly). If $\mathcal{F} \in \gamma$ is not γ-free, then there exists a fixed $\mathcal{F}_0 \in \gamma$ such that $\mathcal{F}_0 \subset \mathcal{F}$. Choose $x \in \bigcap \mathcal{F}_0$. Then $\mathcal{F}_0 \cap \dot{x} = \mathcal{F}_0 \subset \mathcal{F} \cap \dot{x}$ and consequently $\mathcal{F} \cap \dot{x} \in \gamma$, i.e. $(\mathcal{F}, x) \in q_\gamma$.

Let (X, γ) be a Cauchy space:
"(1) $\Longrightarrow$ (3)" (indirectly). If there is some $x \in X$ such that $[\mathcal{F}] = [\dot{x}]$, then $\mathcal{F} \cap \dot{x} \in \gamma$ is a fixed filter. Thus, $\mathcal{F}$ is not γ-free.

"(3) $\Longrightarrow$ (1)" (indirectly). If $\mathcal{F}$ is not γ-free, then there is some $x \in X$ such that $\mathcal{F} \cap \dot{x} \in \gamma$ (cf. "(2) $\Longrightarrow$ (1)"), i.e. $[\mathcal{F}] = [\dot{x}]$.

4.4.5 Theorem. *Let $(X, \mathcal{J}_X)$ be a semiuniform convergence space. Put $Y = X \cup \{[\mathcal{F}] : \mathcal{F}\ \gamma_{\mathcal{J}_X}\text{-free}\}$. If $i : X \longrightarrow Y$ denotes the inclusion map, then an* **SUConv**-*structure $\mathcal{J}_Y$ on Y is defined by*

$$\mathcal{F} \in \mathcal{J}_Y \text{ iff there is some } \mathcal{G} \in \mathcal{B}_Y \text{ with } \mathcal{F} \supset \mathcal{G},$$

where $\mathcal{B}_Y = \{i \times i(\mathcal{H}) : \mathcal{H} \in \mathcal{J}_X\} \cup \{(i(\mathcal{F}) \cap [\dot{\mathcal{F}}]) \times (i(\mathcal{F}) \cap [\dot{\mathcal{F}}]) : \mathcal{F}\ \gamma_{\mathcal{J}_X}\text{-free}\}$, and $(Y, \mathcal{J}_Y)$ is a complete semiuniform convergence space containing $(X, \mathcal{J}_X)$ as a dense subspace (i.e. $\overline{X} = Y$, where the closure is formed with respect to $(Y, q_{\gamma_{\mathcal{J}_Y}})$). The set $\gamma_{\mathcal{J}_Y}$ of all $\mathcal{J}_Y$-Cauchy filters is generated by

$$\{i(\mathcal{F}) : \mathcal{F} \in \gamma_{\mathcal{J}_X} \text{ not } \gamma_{\mathcal{J}_X}\text{-free}\} \cup \{i(\mathcal{F}) \cap [\dot{\mathcal{F}}] : \mathcal{F}\ \gamma_{\mathcal{J}_X}\text{-free}\}.$$

Proof. It is easy to check that $(Y, \mathcal{J}_Y) \in |\mathbf{SUConv}|$. Furthermore, $(X, \mathcal{J}_X)$ is a subspace of $(Y, \mathcal{J}_Y)$: Since $(i \times i)(\mathcal{H}) \in \mathcal{J}_Y$ for each $\mathcal{H} \in \mathcal{J}_X$, $\mathcal{J}_X \subset (\mathcal{J}_Y)_X = \{\mathcal{K} \in F(X \times X) : (i \times i)(\mathcal{K}) \in \mathcal{J}_Y\}$. Conversely, if $\mathcal{K} \in (\mathcal{J}_Y)_X$, then 1) $(i \times i)(\mathcal{K}) \supset (i \times i)(\mathcal{H})$ with $\mathcal{H} \in \mathcal{J}_X$ or 2) $(i \times i)(\mathcal{K}) \supset (i(\mathcal{F}) \cap [\dot{\mathcal{F}}]) \times (i(\mathcal{F}) \cap [\dot{\mathcal{F}}])$ with $\mathcal{F}\ \gamma_{\mathcal{J}_X}$-free. In the first case we obtain $\mathcal{K} = (i \times i)^{-1}((i \times i)(\mathcal{K})) \supset \mathcal{H} = (i \times i)^{-1}((i \times i)(\mathcal{H}))$ and thus $\mathcal{K} \in \mathcal{J}_X$. In the second case we get $\mathcal{K} \supset (i \times i)^{-1}((i(\mathcal{F}) \cap [\dot{\mathcal{F}}]) \times (i(\mathcal{F}) \cap [\dot{\mathcal{F}}])) = i^{-1}(i(\mathcal{F}) \cap [\dot{\mathcal{F}}]) \times i^{-1}(i(\mathcal{F}) \cap [\dot{\mathcal{F}}]) = \mathcal{F} \times \mathcal{F}$. Since $\mathcal{F} \times \mathcal{F} \in \mathcal{J}_X$, $\mathcal{K} \in \mathcal{J}_X$. Next we prove

$$(*) \begin{cases} \gamma_{\mathcal{J}_Y} = \gamma' \text{ with } \gamma' = \{\mathcal{K} \in F(Y) : \text{ there is some } \mathcal{G} \in \mathcal{B} \text{ with } \mathcal{K} \supset \mathcal{G}\}, \\ \text{where } \mathcal{B} = \{i(\mathcal{F}) : \mathcal{F} \in \gamma_{\mathcal{J}_X} \text{ not } \gamma_{\mathcal{J}_X}\text{-free}\} \cup \{i(\mathcal{F}) \cap [\dot{\mathcal{F}}] : \mathcal{F}\ \gamma_{\mathcal{J}_X}\text{-free}\}. \end{cases}$$

a) If $\mathcal{K} \in \gamma_{\mathcal{J}_Y}$, then $\mathcal{K} \times \mathcal{K} \in \mathcal{J}_Y$. Thus, 1) $\mathcal{K} \times \mathcal{K} \supset (i \times i)(\mathcal{H})$ with $\mathcal{H} \in \mathcal{J}_X$ or 2) $\mathcal{K} \times \mathcal{K} \supset (i(\mathcal{F}) \cap [\dot{\mathcal{F}}]) \times (i(\mathcal{F}) \cap [\dot{\mathcal{F}}])$ with $\mathcal{F}\ \gamma_{\mathcal{J}_X}$-free. In the first case X belongs to $\mathcal{K}$ (namely: $X \times X \in \mathcal{H} \subset (i \times i)(\mathcal{H}) \subset \mathcal{K} \times \mathcal{K}$ means $X \times X \supset K \times K$ with $K \in \mathcal{K}$ and consequently $X \supset K$, i.e. $X \in \mathcal{K}$). Thus, $i^{-1}(\mathcal{K}) = \mathcal{F}$ exists and, since $\mathcal{K}$ is a $\mathcal{J}_Y$-Cauchy filter, it is a $\mathcal{J}_X$-Cauchy filter. Furthermore, $i(\mathcal{F}) = \mathcal{K}$ (since $X \in \mathcal{K}$), i.e. $\mathcal{K} \in \gamma'$. In the second case $\mathcal{K} \supset i(\mathcal{F}) \cap [\dot{\mathcal{F}}]$ with $\mathcal{F}\gamma_{\mathcal{J}_X}$-free, i.e. $\mathcal{K} \in \gamma'$.

b) If $\mathcal{K} \in \gamma'$, then 1) $\mathcal{K} \supset i(\mathcal{F})$ with $\mathcal{F} \in \gamma_{\mathcal{J}_X}$ not $\gamma_{\mathcal{J}_X}$-free or 2) $\mathcal{K} \supset i(\mathcal{F}) \cap [\dot{\mathcal{F}}]$ with $\mathcal{F}\ \gamma_{\mathcal{J}_X}$-free. In the first case $\mathcal{K} \times \mathcal{K} \supset i(\mathcal{F}) \times i(\mathcal{F}) = (i \times i)(\mathcal{F} \times \mathcal{F})$, i.e. $\mathcal{K} \times \mathcal{K} \in \mathcal{J}_Y$ since $\mathcal{F} \times \mathcal{F} \in \mathcal{J}_X$. Thus $\mathcal{K} \in \gamma_{\mathcal{J}_Y}$. In the second case $\mathcal{K} \times \mathcal{K} \supset (i(\mathcal{F}) \cap [\dot{\mathcal{F}}]) \times (i(\mathcal{F}) \cap [\dot{\mathcal{F}}])$, i.e. $\mathcal{K} \times \mathcal{K} \in \mathcal{J}_Y$. Hence $\mathcal{K} \in \gamma_{\mathcal{J}_Y}$.

In order to prove that $(Y, \mathcal{J}_Y)$ is complete, let $\mathcal{H} \in \gamma_{\mathcal{J}_Y}$. Then 1) $\mathcal{H} = i(\mathcal{F})$ where $\mathcal{F}$ converges in $(X, q_{\gamma_{\mathcal{J}_X}})$ or 2) $\mathcal{H} = i(\mathcal{F}) \cap [\dot{\mathcal{F}}]$ where $\mathcal{F}$ does not converge in $(X, q_{\gamma_{\mathcal{J}_X}})$. Since $(X, q_{\gamma_{\mathcal{J}_X}})$ is a subspace of $(Y, q_{\gamma_{\mathcal{J}_Y}})$ (cf. 2.3.3.17), $i : X \longrightarrow Y$ is continuous and therefore $\mathcal{H}$ converges in $(Y, q_{\gamma_{\mathcal{J}_Y}})$ in the first case. In the second case $\mathcal{H}$ converges to $[\mathcal{F}]$ in $(Y, q_{\gamma_{\mathcal{J}_Y}})$ because $\mathcal{H} \cap [\dot{\mathcal{F}}] = \mathcal{H} \in \gamma_{\mathcal{J}_Y}$.
It remains to show that X is dense in Y: It suffices to prove $Y \subset \overline{X} = \{y \in Y:$ there is some $\mathcal{G} \in F(Y)$ with $(\mathcal{G}, y) \in q_{\gamma_{\mathcal{J}_Y}}$ and $X \in \mathcal{G}\}$. If $y \in Y$, then 1) $y \in X$ or 2) $y = [\mathcal{F}]$, where $\mathcal{F}$ does not converge in $(X, q_{\gamma_{\mathcal{J}_X}})$. Since $X \in i(\dot{y}) \xrightarrow{q_{\gamma_{\mathcal{J}_Y}}} i(y) = y$, we get $y \in \overline{X}$ in the first case. In the second case we have $X \in i(\mathcal{F}) \xrightarrow{q_{\gamma_{\mathcal{J}_Y}}} [\mathcal{F}] = y$, since $i(\mathcal{F}) \cap [\dot{\mathcal{F}}] \in \gamma_{\mathcal{J}_Y}$. Thus, $y \in \overline{X}$.

4.4.6 Definition. Let $(X, \mathcal{J}_X)$ be a semiuniform convergence space and $(Y, \mathcal{J}_Y)$ the complete semiuniform convergence space constructed in 4.4.5. Then the inclusion map $i : (X, \mathcal{J}_X) \longrightarrow (Y, \mathcal{J}_Y)$ is called the *simple completion* of $(X, \mathcal{J}_X)$. Occasionally, $(Y, \mathcal{J}_Y)$ is already called the simple completion of $(X, \mathcal{J}_X)$.

4.4.7 Theorem. *Let $(X, \mathcal{J}_X)$ be a semiuniform convergence space. Then the following are equivalent:*

(1) *$(X, \mathcal{J}_X)$ is* **Fil***-determined.*
(2) *The simple completion $(Y, \mathcal{J}_Y)$ of $(X, \mathcal{J}_X)$ is* **Fil***-determined.*
(3) *$(X, \mathcal{J}_X)$ is a dense subspace (with respect to* **SUConv***) of a convergence space.*
(4) *$(X, \mathcal{J}_X)$ is a subspace (with respect to* **SUConv***) of a convergence space.*

Proof. (1) $\Longrightarrow$ (2). Since $(X, \mathcal{J}_X)$ is **Fil**–determined, $\mathcal{J}_X = \{\mathcal{F} \in F(X \times X)$: there is some $\mathcal{G} \in \gamma_{\mathcal{J}_X}$ with $\mathcal{F} \supset \mathcal{G} \times \mathcal{G}\}$. We have to prove that $\mathcal{J}_Y \subset \mathcal{J}_{\gamma_{\mathcal{J}_Y}} = \{\mathcal{H} \in F(Y \times Y)$: there is some $\mathcal{K} \in \gamma_{\mathcal{J}_Y}$ with $\mathcal{H} \supset \mathcal{K} \times \mathcal{K}\}$, since $\mathcal{J}_{\gamma_{\mathcal{J}_Y}} \subset \mathcal{J}_Y$ is always valid. Let $\mathcal{H} \in \mathcal{J}_Y$. Then 1) $\mathcal{H} \supset (i \times i)(\mathcal{K})$ with $\mathcal{K} \in \mathcal{J}_X$ or 2) $\mathcal{H} \supset (i(\mathcal{F}) \cap [\dot{\mathcal{F}}]) \times (i(\mathcal{F}) \cap [\dot{\mathcal{F}}])$ with $\mathcal{F}$ $\gamma_{\mathcal{J}_X}$–free. In the first case there is some $\mathcal{G} \in \gamma_{\mathcal{J}_X}$ with $\mathcal{G} \times \mathcal{G} \subset \mathcal{K}$ and thus $(i \times i)(\mathcal{G} \times \mathcal{G}) = i(\mathcal{G}) \times i(\mathcal{G}) \subset (i \times i)(\mathcal{K}) \subset \mathcal{H}$. Since $i(\mathcal{G}) \in \gamma_{\mathcal{J}_Y}$ (note that $i : (X, \gamma_{\mathcal{J}_X}) \longrightarrow (Y, \gamma_{\mathcal{J}_Y})$ is Cauchy continuous), $i(\mathcal{G}) \times i(\mathcal{G}) \in \mathcal{J}_Y$ and consequently $\mathcal{H} \in \mathcal{J}_{\gamma_{\mathcal{J}_Y}}$. In the second case $\mathcal{K} = i(\mathcal{F}) \cap [\dot{\mathcal{F}}]$ with $\mathcal{F}$ $\gamma_{\mathcal{J}_X}$–free belongs to $\gamma_{\mathcal{J}_Y}$ (cf. 4.4.5.) and $\mathcal{K} \times \mathcal{K} \subset \mathcal{H}$, i.e. $\mathcal{H} \in \mathcal{J}_{\gamma_{\mathcal{J}_Y}}$.

(2) $\Longrightarrow$ (3). By assumption $(X, \mathcal{J}_X)$ is a dense subspace of the **Fil**–determined space $(Y, \mathcal{J}_Y)$. Since $(Y, \mathcal{J}_Y)$ is also complete, it follows from 2.3.3.24. that $(Y, \mathcal{J}_Y)$ is a convergence space.

(3) $\Longrightarrow$ (4). Obvious.

(4) $\Longrightarrow$ (1). By 2.3.3.24, each convergence space is **Fil**–determined, and **Fil–D–SUConv** is closed under formation of subspaces (cf. e.g. 2) b) in the proof of 2.3.3.6.).

4.4.8 Corollary. a) *If $(X, \mathcal{J}_X)$ is a complete semiuniform convergence space, then $(X, \mathcal{J}_X) = (Y, \mathcal{J}_Y)$.*

b) *If $(X, \mathcal{J}_X)$ is a convergence space, then $(X, \mathcal{J}_X) = (Y, \mathcal{J}_Y)$.*

Proof. a) If $(X, \mathcal{J}_X)$ is complete, then for each $\mathcal{J}_X$–Cauchy filter $\mathcal{F}$ there is some $x \in X$ with $\mathcal{F} \cap \dot{x} \in \gamma_{\mathcal{J}_X}$, i.e. $\mathcal{F}$ is not $\gamma_{\mathcal{J}_X}$–free. Thus, $X = Y$ and $\mathcal{J}_X = \mathcal{J}_Y$.

b) By 2.3.3.24, each convergence space is complete.

4.4.9 Definitions. 1) A generalized convergence space (X, q) is called
a) a T_0*–space* iff for each pair $(x, y) \in X \times X$, $(\dot{x}, y) \in q$ and $(\dot{y}, x) \in q$ imply $x = y$.
b) a T_1*–space* iff for each pair $(x, y) \in X \times X$, $(\dot{x}, y) \in q$ implies $x = y$.

2) A semiuniform convergence space $(X, \mathcal{J}_X)$ is called a T_0*–space* (resp. T_1*–space*)

iff $(X, q_{\gamma_{\mathcal{J}_X}})$ is a T_0–space (resp. T_1–space).

4.4.10 Remarks. 1) A topological space $(X, \mathcal{X})$ is a T_0–space (resp. T_1–space) in the usual sense iff $(X, q_{\mathcal{X}})$ is a T_0–space (resp. T_1–space) in the sense above, where $(\mathcal{F}, x) \in q_{\mathcal{X}}$ iff $\mathcal{F} \supset \mathcal{U}(x)$.

2) If $(X, \mathcal{J}_X)$ is a semiuniform convergence space, then the following are equivalent:
(1) $(X, \mathcal{J}_X)$ is a T_0–space.
(2) $(X, \mathcal{J}_X)$ is a T_1–space.
If $(X, \mathcal{J}_X)$ is a uniform limit space, then each of the above conditions is equivalent to

(3) $(X, \mathcal{J}_X)$ is a T_2–space.

(Note that the underlying Kent convergence space $(X, q_{\gamma_{\mathcal{J}_X}})$ of $(X, \mathcal{J}_X) \in |\mathbf{ULim}|$ is a T_{2W}–space [cf. 2.3.3.14. 2) b)]. Consequently, if $(X, q_{\gamma_{\mathcal{J}_X}})$ is a T_0–space, it is already a T_2–space).

4.4.11 Proposition. *A semiuniform convergence space $(X, \mathcal{J}_X)$ is a T_1–space iff its simple completion $(Y, \mathcal{J}_Y)$ is a T_1–space.*

Proof. 1) Let $(X, \mathcal{J}_X)$ be a T_1–space and $y_1, y_2 \in Y$ such that $\dot{y}_1 \longrightarrow y_2$. This can occur only if $y_1, y_2 \in X$ or $y_1 = y_2$.

2) Since obviously subspaces of T_1–spaces are T_1–spaces, $(X, \mathcal{J}_X)$ is a T_1–space provided that $(Y, \mathcal{J}_Y)$ is a T_1–space.
Thus, the proposition is proved.

4.4.12 Remarks. 1) Another property which is preserved by the simple completion is connectedness (resp. $\mathcal{E}$–connectedness provided that $\mathcal{E}$ is a class of T_1–spaces) as we will see in the next chapter.

2) Since **Fil** is (concretely) isomorphic to **Fil–D–SUConv**, filter spaces and **Fil**–determined semiuniform convergence spaces may be identified. Thus, it follows from 4.4.5. and 4.4.7. that the simple completion of filter spaces is *Császár's λ-completion of filter spaces* (cf. [34]).

4.4.13 Extension lemma. *Let $(X, \mathcal{J}_X)$ be a semiuniform convergence space, $i : (X, \mathcal{J}_X) \longrightarrow (Y, \mathcal{J}_Y)$ its simple completion, $(X', \mathcal{J}_{X'})$ a separated complete semiuniform convergence space and $f : (X, \mathcal{J}_X) \longrightarrow (X', \mathcal{J}_{X'})$ a uniformly continuous map. Then there is a unique uniformly continuous map $\overline{f} : (Y, \mathcal{J}_Y) \longrightarrow (X', \mathcal{J}_{X'})$ such that $\overline{f} \circ i = f$.*

Proof. A map $\overline{f} : Y \longrightarrow X'$ is defined by

$$\overline{f}(y) = \begin{cases} f(y) & \text{if } y \in X \\ x' & \text{if } y \in Y \setminus X \end{cases}$$

where $(f(\mathcal{F}), x') \in q_{\gamma_{\mathcal{J}_{X'}}}$ with $y = [\mathcal{F}]$ (x' exists, since $(X', \mathcal{J}_{X'})$ is complete, and it is uniquely determined, since $(X', \mathcal{J}_{X'})$ is a T_2-space; furthermore, it is independent of the choice of the representative $\mathcal{F}$). Obviously, $\overline{f} \circ i = f$. In order to prove that $\overline{f}$ is uniformly continuous, let $\mathcal{K} \in \mathcal{J}_Y$. Then 1) $\mathcal{K} \supset (i \times i)(\mathcal{H})$ with $\mathcal{H} \in \mathcal{J}_X$ or 2) $\mathcal{K} \supset (i(\mathcal{F}) \cap [\dot{\mathcal{F}}]) \times (i(\mathcal{F}) \cap [\dot{\mathcal{F}}])$ with $\mathcal{F}$ $\gamma_{\mathcal{J}_X}$-free. In the first case $(\overline{f} \times \overline{f})(\mathcal{K}) \supset (\overline{f} \times \overline{f})(i \times i)(\mathcal{H})) = (\overline{f} \circ i \times \overline{f} \circ i)(\mathcal{H}) = (f \times f)(\mathcal{H}) \in \mathcal{J}_{X'}$, i.e. $(\overline{f} \times \overline{f})(\mathcal{K}) \in \mathcal{J}_{X'}$. In the second case $(\overline{f} \times \overline{f})(\mathcal{K}) \supset (\overline{f} \times \overline{f})((i(\mathcal{F}) \cap [\dot{\mathcal{F}}]) \times (i(\mathcal{F}) \cap [\dot{\mathcal{F}}])) = \overline{f}(i(\mathcal{F}) \cap [\dot{\mathcal{F}}]) \times \overline{f}(i(\mathcal{F}) \cap [\dot{\mathcal{F}}]) = (f(\mathcal{F}) \cap \overline{f}([\dot{\mathcal{F}}])) \times (f(\mathcal{F}) \cap \overline{f}([\dot{\mathcal{F}}])) = (f(\mathcal{F}) \cap \dot{x}') \times (f(\mathcal{F}) \cap \dot{x}') \in \mathcal{J}_{X'}$ since $f(\mathcal{F}) \cap \dot{x}' \in \gamma_{\mathcal{J}_X}$ (i.e. $(f(\mathcal{F}), x') \in q_{\gamma_{\mathcal{J}_X}}$). Thus, $(\overline{f} \times \overline{f})(\mathcal{K}) \in \mathcal{J}_{X'}$. Since now $\overline{f} : Y \longrightarrow X$ is uniformly continuous, it is also continuous with respect to the underlying Kent convergence spaces. Hence, $\overline{f}$ is uniquely determined by $\overline{f} \circ i = f$, since X is dense in Y and X' is a T_2-space.

4.4.14 Theorem. *Let $(X, \mathcal{J}_X)$ be a separated uniform limit space and $(Y, \mathcal{J}_Y)$ its simple completion. Then the bireflective* **SULim***-modification of $(Y, \mathcal{J}_Y)$, namely $(Y, \mathcal{J}_Y^*)$, is a separated complete uniform limit space containing $(X, \mathcal{J}_X)$ as a dense subspace and being characterized (up to isomorphism) by the following universal property: If $f : (X, \mathcal{J}_X) \longrightarrow (X', \mathcal{J}_{X'})$ is a uniformly continuous map from $(X, \mathcal{J}_X)$ into a complete separated uniform limit space $(X', \mathcal{J}_{X'})$, then there is a unique uniformly continuous map $\overline{f} : (Y, \mathcal{J}_Y^*) \longrightarrow (X', \mathcal{J}_{X'})$ such that $\overline{f} \circ i = f$, where $i : X \longrightarrow Y$ denotes the inclusion map.*

Proof. 1) Put $Y' = \{[\mathcal{F}] : \mathcal{F} \in \gamma_{\mathcal{J}_X}\}$. Then $Y' = \{[\mathcal{F}] : \mathcal{F} \in \gamma_{\mathcal{J}_X}$ not $\gamma_{\mathcal{J}_X}$-free$\} \cup \{[\mathcal{F}] : \mathcal{F}$ $\gamma_{\mathcal{J}_X}$-free$\} = \{[\dot{x}] : x \in X\} \cup \{[\mathcal{F}] : \mathcal{F}$ $\gamma_{\mathcal{J}_X}$-free$\}$ (cf. 4.4.4.). The map $j : X \longrightarrow \{[\dot{x}] : x \in X\}$ defined by $j(x) = [\dot{x}]$ for each $x \in X$ is bijective, since $(X, \mathcal{J}_X)$ is a T_1-space. Thus, X and $\{[\dot{x}] : x \in X\}$ may be identified, i.e. $Y = Y'$, where $(Y, \mathcal{J}_Y)$ denotes the simple completion of $(X, \mathcal{J}_X)$.

2) A **ULim**-structure $\mathcal{J}_Y^W$ is defined on Y as follows: $\mathcal{F} \in \mathcal{J}_Y^W$ iff there is some $\mathcal{G} \in \mathcal{B}_Y^W$ with $\mathcal{F} \supset \mathcal{G}$, where $\mathcal{B}_Y^W = \{(i \times i)(\mathcal{H}) \cap ((i(\mathcal{F}_1) \times [\dot{\mathcal{F}}_1]) \cap ([\dot{\mathcal{F}}_1] \times i(\mathcal{F}_1))) \cap \cdots ((i(\mathcal{F}_n) \times [\dot{\mathcal{F}}_n]) \cap ([\dot{\mathcal{F}}_n] \times i(\mathcal{F}_n))) : \mathcal{H} \in \mathcal{J}_X$ and $\mathcal{F}_1, \ldots, \mathcal{F}_n \in \gamma_{\mathcal{J}_X}$ non-convergent$\}$ and $i : X \longrightarrow Y$ is the inclusion map. (Note: If $\mathcal{F} \in \gamma_{\mathcal{J}_X}$ converges to $x \in X$, i.e. $\mathcal{F} \times \dot{x} \in \mathcal{J}_X$, then $(i(\mathcal{F}) \times [\dot{\mathcal{F}}]) \cap ([\dot{\mathcal{F}}] \times i(\mathcal{F})) \in \mathcal{J}_Y^W$, since it is equal to $(i \times i)(\mathcal{H})$ with $\mathcal{H} = (\mathcal{F} \times \dot{x}) \cap (\dot{x} \times \mathcal{F})$). It may be assumed that the filters $\mathcal{F}_\mu$ are pairwise non-equivalent, namely if $\mathcal{F}_\mu \sim \mathcal{F}_\nu$, then $[\mathcal{F}_\mu] = [\mathcal{F}_\nu] = [\mathcal{F}_\mu \cap \mathcal{F}_\nu]$ and consequently, $((i(\mathcal{F}_\mu) \times [\dot{\mathcal{F}}_\mu]) \cap ([\dot{\mathcal{F}}_\mu] \times i(\mathcal{F}_\mu))) \cap ((i(\mathcal{F}_\nu) \times [\dot{\mathcal{F}}_\nu]) \cap ([\dot{\mathcal{F}}_\nu] \times i(\mathcal{F}_\nu))) \supset (i(\mathcal{F}_\mu \cap \mathcal{F}_\nu) \times [\mathcal{F}_\mu \dot{\cap} \mathcal{F}_\nu]) \cap ([\mathcal{F}_\mu \dot{\cap} \mathcal{F}_\nu] \times i(\mathcal{F}_\mu \cap \mathcal{F}_\nu))$. In the following we will make this additional assumption. In order to prove that $\mathcal{J}_Y^W$ is a **ULim**-structure it suffices to show that for filters $\widetilde{\mathcal{U}}, \widetilde{\mathcal{V}} \in \mathcal{B}_Y^W$, $\widetilde{\mathcal{U}}^{-1}$ and $\widetilde{\mathcal{U}} \circ \widetilde{\mathcal{V}}$ belong to $\mathcal{J}_Y^W$ provided that $\widetilde{\mathcal{U}} \circ \widetilde{\mathcal{V}}$ exists: Since $((i(\mathcal{F}_\mu) \times [\dot{\mathcal{F}}_\mu]) \cap ([\dot{\mathcal{F}}_\mu] \times i(\mathcal{F}_\mu)))^{-1} = (i(\mathcal{F}_\mu) \times [\dot{\mathcal{F}}_\mu])^{-1} \cap ([\dot{\mathcal{F}}_\mu] \times i(\mathcal{F}_\mu))^{-1} = ([\dot{\mathcal{F}}_\mu] \times i(\mathcal{F}_\mu)) \cap (i(\mathcal{F}_\mu) \cap [\dot{\mathcal{F}}_\mu])$ and $((i \times i)(\mathcal{H}))^{-1} = (i \times i)(\mathcal{H}^{-1})$, $\widetilde{\mathcal{U}}^{-1} \in \mathcal{B}_Y^W$ provided that $\widetilde{\mathcal{U}} \in \mathcal{B}_Y^W$. If $\mathcal{U}, \mathcal{V} \in \mathcal{J}_X$, and $\mathcal{F}, \mathcal{G} \in \gamma_{\mathcal{J}_X}$ are non-convergent, we get the following formulas:

(1) $(i \times i)(\mathcal{U}) \circ (i \times i)(\mathcal{V}) = (i \times i)(\mathcal{U} \circ \mathcal{V})$ provided that $\mathcal{U} \circ \mathcal{V}$ exists,

(2) $(i \times i)(\mathcal{V}) \circ ((i(\mathcal{F}) \times [\dot{\mathcal{F}}]) \cap ([\dot{\mathcal{F}}]) \times i(\mathcal{F}))) = [\dot{\mathcal{F}}] \times i(\mathcal{V}(\mathcal{F}))$, where $\mathcal{V}(\mathcal{F}) = (\{V[A] : A \in \mathcal{F},\ V \in \mathcal{V}\})$ and $V[A] = \bigcup_{x \in A} V(x)$,

(3) $((i(\mathcal{G}) \times [\dot{\mathcal{G}}]) \cap ([\dot{\mathcal{G}}] \times i(\mathcal{G}))) \circ (i \times i)(\mathcal{U}^{-1}) = i(\mathcal{U}(\mathcal{G})) \times [\dot{\mathcal{G}}]$,

(4) $((i(\mathcal{G}) \times [\dot{\mathcal{G}}]) \cap ([\dot{\mathcal{G}}] \times i(\mathcal{G}))) \circ ((i(\mathcal{F}) \times [\dot{\mathcal{F}}]) \cap ([\dot{\mathcal{F}}] \times i(\mathcal{F})))$ does not exist provided that $[\mathcal{F}] \neq [\mathcal{G}]$,

(5) $((i(\mathcal{G}) \times [\dot{\mathcal{G}}]) \cap ([\dot{\mathcal{G}}] \times i(\mathcal{G}))) \circ ((i(\mathcal{F}) \times [\dot{\mathcal{F}}]) \cap ([\dot{\mathcal{F}}] \times i(\mathcal{F}))) \supset (i \times i)(\mathcal{F} \times \mathcal{G})$ provided that $[\mathcal{F}] = [\mathcal{G}]$.

(e.g. concerning (3), $((i(\mathcal{G}) \times [\dot{\mathcal{G}}]) \cap ([\dot{\mathcal{G}}] \times i(\mathcal{G}))) \circ (i \times i)(\mathcal{U}^{-1})$ is generated by $\{((B \times \{[\mathcal{G}]\}) \cup (\{[\mathcal{G}]\} \times B)) \circ U^{-1}\colon U \in \mathcal{U},\ B \in \mathcal{G}\}$ whereas $i(\mathcal{U}(\mathcal{G})) \times [\dot{\mathcal{G}}]$ is generated by $\{U[B] \times \{[\mathcal{G}]\} : U \in \mathcal{U},\ B \in \mathcal{G}\}$. It is easily verified that both filterbases coincide, since $\mathcal{G}$ is non–convergent, i.e. $[\mathcal{G}] \in Y \setminus X$.)

Since $\mathcal{V} \circ (\mathcal{F} \times \mathcal{F}) = \mathcal{F} \times \mathcal{V}(\mathcal{F})$, $\mathcal{V}(\mathcal{F}) \times \mathcal{V}(\mathcal{F}) = (\mathcal{F} \times \mathcal{V}(\mathcal{F})) \circ (\mathcal{V}(\mathcal{F}) \times \mathcal{F}) \in \mathcal{J}_X$, i.e. $\mathcal{V}(\mathcal{F}) \in \gamma_{\mathcal{J}_X}$, and $(\mathcal{V}(\mathcal{F}) \cap \mathcal{F}) \times (\mathcal{V}(\mathcal{F}) \cap \mathcal{F}) = (\mathcal{F} \times \mathcal{V}(\mathcal{F})) \cap (\mathcal{V}(\mathcal{F}) \times \mathcal{F}) \cap (\mathcal{F} \times \mathcal{F}) \cap (\mathcal{V}(\mathcal{F}) \times \mathcal{V}(\mathcal{F})) \in \mathcal{J}_X$, i.e. $\mathcal{V}(\mathcal{F}) \cap \mathcal{F} \in \gamma_{\mathcal{J}_X}$ or equivalently $[\mathcal{V}(\mathcal{F})] = [\mathcal{F}]$. Thus, the product filter (2) is finer than $(i(\mathcal{V}(\mathcal{F})) \times [\dot{\mathcal{V}(\mathcal{F})}]) \cap ([\dot{\mathcal{V}(\mathcal{F})}] \times i(\mathcal{V}(\mathcal{F}))) \in \mathcal{J}_Y^W$. It is easily checked that the other filters under (1), (2) and (5) belong to $\mathcal{J}_Y^W$. Hence, using 2) c) in the proof of 2.3.2.3., $\widetilde{\mathcal{U}} \circ \widetilde{\mathcal{V}} \in \mathcal{J}_Y^W$.

3) A filter $\mathcal{K}$ on Y is a $\mathcal{J}_Y^W$–Cauchy filter iff there is some $\mathcal{J}_X$–Cauchy filter $\mathcal{F}$ such that $\mathcal{K} \supset i(\mathcal{F}) \cap [\dot{\mathcal{F}}]$:

a) "$\Longleftarrow$". Let $\mathcal{K} \in F(Y)$ such that $\mathcal{K} \supset i(\mathcal{F}) \cap [\dot{\mathcal{F}}]$ for some $\mathcal{F} \in \gamma_{\mathcal{J}_X}$. Then $(i(\mathcal{F}) \cap [\dot{\mathcal{F}}]) \times (i(\mathcal{F}) \cap [\dot{\mathcal{F}}]) = (i \times i)(\mathcal{F} \times \mathcal{F}) \cap ((i(\mathcal{F}) \times [\dot{\mathcal{F}}]) \cap ([\dot{\mathcal{F}}] \times i(\mathcal{F}))) \cap ([\dot{\mathcal{F}}] \times [\dot{\mathcal{F}}]) \in \mathcal{J}_Y^W$ and consequently, $\mathcal{K} \times \mathcal{K} \in \mathcal{J}_Y^W$, i.e. $\mathcal{K} \in \gamma_{\mathcal{J}_Y^W}$.

b) "$\Longrightarrow$". If $\mathcal{K} \in \gamma_{\mathcal{J}_Y^W}$, then $\mathcal{K} \times \mathcal{K} \supset \widetilde{\mathcal{U}}$, where $\widetilde{\mathcal{U}} \in \mathcal{B}_Y^W$ with $\mathcal{H} \in \mathcal{J}_X$ and pairwise non-equivalent filters $\mathcal{F}_\mu$. Put $y_\mu = [\mathcal{F}_\mu]$. Then $\{A_\mu \cup \{y_\mu\} : A_\mu \in \mathcal{F}_\mu\}$ is a base of $i(\mathcal{F}_\mu) \cap [\dot{\mathcal{F}}_\mu]$ and

$$\begin{aligned}\widetilde{\mathcal{B}} = \ &\{H \cup ((A_1 \times \{y_1\}) \cup (\{y_1\} \times A_1)) \cup \cdots \cup ((A_n \times \{y_n\}) \cup (\{y_n\} \times A_n)) : \\ &H \in \mathcal{H} \text{ and } A_\mu \in \mathcal{F}_\mu \text{ for each } \mu \in \{1, \dots, n\}\}\end{aligned}$$

is a base for $\widetilde{\mathcal{U}}$. Thus, each element of $\widetilde{\mathcal{B}}$ contains a set $K \times K$ with $K \in \mathcal{K}$. Let $\widetilde{B} \in \widetilde{\mathcal{B}}$. If $(u, v) \in \widetilde{B}$ and $u \notin X$, then $u = y_\mu$, $v \in A_\mu \in \mathcal{F}_\mu$ for some $\mu \in \{1, \dots, n\}$. If $(u, v) \in \widetilde{B} \cap (X \times X)$, then $(u, v) \in H$. Concerning $\mathcal{K}$, let us consider the following cases:

α) If $\mathcal{K} = [\dot{\mathcal{F}}]$ for some $\mathcal{F} \in \gamma_{\mathcal{J}_X}$, then $\mathcal{K} \supset i(\mathcal{F}) \cap [\dot{\mathcal{F}}]$.
β) Let $\mathcal{K} \neq [\dot{\mathcal{F}}]$ for each $\mathcal{F} \in \gamma_{\mathcal{J}_X}$:

β_1) If $X \in \mathcal{K}$, we may assume that $K \subset X$ (cf. above). Then $K \times K \subset H$. Put $\mathcal{F} = i^{-1}(\mathcal{K})$. Hence,

$$\mathcal{F} \times \mathcal{F} \supset \mathcal{H} \text{ and consequently } \mathcal{F} \times \mathcal{F} \in \mathcal{J}_X, \text{ i.e. } \mathcal{F} \in \gamma_{\mathcal{J}_X}.$$

Since $X \in \mathcal{K}$, $i(\mathcal{F}) = \mathcal{K} \supset i(\mathcal{F}) \cap [\dot{\mathcal{F}}]$.

β_2) If $X \notin \mathcal{K}$, then $K \not\subset X$, i.e. there is some $u \in K$ with $u \notin X$. Obviously, $K \setminus \{u\} \neq \emptyset$, since $\mathcal{K} \neq \dot{y}$ for each $y \in Y$. For each $v \in K \setminus \{u\}$, $(u, v) \in K \times K \subset \widetilde{B}$; thus $u = y_\mu$ for some $\mu \in \{1, \dots, n\}$ and since $\{y_\mu\} \times K \subset K \times K$,

$(y_\mu, v) \in \{y_\mu\} \times A_\mu$ for each $v \in K \setminus \{u\}$. Consequently, $K \subset A_\mu \cup \{y_\mu\}$. For all elements of $\widetilde{\mathcal{B}}$ and the corresponding sets $K \in \mathcal{K}$, we obtain the same point y_μ: namely if there would exist $y_{\mu'} \neq y_\mu$ for some $K' \in \mathcal{K}$, i.e. $K' \subset A_{\mu'} \cup \{y_{\mu'}\}$, $K \cap K' \subset A_\mu \cap A_{\mu'} \subset X$ and thus $X \in \mathcal{K}$ – a contradiction. Hence $\mathcal{K} \supset i(\mathcal{F}_\mu) \cap [\dot{\mathcal{F}_\mu}]$.

4) If $1_Y : (Y, \mathcal{J}_Y) \longrightarrow (Y, \mathcal{J}_Y^*)$ denotes the bireflection of the simple completion $(Y, \mathcal{J}_Y)$ of $(X, \mathcal{J}_X)$ with respect to **SULim**, then $\mathcal{J}_Y^* = \mathcal{J}_Y^W$: Obviously, $\mathcal{J}_Y^*$ is generated by $\{(i \times i)(\mathcal{H}) \cap ((i(\mathcal{F}_1) \cap [\dot{\mathcal{F}_1}]) \times (i(\mathcal{F}_1) \cap [\dot{\mathcal{F}_1}])) \cap \cdots \cap ((i(\mathcal{F}_n) \cap [\dot{\mathcal{F}_n}]) \times (i(\mathcal{F}_n) \cap [\dot{\mathcal{F}_n}]))$: $\mathcal{H} \in \mathcal{J}_X$ and $\mathcal{F}_1 \cdots \mathcal{F}_n \in \gamma_{\mathcal{J}_X}$ non–convergent$\}$. For each $\mathcal{F} \in \gamma_{\mathcal{J}_X}$, the following is valid:

$$(*) \quad (i(\mathcal{F}) \cap [\dot{\mathcal{F}}]) \times (i(\mathcal{F}) \cap [\dot{\mathcal{F}}]) = (i(\mathcal{F}) \times i(\mathcal{F})) \cap (i(\mathcal{F}) \times [\dot{\mathcal{F}}]) \cap ([\dot{\mathcal{F}}] \times i(\mathcal{F})) \cap ([\dot{\mathcal{F}}] \times [\dot{\mathcal{F}}]) \subset (i(\mathcal{F}) \times [\dot{\mathcal{F}}]) \cap ([\dot{\mathcal{F}}] \times i(\mathcal{F}))\,.$$

It follows immediately from $(*)$ that $\mathcal{J}_Y^W \subset \mathcal{J}_Y^*$. According to 3), $i(\mathcal{F}) \cap [\dot{\mathcal{F}}]$ is a $\mathcal{J}_Y^W$–Cauchy filter for each $\mathcal{F} \in \gamma_{\mathcal{J}_X}$, i.e. $(i(\mathcal{F}) \cap [\dot{\mathcal{F}}]) \times (i(\mathcal{F}) \cap [\dot{\mathcal{F}}]) \in \mathcal{J}_Y^W$. Furthermore, for each $\mathcal{H} \in \mathcal{J}_X$, $(i \times i)(\mathcal{H}) \in \mathcal{J}_Y^W$. Since $\mathcal{J}_Y^W$ is a **SULim**–structure (even a **ULim**–structure), $\mathcal{J}_Y^* \subset \mathcal{J}_Y^W$.

5) $(X, \mathcal{J}_X)$ is a dense subspace of $(Y, \mathcal{J}_Y^*)$:
a) Since $(i \times i)(\mathcal{H}) \in \mathcal{J}_Y^*$ for each $\mathcal{H} \in \mathcal{J}_X$, $\mathcal{J}_X \subset (\mathcal{J}_Y^*)_X) = \{\mathcal{K} \in F(X \times X) : (i \times i)(\mathcal{K}) \in \mathcal{J}_Y^*\}$. Conversely, if $\mathcal{K} \in (\mathcal{J}_Y^*)_X$, then $(i \times i)(\mathcal{K}) \supset (i \times i)(\mathcal{H}) \cap ((i(\mathcal{F}_1) \cap [\dot{\mathcal{F}_1}]) \times (i(\mathcal{F}_1) \cap [\dot{\mathcal{F}_n}])) \cdots \cap ((i(\mathcal{F}_n) \cap [\dot{\mathcal{F}_n}]) \times (i[\mathcal{F}_n] \cap [\dot{\mathcal{F}_n}]))$ with $\mathcal{H} \in \mathcal{J}_X$ and $\mathcal{F}_1, \ldots, \mathcal{F}_n \in \gamma_{\mathcal{J}_X}$ non–convergent. Thus, $(i \times i)^{-1}(i \times i)(\mathcal{K}) = \mathcal{K} \supset \mathcal{H} \cap (\mathcal{F}_1 \times \mathcal{F}_1) \cap \cdots \cap (\mathcal{F}_n \times \mathcal{F}_n) \in \mathcal{J}_X$, i.e. $\mathcal{K} \in \mathcal{J}_X$.

b) Let $y \in Y \setminus X$, i.e. $y = [\mathcal{F}]$ with $\mathcal{F} \in \gamma_{\mathcal{J}_X}$ non–convergent. Since $i(\mathcal{F}) \cap [\dot{\mathcal{F}}]$ is a $\mathcal{J}_Y^*$–Cauchy filter, $i(\mathcal{F}) \longrightarrow [\mathcal{F}]$ and $X \in i(\mathcal{F})$. Hence, $y \in \overline{X}$. Since, obviously, $X \subset \overline{X}$, we obtain $Y \subset \overline{X}$ and consequently, $Y = \overline{X}$.

6) $(Y, \mathcal{J}_Y^*)$ is complete and separated:
a) If $\mathcal{K}$ is a $\mathcal{J}_Y^*$–Cauchy filter, then, by 3) and 4), there is some $\mathcal{F} \in \gamma_{\mathcal{J}_X}$ with $\mathcal{K} \supset i(\mathcal{F}) \cap [\dot{\mathcal{F}}]$. Consequently, $\mathcal{K} \cap [\dot{\mathcal{F}}] \in \gamma_{\mathcal{J}_{Y^*}}$, i.e. $\mathcal{K}$ converges to $[\mathcal{F}]$.

b) In order to prove that $(Y, \mathcal{J}_Y^*)$ is separated, it suffices to show that it is a T_1–space (cf. 4.4.10. 2)):

α) Let $u \in Y$ such that $\dot{u} \supset i(\mathcal{F}) \cap [\dot{\mathcal{F}}]$ for some $\mathcal{F} \in \gamma_{\mathcal{J}_X}$. Then either $u = [\mathcal{F}]$ or $u \neq [\mathcal{F}]$. In the second case, we obtain $u \in X$ and $\mathcal{F} \subset \dot{u}$, since $i(\mathcal{F}) \cap [\dot{\mathcal{F}}]$ is generated by $\{F \cup \{[\mathcal{F}]\} : F \in \mathcal{F}\}$. Consequently, $\mathcal{F} \cap \dot{u} = \mathcal{F} \in \gamma_{\mathcal{J}_X}$, i.e. $[\dot{u}] = [\mathcal{F}]$. Since $u \in X$, u and $[\dot{u}]$ may be identified (cf. 1)), i.e. $u = [\mathcal{F}]$ – a contradiction. Thus, the second case cannot occur.

β) Let $u, v \in Y$ such that $\dot{u} \overset{q_{\gamma_{\mathcal{J}_{Y^*}}}}{\longrightarrow} v$. Then $\dot{u} \cap \dot{v} \supset i(\mathcal{F}) \cap [\dot{\mathcal{F}}]$ for some $\mathcal{F} \in \gamma_{\mathcal{J}_X}$. Since $\dot{u}, \dot{v} \supset \dot{u} \cap \dot{v}$, we obtain from α) that $u = v = [\mathcal{F}]$.

7) The universal property of $(Y, \mathcal{J}_Y^*)$ follows immediately from the extension lemma (cf. 4.4.13.) and the universal property of the **SULim**–bireflection:

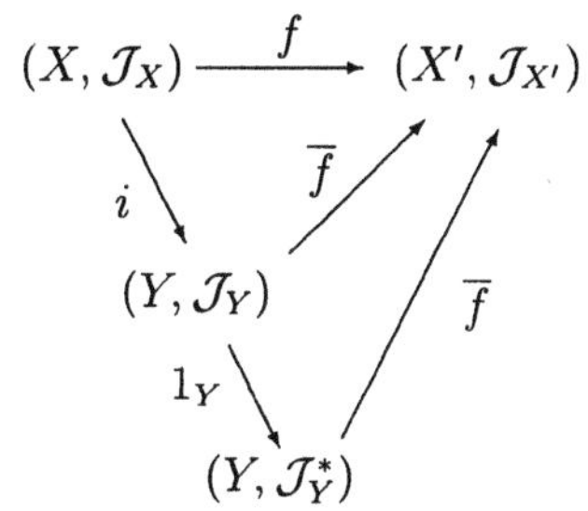

4.4.15 Definition. The completion $i : (X, \mathcal{J}_X) \longrightarrow (Y, \mathcal{J}_Y^*)$ (cf. 4.4.14.) is called the *Wyler completion of the separated uniform limit space* $(X, \mathcal{J}_X)$. Occasionally, $(Y, \mathcal{J}_Y^*)$ is already called the Wyler completion of $(X, \mathcal{J}_X)$.

4.4.16 Remarks. 1) It is easily checked that the construct **SepULim** of separated uniform limit spaces (and uniformly continuous maps) is closed under formation of (weak) subobjects and products (in **ULim**), i.e. **SepULim** is an (extremal) epireflective (full and isomorphism–closed) subconstruct of **ULim** (cf. the theorem under 2.2.4.). If $e_X : (X, \mathcal{J}_X) \longrightarrow (X_S, \mathcal{J}_{X_S})$ denotes the (extremal) epireflection of $(X, \mathcal{J}_X) \in |\mathbf{ULim}|$ w.r.t. **SepULim** and $i_{X_S} : (X_S, \mathcal{J}_{X_S}) \longrightarrow (Y_S, \mathcal{J}_{Y_S^*})$ is the Wyler completion of $(X_S, \mathcal{J}_{X_S})$, then $r_X = i_{X_S} \circ e_X : (X, \mathcal{J}_X) \longrightarrow (Y_S, \mathcal{J}_{Y_S^*})$ is called the *Wyler completion of the uniform limit space* $(X, \mathcal{J}_X)$. Obviously, it is characterized by the following universal property: If $(X, \mathcal{J}_X) \longrightarrow (X', \mathcal{J}_{X'})$ is a uniformly continuous map from $(X, \mathcal{J}_X)$ into a complete separated uniform limit space $(X', \mathcal{J}_{X'})$, then there is a unique uniformly continuous map $\overline{f} : (Y_S, \mathcal{J}_{Y_S}^*) \longrightarrow (X', \mathcal{J}_{X'})$ such that $\overline{f} \circ r_X = f$. Since $(X, \mathcal{J}_X)$ need not be separated, $r_X : (X, \mathcal{J}_X) \longrightarrow (Y_S, \mathcal{J}_{Y_S}^*)$ is generally not an embedding.

2) a) If $(X, \mathcal{W})$ is a uniform space, then each equivalence class $[\mathcal{F}]$ of Cauchy filters in $(X, \mathcal{W})$ contains a *minimal Cauchy filter* (i.e. a minimal element of the set of all Cauchy filters on $(X, \mathcal{W})$ partially ordered by $\subset$), namely $\mathcal{W}(\mathcal{F}) = (\{W[F] : W \in \mathcal{W},\ F \in \mathcal{F}\})$. We know already that $\mathcal{W}(\mathcal{F})$ is a Cauchy filter which is equivalent to $\mathcal{F}$ (cf. the last section of part 2) of the proof of 4.4.14.). Since $\Delta \subset W$ for each $W \in \mathcal{W}$, $\mathcal{W}(\mathcal{F}) \subset \mathcal{F}$. If $\mathcal{G} \subset \mathcal{F}$ is a Cauchy filter in $(X, \mathcal{W})$, then $\mathcal{W}(\mathcal{F}) \subset \mathcal{G}$ (namely, if $W \in \mathcal{W}$ and $F \in \mathcal{F}$, then there is some $G \in \mathcal{G}$ with $G \times G \subset W$ and since $F \cap G \neq \emptyset$, $G \subset W[F]$, i.e. $W[F] \in \mathcal{G}$). Consequently, $\mathcal{W}(\mathcal{F})$ is a minimal Cauchy filter.

b) By a), the set X^* of all minimal Cauchy filters on a uniform space $(X, \mathcal{W})$ is equipotent with the set $\{[\mathcal{F}] : \mathcal{F}$ Cauchy filter on $(X, \mathcal{W}))\}$. Since $\mathcal{W}(\dot{x})$ is the neighborhood filter of $x \in X$ in $(X, \mathcal{X}_{\mathcal{W}})$, $\{\mathcal{W}(\dot{x}) : x \in X\}$ is equipotent with X provided that $(X, \mathcal{W})$ is separated.

c) There is an alternative description of the Hausdorff completion $r_X : (X, \mathcal{W}) \longrightarrow (X^*, \mathcal{W}^*)$ of $(X, \mathcal{W}) \in |\mathbf{Unif}|$ which is a reflection of $(X, \mathcal{W})$ w.r.t. **CSepUnif** (cf. [19]): Let X^* be as under b) and $r_X(x) = \mathcal{W}(\dot{x})$ for each $x \in X$. For each $V = V^{-1} \in \mathcal{W}$, let $\widetilde{V} = \{(\mathcal{F}, \mathcal{G}) \in X^* \times X^*$: there is some

$M \in \mathcal{F} \cap \mathcal{G}$ with $M \times M \subset V\}$. Then $\{\widetilde{V} : V = V^{-1} \in \mathcal{W}\}$ is a base for $\mathcal{W}^*$ and $r_X : (X, \mathcal{W}) \longrightarrow (X^*, \mathcal{W}^*)$ is a dense embedding provided that $(X, \mathcal{W})$ is separated.

d) *The Wyler completion* $(Y, \mathcal{J}_Y^*)$ *of a uniform space* (= *principal uniform limit space*) $(X, \mathcal{J}_X)$ *need not be a uniform space*, e.g. the inclusion map $i : \mathbb{Q}_u \longrightarrow \mathbb{R}_u$ from the (usual) uniform space of rational numbers into the (usual) uniform space of real numbers is the Hausdorff completion of $\mathbb{Q}_u$ (cf. 4.2.2.17. and 4.2.2.19. 2)), but it differs from the Wyler completion of $\mathbb{Q}_u$, since the latter one is not uniform: Let $(X, \mathcal{J}_X) = \mathbb{Q}_u$, i.e. $\mathcal{J}_X = [\mathcal{U}]$, where $\mathcal{U}$ denotes the usual uniformity on $\mathbb{Q}$. If the Wyler completion $(Y, \mathcal{J}_Y^*)$ of $(X, \mathcal{J}_X)$ were a principal uniform limit space, $\mathcal{W} = \bigcap_{\mathcal{H} \in \mathcal{J}_Y^*} \mathcal{H} = (i \times i)(\mathcal{U}) \cap (\bigcap\{i(\mathcal{F}) \cap [\dot{\mathcal{F}}]) \times (i(\mathcal{F}) \cap [\dot{\mathcal{F}}]) : \mathcal{F} \in \gamma_{\mathcal{J}_X}$ is non–convergent $\})$ would belong to $\mathcal{J}_Y^*$, i.e. there were finitely many non–convergent filters $\mathcal{F}_1, \ldots, \mathcal{F}_n \in \gamma_{\mathcal{J}_X}$ such that

$$(*)\quad \mathcal{W} \supset (i \times i)(\mathcal{U}) \cap ((i(\mathcal{F}_1) \cap [\dot{\mathcal{F}}_1]) \times (i(\mathcal{F}_1) \cap [\dot{\mathcal{F}}_1])) \cdots$$
$$\cdots \cap ((i(\mathcal{F}_n) \cap [\dot{\mathcal{F}}_n]) \times (i(\mathcal{F}_n) \cap [\dot{\mathcal{F}}_n]))\,.$$

Choose $U \in \mathcal{U}$ and $F_i \in \mathcal{F}_i$ for each $i \in \{1, \ldots n\}$. Then

$$U \cup ((F_1 \cup \{[\mathcal{F}_1]\}) \times (F_1 \cup \{[\mathcal{F}_1]\})) \cup \cdots ((F_n \cup \{[\mathcal{F}_n]\}) \times (F_n \cup \{[\mathcal{F}_n]\}))$$

would belong to the right side of $(*)$, but it would not belong to $\mathcal{W}$, since it contains only finitely many elements of the diagonal in $(\mathbb{R} \setminus \mathbb{Q}) \times (\mathbb{R} \setminus \mathbb{Q})$, namely $([\mathcal{F}_1], [\mathcal{F}_1]), \cdots, ([\mathcal{F}_n], [\mathcal{F}_n])$, whereas each element of $\mathcal{W}$ contains the diagonal in $(\mathbb{R} \setminus \mathbb{Q}) \times (\mathbb{R} \setminus \mathbb{Q})$ which consists of infinitely many elements. Thus, $\mathcal{W} \notin \mathcal{J}_Y^*$, and consequently, $(Y, \mathcal{J}_Y^*)$ is not uniform.

4.4.17 Definition. A uniform limit space $(X, \mathcal{J}_X)$ is called **Chy**–*determined* provided that $\mathcal{J}_X = \mathcal{J}^*_{\gamma_{\mathcal{J}_X}}$, where

$$\mathcal{J}^*_{\gamma_{\mathcal{J}_X}} = \{\mathcal{F} \in F(X \times X) : \text{ there exist finitely many } \mathcal{F}_1, \cdots, \mathcal{F}_n \in \gamma_{\mathcal{J}_X} \text{ with } \mathcal{F} \supset \textstyle\bigcap_{i=1}^n \mathcal{F}_i \times \mathcal{F}_i\}\,.$$

4.4.18 Theorem. 1) *The construct* **Chy–D–ULim** *of* **Chy**–*determined uniform limit spaces (and uniformly continuous maps) is (concretely) isomorphic to* **Chy**.

2) *If* $(X, \mathcal{J}_X)$ *is a separated* **Chy**–*determined uniform limit space, then its Wyler completion* $(Y, \mathcal{J}_Y^*)$ *is also a separated* **Chy**–*determined uniform limit space.*

Proof. 1) a) Let $(X, \gamma) \in |\,\mathbf{Chy}\,|$ and $\mathcal{J}_\gamma^* = \{\mathcal{F} \in F(X \times X)$: there exist finitely many $\mathcal{F}_1, \ldots, \mathcal{F}_n \in \gamma$ with $\mathcal{F} \supset \bigcap_{i=1}^n \mathcal{F}_i \times \mathcal{F}_i\}$. Then $(X, \mathcal{J}_\gamma^*)$ is a uniform limit space and $\gamma_{\mathcal{J}_\gamma^*} = \gamma$:

α) Since $1_X : (X, \mathcal{J}_\gamma) \longrightarrow (X, \mathcal{J}_\gamma^*)$ is the bireflection of the semiuniform convergence space $(X, \mathcal{J}_\gamma)$ w.r.t. **SULim**, $(X, \mathcal{J}_\gamma^*)$ is a semiuniform limit space. Now let $\mathcal{F}, \mathcal{G} \in \mathcal{J}_\gamma^*$ such that $\mathcal{F} \circ \mathcal{G}$ exists. Then there are finitely many $\mathcal{F}_1 \cdots \mathcal{F}_n \in \gamma$ (resp. $\mathcal{G}_1, \cdots \mathcal{G}_m \in \gamma$) such that $\mathcal{F} \supset \bigcap_{i=1}^n \mathcal{F}_i \times \mathcal{F}_i$ (resp. $\mathcal{G} \supset$

$\bigcap_{j=1}^{m} \mathcal{G}_j \times \mathcal{G}_j)$. Obviously, $\bigcap_{i=1}^{n}(\mathcal{F}_i \times \mathcal{F}_i) \circ \bigcap_{j=1}^{m}(\mathcal{G}_j \times \mathcal{G}_j) \subset \mathcal{F} \circ \mathcal{G}$. In order to prove UC$_4$), it suffices to show that $\bigcap_{i=1}^{n}(\mathcal{F}_i \times \mathcal{F}_i) \circ \bigcap_{j=1}^{m} \mathcal{G}_j \times \mathcal{G}_j \in \mathcal{J}_\gamma^*$. We have $\bigcap_{i=1}^{n}(\mathcal{F}_i \times \mathcal{F}_i) \circ \bigcap_{j=1}^{m}(\mathcal{G}_j \times \mathcal{G}_j) = \bigcap_{(i,j)\in K}(\mathcal{F}_i \times \mathcal{F}_i) \circ (\mathcal{G}_j \times \mathcal{G}_j)$ with $K = \{(i,j) \in \{1,\cdots n\} \times \{1,\cdots m\} : (\mathcal{F}_i \times \mathcal{F}_i) \circ (\mathcal{G}_j \times \mathcal{G}_j)$ exists $\}$ (cf. 2) c) in the proof of 2.3.2.3.). If $(i,j) \in K$, then $\sup\{\mathcal{F}_i, \mathcal{G}_j\}$ exists. Consequently, $\mathcal{H}_{ij} = \mathcal{F}_i \cap \mathcal{G}_j \in \gamma$ for each $(i,j) \in K$. Since $\mathcal{H}_{ij} \times \mathcal{H}_{ij} \subset \mathcal{G}_j \times \mathcal{F}_i = (\mathcal{F}_i \times \mathcal{F}_i) \circ (\mathcal{G}_j \times \mathcal{G}_j)$, $\bigcap_{i=1}^{n}(\mathcal{F}_i \times \mathcal{F}_i) \circ \bigcap_{j=1}^{m}(\mathcal{G}_j \times \mathcal{G}_j) \in \mathcal{J}_\gamma^*$.

β) Obviously, $\gamma \subset \gamma_{\mathcal{J}_\gamma^*}$. Conversely, let $\mathcal{F} \in \gamma_{\mathcal{J}_\gamma^*}$. Then there exist finitely many $\mathcal{F}_1, \ldots, \mathcal{F}_n \in \gamma$ such that $\mathcal{F} \times \mathcal{F} \supset \bigcap_{i=1}^{n} \mathcal{F}_i \times \mathcal{F}_i$. Without loss of generality we may assume that the filters $\mathcal{F}_i$ are pairwise non–equivalent, i.e. $\mathcal{F}_i \cap \mathcal{F}_k \notin \gamma$ provided that $i \neq k$ (namely, if $\mathcal{F}_i \cap \mathcal{F}_k \in \gamma$ for some pair (i,k), then $(\mathcal{F}_i \times \mathcal{F}_i) \cap (\mathcal{F}_k \times \mathcal{F}_k)$ may be substituted by $(\mathcal{F}_i \cap \mathcal{F}_k) \times (\mathcal{F}_i \cap \mathcal{F}_k)$, since $(\mathcal{F}_i \cap \mathcal{F}_k) \times (\mathcal{F}_i \cap \mathcal{F}_k) = (\mathcal{F}_i \times \mathcal{F}_i) \cap (\mathcal{F}_i \times \mathcal{F}_k) \cap (\mathcal{F}_k \times \mathcal{F}_i) \cap (\mathcal{F}_k \times \mathcal{F}_k) \subset (\mathcal{F}_i \times \mathcal{F}_i) \cap (\mathcal{F}_k \times \mathcal{F}_k)$). Thus, $\sup\{\mathcal{F}_i, \mathcal{F}_k\}$ does not exist for $i \neq k$. Consequently, for each $i \in \{1, \ldots, n\}$ one can choose $F_i \in \mathcal{F}_i$ in such a way that $F_i \cap F_k = \emptyset$ provided that $i \neq k$. Furthermore, there is some $F \in \mathcal{F}$ such that

$$\bigcup_{i=1}^{n} F_i \times F_i \supset F \times F .$$

Then $F \subset F_k$ for some $k \in \{1, \ldots, n\}$ (namely, if $x \in F$, then for each $y \in F$, $(x,y) \in F_k \times F_k$ for some $k \in \{1, \ldots, n\}$, which does not depend on the choice of y, since $F_i \cap F_k = \emptyset$ for $i \neq k$, i.e. $y \in F_k$ for each $y \in F$). Hence, $\mathcal{F}_k \subset \mathcal{F}$ (Obviously, $F_k' \in \mathcal{F}_k$ implies $F_k'' = F_k \cap F_k' \in \mathcal{F}_k$; consequently, there is some $F' \in \mathcal{F}$ with $F' \times F' \subset (F_1 \times F_1) \cup \cdots \cup (F_k'' \times F_k'') \cup \cdots \cup (F_n \times F_n)$. Then $F' \subset F_k'' \subset F_k'$; namely if F' were a subset of F_i for some $i \neq k$, $F'' = F \cap F' \subset F' \subset F_i$ and $F'' \subset F \subset F_k$ which is impossible since $F_i \cap F_k = \emptyset$.). Consequently, $\mathcal{F} \in \gamma$.

b) $\mathcal{J}_X = \mathcal{J}^*_{\gamma_{\mathcal{J}_X}}$ for each **Chy**–determined **ULim**–structure $\mathcal{J}_X$.

c) If $f : (X, \gamma) \longrightarrow (X', \gamma')$ is a **Chy**–morphism, then $f : (X, \mathcal{J}_\gamma^*) \longrightarrow (X', \mathcal{J}_{\gamma'}^*)$ is a **Chy–D–ULim**–morphism: Let $\mathcal{F} \in \mathcal{J}_\gamma^*$, i.e. $\mathcal{F} \supset \bigcap_{i=1}^{n} \mathcal{F}_i \times \mathcal{F}_i$ with $\mathcal{F}_i \in \gamma$ for each $i \in \{1, \ldots, n\}$. Then $(f \times f)(\mathcal{F}) \supset (f \times f)(\bigcap_{i=1}^{n} \mathcal{F}_i \times \mathcal{F}_i) = \bigcap_{i=1}^{n}(f \times f)(\mathcal{F}_i \times \mathcal{F}_i) = \bigcap_{i=1}^{n} f(\mathcal{F}_i) \times f(\mathcal{F}_i) \in \mathcal{J}_{\gamma'}^*$ since $f(\mathcal{F}_i) \in \gamma'$ for each $i \in \{1, \ldots, n\}$. Thus, $(f \times f)(\mathcal{F}) \in \mathcal{J}_{\gamma'}^*$.

d) If $f : (X, \mathcal{J}_X) \longrightarrow (X', \mathcal{J}_{X'})$ is a **Chy–D–ULim**–morphism, then $f : (X, \gamma_{\mathcal{J}_X}) \longrightarrow (X', \gamma_{\mathcal{J}_{X'}})$ is a **Chy**–morphism.

e) It follows from a, b), c) and d) that **Chy–D–ULim** is (concretely) isomorphic to **Chy**, i.e. the functor $\mathcal{F}$: **Chy–D–ULim** $\longrightarrow$ **Chy** defined by $\mathcal{F}((X, \mathcal{J}_X)) = (X, \gamma_{\mathcal{J}_X})$ for each **Chy**–determined uniform limit space $(X, \mathcal{J}_X)$ and $\mathcal{F}(f) = f$ for each **Chy–D–ULim**–morphism f is a (concrete) isomorphism.

2) Let $(X, \mathcal{J}_X)$ be a separated uniform limit space such that $\mathcal{J}_X = \mathcal{J}^*_{\gamma_{\mathcal{J}_X}}$. In order to prove $\mathcal{J}_Y^* = \mathcal{J}^*_{\gamma_{\mathcal{J}_{Y^*}}}$ it suffices to show that $\mathcal{J}_Y^* \subset \mathcal{J}^*_{\gamma_{\mathcal{J}_{Y^*}}}$. Let $\mathcal{F} \in \mathcal{J}_Y^*$, i.e.

$$\mathcal{F} \supset (i \times i)(\mathcal{H}) \cap ((i(\mathcal{F}_1) \cap [\dot{\mathcal{F}}_1]) \times i(\mathcal{F}_1) \cap [\dot{\mathcal{F}}_1])) \cap \cdots \cap ((i(\mathcal{F}_n) \cap [\dot{\mathcal{F}}_n]) \times (i(\mathcal{F}_n) \cap [\dot{\mathcal{F}}_n]))$$

with $\mathcal{H} \in \mathcal{J}_X$ and $\mathcal{F}_1 \cdots \mathcal{F}_n \in \gamma_{\mathcal{J}_X}$ non–convergent. By assumption, there are finitely many $\mathcal{H}_1 \cdots \mathcal{H}_m \in \gamma_{\mathcal{J}_X}$ with $\mathcal{H} \supset \bigcap_{j=1}^m \mathcal{H}_j \times \mathcal{H}_j$. Thus,

$$(*) \qquad \mathcal{F} \supset \bigcap_{j=1}^m (i(\mathcal{H}_j) \times i(\mathcal{H}_j)) \cap \bigcap_{k=1}^n (i(\mathcal{F}_k) \cap [\dot{\mathcal{F}}_k]) \times (i(\mathcal{F}_k) \cap [\dot{\mathcal{F}}_k]))).$$

Since $i(\mathcal{G})$ as well as $i(\mathcal{G}) \cap [\dot{\mathcal{G}}]$ is a $\mathcal{J}_Y^*$–Cauchy filter for each $\mathcal{G} \in \gamma_{\mathcal{J}_X}$, it follows from $(*)$ that $\mathcal{F} \in \mathcal{J}^*_{\gamma_{\mathcal{J}_Y^*}}$.

4.4.19 Remarks. 1) Because of 4.4.18. 1) *we need not distinguish between* **Chy**–*determined uniform limit spaces and Cauchy spaces.*

2) It follows from 4.4.18. that for each separated Cauchy space (X, γ), $i : (X, \gamma) \longrightarrow (Y, \gamma_{\mathcal{J}_Y^*})$ is a completion, where $(Y, \mathcal{J}_Y^*)$ is the Wyler completion of $(X, \mathcal{J}_\gamma^*)$. This one is called *the Wyler completion of the separated Cauchy space* (X, γ); it is a completion in the realm of (separated) Cauchy spaces. Obviously, it is characterized (up to isomorphism) by the following universal property: If (X', γ') is a complete separated Cauchy space and $f : (X, \gamma) \longrightarrow (X', \gamma')$ a Cauchy continuous map, then there is a unique Cauchy continuous map $\overline{f} : (Y, \gamma_{\mathcal{J}_Y^*}) \longrightarrow (X', \gamma')$ such that $\overline{f} \circ i = f$.

3) **Chy–D–ULim** *is a bicoreflective (full and isomorphism–closed) subconstruct of* **ULim**, where $1_X : (X, \mathcal{J}^*_{\gamma_{\mathcal{J}_X}}) \longrightarrow (X, \mathcal{J}_X)$ is the bicoreflection of $(X, \mathcal{J}_X) \in |\,\mathbf{ULim}\,|$ w.r.t. **Chy–D–ULim** (similarly to part 1) of the proof of 2.3.3.6.).

4) **Chy** *is a bireflective (full and isomorphism–closed) subconstruct of* **Fil**, where $1_X : (X, \gamma) \longrightarrow (X, \gamma_C)$ is the bireflection of $(X, \gamma) \in |\,\mathbf{Fil}\,|$ w.r.t. **Chy** provided that $\gamma_C = \{\mathcal{F} \in F(X)$: there exist finitely many $\mathcal{F}_1, \ldots, \mathcal{F}_n \in \gamma$ such that every member of $\mathcal{F}_i$ meets every member of $\mathcal{F}_{i+1}$ for each $i \in \{1, \ldots, n-1\}$ and $\mathcal{F} \supset \bigcap_{i=1}^n \mathcal{F}_i\}$. (Obviously, (X, γ_C) is a filter space. In order to prove the Cauchy property, let $\mathcal{F}, \mathcal{G} \in \gamma_C$ such that $\sup\{\mathcal{F}, \mathcal{G}\}$ exists. Then there exist $\mathcal{F}_1, \ldots, \mathcal{F}_n \in \gamma$ such that every member of $\mathcal{F}_i$ meets every member of $\mathcal{F}_{i+1}$ for each $i \in \{1, \ldots, n-1\}$ and $\mathcal{F} \supset \bigcap_{i=1}^n \mathcal{F}_i$, and there exist $\mathcal{G}_1, \ldots, \mathcal{G}_m \in \gamma$ such that every member of $\mathcal{G}_j$ meets every member of $\mathcal{G}_{j+1}$ for each $j \in \{1, \cdots m-1\}$ and $\mathcal{G} \supset \bigcap_{j=1}^m \mathcal{G}_j$. Hence, there exist $i \in \{1, \ldots, n\}$ and $j \in \{1, \ldots, m\}$ such that every member of $\mathcal{F}_i$ meets every member of $\mathcal{G}_j$. It follows from

$$\mathcal{F}_1 \cap \cdots \cap \mathcal{F}_n \cap \mathcal{F}_{n-1} \cap \cdots \cap \mathcal{F}_i \cap \mathcal{G}_j \cap \mathcal{G}_{j-1} \cap \cdots \cap \mathcal{G}_1 \cap \cdots \cap \mathcal{G}_m \subset \mathcal{F} \cap \mathcal{G}$$

that $\mathcal{F} \cap \mathcal{G} \in \gamma_C$. Let $(X', \gamma') \in |\,\mathbf{Chy}\,|$ and let $f : (X, \gamma) \longrightarrow (X', \gamma')$ be a Cauchy continous map. In order to prove that $f : (X, \gamma_C) \longrightarrow (X', \gamma')$ is also Cauchy continuous, let $\mathcal{F} \in \gamma_C$. Then there are $\mathcal{F}_1, \cdots, \mathcal{F}_n \in \gamma$ such that $\sup\{\mathcal{F}_i, \mathcal{F}_{i+1}\}$ exists for each $i \in \{1, \ldots, n-1\}$ and $\mathcal{F} \supset \bigcap_{i=1}^n \mathcal{F}_i$. Consequently, $f(\mathcal{F}) \supset \bigcap_{i=1}^n f(\mathcal{F}_i)$, where every member of $f(\mathcal{F}_i)$ meets every member of $f(\mathcal{F}_{i+1})$ for each $i \in \{1, \ldots, n-1\}$ and $f(\mathcal{F}_k) \in \gamma'$ for each $k \in \{1, \ldots, n\}$, since $f : (X, \gamma) \longrightarrow (X', \gamma')$ is Cauchy continuous. Thus, $f(\mathcal{F}) \in \gamma'$, because $(X', \gamma') \in |\,\mathbf{Chy}\,|$.)

5) **Chy** *is a cartesian closed topological construct, where the power objects in* **Chy** *are formed as in* **Fil**. (It follows from 3) or 4) that **Chy** is a topological construct. It is easily checked that **Chy** is closed under formation of power

objects in **Fil**, namely if Θ and Φ belong to the Cauchy continuous **Fil**–structure $\widehat{\gamma}$ [cf. 3.1.9. ④] on $[\mathbf{X},\mathbf{X}']_{\mathbf{Fil}} = [\mathbf{X},\mathbf{X}']_{\mathbf{Chy}}$, where $\mathbf{X} = (X,\gamma)$ and $\mathbf{X}' = (X',\gamma')$ are Cauchy spaces, and every member of Θ meets every member of Φ, then every member of $\Theta(\mathcal{F})$ meets every member of $\Phi(\mathcal{F})$ provided that $\mathcal{F} \in \gamma$; thus, since $\Theta(\mathcal{F})$ and $\Phi(\mathcal{F})$ belong to γ', $\Theta(\mathcal{F}) \cap \Phi(\mathcal{F}) \in \gamma'$; hence for each $\mathcal{F} \in \gamma$, $(\Theta \cap \Phi)(\mathcal{F}) \in \gamma'$, because $(\Theta \cap \Phi)(\mathcal{F}) \supset \Theta(\mathcal{F}) \cap \Phi(\mathcal{F})$, i.e. $\Theta \cap \Phi \in \widehat{\gamma}$, and consequently, $([\mathbf{X},\mathbf{X}']_{\mathbf{Chy}},\widehat{\gamma})$ is a Cauchy space. Since **Chy** is also bireflective in **Fil** [cf. 4)], the assertion follows from 3.1.8.)

6) Because of 4) initial structures in **Chy** are formed as in **Fil**. By 2.3.3.17. b) initial structures in **Fil** induce initial structures in $\mathbf{KConv_S}$. Thus, it is easily checked that the construct **SepChy** of separated Cauchy spaces (and Cauchy continuous maps) is closed under formation of (weak) subobjects and products in **Chy**, i.e. **SepChy** is an (extremal) epireflective subconstruct of **Chy**. If $e_X : (X,\gamma) \longrightarrow (X_S,\gamma_S)$ denotes the (extremal) epireflection of $(X,\gamma) \in |\mathbf{Chy}|$ w.r.t. **SepChy** and $i_{X_S} : (X_S,\gamma_S) \longrightarrow (Y,\gamma_{\mathcal{J}_Y^*})$ is the Wyler completion of (X_S,γ_S) (cf. 4.4.19. 2)), then $r_X = i_{X_S} \circ e_X : (X,\gamma) \longrightarrow (Y,\gamma_{\mathcal{J}_Y^*})$ is called *the Wyler completion of the Cauchy space* (X,γ), but it is not an embedding in **Chy** unless (X,γ) is separated. It is characterized by the following universal property: If (X',γ') is a complete separated Cauchy space and $f : (X,\gamma) \longrightarrow (X',\gamma')$ a Cauchy continuous map, then there is a unique Cauchy continuous map $\overline{f} : (Y,\gamma_{\mathcal{J}_Y^*}) \longrightarrow (X',\gamma')$ such that $\overline{f} \circ r_X = f$. Furthermore, $r_X[X]$ is dense in $(Y,\gamma_{\mathcal{J}_Y^*})$.

7) The construct **Prox** of proximity spaces may be considered to be a bireflective subconstruct of **Chy**, whereas the construct T_{2W}–**Lim** of weakly Hausdorff limit spaces may be considered to be a bicoreflective subconstruct of **Chy** (cf. chapter 7, section 1). In other words: *Cauchy spaces form a common generalization of proximity spaces and weakly Hausdorff limit spaces.*

4.4.20 Theorem. *If $(X,[\mathcal{V}])$ is a separated principal uniform limit space, then the underlying uniform space of its Wyler completion is (up to isomorphism) the Hausdorff completion of $(X,\mathcal{V})$.*

Proof. Since $(X,[\mathcal{V}])$ is separated, we need not distinguish between the underlying set Y of the Wyler completion of $(X,[\mathcal{V}])$ and $X^* = \{[\mathcal{F}]\colon \mathcal{F}$ Cauchy filter in $(X,\mathcal{V})\}$, and we can identify X with $\{[\dot{x}] : x \in X\}$. Each equivalence class $[\mathcal{F}]$ contains a minimal Cauchy filter, namely $\mathcal{V}(\mathcal{F})$ (cf. 4.4.16. 2) a)): this one does not depend on the choice of the representative $\mathcal{F}$, namely if $\mathcal{F} \sim \mathcal{G}$, then $\mathcal{V}(\mathcal{F}) = \mathcal{V}(\mathcal{F} \cap \mathcal{G}) = \mathcal{V}(\mathcal{G})$, since $\mathcal{V}(\mathcal{F} \cap \mathcal{G}) \subset \mathcal{V}(\mathcal{F})$, $\mathcal{V}(\mathcal{G})$, and $\mathcal{V}(\mathcal{F})$ and $\mathcal{V}(\mathcal{G})$ are minimal. If $i : X \longrightarrow X^*$ denotes the inclusion map and $\mathcal{V}^*$ is the uniformity for X^* generated by $\{\widetilde{V} : V = V^{-1} \in \mathcal{V}\}$, where $\widetilde{V} = \{([\mathcal{F}],[\mathcal{G}]) \in X^* \times X^*\colon$ there is some $M \in \mathcal{V}(\mathcal{F}) \cap \mathcal{V}(\mathcal{G})$ with $M \times M \subset V\}$, then, obviously, $i : (X,\mathcal{V}) \longrightarrow (X^*,\mathcal{V}^*)$ is the Hausdorff completion of $(X,\mathcal{V})$ (cf. 4.4.16. 2) c)). Let $i : (X,[\mathcal{V}]) \longrightarrow (Y,\mathcal{J}_Y^*)$ with $Y = X^*$ be the Wyler completion of $(X,[\mathcal{V}])$ and let $\mathcal{W} = \bigcap\{\mathcal{F} : \mathcal{F} \in \mathcal{J}_Y^*\}$. Then $\mathcal{W} = (i \times i)(\mathcal{V}) \cap \bigcap\{(i(\mathcal{F}) \cap [\dot{\mathcal{F}}]) \times (i(\mathcal{F}) \cap [\dot{\mathcal{F}}])\colon \mathcal{F}$ is a Cauchy filter on

$(X,\mathcal{V})\} = i \times i(\mathcal{V}) \cap \bigcap (i(\mathcal{V}(\mathcal{F})) \cap [\dot{\mathcal{V}(\mathcal{F})}]) \times (i(\mathcal{V}(\mathcal{F})) \cap [\dot{\mathcal{V}(\mathcal{F})}]) : \mathcal{F}$ is a Cauchy filter on $(X,\mathcal{V})\}$ and

(1) $\mathcal{V}^* \subset \mathcal{W}$;

namely, by 4.4.14. there is a uniformly continuous map $h : (X^*, \mathcal{J}_Y^*) \longrightarrow (X^*, [\mathcal{V}^*])$ such that the diagram

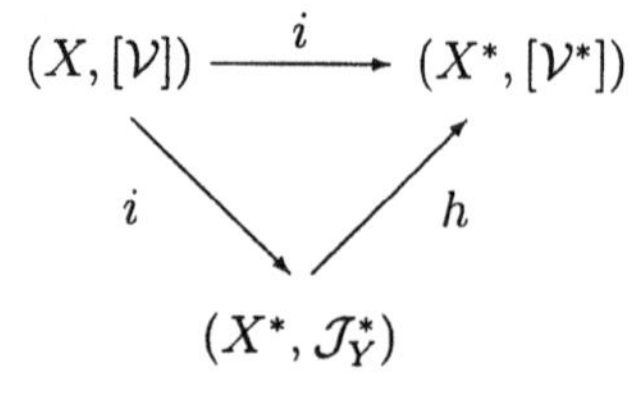

commutes, where for each Cauchy filter $\mathcal{F}$ on $(X, [\mathcal{V}])$, $i(\mathcal{F})$ converges to $h([\mathcal{F}])$ in $(X^*, [\mathcal{V}^*])$ (cf. the proof of 4.4.14. as well as the proof of the extension lemma 4.4.13.); since $i(\mathcal{F})$ converges to $[\mathcal{F}]$ in $(X^*, \mathcal{V}^*)$, i.e. in $(X^*, [\mathcal{V}^*])$, and $(X^*, \mathcal{V}^*)$ is separated, $h([\mathcal{F}]) = [\mathcal{F}]$ for each $[\mathcal{F}] \in X^*$, i.e. $h = 1_{X^*}$; consequently, $\mathcal{J}_Y^* \subset [\mathcal{V}^*]$, which implies $\mathcal{V}^* \subset \bigcap\{\mathcal{F} : \mathcal{F} \in \mathcal{J}_Y^*\} = \mathcal{W}$. Furthermore,

(2) $\mathcal{V}^* = \mathcal{W} \circ \mathcal{W} \circ \mathcal{W}$:

a) Let $W \in \mathcal{W}$. Then there is some $V = V^{-1} \in \mathcal{V}$ and for each $[\mathcal{F}] = [\mathcal{V}(\mathcal{F})] \in X^*$, there is a set $C_{[\mathcal{F}]} \in \mathcal{V}(\mathcal{F})$ such that

(∗) $V \cup \bigcup_{[\mathcal{F}]\in X^*} (C_{[\mathcal{F}]} \cup \{[\mathcal{F}]\}) \times (C_{[\mathcal{F}]} \cup \{[\mathcal{F}]\}) \subset W$.

Let $([\mathcal{F}], [\mathcal{G}]) \in \widetilde{V}$. Then $V \in \mathcal{V}(\mathcal{F}) \times \mathcal{V}(\mathcal{G})$, which implies $V \cap (C_{[\mathcal{F}]} \times C_{[\mathcal{G}]}) \neq \emptyset$, i.e. there is some $([\dot{x}], [\dot{y}]) \in V \cap (C_{[\mathcal{F}]} \times C_{[\mathcal{G}]})$. Hence, it follows from (∗)

$$([\mathcal{F}], [\dot{x}]) \in (C_{[\mathcal{F}]} \cup \{[\mathcal{F}]\}) \times (C_{[\mathcal{F}]} \cup \{[\mathcal{F}]\}) \subset W,$$

$$([\dot{x}], [\dot{y}]) \in V \subset W, \text{ and}$$

$$([\dot{y}], [\mathcal{G}]) \in (C_{[\mathcal{G}]} \cup \{[\mathcal{G}]\}) \times (C_{[\mathcal{G}]} \cup \{[\mathcal{G}]\}) \subset W.$$

Consequently, $([\mathcal{F}], [\mathcal{G}]) \in W^3$, i.e. $\widetilde{V} \subset W^3$. This proves $\mathcal{W}^3 = \mathcal{W} \circ \mathcal{W} \circ \mathcal{W} \subset \mathcal{V}^*$.

b) Since $\mathcal{V}^*$ is a uniformity, $\mathcal{V}^* = \mathcal{V}^* \circ \mathcal{V}^*$. Thus, by (1), $\mathcal{V}^* = \mathcal{V}^* \circ \mathcal{V}^* = \mathcal{V}^* \circ \mathcal{V}^* \circ \mathcal{V}^* \subset \mathcal{W} \circ \mathcal{W} \circ \mathcal{W}$.

Now, let $\mathcal{R}$ be a uniformity on $Y = X^*$ such that $\mathcal{R} \subset \mathcal{W}$. Then $\mathcal{R} = \mathcal{R} \circ \mathcal{R} \circ \mathcal{R} \subset \mathcal{W} \circ \mathcal{W} \circ \mathcal{W} = \mathcal{V}^*$ and since, by (1), $\mathcal{V}^* \subset \mathcal{W}$, $\mathcal{V}^*$ is the finest uniformity which is coarser than $\mathcal{W}$, i.e. $\mathcal{V}^*$ is the finest uniformity which is coarser than each $\mathcal{F} \in \mathcal{J}_Y^*$. Thus, $(X^*, \mathcal{V}^*)$ is the underlying uniform space of $(X^*, \mathcal{J}_Y^*)$ (cf. part 3) of the proof of 2.3.2.3.).

4.4.21 Corollary. *If* **X** *is a separated uniform space, then the underlying uniform space of its simple completion is the Hausdorff completion of* **X**.

Proof. If $\mathbf{X} = (X, \mathcal{J}_X)$ is a separated principal uniform limit space (= separated uniform space), then the underlying uniform space of its simple completion

$(Y, \mathcal{J}_Y)$ is (by definition) the underlying uniform space of the Wyler completion $(Y, \mathcal{J}_Y^*)$ of $\mathbf{X}$, i.e., by 4.4.20, the Hausdorff completion of $\mathbf{X}$.

Chapter 5

Connectedness Properties

A topological space $\mathbf{X} = (X, \mathcal{X})$ is connected iff each continuous map $f : \mathbf{X} \longrightarrow D_2$ from $\mathbf{X}$ into the two–point discrete topological space D_2 is constant. This characterization of the usual concept of connectedness leads to an analogous definition of connectedness for topological constructs, since two–point discrete objects are available, e.g. if D_2^{Δ} denotes the two–point discrete uniform space, then a uniform space $\mathbf{X} = (X, \mathcal{W})$ is called 'connected' (or more exactly: ,uniformly connected) iff each uniformly continuous map $f : \mathbf{X} \longrightarrow D_2^{\Delta}$ is constant.
Uniform connectedness has been studied first by S. Mrowka and W. Pervin [101] in the realm of uniform spaces, but much earlier (and before the now accepted definition of connectedness for topological spaces due to F. Riesz [125], N.J. Lennes [94] and F. Hausdorff [58]) G. Cantor [24] has given a definition of connectedness which is applicable to metric spaces and which coincides with uniform connectedness, whenever metric spaces are regarded as uniform spaces. In this chapter we start with the definition of connectedness for semiuniform convergence spaces which is identical with the usual one for topological spaces provided topological semiuniform convergence spaces (= symmetric topological spaces) are considered. Since it turns out that each uniform space (regarded as a semiuniform convergence space) is connected, it is useful to introduce a concept of uniform connectedness in the realm of semiuniform convergence spaces too. In order to develop a common theory of connectedness, we define for each class $\mathcal{E}$ of semiuniform convergence spaces, $\mathcal{E}$–connected semiuniform convergence spaces as those semiuniform convergence spaces $\mathbf{X} = (X, \mathcal{J}_X)$ such that for each $\mathbf{E} \in \mathcal{E}$, each uniformly continuous map $f : \mathbf{X} \longrightarrow \mathbf{E}$ is constant. Thus, for $\mathcal{E} = \{D_2\}$ we obtain connectedness, whereas for $\mathcal{E} = \{D_2^{\Delta}\}$, $\mathcal{E}$–connectedness means uniform connectedness. Though subspaces of $\mathcal{E}$–connected semiuniform convergence spaces are not $\mathcal{E}$–connected in general, dense subspaces of connected topological semiuniform convergence spaces (resp. uniformly connected uniform spaces) are always connected (resp. uniformly connected). If $C\mathcal{E}$ denotes the class of all $\mathcal{E}$–connected semiuniform convergence spaces, then a class $\mathcal{K}$ of semiuniform convergence spaces is called a connectedness iff there is some $\mathcal{E}$ such that $\mathcal{K} = C\mathcal{E}$. An internal description of connectednesses is given here, where the proof benefits from the fact that **SUConv** has hereditary quotients. Furthermore, a product theorem

for $\mathcal{E}$–connectedness is proved. Since each semiuniform convergence space can be decomposed into maximal $\mathcal{E}$–connected subsets, called $\mathcal{E}$–components, totally $\mathcal{E}$–disconnected semiuniform convergence spaces are studied, too, i.e. semiuniform convergence spaces whose $\mathcal{E}$–components are at most singletons. In particular, the quotient space of a semiuniform convergence space obtained by the decomposition into $\mathcal{E}$–components is totally $\mathcal{E}$–disconnected. This result can be applied to uniform spaces in case $\mathcal{E} = \{D_2^{\Delta}\}$ and leads to a new insight, since its validity depends on the fact that quotients of uniform spaces are formed in **SUConv** and not in **Unif**. If a class $\mathcal{E}$ of semiuniform convergence spaces coincides with the class of $\mathcal{E}$–disconnected semiuniform convergence spaces, it is called a disconnectedness. Disconnectednesses are also characterized internally, and a Galois correspondence between all connectednesses and disconnectednesses is established. The internal characterizations of connectednesses and disconnectednesses of semiuniform convergence spaces are similar to those in the realm of topological spaces due to A.V. Arhangel'skii and R. Wiegandt [6], whose proof is more complicated, since in **Top** quotients are not hereditary.
Local connectedness of topological spaces, originally introduced by H. Hahn [55] for subspaces of the Euclidean plane achieved its present form some years later by H. Hahn [57] and R.L. Moore [100]. In the last part of this chapter the localization of $\mathcal{E}$–connectedness is studied and it turns out that locally $\mathcal{E}$–connected semiuniform convergence spaces form a cartesian closed topological construct. A symmetric topological space which is locally connected in the usual sense is locally connected as a semiuniform convergence space. If subspaces of symmetric topological spaces are formed in **SUConv**, dense subspaces of locally connected symmetric topological spaces (regarded as semiuniform convergence spaces) are locally connected in contrast to the classical situation.

5.1 Connectednesses

5.1.1 Definition. A semiuniform convergence space $(X, \mathcal{J}_X)$ is called *connected* iff each uniformly continuous map $f : (X, \mathcal{J}_X) \longrightarrow (\{0,1\}, \{\dot{0} \times \dot{0}, \dot{1} \times \dot{1}\})$ from $(X, \mathcal{J}_X)$ into the two–point discrete semiuniform convergence space is constant.

5.1.2 Proposition. *A semiuniform convergence space $(X, \mathcal{J}_X)$ is connected provided that the underlying (symmetric) Kent convergence space $(X, q_{\gamma_{\mathcal{J}_X}})$ is connected, i.e. each continuous map $f : (X, q_{\gamma_{\mathcal{J}_X}}) \longrightarrow (\{0,1\}, \{(\dot{0},0), (\dot{1},1)\})$ from $(X, q_{\gamma_{\mathcal{J}_X}})$ into the two–point discrete (symmetric) Kent convergence space is constant.*

Proof. Since each uniformly continuous map between semiuniform convergence spaces is continuous with respect to the underlying (symmetric) Kent convergence spaces, the assertion is obvious.

5.1.3 Remark. 5.1.2. is not reversible. The set $\mathbb{Q}$ of rational numbers with the usual uniformity is a counterexample.

5.1.4 Proposition. *If $(X, \mathcal{J}_X)$ is a convergence space, then $(X, \mathcal{J}_X)$ is connected iff the underlying (symmetric) Kent convergence space $(X, q_{\gamma_{\mathcal{J}_X}})$ is connected.*

Proof. "$\Longrightarrow$". Use 2.3.3.23. 1) b) and c).
"$\Longleftarrow$". cf. 5.1.2.

5.1.5 Definition. Let $(X, \mathcal{J}_X)$ be a semiuniform convergence space. A subset A of X is called a *partition set* (in $(X, \mathcal{J}_X)$) provided that $A \times A \in \mathcal{F}$ or $(X \backslash A) \times (X \backslash A) \in \mathcal{F}$ for each $\mathcal{F} \in \mathcal{J}_X$.

5.1.6 Proposition. *A semiuniform convergence space $(X, \mathcal{J}_X)$ is connected iff the empty set $\emptyset$ and X are the only partition sets in $(X, \mathcal{J}_X)$.*

Proof. "$\Longrightarrow$" (indirect). If A is a partition set in $(X, \mathcal{J}_X)$ with $A \neq \emptyset$ and $A \neq X$, then $f : (X, \mathcal{J}_X) \longrightarrow (\{0,1\}, (\{\dot{0} \times \dot{0}, \dot{1} \times \dot{1}\}))$ defined by

$$f(x) = \begin{cases} 0 & \text{if} \quad x \in A \\ 1 & \text{if} \quad x \in X \backslash A \end{cases}$$

is uniformly continuous and not constant, i.e. $(X, \mathcal{J}_X)$ is not connected.

"$\Longleftarrow$". Let $f : (X, \mathcal{J}_X) \longrightarrow (\{0,1\}, \{\dot{0} \times \dot{0}, \dot{1} \times \dot{1}\})$ be uniformly continuous. Further, let $\mathcal{F} \in \mathcal{J}_X$. Then $(f \times f)(\mathcal{F}) = \dot{0} \times \dot{0}$ or $(f \times f)(\mathcal{F}) = \dot{1} \times \dot{1}$. Consequently, $f^{-1}(0) \times f^{-1}(0) \in \mathcal{F}$ or $f^{-1}(1) \times f^{-1}(1) \in \mathcal{F}$. Hence, $f^{-1}(0)$ is a partition set in $(X, \mathcal{J}_X)$, i.e. $f^{-1}(0) = \emptyset$ or $f^{-1}(0) = X$. Thus, f is constant.

5.1.7 Remark. If $(X, \mathcal{V})$ is a uniform space and $(X, [\mathcal{V}])$ its corresponding principal uniform limit space, then a subset A of X is a partition set in $(X, [\mathcal{V}])$ iff $A \times A \in \mathcal{V}$ or $(X \backslash A) \times (X \backslash A) \in \mathcal{V}$. Obviously, there are no partition sets other than $\emptyset$ and X. Thus, we obtain the following:

5.1.8 Corollary. *Every uniform space (= principal uniform limit space) is connected.*

5.1.9 Proposition. *Let $(X, \mathcal{X})$ be a symmetric topological space (i.e. an R_0-space) and $(X, \mathcal{J}_{\gamma_{q_\mathcal{X}}})$ its corresponding semiuniform convergence space. If $A \subset X$, then the following are equivalent:*
(1) *A is a partition set in $(X, \mathcal{J}_{\gamma_{q_\mathcal{X}}})$.*
(2) *For each $z \in \Delta \subset X \times X$, $A \times A$ is a neighborhood of z or $(X \backslash A) \times (X \backslash A)$ is a neighborhood of z (in the product space $(X, \mathcal{X}) \times (X, \mathcal{X})$).*
(3) A is open–closed in $(X, \mathcal{X})$.

5.1.10 Corollary. *Let $(X, \mathcal{X})$ be a symmetric topological space and $(X, \mathcal{J}_{\gamma_{q_\mathcal{X}}})$ its corresponding semiuniform convergence space. Then $(X, \mathcal{J}_{\gamma_{q_\mathcal{X}}})$ is connected iff* *there are no open–closed subsets in $(X, \mathcal{X})$ other than $\emptyset$ and X (i.e. iff $(X, \mathcal{X})$ is connected in the usual sense).*

5.1.11 Remark. Because of 5.1.8. the above concept of connectedness is not suitable for distinguishing between uniform spaces. But it makes sense to call a uniform space **X** *uniformly connected* iff each uniformly continuous map from **X** into the two–point discrete uniform space is constant. In particular, a metric space (X, d) is uniformly connected as a uniform space iff it is *Cantor–connected* (i.e. for each $\varepsilon > 0$ and any pair of points $x, y \in X$ there is a finite sequence $x_1, \ldots, x_n$ of points of X with $x_1 = x$, $x_n = y$ and $d(x_i, x_{i+1}) < \varepsilon$ for each $i \in \{1, \ldots, n-1\}$). This connectedness concept for metric spaces has been introduced by G. Cantor [24; p. 575]. Obviously, the metric space $\mathbb{Q}$ of rationals is Cantor–connected. The same is true for the following two subsets of the plane:
(1) a hyperbola and its asymptotes,
(2) the complement of a circle.
These aspects are captured by the following definition.

5.1.12 Definition. Let D_2^{Δ} be the two–point discrete uniform space (regarded as a principal uniform limit space), i.e. $D_2^{\Delta} = (\{0, 1\}, [(\Delta)])$, where $[(\Delta)]$ is the set of all filters on $\{0,1\} \times \{0,1\}$ containing the filter (Δ) generated by the diagonal Δ of $\{0,1\} \times \{0,1\}$. A semiuniform convergence space $(X, \mathcal{J}_X)$ is called *uniformly connected* provided that each uniformly continuous map $f : (X, \mathcal{J}_X) \longrightarrow D_2^{\Delta}$ is constant.

5.1.13. In order to develop a common theory of connectedness and uniform connectedness we define the following:

Definition. Let $\mathcal{E}$ be a class of semiuniform convergence spaces. Then a semiuniform convergence space $(X, \mathcal{J}_X)$ is called *$\mathcal{E}$–connected* provided that each uniformly continuous map $f : (X, \mathcal{J}_X) \longrightarrow (E, \mathcal{J}_E)$ is constant for each $(E, \mathcal{J}_E) \in \mathcal{E}$. The class of all $\mathcal{E}$–connected semiuniform convergence spaces is denoted by $C\mathcal{E}$. A subset A of X is called $\mathcal{E}$–connected provided that $(A, \mathcal{J}_A)$ is $\mathcal{E}$–connected, where $\mathcal{J}_A$ denotes the initial **SUConv**–structure on A with respect to the inclusion map $i : A \longrightarrow X$.
Obviously, $\mathcal{E}$–connectedness means connectedness (uniform connectedness) provided that $\mathcal{E} = \{(\{0,1\}, \{\dot{0} \times \dot{0}, \dot{1} \times \dot{1}\})\}$ ($\mathcal{E} = \{D_2^{\Delta}\}$).

5.1.14 Proposition. *Each $\mathcal{E}$–connected semiuniform convergence space is connected provided that $\mathcal{E}$ contains a space with at least two points.*

Proof. If $(X, \mathcal{J}_X)$ is a semiuniform convergence which is not connected, then there is a partition set A in $(X, \mathcal{J}_X)$ with $A \neq \emptyset$ and $A \neq X$. Let $(E, \mathcal{J}_E) \in \mathcal{E}$ with $a, b \in E$ and $a \neq b$. Then $f : (X, \mathcal{J}_X) \longrightarrow (E, \mathcal{J}_E)$ defined by

$$f(x) = \begin{cases} a & \text{if } x \in A \\ b & \text{if } x \in X \setminus A \end{cases}$$

is uniformly continuous, but not constant. Thus $(X, \mathcal{J}_X)$ is not $\mathcal{E}$-connected.

5.1.15 Remarks. 1) By 5.1.14. each uniformly connected semiuniform convergence space is connected, whereas a connected semiuniform converence space need not be uniformly connected, e.g. D_2^Δ is connected but not uniformly connected.

2) It is easily checked that *for* **Fil**-*determined semiuniform convergence spaces there is no difference between connectedness and uniform connectedness*, namely for a **Fil**–determined semiuniform convergence space $(X, \mathcal{J}_X)$ the following are equivalent:
(a) $(X, \mathcal{J}_X)$ is connected,
(b) The underlying filter space $(X, \gamma_{\mathcal{J}_X})$ is connected, i.e. each Cauchy continuous map $f : (X, \gamma_{\mathcal{J}_X}) \longrightarrow (\{0,1\}, \{\dot{0}, \dot{1}\})$ from $(X, \gamma_{\mathcal{J}_X})$ into the two–point discrete filter space $(\{0,1\}, \{\dot{0}, \dot{1}\})$ is constant,
(c) $(X, \mathcal{J}_X)$ is uniformly connected.

5.1.16 Definition. A class $\mathcal{K}$ of semiuniform convergence spaces is called *a connectedness* provided that there is a class $\mathcal{E}$ of semiuniform convergence spaces such that $\mathcal{K}$ is the class of all $\mathcal{E}$–connected semiuniform convergence spaces.

5.1.17 Theorem. *A class $\mathcal{K}$ of semiuniform convergence spaces is a connectedness iff the following are satisfied:*
(1) $\{(X, \mathcal{J}_X) \in |\mathbf{SUConv}| : \text{card}\,(X) \leq 1\} \subset \mathcal{K}$.
(2) *Let $f : (X, \mathcal{J}_X) \longrightarrow (Y, \mathcal{J}_Y)$ be a surjective uniformly continuous map between semiuniform convergence spaces. If $(X, \mathcal{J}_X) \in \mathcal{K}$, then $(Y, \mathcal{J}_Y) \in \mathcal{K}$.*
(3) *Let $(X, \mathcal{J}_X) \in |\mathbf{SUConv}|$ and let $(A_i)_{i\in I}$ be a family of subsets of X with $\bigcap_{i\in I} A_i \neq \emptyset$ such that the subspaces $(A_i, \mathcal{J}_{A_i})$ of $(X, \mathcal{J}_X)$ belong to $\mathcal{K}$ for each $i \in I$. Then $\bigcup_{i\in I} A_i$ (regarded as a subspace of $(X, \mathcal{J}_X)$) belongs to $\mathcal{K}$.*
(4) *Let $f : (X, \mathcal{J}_X) \longrightarrow (Y, \mathcal{J}_Y)$ be a quotient map in* **SUConv**. *Further, let $(Y, \mathcal{J}_Y) \in \mathcal{K}$ and let $f^{-1}(y)$ belong to $\mathcal{K}$ for each $y \in Y$. Then $(X, \mathcal{J}_X) \in \mathcal{K}$.*

Proof. "$\Longrightarrow$". Let $\mathcal{K} = C\mathcal{E}$ for some $\mathcal{E} \subset |\mathbf{SUConv}|$:
(1) is trivial.
(2) Let $(E, \mathcal{J}_E) \in \mathcal{E}$, and let $g : (Y, \mathcal{J}_Y) \longrightarrow (E, \mathcal{J}_E)$ be a uniformly continuous map. Since $g \circ f : (X, \mathcal{J}_X) \longrightarrow (E, \mathcal{J}_E)$ is constant by assumption, g is constant, because f is surjective.
(3) Since $\bigcap_{i\in I} A_i \neq \emptyset$, there is some $x \in \bigcap_{i\in I} A_i$. Thus, if $(E, \mathcal{J}_E) \in \mathcal{E}$ and $f : \bigcup_{i\in I} A_i \longrightarrow E$ is uniformly continuous (w.r.t. the subspace structure of $\bigcup_{i\in I} A_i$ and $\mathcal{J}_E$), $f[A_i] = \{f(x)\}$ for each $i \in I$ and therefore $\bigcup_{i\in I} A_i$ is $\mathcal{E}$–connected.
(4) Let $(E, \mathcal{J}_E) \in \mathcal{E}$ and $g : (X, \mathcal{J}_X) \longrightarrow (E, \mathcal{J}_E)$ be a uniformly continuous map. Then $h : (Y, \mathcal{J}_Y) \longrightarrow (E, \mathcal{J}_E)$ defined by $h \circ f = g$ is well–defined and uniformly continuous. Thus h is constant. Therefore g is constant.

"$\Longleftarrow$". Since $\mathcal{K}$ satisfies (3) and (1), each semiuniform convergence space $(Z, \mathcal{J}_Z)$ may be decomposed into $\mathcal{K}$–components which are defined as follows:

$$K_z = \bigcup\{A \subset Z : A \in \mathcal{K} \text{ and } z \in A\} \text{ for each } z \in Z$$
$$((\mathcal{K})\text{–component of } Z \text{ containing } z).$$

Let $\mathcal{E}$ be the class of all semiuniform convergence spaces whose $(\mathcal{K})$–components are singletons. Obviously, $\mathcal{K} \subset C\mathcal{E}$ (note that $C\mathcal{E}$ is closed under formation of uniformly continuous images).

In order to prove that $C\mathcal{E} \subset \mathcal{K}$, let $(X, \mathcal{J}_X) \in C\mathcal{E}$. Further, let R be the equivalence relation on X corresponding to the decomposition of X into $(\mathcal{K})$–components. We endow X/R with the final **SUConv**–structure with respect to the natural map $\omega : X \longrightarrow X/R$. Since $\mathcal{K}$ satisfies (4), each of the $(\mathcal{K})$–components of X/R is a singleton; namely, if $K \subset X/R$ were a $(\mathcal{K})$–component containing at least two elements, then $(\omega \mid \omega^{-1}[K])' : \omega^{-1}[K] \longrightarrow K$ would be a quotient map (**SUConv** is hereditary!) satisfying (4); thus $\omega^{-1}[K]$ would be a subset of a $(\mathcal{K})$–component of X and should be additionally the union of at least two $(\mathcal{K})$–components of X, which is impossible. That means that the quotient space X/R belongs to $\mathcal{E}$. Since $(X, \mathcal{J}_X)$ is $\mathcal{E}$–connected, $\omega : X \longrightarrow X/R$ is constant. Thus, $(X, \mathcal{J}_X) \in \mathcal{K}$.

5.1.18 Remark. The above theorem implies that each semiuniform convergence space $(X, \mathcal{J}_X)$ may be decomposed into maximal $\mathcal{E}$–connected subsets of X, the so–called *$\mathcal{E}$–components.* The $\mathcal{E}$–component of $(X, \mathcal{J}_X) \in |\mathbf{SUConv}|$ containing $x \in X$ may be described as the union of all $\mathcal{E}$–connected subsets of X containing x. If $\mathcal{E} = (\{0,1\}, \{\dot{0} \times \dot{0}, \dot{1} \times \dot{1}\})$, i.e. if $\mathcal{E}$ consists only of the two–point discrete semiuniform convergence space, then the $\mathcal{E}$–components are called *components.* We say also *uniform component* instead of $\{D_2^{\Delta}\}$–component.

5.1.19 Lemma. *Let $(X, \mathcal{J}_X)$ be a semiuniform convergence space and A a dense subset of X. If $f : (X, \mathcal{J}_X) \longrightarrow (E, \mathcal{J}_E)$ is a uniformly continuous map from $(X, \mathcal{J}_X)$ into a T_1-space $(E, \mathcal{J}_E) \in |\mathbf{SUConv}|$ such that the restriction of f to A is constant, then f is constant.*

Proof. Let $f[A] = \{e_0\}$ with $e_0 \in E$ and let $g : X \longrightarrow E$ be defined by $g(x) = e_0$ for each $x \in X$. In order to prove that $f = g$, let $K = \{x \in X : f(x) = g(x)\}$. Then $cl_{q_{\gamma_{\mathcal{J}_X}}} K = K$: If $x \in cl_{q_{\gamma_{\mathcal{J}_X}}} K$, then there is some $\mathcal{F} \in F(X)$ such that $(\mathcal{F}, x) \in q_{\gamma_{\mathcal{J}_X}}$ and $K \in \mathcal{F}$. Since $f, g : (X, q_{\gamma_{\mathcal{J}_X}}) \longrightarrow (E, q_{\gamma_{\mathcal{J}_E}})$ are continuous, it follows that $(f(\mathcal{F}), f(x)) \in q_{\gamma_{\mathcal{J}_E}}$ and $(g(\mathcal{F}), g(x)) \in q_{\gamma_{\mathcal{J}_E}}$. For each $F \in \mathcal{F}$, $F \cap K \in \mathcal{F}$ and $g[F \cap K] = f[F \cap K]$, which implies $f(\mathcal{F}) = g(\mathcal{F})$. Obviously, $g(\mathcal{F}) = \dot{e}_0$. Thus, $(\dot{e}_0, f(x)) \in q_{\gamma_{\mathcal{J}_X}}$ and $(\dot{e}_0, g(x)) \in q_{\gamma_{\mathcal{J}_X}}$. Since $(E, \mathcal{J}_E)$ is a T_1–space, we obtain $f(x) = e_0 = g(x)$, i.e. $x \in K$.

It follows from $A \subset K \subset X$ that $X = cl_{q_{\gamma_{\mathcal{J}_X}}} A \subset cl_{q_{\gamma_{\mathcal{J}_X}}} K = K \subset X$, i.e. $K = X$. Consequently, f is constant.

5.1.20 Proposition. *The assertion "If a subset M of a semiuniform convergence space $(X, \mathcal{J}_X)$ is $\mathcal{E}$-connected, then $cl_{q_{\gamma_{\mathcal{J}_X}}} M$ is $\mathcal{E}$-connected" is true if and only if $\mathcal{E}$ is a class of T_1-spaces.*

Proof. "$\Longleftarrow$". Let $f : cl_{q_{\gamma \mathcal{J}_X}} M \longrightarrow E$ be uniformly continuous, where $E \in \mathcal{E}$. If $M \neq \emptyset$ (the case $M = \emptyset$ is trivial), then by assumption, $f[M] = \{e_0\}$ with $e_0 \in E$. Since M is dense in $cl_{q_{\gamma \mathcal{J}_X}} M$, then by 5.1.19., f is constant.

"$\Longrightarrow$". Let $E \in \mathcal{E}$ and $x \in E$. If C_x is the $\mathcal{E}$–component of E containing x, then the inclusion map $i : C_x \longrightarrow E$ is uniformly continuous, i.e. $C_x = \{x\}$. Since the assertion " ... " under 5.2.10. is true and C_x is a maximal $\mathcal{E}$–connected subset, we obtain $\{x\} = cl_{q_{\gamma \mathcal{J}_X}} \{x\}$. Thus, E is a T_1–space.

5.1.21 Corollary. *Let $\mathcal{E} \subset |\mathbf{SUConv}|$ be a class of T_1–spaces. Then the following are valid:*
(1) The $\mathcal{E}$–components of each semiuniform convergence space are closed.
(2) If a subset M of a semiuniform convergence space $(X, \mathcal{J}_X)$ is $\mathcal{E}$–connected, then each subset N of X with $M \subset N \subset cl_{q_{\gamma \mathcal{J}_X}} M$ is $\mathcal{E}$–connected.

Proof. (1) Note that $\mathcal{E}$–components are maximal $\mathcal{E}$–connected subsets and use 5.1.20.
(2) M may be regarded as an $\mathcal{E}$–connected subspace of the subspace $(N, \mathcal{J}_N)$ of $(X, \mathcal{J}_X)$. Since $cl_{q_{\gamma \mathcal{J}_N}} M = (cl_{q_{\gamma \mathcal{J}_X}} M) \cap N = N$, the assertion follows from 5.1.20.

5.1.22 Remark. 1) If $\mathcal{E} \subset |\mathbf{SUConv}|$ is the empty class, then $\mathcal{E}$ is a class of T_1–spaces and $C\mathcal{E} = |\mathbf{SUConv}|$.
2) Let $\mathcal{E} \subset |\mathbf{SUConv}|$ be not a class of T_1–spaces: Since $\mathcal{E}$ is not the empty class, there is a space $(E, \mathcal{J}_E) \in \mathcal{E}$ with at least two points such that $(E, \mathcal{J}_E)$ is not a T_1–space, i.e. there are two distinct points $a, b \in E$ with $(\dot{a}, \dot{b}) \in q_{\gamma \mathcal{J}_E}$ or $(\dot{b}, \dot{a}) \in q_{\gamma \mathcal{J}_E}$. Thus, each space $(X, \mathcal{J}_X) \in |\mathbf{SUConv}|$ with at least two points is not $\mathcal{E}$–connected (namely, if $x, y \in X$ with $x \neq y$, then $f : (X, \mathcal{J}_X) \longrightarrow (E, \mathcal{J}_E)$ defined by $f(x) = a$ and $f[X \setminus \{x\}] = \{b\}$ is uniformly continuous [note that $(\dot{a}, \dot{b}) \in q_{\gamma \mathcal{J}_E}$ or $(\dot{b}, \dot{a}) \in q_{\gamma \mathcal{J}_E}$ implies $(\dot{a} \cap \dot{b}) \times (\dot{a} \cap \dot{b}) = (\dot{a} \times \dot{a}) \cap (\dot{a} \times \dot{b}) \cap (\dot{b} \times \dot{a}) \cap (\dot{b} \times \dot{b}) = (\{(a,b),(b,a),(a,a),(b,b)\}) \in \mathcal{J}_E$ and thus $\mathcal{F} \in \mathcal{J}_X$ implies $f \times f(\mathcal{F}) \supset (f \times f[X \times X]) = (\{(a,b),(b,a),(a,a),(b,b)\})$, i.e. $f \times f(\mathcal{F}) \in \mathcal{J}_E]$, but not constant). Consequently, $C\mathcal{E}$ consists of all one–point spaces and the empty space.

5.1.23 Lemma. *Let $((X_i, q_i))_{i \in I}$ be a family of generalized convergence spaces. If $x_{(0)}$ is a point of the product space (X, q) of this family in* **GConv**, *then $U = \{x \in X : p_i(x) = p_i(x_{(0)})$ for all but finitely many $i \in I\}$ is dense in (X, q), where $p_i : X \longrightarrow X_i$ denotes the i–th projection for each $i \in I$.*

Proof. For each $i \in I$, let p_i be the discrete topology on X_i. If (X, p) is the product space of $((X_i, p_i))_{i \in I}$ in **GConv**, then (X, p) is a topological space, since **Top** is bireflective in **GConv**. It is well–known that $cl_p U = X$. Since $1_{X_i} : (X_i, p_i) \longrightarrow (X_i, q_i)$ is continuous for each $i \in I$, $\prod_{i \in I} 1_{X_i} = 1_X : (X, p) \longrightarrow (X, q)$ is continuous, i.e. $p \leq q$. Therefore, $cl_q U \supset cl_p U$, and the lemma is proved.

5.1.24 Theorem. *Let $((X_i, \mathcal{J}_{X_i}))_{i\in I}$ be a family of non-empty semiuniform convergence spaces. Then the product $((X, \mathcal{J}_X), (p_i))$ of this family in* **SUConv** *is $\mathcal{E}$-connected iff $(X_i, \mathcal{J}_{X_i})$ is $\mathcal{E}$-connected for each $i \in I$.*

Proof. "$\Longrightarrow$". Apply 5.1.17. (2) to the projection $p_i : X \longrightarrow X_i$.
"$\Longleftarrow$". a) If $\mathcal{E}$ is not a class of T_1-spaces, then $C\mathcal{E}$ is trivial (cf. 5.1.22.) and the product theorem is valid.
b) Let $\mathcal{E}$ be a class of T_1-spaces and let $x_{(0)} \in X$. If $x_{(n)}$ and $x_{(0)}$ differ by at most $n < \infty$ coordinates, then $x_{(n)}$ and $x_{(0)}$ lie in an $\mathcal{E}$-connected subset of X which is proved by induction on the number n of differing coordinates in the following manner:

α) If $n = 1$ the assertion is correct; namely if $x_{(1)}$ and $x_{(0)}$ differ e.g. in the i-th coordinate, then $Y = X_i \times \prod_{k\neq i}\{p_k(x_{(0)})\} \subset X$ is isomorphic to X_i and hence it is $\mathcal{E}$-connected, and $x_{(1)}$ and $x_{(0)}$ lie in Y.
β) Let the assertion be valid for all $x_{(n-1)}$ $(n \geq 2)$. If $x_{(n)}$ is given, then $x_{(n-1)}$ can be found such that $x_{(n-1)}$ and $x_{(n)}$ differ by one coordinate. By α), $x_{(n)}$ and $x_{(n-1)}$ lie in an $\mathcal{E}$-connected subset C_1, and by the inductive hypothesis, $x_{(n-1)}$ and $x_{(0)}$ lie in an $\mathcal{E}$-connected subset C_2. Since $x_{(n-1)} \in C_1 \cap C_2$, i.e. $C_1 \cap C_2 \neq \emptyset$, 5.1.17. (3) may be applied and $C = C_1 \cup C_2$ is the desired $\mathcal{E}$-connected subset containing $x_{(0)}$ and $x_{(n)}$. Let us denote by $C_{x_{(0)}}$ the $\mathcal{E}$-component of X containing $x_{(0)}$. Thus,

$$U = \{x \in X : x \text{ and } x_{(0)} \text{ differ in at most finitely many coordinates }\} \subset C_{x_{(0)}} \subset X.$$

Since 5.1.23. is also valid for (symmetric) Kent convergence spaces, it follows $cl_{q_\gamma \mathcal{J}_X} U = X$ (cf. 2.3.3.17.). Hence $C_{x_{(0)}}$ is dense in X, which implies $X = C_{x_{(0)}}$ (cf. 5.1.21. 1)), i.e. X is $\mathcal{E}$-connected.

5.1.25 Remark. By 5.1.24., the full subconstruct $\mathcal{A}$ of **SUConv** defined by $|\mathcal{A}| = C\mathcal{E}$ is closed under formation of products, but it is not epireflective in **SUConv**, since it is not closed under formation of subspaces in general, e.g. $I\!R_t$ is connected, but the (closed) subspace $[0,1] \cup [1,2]$ is not connected (note: closed subspaces of symmetric topological spaces are topological). In order to find a necessary and sufficient condition for a subspace of an $\mathcal{E}$-connected semiuniform convergence space to be $\mathcal{E}$-connected we need the following definition.

5.1.26 Definition. Let $(X, \mathcal{J}_X)$ be a semiuniform convergence space, $(S, \mathcal{J}_S)$ a subspace of $(X, \mathcal{J}_X)$ and $\mathcal{E}$ a subclass of $|$**SUConv**$|$. Then $(S, \mathcal{J}_S)$ is called *uniformly $\mathcal{E}$-embedded in* $(X, \mathcal{J}_X)$ provided that for each $(E, \mathcal{J}_E) \in \mathcal{E}$, every uniformly continuous map $f : (S, \mathcal{J}_S) \longrightarrow (E, \mathcal{J}_E)$ has a uniformly continuous extension $\widetilde{f} : (X, \mathcal{J}_X) \longrightarrow (E, \mathcal{J}_E)$.

5.1.27 Proposition. *A subspace $(S, \mathcal{J}_S)$ of an $\mathcal{E}$-connected semiuniform convergence space $(X, \mathcal{J}_X)$ is $\mathcal{E}$-connected iff it is uniformly $\mathcal{E}$-embedded in $(X, \mathcal{J}_X)$.*

Proof. "$\Longrightarrow$". If $(S, \mathcal{J}_S)$ is $\mathcal{E}$-connected, each uniformly continuous map

$f : (S, \mathcal{J}_S) \longrightarrow (E, \mathcal{J}_E)$ is constant for each $(E, \mathcal{J}_E) \in \mathcal{E}$. Since each constant map is extendable, $(S, \mathcal{J}_S)$ is uniformly $\mathcal{E}$–embedded in $(X, \mathcal{J}_X)$.

"$\Longleftarrow$". Let $(E, \mathcal{J}_E) \in \mathcal{E}$. If $f : (S, \mathcal{J}_S) \longrightarrow (E, \mathcal{J}_E)$ is uniformly continuous, then there is a uniformly continuous extension $\tilde{f} : (X, \mathcal{J}_X) \longrightarrow (E, \mathcal{J}_E)$ of f. Since $(X, \mathcal{J}_X)$ is $\mathcal{E}$–connected, $\tilde{f}$ is constant. Thus, f is constant.

5.1.28 Corollary. *Every dense subspace of a uniformly connected uniform space is uniformly connected.*

Proof. Since D_2^{Δ} is a regular, separated and complete semiuniform convergence space, the assertion follows immediately from 5.1.27. and 4.2.1.13.

5.1.29 Remark. Concerning symmetric topological spaces and connectedness a result analogous to 5.1.28. is not valid provided that dense subspaces are formed in $\mathbf{Top_S}$; namely $\mathbb{R}_t$ is connected, but the dense subspace $\mathbb{Q}_t$ of rational numbers is not connected. The situation becomes better provided that subspaces of symmetric topological spaces are formed in **SUConv** instead of $\mathbf{Top_S}$. Obviously, the following extension theorem is valid:

Theorem. *Let $(X, \mathcal{X})$ be a symmetric topological space and $(X, \mathcal{J}_{\gamma_{q_\mathcal{X}}})$ its corresponding semiuniform convergence space. Then each uniformly continuous map $f : (A, \mathcal{J}_A) \longrightarrow (Y, \mathcal{J}_Y)$ of a dense subspace $(A, \mathcal{J}_A)$ of $(X, \mathcal{J}_{\gamma_{q_\mathcal{X}}})$ (formed in* **SUConv**) *into a complete, regular and separated semiuniform convergence space $(Y, \mathcal{J}_Y)$ has a unique uniformly continuous extension $\overline{f} : (X, \mathcal{J}_{\gamma_{q_\mathcal{X}}}) \longrightarrow (Y, \mathcal{J}_Y)$.*

Proof. It follows from the assumptions that $f : (A, q_{\gamma_{\mathcal{J}_A}}) \longrightarrow (Y, q_{\gamma_{\mathcal{J}_Y}})$ is a continuous map from the dense subspace $(A, q_{\gamma_{\mathcal{J}_A}})$ of $(X, q_\mathcal{X})$ into the regular T_2 Kent convergence space $(Y, q_{\gamma_{\mathcal{J}_Y}})$. Since $(Y, \mathcal{J}_Y)$ is compete and separated, we obtain analogously to the first part of the proof of 4.2.1.13. that for each $x \in X$ there is a unique $y \in Y$ such that for each $\mathcal{F} \in F(X)$ with $(\mathcal{F}, x) \in q_\mathcal{X}$ and $A \in \mathcal{F}$, $(f(i^{-1}(\mathcal{F})), y) \in q_{\gamma_{\mathcal{J}_Y}}$, where $i : A \longrightarrow X$ denotes the inclusion map. Thus, the assumptions of 4.2.1.11. are fulfilled and there is a unique continuous extension $\overline{f} : (X, q_\mathcal{X}) \longrightarrow (Y, q_{\gamma_{\mathcal{J}_Y}})$ of $f : (A, q_{\gamma_{\mathcal{J}_A}}) \longrightarrow (Y, q_{\gamma_{\mathcal{J}_Y}})$, i.e. $\overline{f} : (X, \mathcal{J}_{\gamma_{q_\mathcal{X}}}) \longrightarrow (Y, \mathcal{J}_Y)$ is a unique uniformly continuous extension of $f : (A, \mathcal{J}_A) \longrightarrow (Y, \mathcal{J}_Y)$.

This leads to the following:

5.1.30 Corollary. *Every dense subspace (formed in* **SUConv**) *of a connected symmetric topological space (regarded as a semiuniform convergence space) is connected.*

Proof. Since $D_2 = (\{0, 1\}, \{\dot{0} \times \dot{0}, \dot{1} \times \dot{1}\})$ is a complete, regular and separated semiuniform convergence space, the assertion follows immediately from 5.1.27. and the theorem under 5.1.29.

5.1.31 Remark. It follows from 5.1.30. that the set $\mathbb{Q}$ of rational numbers regarded as a subspace (in **SUConv** [or **Fil**]) of $\mathbb{R}_t$, denoted by $\mathbb{Q}_f$, is a connected filter space, but not a topological space.

5.1.32 Theorem. *Let $\mathcal{E}$ be a class of semiuniform convergence spaces and $\mathcal{A}$ a bireflective (full and isomorphism–closed) subconstruct of* **SUConv** *containing $\mathcal{E}$. If $\mathcal{R}$: **SUConv** $\longrightarrow \mathcal{A}$ denotes the bireflector, then a semiuniform convergence space $(X, \mathcal{J}_X)$ is $\mathcal{E}$-connected if and only if $\mathcal{R}((X, \mathcal{J}_X))$ is $\mathcal{E}$-connected.*

Proof. Use 5.1.17 (2) and the defining property of a bireflection.

5.1.33 Remark. It follows from 5.1.32. that *a semiuniform convergence space is uniformly connected iff its underlying uniform space is uniformly connected.*

5.1.34 Corollary. *Let $(X, \mathcal{V})$ be a separated uniform space and $(X, [\mathcal{V}])$ its corresponding principal uniform limit space. Then the following are equivalent:*
(1) *$(X, [\mathcal{V}])$ is uniformly connected, i.e. $(X, \mathcal{V})$ is uniformly connected,*
(2) *The Hausdorff completion of $(X, \mathcal{V})$ is uniformly connected,*
(3) *The Wyler completion $(Y, \mathcal{J}_Y^*)$ of $(X, [\mathcal{V}])$ is uniformly connected,*
(4) *The simple completion $(Y, \mathcal{J}_Y)$ of $(X, [\mathcal{V}])$ is uniformly connected.*

Proof. By 5.1.21. (2) each of the above completions is uniformly connected provided that $(X, [\mathcal{V}])$ (resp. $(X, \mathcal{V})$) is uniformly connected. The underlying uniform space of each of the above completions is the Hausdorff completion of $(X, \mathcal{V})$ (cf. 4.4.20. and 4.4.21.). Thus, by 5.1.28. and 5.1.33., $(X, [\mathcal{V}])$ is uniformly connected provided that one of these completions is uniformly connected.

5.2 Disconnectednesses and their relations to connectednesses

5.2.1 Definition. A semiuniform convergence space $(X, \mathcal{J}_X)$ is called *totally $\mathcal{E}$-disconnected* provided that for each $x \in X$ the $\mathcal{E}$–component of X containing x consists only of x. If $\mathcal{E}$ consists only of the two–point discrete semiuniform convergence space (the two–point discrete principal uniform limit space), then we say *totally disconnected* (*totally uniformly disconnected*) instead of totally $\mathcal{E}$–disconnected.

5.2.2 Proposition. *Let $\mathcal{E} \subset$ |**SUConv**| contain at least one space with more than one point. Then $C\mathcal{E}$ is equal to the class of all connected semiuniform convergence spaces iff $\mathcal{E}$ is a subclass of the class of all totally disconnected spaces.*

Proof. "$\Longrightarrow$ ". Let $\mathbf{E} = (E, \mathcal{J}_E) \in \mathcal{E}$ and $x \in E$. By assumption, the component C_x containing x is $\mathcal{E}$–connected and consequently, it is a singleton (otherwise the inclusion map $i : C_x \longrightarrow E$ were not constant). Thus, $\mathbf{E}$ is totally disconnected.

"$\Longleftarrow$". By 5.1.14. each $\mathbf{X} \in C\mathcal{E}$ is connected. Conversely, let $\mathbf{X} = (X, \mathcal{J}_X)$ be a connected semiuniform convergence space, $\mathbf{E} \in \mathcal{E}$ and $f : \mathbf{X} \longrightarrow \mathbf{E}$ a uniformly continuous map. By 5.1.17. (2), $f[X]$ is a connected subset in $\mathbf{E}$ and by assumption it is a singleton (provided $X \neq \emptyset$), i.e. $f : \mathbf{X} \longrightarrow \mathbf{E}$ is a constant map. Consequently, $\mathbf{X}$ is $\mathcal{E}$-connected.

5.2.3 Proposition. *For each $\mathcal{E} \subset |\mathbf{SUConv}|$, let $D\mathcal{E} = \{(X, \mathcal{J}_X) \in |\mathbf{SUConv}| :$ each uniformly continuous map $f : (E, \mathcal{J}_E) \longrightarrow (X, \mathcal{J}_X)$ is constant for each $(E, \mathcal{J}_E) \in \mathcal{E}\}$. Then $D_H\mathcal{E} = DC\mathcal{E}$ is the class of all totally $\mathcal{E}$-disconnected semiuniform convergence spaces.*

Proof. Let $(X, \mathcal{J}_X) \in D_H\mathcal{E}$ and $x \in X$. If C_x denotes the $\mathcal{E}$-component of $(X, \mathcal{J}_X)$ containing x, then, by assumption, the inclusion map $i : C_x \longrightarrow X$ is constant, i.e. $C_x = \{x\}$. Thus, $(X, \mathcal{J}_X)$ is totally $\mathcal{E}$-disconnected. Conversely, let $(X, \mathcal{J}_X)$ be a totally $\mathcal{E}$-disconnected semiuniform convergence space and $f : (Y, \mathcal{J}_Y) \longrightarrow (X, \mathcal{J}_X)$ a uniformly continuous map from an $\mathcal{E}$-connected semiuniform convergence space $(Y, \mathcal{J}_Y)$ into $(X, \mathcal{J}_X)$. By 5.1.17. (2), $f[Y]$ is an $\mathcal{E}$-connected subset in $(X, \mathcal{J}_X)$ and by assumption, it is a singleton (provided $Y \neq \emptyset$), i.e. $f : (Y, \mathcal{J}_Y) \longrightarrow (X, \mathcal{J}_X)$ is a constant map, which implies $(X, \mathcal{J}_X) \in DC\mathcal{E} = D_H\mathcal{E}$.

5.2.4 Proposition. *Let $\mathcal{E}$ be a subclass of $|\mathbf{SUConv}|$ and $(f_i : (X, \mathcal{J}_X) \longrightarrow (X_i, \mathcal{J}_{X_i}))_{i \in I}$ a mono-source in* **SUConv** *(i.e. each $f_i : (X, \mathcal{J}_X) \longrightarrow (X_i, \mathcal{J}_{X_i})$ is uniformly continuous and for any two distinct points $x, y \in X$ there is some $j \in I$ such that $f_j(x) \neq f_j(y)$). If $(X_i, \mathcal{J}_{X_i})$ belongs to $D\mathcal{E}$ for each $i \in I$, then $(X, \mathcal{J}_X)$ belongs to $D\mathcal{E}$ too.*

Proof. Let $(E, \mathcal{J}_E) \in \mathcal{E}$ and let $f : (E, \mathcal{J}_E) \longrightarrow (X, \mathcal{J}_X)$ be a uniformly continuous map. If f were not constant, there would exist $a, b \in E$ such that $f(a) \neq f(b)$. By assumption, there were some $j \in I$ such that $f_j(f(a)) \neq f_j(f(b))$. Therefore $f_j \circ f : (E, \mathcal{J}_E) \longrightarrow (X_j, \mathcal{J}_{X_j})$ would not be constant which is impossible since $(X_j, \mathcal{J}_{X_j}) \in D\mathcal{E}$.

5.2.5 Corollary. *Let $\mathcal{A}$ be the (full and isomorphism-closed) subconstruct of* **SUConv** *defined by $|\mathcal{A}| = D\mathcal{E}$ [resp. $|\mathcal{A}| = D_H\mathcal{E}$]. Then $\mathcal{A}$ is extremal epireflective.*

Proof. By 5.2.4., $\mathcal{A}$ is closed under formation of products and weak subobjects, which implies that $\mathcal{A}$ is extremal epireflective (cf. the characterization theorem under 2.2.4.).

5.2.6 Proposition. *Let $(X, \mathcal{J}_X)$ be a semiuniform convergence space and R the equivalence relation on X defined by the decomposition of X into $\mathcal{E}$-components. Then the quotient space $(X/R, \mathcal{J}_{X/R})$ (formed in* **SUConv***) is totally $\mathcal{E}$-disconnected, in other words: the natural map $\omega : (X, \mathcal{J}_X) \longrightarrow$*

$(X/R, \mathcal{J}_{X/R})$ is the extremal epireflection of $(X, \mathcal{J}_X)$ w.r.t. the full (and isomorphism–closed) subconstruct $\mathcal{A}$ of **SUConv** *defined by $|\mathcal{A}| = D_H\mathcal{E}$.*

Proof. It follows from the last part of the proof of 5.1.17. that $(X/R, \mathcal{J}_{X/R})$ is totally $\mathcal{E}$–disconnected (put $\mathcal{K} = C\mathcal{E}$ for the present $\mathcal{E}$). Obviously, $\omega : (X, \mathcal{J}_X) \longrightarrow (X/R, \mathcal{J}_{X/R})$ is then the desired extremal epireflection of $(X, \mathcal{J}_X)$ w.r.t. $\mathcal{A}$ with $|\mathcal{A}| = D_H\mathcal{E}$.

5.2.7 Remark. The above proposition is also valid for uniform spaces provided that quotients are formed in **SUConv** and *not* in **Unif**; namely, quotients of uniform spaces formed in **Unif** are rather complicated and they have sometimes unpleasant properties, e.g. the above proposition is false in **Unif** as the following *example* shows:

Let X be the set $\{(0,0),(0,1)\} \cup \bigcup\{\{\frac{1}{n}\} \times [0,1] : n \in I\!N\}$ endowed with the uniformity induced by the uniformity of $I\!R^2_u$. Further, let $Y = \{0\} \cup \{\frac{1}{n} : n \in I\!N\}$ be endowed with the uniformity induced by $I\!R_u$. The canonical projection $\omega : X \longrightarrow Y$ is the restriction of the first projection $p_1 : I\!R^2_u \longrightarrow I\!R_u$ and consequently, it is uniformly continuous and surjective. But it is even a uniform quotient map, since it is a topological quotient map (namely, it is continuous, surjective and [obviously] open) and Y is compact. Furthermore, Y is totally uniformly disconnected (i.e. $Y \in D_H\{D_2^{\Delta}\}$), which is easily seen by means of Cantor's definition of uniform connectedness in metric spaces, and $\omega : X \longrightarrow Y$ is the totally uniformly disconnected reflection of X:
Let Z be a totally uniformly disconnected uniform space and $f : X \longrightarrow Z$ a uniformly continuous map. Then $\overline{f} : Y \longrightarrow Z$ defined by $\overline{f}(\omega(x)) = f(x)$ for each $x \in X$ is well–defined (There is no problem provided that $\omega(x)$ is a singleton; otherwise one may conclude as follows: For each entourage W of Y there is a symmetric entourage V of Y such that $V^2 \subset W$. Since f is uniformly continuous, there is some $\varepsilon > 0$ such that

$(*)$ $\quad d(x,x') < \varepsilon$ implies $(f(x), f(x')) \in V$,

where d is the usual metric inducing the uniformity of X. Choose $n \in N$ such that $\frac{1}{n} < \varepsilon$. Put $a = (0,0)$, $b = (0,1)$, $a' = (\frac{1}{n}, 0)$ and $b' = (\frac{1}{n}, 1)$. It follows from $(*)$

$(**)$ $\quad (f(b), f(b')) \in V$ and $(f(a), f(a')) \in V$.

Furthermore, a' and b' belong to the same uniform component of X and since f is uniformly continuous and Y is totally uniformly disconnected, we obtain

$(***)$ $\quad f(a') = f(b')$.

Thus, $(f(a), f(b)) \in V^2 \subset W$ [since $(**)$ is valid and V is symmetric], i.e. $f(b) \in W(f(a))$ for each entourage W of Y. Since Y is T_1–space, $f(a) = f(b)$.). Since $\omega : X \longrightarrow Y$ is a uniform quotient map, $\overline{f}$ is uniformly continuous and uniquely determined.
Obviously, *the totally uniformly disconnected reflection $\omega : X \longrightarrow Y$ of X has a fibre which is not uniformly connected*, namely $\omega^{-1}(0)$ is a two–point discrete

uniform space and consequently, $\omega^{-1}(0)$ is not uniformly connected. *This implies that the uniform quotient space X/R of X obtained by the decomposition of X into its uniform components is not totally uniformly disconnected* (otherwise the natural map $\nu : X \longrightarrow X/R$ were the totally uniformly disconnected reflection of X and consequently X/R would be isomorphic to Y and the fibres of Y were uniformly connected).

5.2.8 Definition. A subclass $\mathcal{K} \subset |\mathbf{SUConv}|$ is called a *disconnectedness* provided that there is some $\mathcal{F} \subset |\mathbf{SUConv}|$ such that $\mathcal{K} = D\mathcal{F}$.

5.2.9 Remarks. 1) By 5.2.3., the class of all totally $\mathcal{E}$–disconnected semiuniform convergence spaces is a disconnectedness (choose $\mathcal{F} = C\mathcal{E}$).
2) The operators C and D have the following properties:
1. (a) $\mathcal{E} \subset \mathcal{F} \subset |\mathbf{SUConv}|$ implies α) $C\mathcal{E} \supset C\mathcal{F}$
and β) $D\mathcal{E} \supset D\mathcal{F}$
(b) $\mathcal{E} \subset DC\mathcal{E}$ and $\mathcal{E} \subset CD\mathcal{E}$ for each $\mathcal{E} \subset |\mathbf{SUConv}|$,
2. $CDC = C$ and $DCD = D$,
3. $C_H = CD$ and $D_H = DC$ are *hull operators*, i.e. C_H and D_H are *extensive* (cf. 1.(b)), *isotonic* ($\mathcal{E} \subset \mathcal{F} \subset |\mathbf{SUConv}|$ implies $C_H\mathcal{E} \subset C_H\mathcal{F}$ and $D_H\mathcal{E} \subset D_H\mathcal{F}$), *and idempotent* ($C_HC_H = C_H$ and $D_HD_H = D_H$).

5.2.10 Definition. A subclass $\mathcal{K}$ of $|\mathbf{SUConv}|$ is called *C_H–closed* (*D_H–closed*) provided that $\mathcal{K} = C_H\mathcal{K}$ ($\mathcal{K} = D_H\mathcal{K}$).

5.2.11 Proposition. *A subclass $\mathcal{K}$ of* $|\mathbf{SUConv}|$ *is C_H–closed* (*D_H–closed*) *iff it is a connectedness* (*disconnectedness*).

Proof. Apply 2. and 3. under 5.2.9. 2).

5.2.12 Theorem. *There exists a one–one correspondence between the connectednesses and disconnectednesses of* $|\mathbf{SUConv}|$ *which converts the inclusion relation* (**Galois correspondence**) *and is obtained by the operators C and D.*

Proof. By means of C one obtains a one–one correspondence which assigns to each disconnectedness, i.e. D_H–closed subclass of $|\mathbf{SUConv}|$, a connectedness, i.e. a C_H–closed subclass of $|\mathbf{SUConv}|$. The inverse correspondence is obtained by D. It follows from 5.2.9. 1. (a) α) and β) that the inclusion relation is converted.

5.2.13 Theorem. *Let $\mathcal{K}$ be an isomorphism–closed*[1] *subclass of* $|\mathbf{SUConv}|$ *containing a non–empty space. Then the following are equivalent:*
(1) *$\mathcal{K}$ is a disconnectedness,*
(2) (a) *$\mathcal{K}$ is closed under formation of products and subspaces*[1],
(b) *For each surjective uniformly continuous map $f : (X, \mathcal{J}_X) \longrightarrow (Y, \mathcal{J}_Y)$ be-*

---

[1] $\mathcal{K}$ is called isomorphism–closed (resp. closed under formation of products and subspaces) provided that the full subconstruct $\mathcal{A}$ of **SUConv** defined by $|\mathcal{A}| = \mathcal{K}$ has this property.

tween semiuniform convergence spaces such that $(Y, \mathcal{J}_Y) \in \mathcal{K}$ and $f^{-1}(y) \in \mathcal{K}$ for each $y \in Y$, the space $(X, \mathcal{J}_X)$ belongs to $\mathcal{K}$.

Proof. (1) $\Longrightarrow$ (2). Since $\mathcal{K}$ is a disconnectedness, $\mathcal{K} = D_H\mathcal{K}$ (cf. 5.2.11.). By 5.2.5., (a) is valid (cf. the characterization theorem under 2.2.4.). In order to prove (b) let K be a $\mathcal{K}$–component of $(X, \mathcal{J}_X)$ (where we may assume without loss of generality that X is non–empty). Then $f[K] = \{y_0\}$ for some $y_0 \in Y$. Since $f^{-1}(y_0)$ belongs to $\mathcal{K}$ and (a) is valid, K belongs to $\mathcal{K}$. Thus, K is a singleton. This proves (b) (cf. 5.2.3.).

(2) $\Longrightarrow$ (1). Let $\mathcal{A}$ be the full and isomorphism–closed subconstruct of **SUConv** defined by $|\mathcal{A}| = \mathcal{K}$. Since (a) and (b) are valid, $\mathcal{A}$ is extremal epireflective, i.e. $\mathcal{A}$ is closed under formation of products and weak subobjects (let $(Y, \mathcal{J}_Y) \in \mathcal{K}$ and let $f : (X, \mathcal{J}_X) \longrightarrow (Y, \mathcal{J}_Y)$ be a monomorphism in **SUConv**; then $f' : X \longrightarrow f[X]$ defined by $f'(x) = f(x)$ for each $x \in X$ is bijective and $f[X]$ [regarded as a subspace of $(Y, \mathcal{J}_Y)$] belongs to $\mathcal{K}$; thus, $f^{-1}(f(x))$ is a singleton and therefore $f^{-1}(f(x))$ belongs to $\mathcal{K}$ for each $x \in X$; consequently, $(X, \mathcal{J}_X) \in \mathcal{K}$). It suffices to prove that $D_H\mathcal{K} \subset \mathcal{K}$ (cf. 5.2.11. and use 5.2.9. 2) 1. (b)).
Thus, let $\mathbf{X} = (X, \mathcal{J}_X) \in D_H\mathcal{K}$ and let $\omega : \mathbf{X} \longrightarrow \mathbf{X}_{\mathcal{A}}$ be the extremal epireflection of $\mathbf{X}$ w.r.t. $\mathcal{A}$. Since ω is a quotient map, $\mathbf{X}_{\mathcal{A}}$ may be identified with the quotient space $\boldsymbol{X/R_\omega}$, where R_ω denotes the equivalence relation on X defined by

$$x \; R_\omega \; y \Longleftrightarrow \omega(x) = \omega(y)\,.$$

Consequently, ω may be identified with the natural map from $\mathbf{X}$ onto $\boldsymbol{X/R_\omega}$. The proof is finished, if ω is an injective map. This will imply that $\mathbf{X}$ belongs to $\mathcal{K}$. It suffices to show that $\omega(x) \subset X$ is a singleton for each $x \in X$. If there would exist some $x \in X$ such that $\omega(x) \neq \{x\}$, there would be a non–constant uniformly continuous map $f : \boldsymbol{\omega(x)} \longrightarrow \mathbf{Y}$ for some $\mathbf{Y} \in \mathcal{K}$ (by the assumption on $\mathbf{X}$). Since (a) is valid, it may be assumed that f is surjective. Let $Y^* = \{f^{-1}(z) : z \in Y\}$ and let $Z = ((X/R_\omega)\backslash\{\omega(x)\}) \cup Y^*$. Thus, Z is a decomposition of X. Let R_Z be the corresponding equivalence relation on X and let $\omega_1 : X \longrightarrow X/R_Z$ be the natural map. Then $\omega_1 : \mathbf{X} \longrightarrow \boldsymbol{X/R_Z}$ is a uniformly continuous map provided that $\boldsymbol{X/R_Z} = (X/R_Z, \mathcal{J}_{X/R_Z})$, where $\mathcal{J}_{X/R_Z}$ denotes the final **SUConv**–structure w.r.t. ω_1. Let $\omega_2 : X/R_Z \longrightarrow X/R_\omega$ be defined by

$$\omega_2(c) = \begin{cases} c & \text{if } \; c \in (X/R_\omega)\backslash\{\omega(x)\} \\ \omega(x) & \text{if } \; c \in Y^*\,. \end{cases}$$

Obviously, $\omega_2 \circ \omega_1 = \omega$. Thus, $\omega_2 : \boldsymbol{X/R_Z} \longrightarrow \boldsymbol{X/R_\omega}$ is uniformly continuous. Since $\omega_1^{-1}(\omega_1(\omega(x))) = \omega(x)$, the map $(\omega_1 | \; \omega(x))' : \boldsymbol{\omega(x)} \longrightarrow \boldsymbol{\omega_1(\omega(x))}$ is a quotient map (**SUConv** is hereditary!). Therefore, the subspace $\boldsymbol{\omega_1(\omega(x))}$ of $\boldsymbol{X/R_Z}$ may be identified with the quotient space $\boldsymbol{\omega(x)/R_f}$, where R_f denotes the equivalence relation on $\omega(x)$ corresponding to the decomposition Y^*. Consequently, $\overline{f} : \boldsymbol{\omega(x)/R_f} \longrightarrow \mathbf{Y}$ defined by $\overline{f} \circ (\omega_1 | \; \omega(x))' = f$ is a bijective uniformly continuous map. Thus, $\overline{f}$ fulfills condition (b). This implies

$\omega_1(\boldsymbol{\omega}(\boldsymbol{x})) \in \mathcal{K}$. On the other hand $\omega_1(\omega(x)) = \omega_2^{-1}(\omega(x))$. Since $\omega_2^{-1}(c) = c$ for each $c \in (X/R_\omega)\backslash\{\omega(x)\}$, $\omega_2^{-1}(a) \in \mathcal{K}$ for each $a \in X/R_\omega$. But $\boldsymbol{X/R_\omega} \in \mathcal{K}$ and $\mathcal{K}$ fulfills (b). Therefore $\boldsymbol{X/R_Z} \in \mathcal{K}$. Since ω is a reflection, there is a uniformly continuous map $\overline{\omega} : \boldsymbol{X/R_\omega} \longrightarrow \boldsymbol{X/R_Z}$ such that the diagram

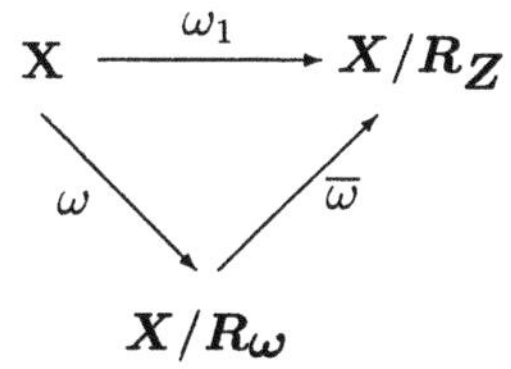

commutes. Consequently, $(\overline{\omega} \circ \omega_2) \circ \omega_1 = \overline{\omega} \circ (\omega_2 \circ \omega_1) = \overline{\omega} \circ \omega = \omega_1 = 1_X \circ \omega_1$ which implies $\overline{\omega} \circ \omega_2 = 1_X$. Hence, ω_2 is injective. This is a contradiction to the fact that $\omega_2^{-1}(\omega(x))$ consists at least of two elements.

5.2.14. At the end of this section let us consider some examples of totally disconnected (resp. totally uniformly disconnected) semiuniform convergence spaces:

Definitions. A semiuniform convergence space $(X, \mathcal{J}_X)$ is called
1) *zerodimensional* provided that it is a subspace (in **SUConv**) of some zerodimensional topological space (regarded as a semiuniform convergence space),
2) *uniformly zerodimensional* provided that it is a subspace of a product of discrete uniform spaces (regarded as semiuniform convergence spaces).

5.2.15 Remarks. 1) *A symmetric topological space is zerodimensional as a semiuniform convergence space iff it is zerodimensional in the usual sense* (cf. 4.3.3.4. ③ for the definition).

2) *The class of all zerodimensional (resp. uniformly zerodimensional) semiuniform convergence spaces is isomorphism–closed and closed under formation of subspaces and products.*

5.2.16 Proposition. 1) *Every zerodimensional semiuniform convergence space is* **Fil**–*determined and totally disconnected.*

2) *Every uniformly zerodimensional semiuniform convergence space is uniform and totally uniformly disconnected.*

Proof. 1) a) Since every symmetric topological space (regarded as a semiuniform convergence space) is **Fil**–determined and **Fil–D–SUConv** is bireflective in **SUConv**, every subspace of a zerodimensional topological space is **Fil**–determined.
b) Every zerodimensional topological space is totally disconnected as a topological space and thus, it is totally disconnected as a semiuniform convergence space. Hence, by 5.2.13. (2) (a), every zerodimensional semiuniform convergence space is totally disconnected.

2) a) Since **Unif** is bireflective in **SUConv**, every uniformly zerodimensional

semiuniform convergence space is uniform.
b) Since every discrete uniform space is totally uniformly disconnected and 5.2.13. (2) (a) is valid, every uniformly zerodimensional semiuniform convergence space is totally uniformly disconnected.

5.2.17 Remark. If $\mathcal{E}$ is a class of semiuniform convergence spaces, then a semiuniform convergence space is called *$\mathcal{E}$–regular* provided that is it a subspace of a product of spaces of $\mathcal{E}$. Let $R\mathcal{E}$ denote the class of all $\mathcal{E}$–regular semiuniform convergence spaces and let $\mathcal{A}$ be the full subconstuct of **SUConv** defined by $|\mathcal{A}| = \mathcal{E}$. The full and isomorphism–closed subconstruct $R\mathcal{A}$ of **SUConv** defined by $|R\mathcal{A}| = R\mathcal{E}$ is the smallest (full and isomorphism–closed) epireflective subconstruct of **SUConv** containing $\mathcal{A}$, the so–called *epireflective hull* of $\mathcal{A}$, where a subconstruct $\mathcal{A}_1$ is called smaller than a subconstruct $\mathcal{A}_2$ provided that $|\mathcal{A}_1| \subset |\mathcal{A}_2|$. If $\mathcal{E}$ consists of all zerodimensional topological spaces (regarded as semiuniform convergence spaces), then $R\mathcal{E}$ is the class of all zerodimensional semiuniform convergence spaces, whereas $R\mathcal{E}$ is the class of all uniformly zerodimensional semiuniform convergence spaces in case $\mathcal{E}$ consists of all discrete uniform spaces (= discrete principal uniform limit spaces). Furthermore, for each epireflective subconstruct $\mathcal{A}$ of **SUConv**, $R\mathcal{A} = \mathcal{A}$, i.e. $R\mathcal{E} = \mathcal{E}$ with $\mathcal{E} = |\mathcal{A}|$. Thus, $R\mathcal{E}$ is the class of all regular semiuniform convergence spaces in case $\mathcal{E}$ is the class of all regular semiuniform convergence spaces. Since $\mathcal{E} \subset D_H\mathcal{E}$ and by 5.2.13. (2) (a), $D_H\mathcal{E}$ is closed under formation of products and subspaces, one obtains that *every $\mathcal{E}$-regular semiuniform convergence space is totally $\mathcal{E}$-disconnected.*

5.3 Local $\mathcal{E}$–connectedness

5.3.1 Definition. A semiuniform convergence space $(X, \mathcal{J}_X)$ is called *locally $\mathcal{E}$-connected* provided that for each $\mathcal{F} \in \mathcal{J}_X$ there is some subfilter $\mathcal{G} \in \mathcal{J}_X$ of $\mathcal{F}$ together with a filter base $\mathcal{B}$ for $\mathcal{G}$ consisting of $\mathcal{E}$-connected subsets of $(X, \mathcal{J}_X) \times (X, \mathcal{J}_X)$. One says *locally connected* (resp. *locally uniformly connected*) instead of locally $\mathcal{E}$–connected in case $\mathcal{E}$ consists of the two–point discrete semiuniform convergence space (resp. $\mathcal{E} = \{D_2^{\Delta}\}$).

5.3.2 Remark. A **Fil**–determined semiuniform convergence space $(X, \mathcal{J}_X)$ is locally connected iff it is locally uniformly connected (cf. 5.1.15. 2)).

5.3.3 Proposition. *Let $(X, \mathcal{X})$ be a symmetric topological space and $(X, \mathcal{J}_{\gamma_{q\mathcal{X}}})$ the corresponding semiuniform convergence space. Then $(X, \mathcal{J}_{\gamma_{q\mathcal{X}}})$ is locally $\mathcal{E}$-connected iff for each $x \in X$ the neighborhood filter $\mathcal{U}(x)$ of x (in $(X, \mathcal{X})$) has a filter base of $\mathcal{E}$-connected subsets of $(X, \mathcal{J}_{\gamma_{q\mathcal{X}}})$.*

Proof. "$\Longrightarrow$". By assumption, for each $x \in X$, there is a subfilter $\mathcal{G} \in \mathcal{J}_{\gamma_{q\mathcal{X}}}$ of $\mathcal{U}(x) \times \mathcal{U}(x)$ with a filter base $\mathcal{B}$ of $\mathcal{E}$–connected subsets of $(X, \mathcal{J}_{\gamma_{q\mathcal{X}}}) \times (X, \mathcal{J}_{\gamma_{q\mathcal{X}}})$. Since $(X, \mathcal{X})$ is symmetric, $\mathcal{G} = \mathcal{U}(x) \times \mathcal{U}(x)$. If $p_1 : X \times X \longrightarrow X$ denotes the first projection, $\{p_1[B] : B \in \mathcal{B}\}$ is a filter base of $\mathcal{U}(x)$ consisting of $\mathcal{E}$–connected subsets of $(X, \mathcal{J}_{\gamma_{q\mathcal{X}}})$.

"$\Longleftarrow$". Let $\mathcal{F} \in \mathcal{J}_{\gamma_{q\mathcal{X}}}$, i.e. there is some $x \in X$ with $\mathcal{F} \supset \mathcal{U}(x) \times \mathcal{U}(x)$. By assumption, there is a filter base $\mathcal{B}$ for $\mathcal{U}(x)$ consisting of $\mathcal{E}$-connected subsets of $(X, \mathcal{J}_{\gamma_{q\mathcal{X}}})$. Then $\mathcal{B} \times \mathcal{B} = \{B \times B' : B, B' \in \mathcal{B}\}$ is a filter base for $\mathcal{U}(x) \times \mathcal{U}(x)$ consisting of $\mathcal{E}$-connected subsets of $(X, \mathcal{J}_{\gamma_{q\mathcal{X}}}) \times (X, \mathcal{J}_{\gamma_{q\mathcal{X}}})$. Obviously, $\mathcal{U}(x) \times \mathcal{U}(x) \in \mathcal{J}_{\gamma_{q\mathcal{X}}}$. Thus, $(X, \mathcal{J}_{\gamma_{q\mathcal{X}}})$ is locally $\mathcal{E}$-connected.

5.3.4 Corollary. 1) *Let $(X, \mathcal{X})$ be a symmetric topological space which is locally connected in the usual sense. Then $(X, \mathcal{X})$ is locally connected as a semiuniform convergence space, i.e. $(X, \mathcal{J}_{\gamma_{q\mathcal{X}}})$ is locally connected.*
2) *Let $(X, \mathcal{X})$ be a regular topological space. If $(X, \mathcal{X})$ is locally connected as a semiuniform convergence space, then $(X, \mathcal{X})$ locally connected in the usual sense.*

Proof. 1) follows from 5.1.2.
2) Let $x \in X$ and $U_x \in \mathcal{U}(x)$. Since $(X, \mathcal{X})$ is regular, there is a closed neighborhood V_x of x such that $V_x \subset U_x$. By assumption, there is a neighborhood $C_x \subset V_x$ of x which is connected as a subspace of $(X, \mathcal{J}_{\gamma_{q\mathcal{X}}})$. By 5.1.20., the closure $\overline{C_x}$ of C_x is also connected as a subspace of $(X, \mathcal{J}_{\gamma_{q\mathcal{X}}})$. Since closed subspaces of symmetric topological spaces are formed as in **SUConv** and 5.1.10. is valid, $\overline{C_x}$ is a connected neighborhood of x in the usual sense, which is contained in U_x. Thus, $(X, \mathcal{X})$ is locally connected in the usual sense.

5.3.5 Proposition. *Let $(X, \mathcal{V})$ be a uniform space and $(X, [\mathcal{V}])$ its corresponding principal uniform limit space. If $(X, [\mathcal{V}])$ is locally $\mathcal{E}$-connected, then $(X, [\mathcal{V}])$ is $\mathcal{E}$-connected.*

Proof. By assumption, there is a filter base $\mathcal{B}$ for $\mathcal{V}$ consisting of $\mathcal{E}$-connected subsets of $(X, [\mathcal{V}]) \times (X, [\mathcal{V}])$. Let $B \in \mathcal{B}$ and let $p_1 : X \times X \longrightarrow X$ be the first projection. Then, since $\Delta \subset B$,

$$X = p_1[\Delta] \subset p_1[B] \subset X,$$

i.e. $(X, [\mathcal{V}])$ is $\mathcal{E}$-connected (note 5.1.17. (2)).

5.3.6 Proposition. *Let $(f_i : (X_i, \mathcal{J}_{X_i}) \longrightarrow (X, \mathcal{J}_X))_{i \in I}$ be a final epi-sink in **SUConv**. If all $(X_i, \mathcal{J}_{X_i})$ are locally $\mathcal{E}$-connected, then $(X, \mathcal{J}_X)$ is locally $\mathcal{E}$-connected.*

Proof. Let $\mathcal{F} \in \mathcal{J}_X$. Then there are some $i \in I$ and some $\mathcal{F}_i \in \mathcal{J}_{X_i}$ such that $(f_i \times f_i)(\mathcal{F}_i) \subset \mathcal{F}$. Furthermore, there is some $\mathcal{G}_i \in \mathcal{J}_{X_i}$ such that $\mathcal{G}_i \subset \mathcal{F}_i$ as well as a filter base $\mathcal{B}_i$ for $\mathcal{G}_i$ consisting of $\mathcal{E}$-connected subsets of $(X_i, \mathcal{J}_{X_i}) \times (X_i, \mathcal{J}_{X_i})$. Thus, $(f_i \times f_i)(\mathcal{G}_i) \in \mathcal{J}_X$ is a subfilter of $\mathcal{F}$ and $\{(f_i \times f_i)[B_i] : B_i \in \mathcal{B}_i\}$ is a filter base for $(f_i \times f_i)(\mathcal{G}_i)$ consisting of $\mathcal{E}$-connected subsets of $(X, \mathcal{J}_X) \times (X, \mathcal{J}_X)$ (note: $f_i \times f_i$ is uniformly continuous and B_i is $\mathcal{E}$-connected).

5.3.7 Corollary. *Let $\mathbf{LCon}_{\mathcal{E}}$ be the construct of all locally $\mathcal{E}$-connected semiuniform convergence spaces (and uniformly continuous maps). Then $\mathbf{LCon}_{\mathcal{E}}$ is a* *bicoreflective (full and isomorphism-closed) subconstruct of* **SUConv**.

5.3.8 Proposition. *Let $((X_i, \mathcal{J}_{X_i}))_{i\in I}$ be a family of non-empty semiuniform convergence spaces. Then the product space $(\prod_{i\in I} X_i, \mathcal{J}_X)$ of this family is locally $\mathcal{E}$-connected iff $(X_i, \mathcal{J}_{X_i})$ is locally $\mathcal{E}$-connected for each $i \in I$ and $\mathcal{E}$-connected for all but finitely many $i \in I$.*

Proof. "$\Longleftarrow$". Let $\mathcal{F} \in \mathcal{J}_X$. Then $(p_i \times p_i)(\mathcal{F}) \in \mathcal{J}_{X_i}$ for each $i \in I$, where $p_i : \prod_{i\in I} X_i \longrightarrow X_i$ denotes the i-th projection. By assumption, there are finitely many elements $i_1, \dots, i_n$ of I such that $(X_i, \mathcal{J}_{X_i})$ is $\mathcal{E}$-connected for each $i \in I\backslash\{i_1, \dots, i_n\}$, and for each $i \in \{i_1, \dots, i_n\}$ there is a uniform subfilter $\mathcal{G}_i$ of $(p_i \times p_i)(\mathcal{F})$ together with a base $\mathcal{B}_i$ for $\mathcal{G}_i$ consisting of $\mathcal{E}$-connected subsets. If $j : \prod_{i\in I}(X_i \times X_i) \longrightarrow \prod_{i\in I} X_i \times \prod_{i\in I} X_i$ denotes the canonical isomorphism, then

$$j(\textstyle\prod_{i\in I} \mathcal{G}_i) \subset j(\prod_{i\in I}(p_i \times p_i)(\mathcal{F})) \subset \mathcal{F}$$

and $\mathcal{B} = \{j[\prod_{i\in I} M_i] : M_i = X_i \times X_i$ for each $i \in I\backslash\{i_1, \dots, i_n\}$, and $M_i \in \mathcal{B}_i$ for each $i \in \{i_1, \dots, i_n\}\}$ is a filter base for $j(\prod_{i\in I} \mathcal{G}_i) \in \mathcal{J}_X$ consisting of $\mathcal{E}$-connected subsets of $(\prod_{i\in I} X_i, \mathcal{J}_X) \times (\prod_{i\in I} X_i, \mathcal{J}_X)$.

"$\Longrightarrow$". Let $\mathcal{F}_i \in \mathcal{J}_{X_i}$ for each $i \in I$. Then the product filter $\prod_{i\in I} \mathcal{F}_i$ is a filter on $\prod_{i\in I}(X_i \times X_i)$ and consequently $j(\prod_{i\in I} \mathcal{F}_i)$ is a filter on $\prod_{i\in I} X_i \times \prod_{i\in I} X_i$. Let $p'_i : \prod_{i\in I}(X_i \times X_i) \longrightarrow X_i \times X_i$ denote the i-th projection for each $i \in I$. Since $(p_i\times p_i)(j(\prod_{i\in I} \mathcal{F}_i)) = p'_i(\prod_{i\in I} \mathcal{F}_i) = \mathcal{F}_i \in \mathcal{J}_{X_i}$, $j(\prod_{i\in I} \mathcal{F}_i) \in \mathcal{J}_X$. By assumption, there is a subfilter $\mathcal{G} \in \mathcal{J}_X$ of $j(\prod_{i\in I} \mathcal{F}_i)$ with a filter base $\mathcal{B}$ of $\mathcal{E}$-connected subsets. Hence, $(p_i \times p_i)(\mathcal{G}) \in \mathcal{J}_{X_i}$ is a subfilter of $\mathcal{F}_i$ and $\{(p_i \times p_i)[B] : B \in \mathcal{B}\}$ is filter a base for $(p_i \times p_i)(\mathcal{G})$ consisting of $\mathcal{E}$-connected subsets. Consequently, $(X_i, \mathcal{J}_{X_i})$ is locally $\mathcal{E}$-connected for each $i \in I$. Let $B \in \mathcal{B}$. Since $B \supset j(\prod_{i\in I} F_i)$, where $F_i \in \mathcal{F}_i$ for each $i \in I$ and $F_i = X_i \times X_i$ for all but finitely many $i \in I$, one obtains for all but finitely many $i \in I$,

$$X_i \times X_i = F_i = p'_i(\textstyle\prod_{i\in I} F_i) = (p_i \times p_i)(j(\prod_{i\in I} F_i)) \subset (p_i \times p_i)[B] \subset X_i \times X_i\,,$$

i.e. $X_i \times X_i = (p_i \times p_i)[B]$ is $\mathcal{E}$-connected, which implies that $(X_i, \mathcal{J}_{X_i})$ is $\mathcal{E}$-connected for all but finitely many $i \in I$.

5.3.9 Theorem. **LCon**$_{\mathcal{E}}$ *is a cartesian closed topological construct.*

Proof. Since **SUConv** is cartesian closed and **LCon**$_{\mathcal{E}}$ is a bicoreflective subconstruct (cf. 5.3.7) which is closed under formation of finite products (cf. 5.3.8.), it follows from 3.1.7. that **LCon**$_{\mathcal{E}}$ is cartesian closed.

5.3.10 Remark. Consider the following subset of the plane $I\!R^2$:

$$X = \{(x,y) : y = \sin\frac{1}{x},\ 0 < x \leq 1\} \cup \{(0,y) : -1 \leq y \leq 1\}\,.$$

If $I\!R^2$ is endowed with the usual topology, then it is locally connected (in the usual sense), whereas the closed subspace determined by X is not locally connected (in the usual sense). This implies that *local $\mathcal{E}$-connectedness is not hereditary in* **SUConv** *in general* (*cf.* 5.3.4.).

5.3.11 Proposition. *Let $(X, \mathcal{X})$ be a symmetric topological space and $(X, \mathcal{J}_{\gamma_{q\mathcal{X}}})$ its corresponding semiuniform convergence space. Then every dense subspace $(A, \mathcal{J}_A)$ of $(X, \mathcal{J}_{\gamma_{q\mathcal{X}}})$ is locally connected provided that $(X, \mathcal{J}_{\gamma_{q\mathcal{X}}})$ is locally connected.*

Proof. 1) Consider $(X, \mathcal{J}_{\gamma_{q\mathcal{X}}})$, where $(X, \mathcal{X})$ is an R_0-space. Let A be dense in $(X, \mathcal{J}_{\gamma_{q\mathcal{X}}})$ and O an open connected subset in $(X, \mathcal{J}_{\gamma_{q\mathcal{X}}})$. Then $A \cap O$ is connected in $(X, \mathcal{J}_{\gamma_{q\mathcal{X}}})$ [note $cl_{q\mathcal{X}} O = cl_{q\mathcal{X}}(A \cap O)$ and use 5.1.20. and 5.1.30.].
2) Let $(X, \mathcal{J}_{\gamma_{q\mathcal{X}}})$ be locally connected and $(A, \mathcal{J}_A)$ a dense subspace. In order to prove that $(A, \mathcal{J}_A)$ is locally connected let $\mathcal{F} \in \mathcal{J}_A$. Then $(i \times i)(\mathcal{F}) \in \mathcal{J}_{\gamma_{q\mathcal{X}}}$, where $i : A \longrightarrow X$ denotes the inclusion map. Hence, there is some $x \in X$ such that

$$(*) \qquad (i \times i)(\mathcal{F}) \supset \mathcal{U}(x) \times \mathcal{U}(x)\,,$$

where $\mathcal{U}(x)$ denotes the neighborhood filter of x in $(X, \mathcal{X})$. Since A is dense in $(X, \mathcal{X})$, $i^{-1}(\mathcal{U}(x))$ exists and it follows from $(*)$

$$(**) \qquad \mathcal{F} \supset (i \times i)^{-1}(\mathcal{U}(x) \times \mathcal{U}(x)) = i^{-1}(\mathcal{U}(x) \times i^{-1}(\mathcal{U}(x))\,.$$

By assumption, the open connected neighborhoods of x form a base for $\mathcal{U}(x)$ (note that in locally connected spaces the components of open neighborhoods are open). Thus, $\mathcal{B} = \{A \cap O_x : O_x \text{ is an open connected neighborhood of } x\}$ is a base for $i^{-1}(\mathcal{U}(x))$ consisting of connected subsets of $(X, \mathcal{J}_{\gamma_{q\mathcal{X}}})$ (resp. $(A, \mathcal{J}_A)$) [cf. 1)]. Hence, $\mathcal{G} = i^{-1}(\mathcal{U}(x)) \times i^{-1}(\mathcal{U}(x))$ has a base of connected subsets of $(A, \mathcal{J}_A) \times (A, \mathcal{J}_A)$ and belongs to $\mathcal{J}_A$. This implies that $(A, \mathcal{J}_A)$ is locally connected.

5.3.12 Remarks. 1) Let $(X, \mathcal{X})$ be a locally connected R_0-space in the usual sense. Then every dense subspace of $(X, \mathcal{J}_{\gamma_{q\mathcal{X}}})$ is locally connected (cf. 5.3.11. and 5.3.4.). Thus, $\mathbb{Q}_f$ (cf. 5.1.31.) is locally connected, since $\mathbb{R}_t$ is locally connected and $\mathbb{Q}$ is dense in $\mathbb{R}_t$.
2) a) Consider the following Hausdorff space $(X, \mathcal{X})$ which is connected but neither locally connected nor regular: Let X be the set $\mathbb{R}$ of real number and $\mathcal{T}$ the usual topology on X. If $\mathcal{X}$ denotes the topology on X generated by $\mathcal{T} \cup \{\mathbb{R} \backslash A\}$, where $A = \{\frac{1}{n} : n \in \mathbb{N}\}$, then $(X, \mathcal{X})$ is the desired space.
b) The semiuniform convergence space $(X, \mathcal{J}_{\gamma_{q\mathcal{X}}})$ with $(X, \mathcal{X})$ as under a) is locally connected (use 5.1.30.) but $(X, \mathcal{X})$ is not locally connected, i.e. the inversion of 5.3.4. 1) is not valid in general.

Chapter 6

Function Spaces

Simple convergence (= pointwise convergence) and uniform convergence known from Analysis are studied first in the realm of classical General Topology. Also continuous convergence introduced by H. Hahn [56] is considered in this context. Since pointwise convergence can be described by means of the product topology, which was first observed by A. Tychonoff [143], uniform spaces are needed for uniform convergence (the uniformity of uniform convergence was first explicitly defined by J.W. Tukey [141]). In order to study continuous convergence in the realm of topological spaces, the restriction to locally compact Hausdorff spaces is necessary (cf. theorem 6.1.31.), i.e. for infinite–dimensional analysis continuous convergence cannot be described in this framework. Since the complex plane is a locally compact Hausdorff space, in classical Function Theory continuous convergence is available and according to C. Carathéodory [25], it is often useful to substitute uniform convergence on compacta by continuous convergence. By the way, if we consider locally compact Hausdorff spaces, then the topology describing continuous convergence is the compact–open topology, introduced and studied first by R.H. Fox [45] and R. Arens [4]. By the introduction of this book, the reason why topological spaces are not sufficient for studying continuous convergence can also be formulated as follows: **Top** is not cartesian closed.

Since the structure of simple convergence, uniform convergence and continuous convergence can be derived from the natural function space structure in **SUConv**, we study in the second and third part of this chapter function space structures in the construct **SUConv** and in suitable subconstructs. We have seen already that the localization of connectedness leads to a cartesian closed subconstuct of **SUConv**. Now local compactness and local precompactness are investigated. It turns out that the localization of compactness leads to a cartesian closed topological construct which coincides with the construct of compactly generated semi-uniform convergence spaces – a result which is not valid in **Top**. Compactly generated Hausdorff spaces are usually called k–spaces, and R. Brown ([21] and [22]) proved that they even form a cartesian closed topological construct (where the Hausdorff axiom is essential according to H. Breger [20] and J. Činčura [31]), but not every k–space is locally compact. It is proved that a Hausdorff space is a

k–space iff it is the induced topological space (= underlying topological space) of some locally compact semiuniform convergence space (the class of k–spaces was introduced by D. Gale [53], where he attributed the notion 'k–space' to Hurewicz without giving a reference). Furthermore, Alexandroff's one–point compactification of locally compact Hausdorff spaces introduced in [2] is studied for those Kent convergence spaces which are locally compact T_2. The localization of precompactness results even in a topological universe.
At the end of section 2 the relations between compactness, precompactness, local compactness and local precompactness are studied and it turns out that the localization of compactness (or precompactness) is a more topological than a uniform procedure, since in uniform spaces there is no difference between precompactness (resp. compactness) and local precompactness (resp. local compactness), whereas in regular topological spaces local precompactness (= local compactness) differs from precompactness (= compactness).
In the third section of the present chapter precompactness and compactness in the natural function spaces of **SUConv** are characterized, where some results of the preceding two sections are used. Finally, the classical Ascoli theorem is obtained, which goes back to G. Ascoli [8] and C. Arzelà [7]. We have profited much in this section from a paper by O. Wyler [153].

6.1 Simple convergence, uniform convergence and continuous convergence in the realm of classical General Topology

6.1.1. In Analysis several kinds of convergence for sequences of functions are studied, e.g. simple convergence, uniform convergence or continuous convergence. Remember: If $f : I\!R \longrightarrow I\!R$ is a map and $(f_n : I\!R \longrightarrow I\!R)$ a sequence of maps, then (f_n) is said to *converge*

(1) *simply* to f provided that for each $x \in I\!R$ the sequence $(f_n(x))$ converges to $f(x)$ in $I\!R$, i.e. provided that for each $x \in X$ and each $\varepsilon > 0$ there is some $N(x,\varepsilon) \in I\!N$ such that for all $n \geq N(x,\varepsilon)$, $|f_n(x) - f(x)| < \varepsilon$,

(2) *uniformly* to f provided that for each $\varepsilon > 0$ there is some $N(\varepsilon) \in I\!N$ such that for all $n \geq N(\varepsilon)$, $|f_n(x) - f(x)| < \varepsilon$ for each $x \in X$,

(3) *continuously* to f provided that for each $x \in X$ and each sequence (x_n) in $I\!R$ converging to x the sequence $(f_n(x_n))$ converges to $f(x)$ in $I\!R$.

In the following concepts of convergence for suitable function sets are defined such that the above examples occur as special cases.

6.1.2 Definition. Let $\mathbf{X} = (X, \mathcal{X})$, $\mathbf{Y} = (Y, \mathcal{Y})$ be topological spaces, $C(\mathbf{X}, \mathbf{Y})$ the set of all continuous maps from $\mathbf{X}$ to $\mathbf{Y}$ and $F(X,Y)$ the set of all maps from X to Y, i.e. $F(X,Y) = Y^X$. Then the product topology on $F(X,Y)$, i.e. the initial topology on $F(X,Y)$ with respect to the family $(p_x : F(X,Y) \longrightarrow Y)_{x \in X}$ of

all projections $p_x : F(X,Y) \longrightarrow Y$ defined by $p_x(f) = f(x)$, is called *the topology of simple convergence* (or *the topology of pointwise convergence*), denoted by $\mathcal{T}_s$. One writes $F_s(X,Y)$ instead of $(F(X,Y),\mathcal{T}_s)$, and $C_s(\mathbf{X},\mathbf{Y})$ for the subspace of $F_s(X,Y)$ determined by the subset $C(\mathbf{X},\mathbf{Y})$.
A filter on $F(X,Y)$ is said to *converge simply* (*or pointwise*) to $f \in F(X,Y)$ provided that it converges to f in $F_s(X,Y)$.

6.1.3 Remark. It follows immediately from the characterization of filter convergence in (topological) product spaces (cf. chapter 0) that a *filter $\mathcal{F}$ on $F(X,Y)$ converges simply to $f \in F(X,Y)$ iff for each $x \in X$, $p_x(\mathcal{F})$ converges to $f(x)$ in* $\mathbf{Y}$. Consequently, *a sequence (f_n) in $F(X,Y)$ converges simply to $f \in F(X,Y)$ iff for each $x \in X$, the sequence $(f_n(x))$ converges to $f(x)$ in* $\mathbf{Y}$. Note further, that the topology on X is not needed for defining the topology of simple convergence on $F(X,Y)$.

6.1.4 Proposition. 1) *If* $\mathbf{Y}$ *has one of the properties T_0, T_1, T_2, regular, completely regular or totally disconnected, then $F_s(X,Y)$ and $C_s(\mathbf{X},\mathbf{Y})$ have these properties, too.*
2) *If* $\mathbf{Y}$ *is compact (resp. connected), then $F_s(X,Y)$ is compact (resp. connected).*

Proof. 1) All mentioned properties are closed under formation of products and subspaces.
2) Compactness and connectedness are closed under formation of products.

6.1.5 Remark. The topology of simple convergence is easy to handle and has pleasant properties (cf. 6.1.4.). Its disadvantage lies in the fact that it is too coarse, i.e. too many filters (or sequences) converge simply. Thus, continuity may be destroyed as the following *example* shows: If $X = [0,1]$ and $Y = \mathbb{R}$ are endowed with the usual topologies, then the sequence $(f_n : \mathbf{X} \longrightarrow \mathbf{Y})$ of continuous functions, defined by $f_n(x) = x^n$ for each $x \in X$, converges simply to the non-continuous function $f : \mathbf{X} \longrightarrow \mathbf{Y}$ defined by

$$f(x) = \begin{cases} 0 & \text{if } x \neq 1 \\ 1 & \text{if } x = 1 \,. \end{cases}$$

In order to avoid such an unpleasant situation, in Analysis the topology of uniform convergence is studied on $F(X,\mathbb{R})$.

6.1.6 Proposition. *Let X be a set and $(Y,\mathcal{V})$ a uniform space. For each $V \in \mathcal{V}$, let*

$$W(V) = \{(f,g) \in F(X,Y) \times F(X,Y) : (f(x),g(x)) \in V \text{ for each } x \in X\}\,.$$

Then $\{W(V) : V \in \mathcal{V}\}$ is a base for a uniformity $\mathcal{U}$ on $F(X,Y)$.

Proof. This follows from the following observations:
1. For each $f \in F(X,Y)$ and each $W(V)$, $(f,f) \in W(V)$, i.e. the diagonal of

$F(X,Y) \times F(X,Y)$ is contained in each $W(V)$.
2. $(W(V))^{-1} = W(V^{-1})$
3. $(W(V))^2 \subset W(V^2)$
4. Since $V \subset V'$ implies $W(V) \subset W(V')$, $W(V_1 \cap V_2) \subset W(V_1) \cap W(V_2)$.

6.1.7 Definition. The uniformity $\mathcal{U}$ defined in 6.1.6. is called *the uniform structure of uniform convergence.* We write $F_u(X,Y)$ instead of $(F(X,Y),\mathcal{U})$. The topology induced by $\mathcal{U}$ is called the *topology of uniform convergence.* A filter $\mathcal{F}$ on $F(X,Y)$ is said to *converge uniformly* to $f \in F(X,Y)$ provided that it converges to f with respect to this topology.

6.1.8 Remarks. 1) It follows immediately from the definition of $\mathcal{U}$ that $p_x : F_u(X,Y) \longrightarrow (Y,\mathcal{V})$ is uniformly continuous for each $x \in X$. If X is a set and $(Y,\mathcal{V})$ a uniform space, the product uniformity on $F(X,Y)$ is called *the uniform structure of simple convergence.* The induced topology is then the topology of simple convergence. Since the product uniformity on $F(X,Y)$ is the coarsest uniformity such that all projections p_x are uniformly continuous, it follows that *the uniform structure of simple convergence is coarser than the uniform structure of uniform convergence*, which implies that *the topology of uniform convergence is finer than the topology of simple convergence, i.e. each uniformly convergent filter $\mathcal{F}$ on $F(X,Y)$ is simply convergent.*

2) If (Y,d) is a metric space and $\mathcal{V}_d$ the uniformity on Y induced by d, then a sequence (f_n) in $F(X,Y)$ converges uniformly to $f \in F(X,Y)$ iff for each $\varepsilon > 0$ there is some $N(\varepsilon) \in \mathbb{N}$ such that for all $n \geq N(\varepsilon)$, $d(f_n(x), f(x)) < \varepsilon$ for each $x \in X$ (note: $\{V_\varepsilon : \varepsilon > 0\}$ is a base for $\mathcal{V}_d$ where $V_\varepsilon = \{(y,y') \in Y \times Y : d(y,y') < \varepsilon\}$). Then $\{W(V_\varepsilon) \mid \varepsilon > 0\}$ is a base for $\mathcal{U}$ and $\{W(V_\varepsilon)(f) \mid \varepsilon > 0\}$ with $W(V_\varepsilon)(f) = \{g \in F(X,Y) : (f,g) \in W(V_\varepsilon)\}$ is a neighborhood base at f with respect to the topology of uniform convergence.)

6.1.9 Proposition. *If $(Y,\mathcal{V})$ is a metrizable uniform space, i.e. $\mathcal{V}$ is induced by a metric, then $F_u(X,Y)$ is metrizable.*

Proof. Since $(Y,\mathcal{V})$ is metrizable, there is a bounded metric d on Y inducing $\mathcal{V}$ (namely if d' is a metric on Y inducing $\mathcal{V}$, then $d = \min\{d', 1\}$ is a bounded metric inducing $\mathcal{V}$). A metric $\tilde{d}$ on $F(X,Y)$ is defined by

$$\tilde{d}(f,g) = \sup\{d(f(x), g(x)) : x \in X\}.$$

Obviously, $\tilde{d}(f,g) \leq \varepsilon$ is equivalent to $d(f(x),g(x)) \leq \varepsilon$ for each $x \in X$. Thus, the uniformity induced by $\tilde{d}$ has the same base as $\mathcal{U}$, i.e. it coincides with $\mathcal{U}$. Consequently, $F_u(X,Y)$ is metrizable by means of $\tilde{d}$.

6.1.10 Definition. Let $\mathbf{X} = (X,\mathcal{X})$ be a topological space and $\mathbf{Y} = (Y,\mathcal{V})$ a uniform space. Then the subspace of $F_u(X,Y)$ determined by the subset $C(\mathbf{X},\mathbf{Y})$ of all continuous maps from $\mathbf{X}$ into $\mathbf{Y}$ is denoted by $C_u(\mathbf{X},\mathbf{Y})$.

6.1.11 Proposition. *If $\mathbf{X} = (X, \mathcal{X})$ is a topological space and $\mathbf{Y} = (Y, \mathcal{V})$ a uniform space, then $C_u(\mathbf{X}, \mathbf{Y})$ is a closed subspace of $F_u(X, Y)$.*

Proof. It will be shown that $F(X,Y)\backslash C(\mathbf{X}, \mathbf{Y})$ is open with respect to the topology of uniform convergence on $F(X,Y)$. If f is not continuous at x there is some $U \in \mathcal{V}$ such that $f^{-1}[U(f(x))]$ is not a neighborhood of x. Choose some $V = V^{-1} \in \mathcal{V}$ with $V^3 \subset U$. Let $g \in W(V)(f)$, i.e. $(f(x'), g(x')) \in V$ for each $x' \in X$. Hence, $g^{-1}[V(g(x))] \subset f^{-1}[U(f(x))]$ (namely, $x' \in g^{-1}[V(g(x))]$ implies $g(x') \in V(g(x))$, i.e. $(g(x), g(x')) \subset V$; since $(f(x), g(x)), (f(x'), g(x')) \in V$ and $V = V^{-1}$, it follows that $(f(x), f(x')) \in V^3 \subset U$, i.e. $f(x') \in U(f(x))$ or equivalently, $x' \in f^{-1}[U(f(x))]$). Consequently, $g^{-1}[V(g(x))]$ is not a neighborhood of x, i.e. g is not continuous at x. Thus, $W(V)(f)$ is a neighborhood of f with respect to the topology of uniform convergence each member of which is a non–continuous function.

6.1.12 Remarks. 1) If $i : C_u(\mathbf{X}, \mathbf{Y}) \longrightarrow F_u(X, Y)$ denotes the inclusion map and $\mathcal{F}$ is a filter on $C_u(\mathbf{X}, \mathbf{Y})$ such that $i(\mathcal{F})$ converges uniformly to $f \in F(X, Y)$, then the above proposition says that f is continuous. In particular, *if (f_n) is a sequence in $C_u(\mathbf{X}, \mathbf{Y})$ which converges uniformly to $f \in F(X, Y)$, then f is continuous.*
In other words: Uniform convergence does not destroy continuity.

2) It is known from Analysis that a sequence (f_n) of continuous functions $f_n : \mathbb{R} \longrightarrow \mathbb{R}$ which converges uniformly only on each compact subset C of $\mathbb{R}$ has a continuous limit function, i.e. uniform convergence is not necessary in order to obtain a continuous limit function. Thus, it will be possible to substitute the uniform structure of uniform convergence in 6.1.11. by a coarser one as one can see in the following.

6.1.13 Definition. Let X be a set, $(Y, \mathcal{V})$ a uniform space and $\mathcal{S} \subset \mathcal{P}(X)$. Furthermore, for each $S \in \mathcal{S}$ let $\varphi_S : F(X,Y) \longrightarrow F_u(S, Y)$ be defined by $\varphi_S(f) = f \mid S$. The initial uniformity on $F(X, Y)$ w.r.t. $(\varphi_S)_{S \in \mathcal{S}}$, i.e. the coarsest uniformity on $F(X,Y)$ which makes each φ_S uniformly continuous, is called *the uniform structure of uniform convergence on members of $\mathcal{S}$*, denoted by $\mathcal{U} \mid \mathcal{S}$. One writes $F_{\mathcal{S}}(X,Y)$ instead of $(F(X,Y), \mathcal{U}|\mathcal{S})$. If X is endowed with a topology $\mathcal{X}$ one writes $C_{\mathcal{S}}(\mathbf{X}, \mathbf{Y})$ for the (uniform) subspace of $F_{\mathcal{S}}(X, Y)$ determined by $C(\mathbf{X}, \mathbf{Y})$, where $\mathbf{X} = (X, \mathcal{X})$ and $\mathbf{Y} = (Y, \mathcal{Y}_{\mathcal{V}})$ ($\mathcal{Y}_{\mathcal{V}}$ denotes the topology induced by $\mathcal{V}$).

6.1.14 Proposition. *Let X be a set, $(Y, \mathcal{V})$ a uniform space and $\mathcal{S} \subset \mathcal{P}(X)$. For each $A \subset X$ and each $V \in \mathcal{V}$ let*

$$W(A,V) = \{(f,g) \in F(X,Y) \times F(X,Y) : (f(x), g(x)) \in V \text{ for each } x \in A\}.$$

Then $\mathcal{R} = \{W(S,V) : S \in \mathcal{S},\ V \in \mathcal{V}\}$ is a subbase for $\mathcal{U} \mid \mathcal{S}$, i.e. the set of all finite intersections of elements of $\mathcal{R}$ is a base for $\mathcal{U} \mid \mathcal{S}$.

Proof. For each $V \in \mathcal{V}$ and each $S \in \mathcal{S}$ let $W_S(V) = \{(f,g) \in F(S,Y) \times$

$F(S,Y) : (f(x), g(x)) \in V$ for each $x \in S\}$. By 6.1.6., $\{W_S(V) : V \in \mathcal{V}\}$ is a base for the uniform structure of uniform convergence on $F(S,Y)$. Obviously,

$$W(S,V) = (\varphi_S \times \varphi_S)^{-1}[W_S(V)] .$$

Thus, the above proposition is evident (cf. the last section of 1.1.6. ②).

6.1.15 Proposition. *Let $\mathcal{E}$ and $\mathcal{S}$ be non-empty sets of subsets of X. Then the following are valid:*
1) *If each element of $\mathcal{E}$ is contained in a finite union of elements of $\mathcal{S}$, then the uniform structure of $F_{\mathcal{E}}(X,Y)$ is coarser than the uniform structure of $F_{\mathcal{S}}(X,Y)$.*
2) *The uniform structure of $F_{\mathcal{E}}(X,Y)$ is coarser than the uniform structure of $F_u(X,Y)$.*
3) *If $\mathcal{S}$ is a cover of X, then the uniform structure of simple convergence on $F(X,Y)$ is coarser than the uniform structure of $F_{\mathcal{S}}(X,Y)$.*

Proof. 1) Let $E \in \mathcal{E}$. Then there exist finitely many $S_1, \ldots, S_n \in \mathcal{S}$ such that $E \subset \bigcup_{i=1}^n S_i$. Obviously, for each $V \in \mathcal{V}$,

$$(*) \qquad \bigcap_{i=1}^n W(S_i, V) = W(\bigcup_{i=1}^n S_i, V) \subset W(E,V) .$$

The assertion follows immediately from $(*)$.
2) Choose $\mathcal{S} = \{X\}$ and apply 1).
3) Choose $\mathcal{E} = \{\{x\} : x \in X\}$ and apply 1).

6.1.16 Proposition. *If $\mathcal{S}$ is a cover of X, then a filter $\mathcal{F}$ on $F_{\mathcal{S}}(X,Y)$ converges iff $\mathcal{F}$ is a Cauchy filter and $\mathcal{F}$ converges pointwise.*

Proof. If $\mathcal{F}$ converges, it is a Cauchy filter and since $p_x : F_{\mathcal{S}}(X,Y) \longrightarrow (Y, \mathcal{V})$ is uniformly continuous (use part 3) of the above proposition) and thus continuous for each $x \in X$, $p_x(\mathcal{F})$ converges for each $x \in X$. Conversely, if $\mathcal{F}$ is a Cauchy filter which converges pointwise to f, it must be shown that $\mathcal{F}$ converges to f. Without loss of generality we may assume that $\mathcal{S}$ is closed under formation of finite unions (namely, if $\mathcal{S}$ is not closed under formation of finite unions it may be replaced by $\mathcal{S}' = \{\bigcup_{A \in \mathcal{U}} A : \mathcal{U} \subset \mathcal{S}$ is finite$\}$ because $\mathcal{U} \mid \mathcal{S} = \mathcal{U} \mid \mathcal{S}'$ [use $(*)$ in the first part of the proof of 6.1.15.]). Then $\{W(S,V) : S \in \mathcal{S}$ and $V \in \mathcal{B}\}$ is a base for $\mathcal{U} \mid \mathcal{S}$ if $\mathcal{B}$ is a base for $\mathcal{V}$. Consequently, $\{W(S,V)(f) : S \in \mathcal{S}$ and $V \in \mathcal{B}\}$ is a neighborhood base at f. Let $\mathcal{B} = \{V \in \mathcal{V} : V = \overline{V}\}$ (cf. 4.2.1.4.), $V \in \mathcal{B}$ and $S \in \mathcal{S}$. Since $\mathcal{F}$ is a Cauchy filter, there is some $F \in \mathcal{F}$ such that $F \times F \subset W(S,V)$. Thus, $p_x[F] \times p_x[F] = (p_x \times p_x)[F \times F] \subset V$ for each $x \in S$. Since $p_x(\mathcal{F})$ converges to $f(x)$, $f(x)$ is an adherence point of $p_x(\mathcal{F})$ and consequently, $f(x) \in \overline{p_x[F]}$. Obviously, $\overline{p_x[F]} \times \overline{p_x[F]} = \overline{p_x[F] \times p_x[F]} \subset \overline{V} = V$. Hence, $(f(x), g(x)) \in V$ for all $x \in S$ and $g \in F$, i.e. $F \subset W(S,V)(f)$. It follows that $W(S,V)(f) \in \mathcal{F}$, i.e. $\mathcal{F}$ converges to f.

6.1.17 Definition. If X is a set and $(Y, \mathcal{V})$ a uniform space, a filter $\mathcal{F}$ on $F(X,Y)$ is said to *converge uniformly on a subset S of X* provided that $\varphi_S(\mathcal{F})$ converges in $F_u(S,Y)$.

6.1.18 Corollary. *Let $\mathcal{S}$ be a cover of X. A filter $\mathcal{F}$ on $F_{\mathcal{S}}(X,Y)$ converges iff $\mathcal{F}$ converges uniformly on each $S \in \mathcal{S}$.*

Proof. If $\mathcal{F}$ converges on $F_{\mathcal{S}}(X,Y)$, then $\varphi_S(\mathcal{F})$ converges too, because φ_S is continuous. Conversely, let $\varphi_S(\mathcal{F})$ be convergent for each $S \in \mathcal{S}$. For each $x \in S$, $p_x(\varphi_S(\mathcal{F})) = p_x(\mathcal{F})$ converges. It remains to prove that $\mathcal{F}$ is a Cauchy filter (cf. 6.1.16.). If $\mathcal{S}' = \{\bigcup_{A\in\mathcal{A}} A : \mathcal{A} \subset \mathcal{S}$ is finite $\}$, then $\mathcal{B} = \{W(S',V) : S' \in \mathcal{S}'$ and V is an entourage of $Y\}$ is a base for $\mathcal{U} \mid \mathcal{S}$. Let V be an entourage of Y and $S_1, \ldots, S_n \in \mathcal{S}$. Then for each $i \in \{1, \ldots, n\}$ there is some $F_i \in \mathcal{F}$ such that

$$(*) \qquad \varphi_{S_i}[F_i] \times \varphi_{S_i}[F_i] = \varphi_{S_i} \times \varphi_{S_i}[F_i \times F_i] \subset W_{S_i}(V)$$

(cf. the proof of 6.1.14. concerning the notation). Because of $(\varphi_{S_i} \times \varphi_{S_i})^{-1}[W_{S_i}(V)] = W(S_i,V)$ one obtains from $(*)$, $F_i \times F_i \subset W(S_i,V)$. Put $F = \bigcap_{i=1}^n F_i$. Thus, $F \in \mathcal{F}$ and $F \times F \subset \bigcap_{i=1}^n (F_i \times F_i) \subset \bigcap_{i=1}^n W(S_i,V) = W(\bigcup_{i=1}^n S_i, V)$. Consequently, $\mathcal{F}$ is a Cauchy filter on $F_{\mathcal{S}}(X,Y)$.

6.1.19 Remark. 6.1.18. is also valid if $\mathcal{S}$ is not a cover of X (cf. exercise 46)). In the following we will consider mainly the case that $\mathcal{S}$ is the set of all compact subsets of a topological space $\mathbf{X} = (X, \mathcal{X})$, i.e. $\mathcal{S}$ is a cover of X.

6.1.20 Definition. Let $\mathbf{X}$ be a topological space and $\mathbf{Y}$ a uniform space. If $\mathcal{S}_c$ denotes the set of all compact subsets of $\mathbf{X}$, then $\mathcal{U} \mid \mathcal{S}_c$ is called *the uniform structure of uniform convergence on compacta* (or shortly: *the uniform structure of compact convergence*). Instead of $F_{\mathcal{S}_c}(X,Y)$ (resp. $C_{\mathcal{S}_c}(\mathbf{X},\mathbf{Y})$) one writes $F_c(X,Y)$ (resp. $C_c(\mathbf{X},\mathbf{Y})$). The induced topology is called *the topology of uniform convergence on compacta* (or shortly: *the topology of compact convergence*).

6.1.21 Corollary. *If $\mathbf{X}$ is a compact topological space, then $F_c(X,Y) = F_u(X,Y)$ and $C_c(\mathbf{X},\mathbf{Y}) = C_u(\mathbf{X},\mathbf{Y})$.*

Proof. Choose $\mathcal{E} = \{X\}$ and $\mathcal{S} = \mathcal{S}_c$ in 6.1.15. 1) and $\mathcal{E} = \mathcal{S}_c$ in 6.1.15. 2).

6.1.22 Corollary. *If $\mathbf{X} = (X, \mathcal{X})$ is a compactly generated topological space (i.e. $O \in \mathcal{X}$ iff $O \cap K \in \mathcal{X}_K$ for each $K \in \mathcal{S}_c$), then $C_c(\mathbf{X},\mathbf{Y})$ is a closed subspace of $F_c(X,Y)$.*

Proof. Since $\mathbf{X}$ is compactly generated, a map $f : \mathbf{X} \longrightarrow \mathbf{Y}$ is continuous iff $f|K : \mathbf{K} \longrightarrow \mathbf{Y}$ is continuous for each compact $K \subset X$ (note: $\mathcal{X}$ is the final topology on X w.r.t. the family $(j_i : K_i \longrightarrow X)_{i\in I}$ of the inclusions of all compact subsets of X). Thus,

$$(*) \qquad C(\mathbf{X},\mathbf{Y}) = \bigcap_{K\in\mathcal{S}_c} \varphi_K^{-1}[C(\mathbf{K},\mathbf{Y})].$$

Since for each $K \in \mathcal{S}_c$, $C(\mathbf{K},\mathbf{Y})$ is closed in $F_u(K,Y)$ and $\varphi_K : F_c(X,Y) \longrightarrow F_u(K,Y)$ is continuous, it follows from $(*)$ that $C(\mathbf{X},\mathbf{Y})$ is closed in $F_c(X,Y)$.

6.1.23 Remarks. 1) Since each locally compact Hausdorff space as well as each topological space satisfying the first axiom of countability is compactly generated (cf. [149; 43.9]), the above corollary is applicable to classical Real (or Complex) Analysis and says that compact convergence does not destroy continuity (cf. 6.1.12. 2)).

2) There is an alternative description of the topology of compact convergence on $C(\mathbf{X}, \mathbf{Y})$ provided that $\mathbf{X}$ is a topological space and $\mathbf{Y}$ a uniform space which will be considered in the following.

6.1.24 Definition. Let $\mathbf{X}$ and $\mathbf{Y}$ be topological spaces. The *compact–open topology* on $C(\mathbf{X}, \mathbf{Y})$ is the topology having for a subbase the sets $(K, O) = \{f \in C(\mathbf{X}, \mathbf{Y}) : f[K] \subset O\}$, for K compact in $\mathbf{X}$ and O open in $\mathbf{Y}$. $C(\mathbf{X}, \mathbf{Y})$ endowed with the compact–open topology is denoted by $C_{co}(\mathbf{X}, \mathbf{Y})$.

6.1.25. In order to prove that the compact–open topology is the right one for the desired alternative description of the topology of compact convergence on $C(\mathbf{X}, \mathbf{Y})$ the following is needed:

Proposition. *Let $(Y, \mathcal{V})$ be a uniform space and A a compact subset of Y. Then for each neighborhood N of A there is some $V \in \mathcal{V}$ such that $V[A] \subset N$.*

Proof. Let N be a neighborhood of A. Then there is some open set O such that $A \subset O \subset N$. Hence, for each $x \in A$ there is some $U_x \in \mathcal{V}$ such that $U_x(x) \subset O$. Since for each U_x there is some $W_x \in \mathcal{V}$ with $W_x^2 \subset U_x$, one obtains $W_x^2(x) \subset O$ for each $x \in A$. Furthermore, there are finitely many $x_1, \ldots, x_n \in A$ such that the sets $W_{x_i}(x_i)$ cover A and for each $i \in \{1, \ldots, n\}$, $W_{x_i}^2(x_i) \subset O$. Let $V = \bigcap_{i=1}^n W_{x_i}$. Then $V[A] = \bigcup_{y \in A} V(y) \subset O \subset N$ (namely, if $y \in A$ there exists $i \in \{1, \ldots, n\}$ with $y \in W_{x_i}(x_i)$, and consequently $V(y) \subset V[W_{x_i}(x_i)] \subset W_{x_i}^2(x_i) \subset O$).

6.1.26 Theorem. *Let $\mathbf{X} = (X, \mathcal{X})$ be a topological space and $\mathbf{Y} = (Y, \mathcal{V})$ a uniform space. Then the topology of compact convergence on $C(\mathbf{X}, \mathbf{Y})$ is the compact–open topology.*

Proof. 1) In order to prove that the compact–open topology is coarser than the topology of compact convergence, it suffices to show that each set of the form (K, O) is open w.r.t. the topology of compact convergence. Let $f \in (K, O)$. Then $f[K]$ is compact and by 6.1.25. there is some $V \in \mathcal{V}$ such that $V[f[K]] \subset O$. Thus, $W(K, V)(f) \subset (K, O)$ (namely, if $g \in W(K, V)(f)$, then $(f(x), g(x)) \in V$ for each $x \in K$, i.e. $g(x) \in V(f(x))$ for each $x \in K$, and therefore $g(x) \in O$, i.e. $g[K] \subset O$ or equivalently $g \in (K, O)$). Consequently, (K, O) is open in $C_c(\mathbf{X}, \mathbf{Y})$.

2) Conversely, since $\mathcal{B} = \{W(K, V)(f) : K \in \mathcal{S}_c \text{ and } V \in \mathcal{V}\}$ is a neighborhood base at $f \in C(\mathbf{X}, \mathbf{Y})$ w.r.t. the topology of compact convergence (note that $\mathcal{S}_c$ is closed under formation of finite unions), it must be shown that each element of $\mathcal{B}$ contains an element of the neighborhood base at f w.r.t. the compact–open topology. Indeed, it will be shown that for each compact subset K of X, each

$V \in \mathcal{V}$, and each $f \in C(\mathbf{X}, \mathbf{Y})$ there are compact subsets $K_1, \ldots, K_n$ of X and open subsets $O_1, \ldots, O_n$ of Y such that $f[K_i] \subset O_i$, i.e. $f \in \bigcap_{i=1}^{n}(K_i, O_i)$, and if $g \in \bigcap_{i=1}^{n}(K_i, O_i)$, i.e. $g[K_i] \subset O_i$ for each $i \in \{1, \ldots, n\}$, then $g \in W(K, V)(f)$, i.e. $(f(x), g(x)) \in V$ for each $x \in K$: Let U be a closed symmetric entourage of Y such that $U^3 \subset V$, choose $x_1, \ldots, x_n \in K$ such that the sets $U(f(x_i))$ cover $f[K]$, let $K_i = K \cap f^{-1}[U(f(x_i))]$ and let O_i be the interior of $U^2(f(x_i))$. Obviously, $f[K_i] \subset U(f(x_i)) \subset (U^2(f(x_i)))^0 = O_i$. Furthermore, if $g[K_i] \subset O_i$ for each $i \in \{1, \ldots, n\}$, then one obtains: For each $x \in K$ there is some $i \in \{1, \ldots, \}$ such that $x \in K_i$. Hence, $g(x) \in U^2(f(x_i))$ and $f(x) \in U(f(x_i))$. Consequently, $(f(x), g(x)) \in U^3 \subset V$.

6.1.27. The compact–open topology on $C(\mathbf{X}, \mathbf{Y})$ is closely related to the question whether there is a topology on $C(\mathbf{X}, \mathbf{Y})$ describing continuous convergence which can be defined as follows:

Definition. Let $\mathbf{X}$ and $\mathbf{Y}$ be topological spaces. A filter $\mathcal{F}$ on a non-empty subset M of $F(X, Y)$ *converges continuously* to $f \in F(X, Y)$ provided that for each $x \in X$ and each filter $\mathcal{G}$ on X converging to x in $\mathbf{X}$ the filter $e_{X,Y}(\mathcal{G} \times \mathcal{F})$ converges to $f(x)$ in $\mathbf{Y}$, where $\mathcal{G} \times \mathcal{F}$ denotes the product filter and $e_{X,Y} : X \times M \longrightarrow Y$ is the evaluation map (i.e. $e_{X,Y}(x, g) = g(x)$ for each $(x, g) \in X \times M$).

6.1.28 Remark. The filter $e_{X,Y}(\mathcal{G} \times \mathcal{F})$ is also denoted by $\mathcal{F}(\mathcal{G})$; it has the filter base $\{F(G) : F \in \mathcal{F}, G \in \mathcal{G}\}$ where $F(G) = \{f(x) : f \in F, x \in G\}$.

6.1.29 Proposition. *Let $\mathbf{X}, \mathbf{Y}$ be topological spaces. If a filter $\mathcal{F}$ on a non–empty subset M of $F(X, Y)$ converges continuously to $f \in F(X, Y)$, then $f : \mathbf{X} \longrightarrow \mathbf{Y}$ is continuous provided that $\mathbf{Y}$ is regular.*

Proof. Let $\mathcal{G}$ be a filter on $\mathbf{X}$ converging to $x \in X$. Put $\mathcal{F}(\mathcal{G}) = \mathcal{H}$. Then $\mathcal{H}$ converges to $f(x)$ in $\mathbf{Y}$ and for each $H \in \mathcal{H}$ there exist $F_H \in \mathcal{F}$ and $G_H \in \mathcal{G}$ such that $F_H(G_H) \subset H$. Since $\mathcal{F}$ converges continuously to f, $\mathcal{F}(\dot{x})$ converges to $f(x)$ for each $x \in G_H$. Furthermore, for each $x \in G_H$ and each $F \in \mathcal{F}$, $F(\{x\}) \cap F_H(G_H) \neq \emptyset$. Consequently, $f[G_H] \subset \overline{F_H(G_H)} \subset \overline{H}$. Thus, $f(\mathcal{G}) \supset \overline{\mathcal{H}}$. Since $\mathbf{Y}$ is regular, $\overline{\mathcal{H}}$ converges to $f(x)$. Hence, $f(\mathcal{G})$ converges to $f(x)$, i.e. f is continuous.

6.1.30 Remarks. 1) It follows from the above proposition that on subsets M of $F(X, Y)$ containing non–continuous functions there does not exist any topology describing continuous convergence provided that $\mathbf{Y}$ is regular (consider the filter $\mathcal{F} = \dot{f}$ for a non–continuous f!). Thus, in the following the case that $M = C(\mathbf{X}, \mathbf{Y})$ is mainly considered.

2) *A sequence (f_n) in $C(\mathbf{X}, \mathbf{Y})$ is said to converge continuously to $f \in C(\mathbf{X}, \mathbf{Y})$ provided that the elementary filter of (f_n) converges continuously to f. If $\mathbf{X}$ fulfills the first axiom of countability, then a sequence (f_n) in $C(\mathbf{X}, \mathbf{Y})$ converges continuously to $f \in C(\mathbf{X}, \mathbf{Y})$ iff for each $x \in X$ and each sequence (x_n) in $\mathbf{X}$*

converging to x the sequence $(f_n(x_n))$ converges to $f(x)$ in $\mathbf{Y}$ (exercise!).

3) In the following it will be examined under what conditions a topology on $C(\mathbf{X},\mathbf{Y})$ exists which describes continuous convergence. In this context local compactness will play an essential role. In Hausdorff spaces local compactness can be characterized as follows:
A Hausdorff space $\mathbf{X}$ *is locally compact iff for each quotient map* $f:\mathbf{Y}\longrightarrow\mathbf{Z}$ *between Hausdorff spaces,* $1_X\times f:\mathbf{X}\times\mathbf{Y}\longrightarrow\mathbf{X}\times\mathbf{Z}$ *is a quotient map.* (cf. Whitehead [148] for "$\Longrightarrow$" and Michael [99] for "$\Longleftarrow$" as well as Herrlich [67] for both directions).

6.1.31 Theorem. *If* $\mathbf{X}$ *is a Hausdorff space, then the following are equivalent:*
(1) *The evaluation map* $e_{X,Y}:\mathbf{X}\times C_{co}(\mathbf{X},\mathbf{Y})\longrightarrow\mathbf{Y}$ *is continuous for each topological space* $\mathbf{Y}$.
(2) *for each topological space* $\mathbf{Y}$ *there is a coarsest topology* $\mathcal{Z}$ *on* $C(\mathbf{X},\mathbf{Y})$ *such that the evaluation map* $e_{X,Y}:\mathbf{X}\times(C(\mathbf{X},\mathbf{Y}),\mathcal{Z})\longrightarrow\mathbf{Y}$ *is continuous,*
(3) *for each topological space* $\mathbf{Y}$ *there is a topology* $\mathcal{Z}$ *on* $C(\mathbf{X},\mathbf{Y})$ *describing continuous convergence,*
(4) *for each topological space* $\mathbf{Y}$ *there is a topology* $\mathcal{Z}$ *on* $C(\mathbf{X},\mathbf{Y})$ *such that for each topological space* $\mathbf{Z}$ *the map* $*:C(\mathbf{X}\times\mathbf{Z},\mathbf{Y})\longrightarrow C(\mathbf{Z},(C(\mathbf{X},\mathbf{Y}),\mathcal{Z}))$ *defined by* $(f^*(z))(x)=f(x,z)$ *is bijective,*
(5) $\mathbf{X}$ *is locally compact.*

If the above conditions are fulfilled, then the topology $\mathcal{Z}$, *uniquely determined by each of the conditions* (2), (3) *or* (4), *is the compact-open topology.*

Proof. (1) $\Longrightarrow$ (2). Let $\mathcal{Z}$ be a topology on $C(\mathbf{X},\mathbf{Y})$ such that $e_{X,Y}:\mathbf{X}\times(C(\mathbf{X},\mathbf{Y}),\mathcal{Z})\longrightarrow\mathbf{Y}$ is continuous. Since (1) is valid it suffices to prove that the compact-open topology is coarser than $\mathcal{Z}$. Let K be compact in $\mathbf{X}$, O open in $\mathbf{Y}$ and $f\in(K,O)$. Thus, for each $x\in K$, $e_{X,Y}(x,f)=f(x)\in O$. It follows from the continuity of $e_{X,Y}$ that there are neighborhoods U_x of x in $\mathbf{X}$ and V_f^x of f in $(C(\mathbf{X},\mathbf{Y}),\mathcal{Z})$ such that $e_{X,Y}[U_x\times V_f^x]\subset O$. Since K is compact there are finitely many $x_1,\dots,x_n\in K$ such that $K\subset\bigcup_{i=1}^n U_{x_i}$. Hence, $V=\bigcap_{i=1}^n V_f^{x_i}$ is a neighborhood of f w.r.t. $\mathcal{Z}$ such that $f\in V\subset(K,O)$. Consequently, (K,O) is a neighborhood of f w.r.t. $\mathcal{Z}$, i.e. $(K,O)\in\mathcal{Z}$.

(2) $\Longrightarrow$ (3). Let $\mathcal{Z}$ be the coarsest topology on $C(\mathbf{X},\mathbf{Y})$ such that $e_{X,Y}:\mathbf{X}\times(C(\mathbf{X},\mathbf{Y}),\mathcal{Z})\longrightarrow\mathbf{Y}$ is continuous. If a filter $\mathcal{F}$ on $C(\mathbf{X},\mathbf{Y})$ converges to $f\in C(\mathbf{X},\mathbf{Y})$ w.r.t. $\mathcal{Z}$, then it follows from the continuity of $e_{X,Y}$ that $\mathcal{F}$ converges continuously to f. Conversely, let $\mathcal{F}$ be a filter on $C(\mathbf{X},\mathbf{Y})$ converging continuously to $f\in C(\mathbf{X},\mathbf{Y})$. Then $\mathcal{Z}_f=\{M\subset C(\mathbf{X},\mathbf{Y}):f\in M\Longrightarrow M\in\mathcal{F}\}$ is a topology on $C(\mathbf{X},\mathbf{Y})$ such that $e_{X,Y}:\mathbf{X}\times(C(\mathbf{X},\mathbf{Y}),\mathcal{Z}_f)\longrightarrow\mathbf{Y}$ is continuous (Let $(x,g)\in X\times C(\mathbf{X},\mathbf{Y})$ and $O_{g(x)}$ an open neighborhood of $g(x)$ in $\mathbf{Y}$.

1. $g\neq f$:
Since g is continuous, there is a neighborhood U_x of x in $\mathbf{X}$ such that $g[U_x]\subset O_{g(x)}$. Further, $\{g\}$ is an open neighborhood of g w.r.t. $\mathcal{Z}_f$. Then $e_{X,Y}[U_x\times\{g\}]\subset O_{g(x)}$, i.e. $e_{X,Y}$ is continuous at (x,g).

2. $g = f$:
Obviously, $\overset{\circ}{\mathcal{U}}(f) = \{F \in \mathcal{F} : f \in F\}$. Since $\mathcal{F}$ converges continuously to f, $\mathcal{F}(\mathcal{U}(x)) \supset \mathcal{U}(f(x))$ and consequently $O_{f(x)} \in \mathcal{F}(\mathcal{U}(x))$, i.e. there is some $F \in \mathcal{F}$ and some $U_x \in \mathcal{U}(x)$ such that $F[U_x] \subset O_{f(x)}$. Since f is continuous, there is some $V_x \in \mathcal{U}(x)$ such that $f[V_x] \subset O_{f(x)}$. Put $W_x = U_x \cap V_x$. Since $F \cup \{f\} \in \mathcal{F}$, $F \cup \{f\} \in \overset{\circ}{\mathcal{U}}(f)$ and $e_{X,Y}[W_x \times (F \cup \{f\})] \subset O_{f(x)}$, i.e. $e_{X,Y}$ is continuous at (x, f).).

Consequently, $\mathcal{Z} \subset \mathcal{Z}_f$. Obviously, $\mathcal{F}$ converges to f w.r.t. $\mathcal{Z}_f$. Thus, $\mathcal{F}$ converges to f w.r.t. $\mathcal{Z}$.

(3) $\Longrightarrow$ (4). Let $\mathcal{Z}$ be a topology on $C(\mathbf{X}, \mathbf{Y})$ describing continuous convergence. Then $e_{X,Y} : \mathbf{X} \times (C(\mathbf{X}, \mathbf{Y}), \mathcal{Z}) \longrightarrow \mathbf{Y}$ is continuous [namely, if p_x (resp. p_c) denotes the projection on the first (resp. second) factor of $X \times C(\mathbf{X}, \mathbf{Y})$, then $p_c(\mathcal{F})$ converges continuously to $f \in C(\mathbf{X}, \mathbf{Y})$ and $p_x(\mathcal{F})$ converges to $x \in X$ provided that $\mathcal{F} \in F(X \times C(\mathbf{X}, \mathbf{Y}))$ converges to (x, f) in $\mathbf{X} \times C(\mathbf{X}, \mathbf{Y}), \mathcal{Z})$, i.e. $e_{X,Y}(p_x(\mathcal{F}) \times p_c(\mathcal{F})) \subset e_{X,Y}(\mathcal{F})$ converges to $f(x)$].
If $f \in C(\mathbf{X} \times \mathbf{Z}, \mathbf{Y})$, then $f^* : \mathbf{Z} \longrightarrow (C(\mathbf{X}, \mathbf{Y}), \mathcal{Z})$ is a map. In order to show that f^* is continuous, let $\mathcal{F}$ be a filter on Z converging to $z \in Z$ in $\mathbf{Z}$. Furthermore, let x be an element of X and $\mathcal{G}$ a filter on X converging to x in $\mathbf{X}$. Since f is continuous, $f(\mathcal{G} \times \mathcal{F})$ converges to $f(x, z)$ in $\mathbf{Y}$. Thus, $e_{X,Y}[\mathcal{G} \times f^*[\mathcal{F}]]$ converges to $f^*(z)(x) = f(x, z)$, since obviously, $e_{X,Y}[G \times f^*[F]] = f[G \times F]$ for each $G \in \mathcal{G}$ and each $F \in \mathcal{F}$. Consequently, $f^*(\mathcal{F})$ converges continuously to $f^*(z)$. By assumption, $f^*(\mathcal{F})$ converges to $f^*(z)$ in $(C(\mathbf{X}, \mathbf{Y}), \mathcal{Z})$, i.e. f^* is continuous. Thus, $* : C(\mathbf{X} \times \mathbf{Z}, \mathbf{Y}) \longrightarrow C(\mathbf{Z}, (C(\mathbf{X}, \mathbf{Y}), \mathcal{Z}))$ is a map. By definition it is injective. If $g : \mathbf{Z} \longrightarrow (C(\mathbf{X}, \mathbf{Y}), \mathcal{Z})$ is continuous, then $f = e_{X,Y} \circ (1_X \times g) : \mathbf{X} \times \mathbf{Z} \longrightarrow \mathbf{Y}$ is continuous (note, that $e_{X,Y}$ is continuous) such that $f^*(z)(x) = f(x, z) = g(z)(x)$, i.e. $f^* = g$. Hence, $*$ is surjective and thus, it is bijective.

(4) $\Longrightarrow$ (5). By 6.1.30. 3) it suffices to prove that for each quotient map $f : \mathbf{Z}' \longrightarrow \mathbf{Z}''$ between Hausdorff spaces, $1_X \times f$ is a quotient map. In order to do so let $f : \mathbf{Z}' \longrightarrow \mathbf{Z}''$ be a quotient map between Hausdorff spaces and $g : \mathbf{X} \times \mathbf{Z}'' \longrightarrow \mathbf{Y}$ a map such that $h = g \circ (1_X \times f) : \mathbf{X} \times \mathbf{Z}' \longrightarrow \mathbf{Y}$ is continuous. It suffices to show that g is continuous. By assumption there is a topology $\mathcal{Z}$ on $C(\mathbf{X}, \mathbf{Y})$ such that

$$* : C(\mathbf{X} \times \mathbf{Z}, \mathbf{Y}) \longrightarrow C(\mathbf{Z}, (C(\mathbf{X}, \mathbf{Y}), \mathcal{Z}))$$

is bijective for $\mathbf{Z} = \mathbf{Z}'$ or $\mathbf{Z} = \mathbf{Z}''$. In particular, $h^* : \mathbf{Z}' \longrightarrow (C(\mathbf{X}, \mathbf{Y}), \mathcal{Z})$ is continuous. Since $(h^*(z'))(x) = h(x, z') = g(x, f(z')) = g^*(f(z'))(x) = ((g^* \circ f)(z'))(x)$, $h^* = g^* \circ f$ and consequently, $g^* \circ f$ is continuous. Thus, g^* is continuous because f is a quotient map. Since $*$ is bijective, g is continuous. Consequently, $1_X \times f : \mathbf{X} \times \mathbf{Z}' \longrightarrow \mathbf{X} \times \mathbf{Z}''$ is a quotient map and $\mathbf{X}$ is locally compact.

(5) $\Longrightarrow$ (1). Let $\mathbf{X}$ be locally compact and O an open neighborhood of $e_{X,Y}(x, f) = f(x)$ in $\mathbf{Y}$. Since f is continuous, $f^{-1}[O]$ is an open neighborhood of x in $\mathbf{X}$ which contains a compact neighborhood K of x because $\mathbf{X}$ is a locally compact Hausdorff space. Obviously, $f[K] \subset O$. Thus, $V = K \times (K, O)$ is

a neighborhood of (x, f) in $\mathbf{X} \times C_{co}(\mathbf{X}, \mathbf{Y})$ such that $e_{X,Y}[V] \subset O$. Consequently, $e_{X,Y} : \mathbf{X} \times C_{co}(\mathbf{X}, \mathbf{Y}) \longrightarrow \mathbf{Y}$ is continuous.

The topology $\mathcal{Z}$ is uniquely determined by each of the conditions (2), (3) or (4) and coincides with the compact–open topology:
ad (2). Since (2) implies (1) the unique topology $\mathcal{Z}$ is the compact–open topology according to the proof of '(1) $\Longrightarrow$ (2)'.

ad (3). By ① a) (A) and (B) of the introduction of this book $\mathcal{Z}$ is the coarsest topology on $C(\mathbf{X}, \mathbf{Y})$ such that the evaluation map $e_{X,Y} : \mathbf{X} \times (C(\mathbf{X}, \mathbf{Y}), \mathcal{Z}) \longrightarrow \mathbf{Y}$ is continuous and this one is the compact–open topology (cf. 'ad (2)').

ad (4). $\mathcal{Z}$ is the coarsest topology such that the evaluation map $e_{X,Y} : \mathbf{X} \times (C(\mathbf{X}, \mathbf{Y}), \mathcal{Z}) \longrightarrow \mathbf{Y}$ is continuous (and consequently, the compact–open topology), namely: $e^*_{X,Y} : (C(\mathbf{X}, \mathbf{Y}), \mathcal{Z}) \longrightarrow (C(\mathbf{X}, \mathbf{Y}), \mathcal{Z})$ is the identity on $(C(\mathbf{X}, \mathbf{Y}), \mathcal{Z})$ and thus it is continuous, which implies the continuity of $e_{X,Y}$; furthermore, if $\mathcal{Z}'$ is a topology on $C(\mathbf{X}, \mathbf{Y})$ such that the evaluation map $e_{X,Y} : \mathbf{X} \times (C(\mathbf{X}, \mathbf{Y}), \mathcal{Z}') \longrightarrow \mathbf{Y}$ is continuous, $e^*_{X,Y} = 1_{C(\mathbf{X},\mathbf{Y})} : (C(\mathbf{X}, \mathbf{Y}), \mathcal{Z}') \longrightarrow (C(\mathbf{X}, \mathbf{Y}), \mathcal{Z})$ is continuous and then, $\mathcal{Z} \subset \mathcal{Z}'$.

6.1.32 Corollary. *Let* $\mathbf{X}$ *be a locally compact Hausdorff space,* $\mathbf{Y}$ *a uniform space and* $f : \mathbf{X} \longrightarrow \mathbf{Y}$ *a continuous map. Then a filter* $\mathcal{F}$ *on* $C(\mathbf{X}, \mathbf{Y})$ *converges continuously to* f *iff it converges uniformly to* f *on each compact subset* K *of* $\mathbf{X}$ *(i.e. iff* $\mathcal{F}$ *converges to* f *in* $C_c(\mathbf{X}, \mathbf{Y})$*).*

Proof. By 6.1.26., the topology of compact convergence on $C(\mathbf{X}, \mathbf{Y})$ is the compact–open topology and by 6.1.31., the compact–open topology is the unique topology describing continuous convergence. Thus, the above corollary is obvious.

6.1.33 Remarks. 1) It follows from the above corollary that e.g. *in Function Theory we need not distinguish between uniform convergence on compacta and continuous convergence*, because regions or even non–empty open subsets of the complex plane are locally compact Hausdorff. Since continuous convergence is much easier to handle, Carathéodory [25] proposed already in 1929 to substitute uniform convergence (on compacta) by continuous convergence whenever it is favorable.

2) It is well–known that if I is a set, then the product space $\mathbb{R}^I_t$ is locally compact iff I is finite. Thus, by 6.1.23., *for infinite–dimensional Analysis continuous convergence cannot be described by means of a topology in general.* In order to describe continuous convergence also for infinite–dimensional Analysis more general structures than topological structures are needed: By 3.1.9. ⑥ b) (resp. c)) *the construct* **Lim** (resp. $\mathbf{Lim_S}$) *is cartesian closed and the natural function space structure is the structure of continuous convergence.* The same is true for the smaller construct **PsTop** (resp. **Ps** $\mathbf{Top_S}$) [cf. exercise 54)]. It has been shown earlier that $\mathbf{Lim_S}$ (resp. $\mathbf{PsTop_S}$) is embedded into **SUConv** and that the $\mathbf{Lim_S}$–structure of continuous convergence can be derived from the natural

function space structure in **SUConv** (cf. 3.1.11.). Since the natural function space structure in $\mathbf{PsTop_S}$ is also the structure of continuous convergence the latter result remains true provided that $\mathbf{Lim_S}$ is substituted by $\mathbf{PsTop_S}$. Thus, it makes sense to concentrate upon **SUConv** and some suitable subconstructs of **SUConv** in the following.

6.2 Local compactness and local precompactness in semiuniform convergence spaces

6.2.1 Local compactness

6.2.1.1 Definitions. 1) *A subset* A of a semiuniform convergence space $(X, \mathcal{J}_X)$ *is* called *compact* provided that $(A, \mathcal{J}_A)$ is compact where $\mathcal{J}_A$ is the initial **SUConv**–structure on A with respect to the inclusion map $i : A \longrightarrow X$.
2) A semiuniform convergence space $(X, \mathcal{J}_X)$ is called
a) *locally compact* provided that each $\mathcal{F} \in \mathcal{J}_X$ contains a compact subset of the product space $(X, \mathcal{J}_X) \times (X, \mathcal{J}_X)$,
b) *diagonal* provided that the filter (Δ_X) generated by the diagonal $\Delta_X = \{(x, x) : x \in X\}$ of $X \times X$ belongs to $\mathcal{J}_X$, if the filter (Δ_X) exists.

6.2.1.2 Remark. Obviously, every uniform space (regarded as a semiuniform convergence space) is a diagonal uniform limit space. Diagonal uniform limit spaces are also called *Cook–Fischer spaces.*

6.2.1.3 Proposition. 1) *Let $(X, \mathcal{J}_X)$ be a compact semiuniform convergence space. Then $(X, \mathcal{J}_X)$ is locally compact.*
2) *A diagonal semiuniform convergence space $(X, \mathcal{J}_X)$ is compact provided that it is locally compact.*

Proof. 1) Let $\mathcal{F} \in \mathcal{J}_X$. Then $X \times X \in \mathcal{F}$ and by 4.3.2.11. $X \times X$ is a compact subset of $(X, \mathcal{J}_X) \times (X, \mathcal{J}_X)$.
2) Since $(\Delta_X) \in \mathcal{J}_X$, there is, by assumption, some $M \supset \Delta_X$ such that M is a compact subset of $(X, \mathcal{J}_X) \times (X, \mathcal{J}_X)$. If $p_1 : (X, \mathcal{J}_X) \times (X, \mathcal{J}_X) \longrightarrow (X, \mathcal{J}_X)$ denotes the first projection, then it follows from $X = p_1[\Delta_X] \subset p_1[M] \subset X$ that $(X, \mathcal{J}_X)$ is compact (cf. 4.3.2.10.).

6.2.1.4 Corollary. *A principal uniform limit space (= uniform space) is locally compact iff it is compact.*

Proof. Note 6.2.1.2. and apply 6.2.1.3.

6.2.1.5 Proposition. *Let $(X, \mathcal{J}_X)$ be a locally compact semiuniform convergence space. Then the underlying Kent convergence space $(X, q_{\gamma_{\mathcal{J}_X}})$ is locally compact in the usual sense, i.e. for each $(\mathcal{F}, x) \in q_{\gamma_{\mathcal{J}_X}}$ there is some compact subset K of $(X, q_{\gamma_{\mathcal{J}_X}})$ such that $K \in \mathcal{F}$.*

Proof. If $(\mathcal{F},x) \in q_{\gamma_{\mathcal{J}_X}}$, then $\mathcal{F} \cap \dot{x} \in \gamma_{\mathcal{J}_X}$, and consequently $\mathcal{F} \times \mathcal{F} \in \mathcal{J}_X$. Thus, by assumption, $\mathcal{F} \times \mathcal{F}$ contains a compact set M. Hence $\mathcal{F}$ contains a compact set.

6.2.1.6 Proposition. *Let $(X,\mathcal{J}_X)$ be a convergence space and $(X,q_{\gamma_{\mathcal{J}_X}})$ the corresponding symmetric Kent convergence space. Then the following are equivalent:*
(1) *$(X,\mathcal{J}_X)$ is locally compact.*
(2) *$(X,q_{\gamma_{\mathcal{J}_X}})$ is locally compact in the usual sense.*

Proof. (1) $\Longrightarrow$ (2). See 6.2.1.5.
(2) $\Longrightarrow$ (1). Let $\mathcal{H} \in \mathcal{J}_X$. Then there are $x \in X$ and $\mathcal{F} \in F(X)$ such that $(\mathcal{F},x) \in q_{\gamma_{\mathcal{J}_X}}$ and $\mathcal{H} \supset \mathcal{F} \times \mathcal{F}$. By assumption, there is a compact $F \in \mathcal{F}$ and, by Tychonoff, $F \times F \in \mathcal{H}$ is compact.

6.2.1.7 Proposition., *The construct* **LC–SUConv** *of locally compact semiuniform convergence spaces (and uniformly contiuous maps) is a bicoreflective (full and isomorphism–closed) subconstruct of* **SUConv**.

Proof. Let $(X,\mathcal{J}_X) \in |\mathbf{SUConv}|$. Put $\mathcal{J}_X^* = \{\mathcal{F} \in \mathcal{J}_X$: there is some $F \in \mathcal{F}$ such that F is a compact subset of $(X,\mathcal{J}_X) \times (X,\mathcal{J}_X)\}$. Then $\mathcal{J}_X^* \subset \mathcal{J}_X$ and $(X,\mathcal{J}_X^*)$ is a semiuniform convergence space (note: H^{-1} is compact provided that $H \subset X \times X$ is compact) which is locally compact, namely if $\mathcal{F} \in \mathcal{J}_X^*$ then $\mathcal{F} \in \mathcal{J}_X$ and there is some $F \in \mathcal{F}$ such that F is a compact subset of $(X,\mathcal{J}_X) \times (X,\mathcal{J}_X)$, but F is also a compact subset of $(X,\mathcal{J}_X^*) \times (X,\mathcal{J}_X^*)$ (note: $(\mathcal{J}_X \times \mathcal{J}_X)^* = \mathcal{J}_X^* \times \mathcal{J}_X^*$ where $\times$ stands for forming the product structure; then it follows from the following part of this proof that the inclusion map $i : (F,(\mathcal{J}_X \times \mathcal{J}_X)_F) \longrightarrow (X \times X,(\mathcal{J}_X \times \mathcal{J}_X)^*)$ is uniformly continuous and thus $i[F] = F$ is a compact subset of $(X \times X,(\mathcal{J}_X \times \mathcal{J}_X)^*) = (X,\mathcal{J}_X^*) \times (X,\mathcal{J}_X^*)$). Then $1_X : (X,\mathcal{J}_X^*) \longrightarrow (X,\mathcal{J}_X)$ is the desired bicoreflection: Let $(Y,\mathcal{J}_Y) \in |\mathbf{LC\text{–}SUConv}|$ and let $f : (Y,\mathcal{J}_Y) \longrightarrow (X,\mathcal{J}_X)$ be a uniformly continuous map. If $\mathcal{F} \in \mathcal{J}_Y$ there is some $F \in \mathcal{F}$ such that F is a compact subset of $(Y,\mathcal{J}_Y) \times (Y,\mathcal{J}_Y)$ and $(f \times f)(\mathcal{F}) \in \mathcal{J}_X^*$ because $(f \times f)[F] \in (f \times f)(\mathcal{F})$ is compact (since $f \times f$ is uniformly continuous, it is continuous!), i.e. $f : (Y,\mathcal{J}_Y) \longrightarrow (X,\mathcal{J}_X^*)$ is uniformly continuous.

6.2.1.8 Corollary. *Let $(f_i : (X_i,\mathcal{J}_{X_i}) \longrightarrow (X,\mathcal{J}_X))_{i \in I}$ be a final sink in* **SUConv** *such that each $(X_i,\mathcal{J}_{X_i})$ is locally compact. Then $(X,\mathcal{J}_X)$ is locally compact.*

Proof. Use 6.2.1.7. and apply 2.2.12.

6.2.1.9 Proposition. *Let $((X_i,\mathcal{J}_{X_i}))_{i \in I}$ be a family of non-empty semiuniform convergence spaces. Then the product space $(\prod_{i \in I} X_i,\mathcal{J}_X)$ of this family is locally compact iff $(X_i,\mathcal{J}_{X_i})$ is locally compact for all $i \in I$ and compact for all but finitely many $i \in I$.*

Proof. 1) '$\Longleftarrow$'. Let $\mathcal{F} \in \mathcal{J}_X$. Then $(p_i \times p_i)(\mathcal{F}) \in \mathcal{J}_{X_i}$ for each $i \in I$, where $p_i : \prod_{i\in I} X_i \longrightarrow X_i$ denotes the i–th projection. By assumption, there are finitely many elements $i_1, \dots, i_n$ of I such that $(X_i, \mathcal{J}_{X_i})$ is compact for each $i \in I\setminus\{i_1, \dots, i_n\}$, and for each $i \in \{i_1, \dots, i_n\}$ there is some compact $H_i \in (p_i \times p_i)(\mathcal{F})$. Put $H_i = X_i \times X_i$ for each $i \in I\setminus\{i_1, \dots, i_n\}$. If $j : \prod_{i\in I}(X_i \times X_i) \longrightarrow (\prod_{i\in I} X_i) \times (\prod_{i\in I} X_i)$ denotes the canonical isomorphism, then $j(\prod_{i\in I} H_i) \in j(\prod_{i\in I}(p_i \times p_i)(\mathcal{F})) \subset \mathcal{F}$ is compact.
2) '$\Longrightarrow$'. Let $\mathcal{F}_i \in \mathcal{J}_{X_i}$ for each $i \in I$. Then the product filter $\prod_{i\in I} \mathcal{F}_i$ is a filter on $\prod_{i\in I} X_i \times X_i$ and consequently, $j(\prod_{i\in I} \mathcal{F}_i)$ is a filter on $\prod_{i\in I} X_i \times \prod_{i\in I} X_i$. Let $p_i' : \prod_{i\in I} X_i \times X_i \longrightarrow X_i \times X_i$ denote the i–th projection for each $i \in I$. Since $(p_i \times p_i)(j(\prod_{i\in I} \mathcal{F}_i)) = p_i'(\prod_{i\in I} \mathcal{F}_i) = \mathcal{F}_i \in \mathcal{J}_{X_i}$, $j(\prod_{i\in I} \mathcal{F}_i) \in \mathcal{J}_X$. By assumption, there is some compact $K \in j(\prod_{i\in I} \mathcal{F}_i)$. Hence, for each $i \in I$, $(p_i \times p_i)[K]$ is compact and belongs to $(p_i \times p_i)(j(\prod_{i\in I} \mathcal{F}_i)) = p_i'(\prod_{i\in I} \mathcal{F}_i) = \mathcal{F}_i$. Consequently, $(X_i, \mathcal{J}_{X_i})$ is locally compact for each $i \in I$. Since $K \supset j(\prod_{i\in I} F_i)$, where $F_i \in \mathcal{F}_i$ for each $i \in I$ and $F_i = X_i \times X_i$ for all but finitely many $i \in I$, one obtains for all but finitely many $i \in I$:

$$X_i \times X_i = F_i = p_i'(\textstyle\prod_{i\in I} F_i) = (p_i \times p_i)(j(\prod_{i\in I} F_i)) \subset (p_i \times p_i)[K] \subset X_i \times X_i\,,$$

i.e. $X_i \times X_i = (p_i \times p_i)[K]$ is compact, which implies that $(X_i, \mathcal{J}_{X_i})$ is compact for all but finitely many $i \in I$.

6.2.1.10 Proposition. *Let $(X, \mathcal{J}_X)$ be a locally compact semiuniform convergence space and A a closed subset of X. Then $(A, \mathcal{J}_A)$ is locally compact provided that $\mathcal{J}_A$ denotes the initial* **SUConv***–structure on A w.r.t. the inclusion map $i : A \longrightarrow X$.*

Proof. Let $\mathcal{F} \in \mathcal{J}_A$. Then $(i \times i)(\mathcal{F}) \in \mathcal{J}_X$ and by assumption, there is a compact set $K \in (i \times i)(\mathcal{F})$. Obviously, $K \cap (A \times A) \in \mathcal{F}$ is compact, since it is a closed subset of K.

6.2.1.11 Theorem. *The construct* **LC–SUConv** *(of locally compact semiuniform convergence spaces) is cartesian closed and topological.*

Proof. Since **SUConv** is a cartesian closed topological construct (cf. 3.1.9. ③) and, by 6.2.1.7. and 6.2.1.9. **LC–SUConv** is a bicoreflective (full and isomorphism–closed) subconstruct being closed under formation of finite products, it follows from 3.1.7. (resp. 2.2.12.) that **LC–SUConv** is cartesian closed (resp. topological).

6.2.1.12 Remark. The natural function spaces in **LC–SUConv** are formed as follows: *Let $\mathbf{X} = (X, \mathcal{J}_X)$ and $\mathbf{Y} = (Y, \mathcal{J}_Y)$ be locally compact semiuniform convergence spaces. Then the natural function space $\mathbf{Y}^{\mathbf{X}}$ is the set $[\mathbf{X}, \mathbf{Y}]$ of all uniformly continuous maps from $\mathbf{X}$ into $\mathbf{Y}$ endowed with the* **LC–SUConv***–structure $(\mathcal{J}_{X,Y})_{LC} = \{\Phi \in \mathcal{J}_{X,Y}$: there is some $K \in \Phi$ such that K is a compact subset of $([\mathbf{X}, \mathbf{Y}], \mathcal{J}_{X,Y}) \times ([\mathbf{X}, \mathbf{Y}], \mathcal{J}_{X,Y})\}$,* where $\mathcal{J}_{X,Y} = \{\Phi \in F([\mathbf{X}, \mathbf{Y}] \times [\mathbf{X}, \mathbf{Y}]) : \Phi(\mathcal{F}) \in \mathcal{J}_Y$ for each $\mathcal{F} \in \mathcal{J}_X\}$ and $\Phi(\mathcal{F})$ is the filter generated by $\{A(F) : A \in \Phi,$

$F \in \mathcal{F}\}$ with $A(F) = \{(f(a), g(b)) : (f,g) \in A$ and $(a,b) \in F\}$ (see 3.1.7. and use 6.2.1.7. and 6.2.1.9.).

6.2.1.13 Definition. A semiuniform convergence space $(X, \mathcal{J}_X)$ is called *compactly generated* provided that $\mathcal{J}_X$ is the final **SUConv**–structure with respect to the family $(j_i : (K_i, \mathcal{J}_{K_i}) \longrightarrow (X, \mathcal{J}_X))_{i \in I}$ of the inclusions of all compact subspaces of $(X, \mathcal{J}_X)$.

6.2.1.14 Theorem. *A semiuniform convergence space $(X, \mathcal{J}_X)$ is compactly generated if and only if it is locally compact.*

Proof. 1) '$\Longrightarrow$'. Since $(X, \mathcal{J}_X)$ is compactly generated, $\mathcal{J}_X = \{\mathcal{F} \in F(X \times X)$: there exist $i \in I$ and $\mathcal{F}_i \in \mathcal{J}_{K_i}$ such that $(j_i \times j_i)(\mathcal{F}_i) \subset \mathcal{F}\}$. Thus, for each $\mathcal{F} \in \mathcal{J}_X$ there are some $i \in I$ and some $\mathcal{F}_i \in \mathcal{J}_{K_i}$ such that $\mathcal{F} \supset (j_i \times j_i)(\mathcal{F}_i)$, which implies that the compact set $K_i \times K_i$ belongs to $\mathcal{F}$.
2) '$\Longleftarrow$'. Let $(X, \mathcal{J}_X) \in |\mathbf{LC\text{–}SUConv}|$ and let $\mathcal{J}'_X$ be the final **SUConv**–structure with respect to the family $(j_i : (K_i, \mathcal{J}_{K_i}) \longrightarrow (X, \mathcal{J}_X))$ of the inclusions of all compact subspaces of $(X, \mathcal{J}_X)$. In order to prove that $\mathcal{J}_X = \mathcal{J}'_X$, let $\mathcal{F} \in \mathcal{J}_X$. By assumption, there is some compact $\widehat{K} \in \mathcal{F}$. Without loss of generality, let $\widehat{K}$ be symmetric, i.e. $\widehat{K} = \widehat{K}^{-1}$ (note: If $\widehat{K} \in \mathcal{F}$ is compact, then $\widehat{K}^{-1}$ is compact and therefore $\widehat{K} \cup \widehat{K}^{-1} = (\widehat{K} \cup \widehat{K}^{-1})^{-1} \in \mathcal{F}$ is compact). Put $K = p_1[\widehat{K}]$ where p_1 denotes the first projection. Then K is compact. Let $j_K : K \longrightarrow X$ be the corresponding inclusion. $\mathcal{F}_K = (j_K \times j_K)^{-1}(\mathcal{F})$ exists (note: $\widehat{K} \cap F \subset (K \times K) \cap F$ belongs to $\mathcal{F}$ for each $F \in \mathcal{F}$) and $(j_K \times j_K)(\mathcal{F}_K) = \mathcal{F}$. Thus, $\mathcal{F}_K \in \mathcal{J}_K$ and $\mathcal{F} \in \mathcal{J}'_X$. Consequently, $\mathcal{J}_X \subset \mathcal{J}'_X$ and, since $\mathcal{J}'_X \subset \mathcal{J}_X$ is trivial, even $\mathcal{J}_X = \mathcal{J}'_X$.

6.2.1.15 Corollary. *Let $(X, \mathcal{J}_X)$ be a locally compact semiuniform convergence space. Then a map $f : (X, \mathcal{J}_X) \longrightarrow (Y, \mathcal{J}_Y)$ from $(X, \mathcal{J}_X)$ into a semiuniform convergence space $(Y, \mathcal{J}_Y)$ is uniformly continuous iff it is uniformly continuous on compacta, i.e. for each compact subspace $(K, \mathcal{J}_K)$ of $(X, \mathcal{J}_X)$ the restriction $f|_K : (K, \mathcal{J}_K) \longrightarrow (Y, \mathcal{J}_Y)$ is uniformly continuous.*

Proof. Use 6.2.1.14. and the defining property of final structures.

6.2.1.16 Remarks. 1) The above definition of compactly generated semiuniform convergence spaces is similar to the definition of compactly generated topological spaces, but a compactly generated topological space need not be locally compact, e.g. the topological space $\mathbb{R}_t^{\mathbb{N}}$ of sequences of real numbers is metrizable and thus a compactly generated space but not locally compact.

2) a) Compactly generated topological spaces form a bicoreflective (full and isomorphism–closed) subconstruct of **Top**; namely, it is easily checked that *quotients and coproducts of compactly generated topological spaces are compactly generated* (exercise!). Furthermore, every compactly generated topological space is a quotient (space) of a coproduct of compact topological spaces, i.e. it is a quotient of a locally compact topological space (Let **X** be a compactly generated topo-

logical space and $(i_\ell : \mathbf{K}_\ell \longrightarrow \mathbf{X})_{\ell \in L}$ the family of the inclusions of all compact subspaces of $\mathbf{X}$. Let $((j_\ell)_{\ell \in L}, \coprod_{\ell \in L} \mathbf{K}_\ell)$ be the coproduct of $(\mathbf{K}_\ell)_{\ell \in L}$. Then a map $f : \coprod_{\ell \in L} \mathbf{K}_\ell \longrightarrow \mathbf{X}$ is defined by $f \circ j_\ell = i_\ell$ for each $\ell \in L$, which is surjective and continuous. Since $\mathbf{X}$ is compactly generated it is easily checked that f is a quotient map.).
Usually, a *k–space* means a compactly generated Hausdorff space. It has already been mentioned earlier that every locally compact Hausdorff space is a k–space. Thus, *a Hausdorff* $\mathbf{X}$ *is a k–space iff it is a quotient (space) of a locally compact Hausdorff space.* Thus, it follows immediately from the characterization of local compactness in Hausdorff spaces by means of quotient maps (cf. 6.1.30. 3)) that *the product of a locally compact Hausdorff space and a k–space is a k–space*; it has been proved by Dowker [36] in 1952 that the product of two k–spaces is not a k space in general.

b) *In contrast to the situation for semiuniform convergence spaces the construct* **CGTop** *of compactly generated topological spaces is not cartesian closed* as the following example shows.

Example. Let $[0,1]$ be endowed with the usual topology and let $I\!N$ be the discrete topological space of natural numbers. Let K be the one–point compactification of $I\!N \times [0,1]$ and $\mathbb{Q}^*$ the one–point compactification of the usual topological space $\mathbb{Q}$ of rational numbers (concerning the definition of the one–point compactification see 6.2.1.27.3)). $I\!N \times [0,1]$ is locally compact Hausdorff and by 6.1.23. 1), $\mathbb{Q}$ is a k–space. Consider the natural (quotient) map $\omega : K \longrightarrow K/I\!N \times \{0\}$, where $K/I\!N \times \{0\}$ denotes the space K with $I\!N \times \{0\}$ identified to a point. Then $\omega \times 1_{\mathbb{Q}^*} : K \times \mathbb{Q}^* \longrightarrow (K/I\!N \times \{0\}) \times \mathbb{Q}^*$ is not a quotient map (If $\omega \times 1_{\mathbb{Q}^*}$ were a quotient map, then the restriction to an open inverse image

$$\omega \times 1_{\mathbb{Q}^*}|_{I\!N \times [0,1] \times \mathbb{Q}} : I\!N \times [0,1] \times \mathbb{Q} \longrightarrow (I\!N \times [0,1]/I\!N \times \{0\}) \times \mathbb{Q}$$

would be a quotient map, too. Since $(I\!N \times [0,1]) \times \mathbb{Q}$ is compactly generated, $(I\!N \times [0,1]/I\!N \times \{0\}) \times \mathbb{Q}$ would also be compactly generated. But this is not true; namely, if X denotes the space $I\!N \times [0,1]/I\!N \times \{0\}$, then the set

$$M = \{((n,0), \frac{1}{n}) : n \in I\!N\}$$

is not closed in $X \times \mathbb{Q}$ (because each neighborhood of $((1,0),0)$ meets M), but for each compact subset C of $X \times \mathbb{Q}$, $M \cap C$ is closed in C, which can be verified as follows:

Since each compact subset of $X \times \mathbb{Q}$ is contained in the product of its projections, it suffices to prove that $M \cap (K_1 \times K_2)$ is closed in $X \times \mathbb{Q}$ provided that K_1 (resp. K_2) is compact in X (resp. $\mathbb{Q}$); and since each compact subset of X meets only finitely many $\{n\} \times [0,1]$, it suffices even to prove that for each compact subset K_2 of $\mathbb{Q}$,

$$M \cap \Big((\bigcup_{i=1}^{k} \{n_i\} \times [0,1]) \times K_2 \Big)$$

is closed in $X \times \mathbb{Q}$ for each $k \in \mathbb{N}$. But this intersection is equal to

$$\left\{\left((n_i,0),\frac{1}{n_i}\right) : 1 \leq i \leq k \text{ and } \frac{1}{n_i} \in K_2\right\}$$

which is a closed subset of $X \times \mathbb{Q}$.).
Since **CGTop** is bicoreflective in **Top**, quotients in **CGTOP** are formed as in **Top**. Furthermore, K, $K/\mathbb{N}\times\{0\}$ and $\mathbb{Q}^*$ as well as $K\times\mathbb{Q}^*$ and $(K/\mathbb{N}\times\{0\})\times\mathbb{Q}^*$ are compact and consequently compactly generated. Thus, it follows from the above result that **CGTop** is not cartesian closed (otherwise $\omega \times 1_{\mathbb{Q}^*}$ would be a quotient map).

3) a) Let $(X,\mathcal{X})$ be a topological space and $\mathcal{X}^*$ the final topology on X with respect to the family of the inclusions of all compact subspaces of $(X,\mathcal{X})$. Then $1_X : (X,\mathcal{X}^*) \longrightarrow (X,\mathcal{X})$ is the bicoreflection of $(X,\mathcal{X})$ with respect to **CGTop**.
b) Let **CGHaus** be the construct of all compactly generated Hausdorff spaces (= k–spaces) [and continuous maps]. Then **CGHaus** is a bicoreflective subconstruct of the construct **Haus** of Hausdorff spaces (and continuous maps) where the bicoreflection of $(X,\mathcal{X}) \in |\mathbf{Haus}|$ with respect to **CGHaus** is constructed as under a) [note: $(X,\mathcal{X}^*)$ is Hausdorff provided that $(X,\mathcal{X})$ is Hausdorff].
4) **CGHaus** *is a cartesian closed construct*, where the natural function spaces are formed as follows: Let $\mathbf{X}$ and $\mathbf{Y}$ be k–spaces. If $\mathcal{C}$ denotes the compact–open topology on $C(\mathbf{X},\mathbf{Y})$, then $(C(\mathbf{X},\mathbf{Y}),\mathcal{C}^*)$ is the power object $\mathbf{Y}^{\mathbf{X}}$ in **CGHaus** (cf. [96; VII. 8.] for more detailed information and note that $(C(\mathbf{X},\mathbf{Y}),\mathcal{C})$ is Hausdorff whenever $\mathbf{Y}$ is Hausdorff). Obviously, **CGHaus** is not topological (the inclusion map $i : \mathbb{Q} \longrightarrow \mathbb{R}$ is an epimorphism in **CGHaus**, which is not surjective!).

4) Compactly generated topological spaces have not been described by means of suitable axioms in contrast to the situation for compactly generated semiuniform convergence spaces (cf. 6.2.1.14.).

6.2.1.17 Definition. Let $(X,\mathcal{J}_X)$ be a semiuniform convergence space. Then the topological space with underlying set X whose closed subsets are exactly the closed subsets of the Kent convergence space $(X,q_{\gamma_{\mathcal{J}_X}})$ is called the *induced topological space of* $(X,\mathcal{J}_X)$.

6.2.1.18 Theorem. *Let $(X,\mathcal{X})$ be an R_0–space (= symmetric topological space) and $(X,\mathcal{J}_X^*)$ the locally compact modification* (cf. 6.2.1.7.) *of its corresponding semiuniform convergence space $(X,\mathcal{J}_X)$ defined by $\mathcal{F} \in \mathcal{J}_X$ iff there is some $x \in X$ such that $\mathcal{F} \supset \mathcal{U}_{\mathcal{X}}(x)\times\mathcal{U}_{\mathcal{X}}(x)$, where $\mathcal{U}_{\mathcal{X}}(x)$ denotes the neighborhood filter of $x \in X$ with respect to $\mathcal{X}$. Then the induced topological space $(X,\mathcal{X}^*)$ of $(X,\mathcal{J}_X^*)$ is the compactly generated modification of $(X,\mathcal{X})$ in the usual topological sense (i.e. $1_X : (X,\mathcal{X}^*) \longrightarrow (X,\mathcal{X})$ is the bicoreflection of $(X,\mathcal{X})$ with respect to* **CGTop**).

Proof. Put $q = q_{\mathcal{X}}$ and $q^* = q_{\gamma_{\mathcal{J}_X^*}}$. Obviously, $(\mathcal{F},x) \in q^*$ iff $(\mathcal{F},x) \in q$ and $\mathcal{F}$ contains a compact subset K of (X,q). Furthermore, the Kent convergence space

(X, q^*) is even a limit space (use that finite unions of compact sets are compact). It is easily verified that the following are valid:
1. (X, q^*) and (X, q) have the same compact subsets.
2. The **Lim**–structure q^* is final w.r.t. the family $(i_\ell : \mathbf{K}_\ell \longrightarrow (X, q))$ of the inclusions of all compact subspaces of the (symmetric) topological space (X, q). Let $\mathcal{R} : \mathbf{Lim} \longrightarrow \mathbf{Top}$ denote the bireflector. Then $(\mathcal{R}(i_\ell) : \mathcal{R}(\mathbf{K}_\ell) \longrightarrow \mathcal{R}((X, q^*))$ is a final sink in **Top**, where $\mathcal{R}(i_\ell) = i_\ell$ and $\mathcal{R}((X, q^*)) = (X, \mathcal{X}^*)$ (note that $\mathcal{R}$ preserves coproducts and quotients by means of the dualization of 2.1.10.). Since $\mathbf{K}_\ell$ is topological (as a subspace in **Lim** of a topological space), $\mathcal{R}(\mathbf{K}_\ell) = \mathbf{K}_\ell$, and $\mathcal{X}^*$ is the final **Top**–structure on X w.r.t. the familiy of the inclusions of all compact subspaces of $(X, \mathcal{X})$, i.e. $1_X : (X, \mathcal{X}^*) \longrightarrow (X, \mathcal{X})$ is the bicoreflection of $(X, \mathcal{X})$ w.r.t. **CGTop**.

6.2.1.19 Remarks. 1) *The induced topological space of a locally compact semiuniform convergence space is generally not locally compact whereas the underlying Kent convergence space is locally compact* (cf. 6.2.1.5.), namely if $(X, \mathcal{J}_X)$ is the corresponding semiuniform convergence space of the (symmetric) topological space $\mathbb{R}_t^{\mathbb{N}}$, then $(X, \mathcal{J}_X^*)$ is a locally compact semiuniform convergence space whose induced topological space is $\mathbb{R}_t^{\mathbb{N}}$ which is not locally compact (apply the above theorem!).
2) The induced topological space of a locally compact semiuniform convergence space $(X, \mathcal{J}_X)$ is compactly generated (namely since $(X, q_{\gamma_{\mathcal{J}_X}})$ is locally compact, it is a compactly generated Kent convergence space, i.e. $q_{\gamma_{\mathcal{J}_X}}$ is the final **KConv**–structure on X w.r.t. the family $(i_\ell : \mathbf{K}_\ell \longrightarrow (X, q_{\gamma_{\mathcal{J}_X}}))$ of the inclusions of all compact subspaces of $(X, q_{\gamma_{\mathcal{J}_X}})$; then $(\mathcal{R}(i_\ell) : \mathcal{R}(\mathbf{K}_\ell) \longrightarrow \mathcal{R}((X, q_{\gamma_{\mathcal{J}_X}})))$ is a final sink in **Top** provided that $\mathcal{R} : \mathbf{KConv} \longrightarrow \mathbf{Top}$ denotes the bireflector, which implies that $\mathcal{R}((X, q_{\gamma_{\mathcal{J}_X}}))$, i.e. the induced topological space of $(X, \mathcal{J}_X)$, is compactly generated). Thus, using the above theorem, *a Hausdorff space is a k–space iff it is the induced topological space of some locally compact semiuniform convergence space.*
3) If the underlying Kent convergence space of some semiuniform convergence space $(X, \mathcal{J}_X)$ is locally compact, then $(X, \mathcal{J}_X)$ need not be locally compact, e.g. the uniform space $\mathbb{R}_u$ of real numbers is not locally compact, but its underlying Kent convergence space, namely $\mathbb{R}_t$, is locally compact.

6.2.1.20 Definitions. A semiuniform convergence space $(X, \mathcal{J}_X)$ is called
1) *strongly locally compact* provided that for each $\mathcal{F} \in \mathcal{J}_X$ there is some subfilter $\mathcal{G} \in \mathcal{J}_X$ of $\mathcal{F}$ together with a filter base $\mathcal{B}$ for $\mathcal{G}$ consisting of compact subsets of $(X, \mathcal{J}_X) \times (X, \mathcal{J}_X)$.
2) *t–regular* provided that for each $\mathcal{F} \in \mathcal{J}_X$ the subfilter $\overline{\mathcal{F}}^t$, generated by the filter base of all closed elements of $\mathcal{F}$, belongs to $\mathcal{J}_X$.

6.2.1.21 Remark. The following facts are easily verified (cf. the corresponding results on regularity under 4.2.1.):
a) Every t–regular semiuniform convergence space is regular (note: $\overline{\mathcal{F}}^t \subset \overline{\mathcal{F}}$).

b) Every uniform space (= principal uniform limit space) is t–regular.
c) Let $(X, \mathcal{X})$ be a symmetric topological space (= R_0–space) and $(X, \mathcal{J}_X)$ its corresponding semiuniform convergence space. Then $(X, \mathcal{X})$ is regular (in the usual topological sense) iff $(X, \mathcal{J}_X)$ is t–regular.
d) The construct **T–Reg** of t–regular semiuniform convergence spaces (and uniformly continuous maps) is a bireflective (full and isomorphism–closed) subconstruct of **SUConv** and thus a topological construct.

6.2.1.22 Proposition. *Let $(X, \mathcal{J}_X)$ be a convergence space. Then the following are equivelent:*
(1) $(X, \mathcal{J}_X)$ is t–regular,
(2) For each $(\mathcal{F}, x) \in q_{\gamma_{\mathcal{J}_X}}$, $(\overline{\mathcal{F}}^t, x) \in q_{\gamma_{\mathcal{J}_X}}$.

Proof. a) Obviously, for each $\mathcal{F} \in F(X)$, $\overline{\mathcal{F} \times \mathcal{F}}^t = \overline{\mathcal{F}}^t \times \overline{\mathcal{F}}^t$.
b) Let (2) be satisfied and $\mathcal{F} \in \mathcal{J}_X$. By assumption there is some $(\mathcal{G}, x) \in q_{\gamma_{\mathcal{J}_X}}$ such that $\mathcal{G} \times \mathcal{G} \subset \mathcal{F}$ and by (2), $(\overline{\mathcal{G}}^t, x) \in q_{\gamma_{\mathcal{J}_X}}$. Using a), $\overline{\mathcal{G}}^t \times \overline{\mathcal{G}}^t \subset \overline{\mathcal{F}}^t$, i.e. $\overline{\mathcal{F}}^t \in \mathcal{J}_X$. Thus, (1) is satisfied.
c) Let $(X, \mathcal{J}_X)$ be t–regular and $(\mathcal{F}, x) \in q_{\gamma_{\mathcal{J}_X}}$, i.e. $(\mathcal{F} \cap \dot{x}) \times (\mathcal{F} \cap \dot{x}) \in \mathcal{J}_X$. Put $\mathcal{G} = \mathcal{F} \cap \dot{x}$. Since $(X, \mathcal{J}_X)$ is t–regular, $\overline{\mathcal{G} \times \mathcal{G}}^t = \overline{\mathcal{G}}^t \times \overline{\mathcal{G}}^t \in \mathcal{J}_X$. By assumption there is some $(\mathcal{H}, z) \in q_{\gamma_{\mathcal{J}_X}}$ such that $\overline{\mathcal{G}}^t \times \overline{\mathcal{G}}^t \supset \mathcal{H} \times \mathcal{H}$ which implies $\overline{\mathcal{G}}^t \supset \mathcal{H}$ and consequently $(\overline{\mathcal{G}}^t, z) \in q_{\gamma_{\mathcal{J}_x}}$. Since $(X, q_{\gamma_{\mathcal{J}_X}})$ is symmetric and $x \in \bigcap_{K \in \overline{\mathcal{G}}^t} K$, $(\overline{\mathcal{G}}^t, x) \in q_{\gamma_{\mathcal{J}_X}}$ which implies $(\overline{\mathcal{F}}^t, x) \in q_{\gamma_{\mathcal{J}_X}}$. Thus, (2) is satisfied.

6.2.1.23 Proposition. *Let $(X, \mathcal{J}_X)$ be a t–regular semiuniform convergence space. Then $(X, \mathcal{J}_X)$ is locally compact iff it is strongly locally compact.*

Proof. Let $(X, \mathcal{J}_X)$ be locally compact and $\mathcal{F} \in \mathcal{J}_X$. Since $(X, \mathcal{J}_X)$ is also t–regular, the subfilter $\overline{\mathcal{F}}^t$ belongs to $\mathcal{J}_X$ and contains a compact set K. Then $\overline{\mathcal{F}}^t$ is generated by the filter base $\mathcal{B} = \{F \in \mathcal{F} : F = \overline{F} \subset K\}$ consisting of compact subsets, i.e. $(X, \mathcal{J}_X)$ is strongly locally compact. The inverse implication is obvious.

6.2.1.24 Remarks. 1) It follows from 6.2.1.4., 6.2.1.21. b) and the above proposition that a uniform space is strongly locally compact iff it is compact.
2) If $(X, \mathcal{X})$ is a symmetric topological space and $(X, \mathcal{J}_X)$ its corresponding semiuniform convergence space, then $(X, \mathcal{J}_X)$ is strongly locally compact iff for each $x \in X$ the neighborhood filter $\mathcal{U}_{\mathcal{X}}(x)$ has a base of compact sets (cf. the corresponding proof for locally $\mathcal{E}$–connected spaces in chapter 5).
3) The construct **SLC–SUConv** of strongly locally compact semiuniform convergence spaces (and uniformly continuous maps) is a bicoreflective cartesian closed subconstruct of **SUConv** (cf. the corresponding proof for locally $\mathcal{E}$–connected spaces in chapter 5).

6.2.1.25 Proposition. *Every open subspace (in* **Conv***) of a t–regular locally* *compact convergence space* $(X, \mathcal{J}_X)$ *is strongly locally compact (and thus locally compact).*

Proof. Let O be an open subset of $(X, q_{\gamma_{\mathcal{J}_X}})$ and $(O, \mathcal{J}_O)$ the subspace of $(X, \mathcal{J}_X)$ in **SUConv**. Then $(O, \mathcal{J}_{\gamma_{q_{\gamma_{\mathcal{J}_O}}}})$ is the (open) subspace of $(X, \mathcal{J}_X)$ in **Conv**. Since $(X, \mathcal{J}_X)$ is a t–regular convergence space, $(X, q_{\gamma_{\mathcal{J}_X}})$ fulfills condition (2) of 6.2.1.22. Hence $(O, q_{\gamma_{\mathcal{J}_O}})$ fulfills this condition too, i.e. $(O, \mathcal{J}_{\gamma_{q_{\gamma_{\mathcal{J}_O}}}})$ is a t–regular convergence space. Thus, by 6.2.1.23., it suffices to prove that $(O, \mathcal{J}_{\gamma_{q_{\gamma_{\mathcal{J}_O}}}})$ is locally compact. Now let $\mathcal{F} \in \mathcal{J}_{\gamma_{q_{\gamma_{\mathcal{J}_O}}}}$. Then there is some $x \in O$ and some $\mathcal{G} \in F(O)$ such that $\mathcal{F} \supset \mathcal{G} \times \mathcal{G}$ and $(\mathcal{G}, x) \in q_{\gamma_{\mathcal{J}_O}}$. Furthermore, $(i(\mathcal{G}), x) \in q_{\gamma_{\mathcal{J}_X}}$ provided that $i : O \longrightarrow X$ denotes the inclusion map. By assumption, $(X, q_{\gamma_{\mathcal{J}_X}})$ is locally compact (cf. 6.2.1.5.) and t–regular (i.e. it satisfies the condition (2) in 6.2.1.22.). Thus, there is a subfilter $\widehat{\mathcal{F}} \subset i(\mathcal{G})$ together with a filter base $\widehat{\mathcal{B}}$ consisting of compact subsets of $(X, q_{\gamma_{\mathcal{J}_X}})$ such that $(\widehat{\mathcal{F}}, x) \in q_{\gamma_{\mathcal{J}_X}}$ (similarly to 6.2.1.23.). Since O is open, $O \in \widehat{\mathcal{F}}$ and consequently there is a compact set $K \in \widehat{\mathcal{F}}$ such that $O \supset K$. Then $K \in i^{-1}(\widehat{\mathcal{F}}) \subset i^{-1}(i(\mathcal{G})) = \mathcal{G}$. By Tychonoff $K \times K \in \mathcal{G} \times \mathcal{G} \subset \mathcal{F}$ is compact.

6.2.1.26 Theorem (*One–point compactification*). *Let (X, q) be a non–compact Kent convergence space. Put $X^* = X \cup \{\infty\}$ with $\infty \notin X$ and $q^* = \{(\mathcal{F}^*, x) \in F(X^*) \times X$: there is some $(\mathcal{F}, x) \in q$ with $\mathcal{F} \subset \mathcal{F}^*\} \cup \{(\mathcal{F}^*, \infty) : \mathcal{F}^* \in F(X^*)$ and $\mathcal{B} \subset \mathcal{F}^*\}$ where $\mathcal{B} = \{O^* \subset X^* : \infty \in O^*$ and $X^* \backslash O^*$ is a compact subset of $(X, q)\}$. Then (X^*, q^*) is a compact Kent convergence space containing (X, q) as a dense subspace. Furthermore, (X^*, q^*) is T_2 provided that (X, q) is locally compact T_2.*

Proof. I) It is easily checked that (X^*, q^*) is a Kent convergence space.
II) (X, q) is a subspace of (X^*, q^*): Let $q' = \{(\mathcal{F}, x) \in F(X) \times X$: $(i(\mathcal{F}), x) \in q^*\}$, where $i : X \longrightarrow X^*$ denotes the inclusion map. In order to prove that $q' = q$, let $(\mathcal{F}, x) \in q$. Put $i(\mathcal{F}) = \mathcal{F}^*$. Then $(\mathcal{F}^*, x) \in q^*$, i.e. $(\mathcal{F}, x) \in q'$, since $\mathcal{F} \subset \mathcal{F}^*$. Conversely, let $(\mathcal{F}, x) \in q'$, i.e. $(i(\mathcal{F}), x)) \in q^*$. Since $x \neq \infty$, there is some $(\mathcal{G}, x) \in q$ such that $\mathcal{G} \subset \mathcal{F}^*$. Hence $\mathcal{G} \subset \mathcal{F}$ and consequently $(\mathcal{F}, x) \in q$.

III) X is dense in (X^*, q^*), i.e. $X^* = cl_{q^*} X = \{x^* \in X^*$: there is some $\mathcal{G}^* \in F(X^*)$ such that $(\mathcal{G}^*, x^*) \in q^*$ and $X \in \mathcal{G}^*\}$: Since it suffices to prove that $X^* \subset cl_{q^*} X$, let $x^* \in X^*$. Then $x^* = x \in X$ or $x^* = \infty$. In the first case it follows from $(\dot{x}, x) \in q$ that $(\dot{x}^*, x) \in q^*$ (since $\dot{x} \subset \dot{x}^*$) and thus $x^* \in cl_{q^*} X$ since $X \in \dot{x}^*$. In the second case one obtains:

a) $\mathcal{B}' = \{O^* \cap X : O^* \in \mathcal{B}\}$ is a filter base on X: Since $\mathcal{B}$ is closed under formation of finite intersections, $\mathcal{B}'$ is also closed under formation of finite intersections, and $\emptyset \notin \mathcal{B}'$ since otherwise X were compact.
b) $X \in i((\mathcal{B}')) = \mathcal{F}^*$ and by a), $\mathcal{F}^*$ is a filter on X^*. Since obviously $\mathcal{B} \subset \mathcal{F}^*$, $(\mathcal{F}^*, \infty) \in q^*$ which implies $\infty \in cl_{q^*} X$.
IV) (X^*, q^*) is compact: Let $\mathcal{U}^*$ be an ultrafilter on X^*. Then $(\mathcal{U}^*, \infty) \in q^*$ or

$(\mathcal{U}^*, \infty) \notin q^*$. In the latter case there is some $O^* \in \mathcal{B}$ such that $O^* \notin \mathcal{U}^*$. Since $\mathcal{U}^*$ is an ultrafilter, $X^* \backslash O^* \in \mathcal{U}^*$. Let $j : X^* \backslash O^* \longrightarrow X^*$ be the inclusion map. Then $j^{-1}(\mathcal{U}^*)$ exists, and it has an adherence point $x \in X^* \backslash O^* \subset X$ since $X^* \backslash O^*$ is compact. Hence, $(\mathcal{U}^*, x) \in q^*$.

V) Let (X, q) be locally compact T_2. In order to prove that (X^*, q^*) is T_2, let $\mathcal{F}^* \in F(X^*)$ and $x^*, y^* \in X^*$ such that $(\mathcal{F}^*, x^*) \in q^*$ and $(\mathcal{F}^*, y^*) \in q^*$. Let us consider two cases:
1. $x^* = x \in X$ and $y^* = y \in X$,
2. $x^* = x \in X$ and $y^* = \infty$.

In the first case there are $(\mathcal{G}, x) \in q$ and $(\mathcal{H}, y) \in q$ such that $\mathcal{G} \subset \mathcal{F}^*$ and $\mathcal{H} \subset \mathcal{F}^*$. Consequently $\mathcal{K} = \sup\{\mathcal{G}, \mathcal{H}\}$ exists and converges to x and y in (X, q) which implies $x = y$ since (X, q) is T_2. In the second case there is some $(\mathcal{F}, x) \in q$ such that $\mathcal{F} \subset \mathcal{F}^*$. Since (X, q) is locally compact there exists some compact $K \in \mathcal{F}$ and since $(\mathcal{F}^*, \infty) \in q^*$, $O^* = X^* \backslash K \in \mathcal{F}^*$, i.e. the second case cannot occur (otherwise $\emptyset$ would belong to $\mathcal{F}^*$). Consequently, (X^*, q^*) is T_2.

6.2.1.27 Remarks. 1) There are other possibilities to construct a one–point compactification of a non-compact Kent convergence space, e.g. q^* can be replaced by q^{**} where q^{**} arises from q^* by substituting $\mathcal{B}' = \{O^* \subset X^* : \infty \in O^*$ and $X^* \backslash O^*$ is a closed and compact subset of $(X, q)\}$ for $\mathcal{B}$.

2) a) (X^*, q^*) has the property P iff (X, q) has the property P where P stands for limit space, pseudotopological space, pretopological space or T_1 respectively.
b) (X^*, q^{**}) is topological iff (X, q) is topological.

3) If (X, q) is a locally compact topological T_2–space, then $q^* = q^{**}$, i.e. (X^*, q^*) is a compact topological T_2–space; in other words: (X^*, q^*) is the usual one–point compactification (= *Alexandroff compactification*) of (X, q).

4) A locally compact Hausdorff space (= locally compact topological T_2–space) is a Tychonoff space since it is a subspace of its Alexandorff compactification. Thus, it is regular (= t–regular) and it follows from 6.2.1.23. the well–known fact that *for a Hausdorff* $(X, \mathcal{X})$ *the following are equivalent:*
(1) $(X, \mathcal{X})$ *is locally compact, i.e. each point* $x \in X$ *has a compact neighborhood,*
(2) $(X, \mathcal{X})$ *is strongly locally compact, i.e. each point* $x \in X$ *has a neighborhood base consisting of compact sets.*

5) The axiom C_3) for Kent convergence spaces is only used in the above theorem in order to prove that (X^*, q^*) fulfills C_3). Thus, the above theorem is also valid for generalized convergence spaces instead of Kent convergence spaces.

6.2.2 Local precompactness

6.2.2.1 Definitions. 1) *A subset* A of a semiuniform convergence space $(X, \mathcal{J}_X)$ *is* called *precompact* provided that $(A, \mathcal{J}_A)$ is precompact, where $\mathcal{J}_A$ is the initial **SUConv**–structure on A with respect to the inclusion map $i : A \longrightarrow X$.
2) A semiuniform convergence space is called *locally precompact* provided that

each $\mathcal{F} \in \mathcal{J}_X$ contains a precompact subset of the product space $(X, \mathcal{J}_X) \times (X, \mathcal{J}_X)$.

6.2.2.2 Proposition. *Every locally compact semiuniform convergence space is locally precompact.*

Proof. Use 4.3.2.4. 1).

6.2.2.3 Proposition. *Let $(X, \mathcal{J}_X)$ be a t-regular weakly complete semiuniform convergence space. If $(X, \mathcal{J}_X)$ is locally precompact, then $(X, \mathcal{J}_X)$ is locally compact.*

Proof. Let $\mathcal{F} \in \mathcal{J}_X$. Since $(X, \mathcal{J}_X)$ is t–regular, $\overline{\mathcal{F}}^t \in \mathcal{J}_X$. By assumption there is a precompact subset H of $(X, \mathcal{J}_X) \times (X, \mathcal{J}_X)$ such that $H \in \overline{\mathcal{F}}^t$. Thus, there is a closed subset F of $X \times X$ such that $F \in \mathcal{F}$ and $F \subset H$. Consequently, F is precompact (cf. 4.3.2.2.) and weakly complete (cf. 4.3.2.12. 3) b)), i.e. F is compact.

6.2.2.4 Corollary. *A t–regular convergence space (in particular a regular topological space) is locally precompact iff it is locally compact.*

Proof. Apply 6.2.2.2. and 6.2.2.3. Note that each convergence space is complete (cf. 2.3.3.24.) and thus weakly complete.

6.2.2.5 Proposition. 1) *Let $(X, \mathcal{J}_X)$ be a precompact semiuniform convergence space. Then $(X, \mathcal{J}_X)$ is locally precompact.*
2) *A diagonal semiuniform convergence space $(X, \mathcal{J}_X)$ is precompact provided that it is locally precompact.*

Proof. Analogously to 6.2.1.3.

6.2.2.6 Corollary. *A principal uniform limit space (= uniform space) is locally precompact iff it is precompact.*

6.2.2.7 Remarks. 1) If a diagonal semiuniform convergence space $(X, \mathcal{J}_X)$ is locally compact, then it is locally precompact. The inverse implication is valid provided that $(X, \mathcal{J}_X)$ is weakly complete.
2) If (X, γ) is a filter space, then a subset C of X is called precompact provided that (C, γ_C) is precompact, where γ_C denotes the initial **Fil**–structure on X with respect to the inclusion map $i : C \longrightarrow X$; equivalently: C is a precompact subset of (X, γ) iff each ultrafilter $\mathcal{U}$ on X with $C \in \mathcal{U}$ belongs to γ. A subset C of a semiuniform convergence space $(X, \mathcal{J}_X)$ is precompact (in the sense of 6.2.2.1.) iff it is precompact in $(X, \gamma_{\mathcal{J}_X})$.

6.2.2.8 Definition. A filter space (X, γ) is called *locally precompact* provided that each $\mathcal{F} \in \gamma$ contains a precompact subset of (X, γ).

6.2.2.9 Proposition. *Let $(X, \mathcal{J}_X)$ be a locally precompact semiuniform convergence space. Then the underlying filter space $(X, \gamma_{\mathcal{J}_X})$ is locally precompact.*

Proof. Let $\mathcal{F} \in \gamma_{\mathcal{J}_X}$. Then $\mathcal{F} \times \mathcal{F} \in \mathcal{J}_X$, and (by assumption) there is some precompact $H \in \mathcal{F} \times \mathcal{F}$,

i.e. $H \supset F \times F$ for some $F \in \mathcal{F}$. Since the first projection

$$p_1 : (X, \mathcal{J}_X) \times (X, \mathcal{J}_X) \longrightarrow (X, \mathcal{J}_X)$$

is uniformly continuous and therefore Cauchy continuous, it follows from 4.3.2.10. that $p_1[H] \supset F$ is precompact in $(X, \mathcal{J}_X)$ and thus, F is precompact in $(X, \mathcal{J}_X)$ (cf. 4.3.2.2), i.e. F is precompact in $(X, \gamma_{\mathcal{J}_X})$.

6.2.2.10 Proposition. *Let (X, γ) be a filter space and $(X, \mathcal{J}_\gamma)$ its corresponding semiuniform convergence space. Then the following are equivalent:*
(1) *The semiuniform convergence space $(X, \mathcal{J}_\gamma)$ is locally precompact,*
(2) *The filter space (X, γ) is locally precompact.*

Proof. (1) $\Longrightarrow$ (2). Since $\gamma_{\mathcal{J}_\gamma} = \gamma$ this implication follows from 6.2.2.9.
(2) $\Longrightarrow$ (1). Let $\mathcal{G} \in \mathcal{J}_\gamma$. Then there exists some $\mathcal{F} \in \gamma$ with $\mathcal{F} \times \mathcal{F} \subset \mathcal{G}$. By assumption, there is some precompact $F \in \mathcal{F}$ in (X, γ) and thus in $(X, \mathcal{J}_\gamma)$. Consequently, $F \times F$ is precompact in $(X, \mathcal{J}_\gamma) \times (X, \mathcal{J}_\gamma)$ (cf. 4.3.2.2) and belongs to $\mathcal{G}$.

6.2.2.11 Proposition. *The construct* **LPC–SUConv** *of locally precompact semiuniform convergence spaces (and uniformly continuous maps) is a bicoreflective (full and isomorphism-closed) subconstruct of* **SUConv**, *where $1_X : (X, \mathcal{J}_X^*) \longrightarrow (X, \mathcal{J}_X)$ is the bicoreflection of $(X, \mathcal{J}_X) \in |$***SUConv***$|$ with respect to* **LPC–SUConv** *provided that*

$\mathcal{J}_X^* = \{\mathcal{F} \in \mathcal{J}_X :$ *there is some $F \in \mathcal{F}$ such that F is a precompact subset of $(X, \mathcal{J}_X) \times (X, \mathcal{J}_X)\}$.*

Proof. Analogously to 6.2.1.7.

6.2.2.12 Corollary. *Let $(f_i : (X_i, \mathcal{J}_{X_i}) \longrightarrow (X, \mathcal{J}_X))_{i \in I}$ be a final sink in* **SUConv** *such that each $(X_i, \mathcal{J}_{X_i})$ is locally precompact. Then $(X, \mathcal{J}_X)$ is locally precompact.*

Proof. Analogously to 6.2.1.8.

6.2.2.13 Proposition. *Let $((X_i, \mathcal{J}_{X_i}))_{i \in I}$ be a family of non–empty semiuniform convergence spaces. Then the product space $(\prod_{i \in I} X_i, \mathcal{J}_X)$ of this family is locally precompact iff $(X_i, \mathcal{J}_{X_i})$ is locally precompact for all $i \in I$ and precompact for all but finitely many $i \in I$.*

Proof. Analogously to 6.2.1.9.

6.2.2.14 Proposition. **LPC–SUConv** is closed under formation of subspaces.

Proof. Let $(X, \mathcal{J}_X) \in |\mathbf{SUConv}|$ and $(A, \mathcal{J}_A)$ a subspace of $(X, \mathcal{J}_X)$ where one may assume that $A \subset X$ and $\mathcal{J}_A$ is the initial **SUConv**–structure on A w.r.t. the inclusion map $i : A \longrightarrow X$. Further, let $\mathcal{F} \in \mathcal{J}_A$. Then $(i \times i)(\mathcal{F}) \in \mathcal{J}_X$ and by assumption there is a precompact set $C \in (i \times i)(\mathcal{F})$. Obviously, $C \cap (A \times A) \in \mathcal{F}$ is precompact since it is a subset of C.

6.2.2.15 Theorem. **LPC–SUConv** *is a topological universe.*

Proof. 1) The fact that **LPC–SUConv** is cartesian closed and topological is proved analogously to 6.2.1.11.
2) Since **SUConv** is extensional and **LPC–SUConv** is a bicoreflective subconstruct of **SUConv** which is closed under formation of subspaces in **SUConv**, it follows from 3.2.5. that **LPC–SUConv** is extensional.

6.2.2.16 Remarks. 1) The natural function spaces (= power objects) in **LPC–SUConv** are formed as follows: *Let* $\mathbf{X} = (X, \mathcal{J}_X)$ *and* $\mathbf{Y} = (Y, \mathcal{J}_Y)$ *be locally precompact semiuniform convergence spaces. Then the natural function space* $\mathbf{Y}^{\mathbf{X}}$ *is the set* $[\mathbf{X}, \mathbf{Y}]$ *of all uniformly continuous maps from* $\mathbf{X}$ *into* $\mathbf{Y}$ *endowed with the* **LPC–SUConv***-structure* $(\mathcal{J}_{X,Y})_{LPC} = \{\Phi \in \mathcal{J}_{X,Y}$: *there is some* $C \in \Phi$ *such that* C *is a precompact subset of* $([\mathbf{X}, \mathbf{Y}], \mathcal{J}_{X,Y}) \times ([\mathbf{X}, \mathbf{Y}], \mathcal{J}_{X,Y})\}$ *provided that* $\mathcal{J}_{X,Y}$ *denotes the uniformly continuous* **SUConv**–*structure* (cf. 3.1.7. and use 6.2.2.11. and 6.2.2.13.).
2) The one–point extensions in **LPC–SUConv** are formed as follows: Let $(X, \mathcal{J}_X)$ be a locally precompact semiuniform convergence space and $X^* = X \cup \{\infty_X\}$ with $\infty_X \notin X$; then *the one-point extension of* $(X, \mathcal{J}_X)$ *in* **LPC–SUConv** *is* $(X^*, (\mathcal{J}_X^{\#})_{LPC})$, where

$(\mathcal{J}_X^{\#})_{LPC} = \{\mathcal{F} \in \mathcal{J}_X^{\#}$: there is some $\mathcal{F} \in \mathcal{F}$ such that F is a precompact subset of $(X^*, \mathcal{J}_X^{\#}) \times (X^*, \mathcal{J}_X^{\#})\}$

and

$$\mathcal{J}_X^{\#} = \{\mathcal{H} \in F(X^* \times X^*) : \text{ the trace of } \mathcal{H} \text{ on } X \times X \text{ exists and belongs to } \mathcal{J}_X \text{ or } \{(\infty_X, \infty_X)\} \cup (X^* \times \{\infty_X\}) \cup (\{\infty_X\} \times X^*) \in \mathcal{H}\}$$

(cf. 3.2.5., 3.2.7. ③ and 6.2.2.11.).

6.2.2.17 Definition. A semiuniform convergence space $(X, \mathcal{J}_X)$ is called *precompactly generated* provided that $\mathcal{J}_X$ is the final **SUConv**–structure with respect to the family $(j_i : (C_i, \mathcal{J}_{C_i}) \longrightarrow (X, \mathcal{J}_X))_{i \in I}$ of the inclusions of all precompact subspaces of $(X, \mathcal{J}_X)$.

6.2.2.18 Theorem. *A semiuniform convergence space* $(X, \mathcal{J}_X)$ *is precompactly generated if and only if it is locally precompact.*

Proof. Analogously to 6.2.1.14.

6.2.2.19 Remarks. 1) The relationships between compactness, precompactness, local compactness and local precompactness can be summarized by means of the following implication scheme:

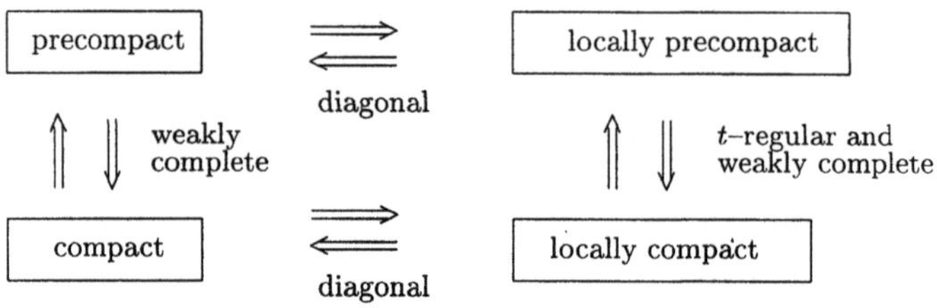

2) *The localization of precompactness (resp. compactness) is a more topological procedure than a uniform one*, namely in uniform spaces there is no difference between precompactness (resp. compactness) and local precompactness (resp. local compactness) whereas in regular topological spaces local precompactness (= local compactness) differs from precompactness (= compactness) [cf. the above implication scheme].

6.3 Precompactness and compactness in the natural function spaces of the construct of semiuniform convergence spaces

6.3.1 Definitions. 1) A semiuniform convergence space $(X, \mathcal{J}_X)$ is called a *semi–pseudouniform space* provided that the following is satisfied:

Ps) $\mathcal{F} \in \mathcal{J}_X$ whenever $\mathcal{U} \in \mathcal{J}_X$ for each ultrafilter $\mathcal{U} \supset \mathcal{F}$.

2) A uniform limit space which is semi–pseudouniform is called a *pseudouniform space.*

3) The construct of semi–pseudouniform spaces (resp. pseudouniform spaces) [and uniformly continuous maps] is denoted by **SPsU** (resp. **PsU**).

6.3.2 Remarks. 1) Every semi–pseudouniform space $(X, \mathcal{J}_X)$ is a semiuniform limit space (namely, if $\mathcal{F}, \mathcal{G} \in F(X \times X)$ and $\mathcal{F} \cap \mathcal{G} \notin \mathcal{J}_X$, then by (Ps) there is an ultrafilter $\mathcal{U}$ on $X \times X$ such that $\mathcal{U} \notin \mathcal{J}_X$ and $\mathcal{F} \cap \mathcal{G} \subset \mathcal{U}$, which implies $\mathcal{F} \subset \mathcal{U}$ or $\mathcal{G} \subset \mathcal{U}$; thus, $\mathcal{F} \notin \mathcal{J}_X$ or $\mathcal{G} \notin \mathcal{J}_X$, i.e. UC_4) is satisfied).

2) a) A *semiuniform space* is a pair $(X, \mathcal{W})$ where X is a set and $\mathcal{W}$ a non–empty subset of $\mathcal{P}(X \times X)$ satisfying the conditions F_2) and F_3) for a filter as well as the axioms U_1) and U_2) for uniform spaces (but not necessarily U_3).). Semiuniform spaces (and uniformly continuous maps) have been studied intensively in Čech's book [29]. The construct **SUnif** of semiuniform spaces is concretely isomorphic to the construct **PrSULim** of principal semiuniform limit spaces (and uniformly

continuous maps), where a semiuniform limit space $(X, \mathcal{J}_X)$ is called *principal* provided that there is a non–empty subset $\mathcal{F}$ of $\mathcal{P}(X \times X)$ satisfying the conditions F_2) and F_3) for a filter such that $\mathcal{J}_X = [\mathcal{F}]$ $(= \{\mathcal{G} \in F(X \times X) : \mathcal{G} \supset \mathcal{F}\})$. Thus, *we need not distinguish between semiuniform spaces and principal semiuniform limit spaces.*
b) **SUnif** can be embedded into **SPsU** as a bireflective subconstruct:
α) $(X, \mathcal{W}) \in |\mathbf{SUnif}|$ implies $(X, [\mathcal{W}]) \in |\mathbf{SPsU}|$: Let $\mathcal{F} \in F(X \times X)$ and $\mathcal{W} \not\subset \mathcal{F}$, i.e. there is some $W \in \mathcal{W}$ such that $W \notin \mathcal{F}$. Thus, $((X \times X)\backslash W) \cap F \neq \emptyset$ for all $F \in \mathcal{F}$. Hence $\{((X \times X)\backslash W) \cap F : F \in \mathcal{F}\}$ is a base for a filter $\mathcal{F}'$ on $X \times X$. Let $\mathcal{U}$ be an ultrafilter on $X \times X$ which is finer that $\mathcal{F}'$. Then $\mathcal{F} \subset \mathcal{F}' \subset \mathcal{U}$ and $\mathcal{U} \notin [\mathcal{W}]$ since $(X \times X)\backslash W \in \mathcal{U}$. Thus, $(X, [\mathcal{W}])$ fulfills (Ps).
β) Let $(X, \mathcal{J}_X) \in |\mathbf{SPsU}|$ and put $\mathcal{W}_{\mathcal{J}_X} = \bigcap_{\mathcal{F} \in \mathcal{J}_X} \mathcal{F}$. Then $(X, [\mathcal{W}_{\mathcal{J}_X}])$ is a principal semiuniform limit space (= semiuniform space) and $1_X : (X, \mathcal{J}_X) \longrightarrow (X, [\mathcal{W}_{\mathcal{J}_X}])$ is the desired bireflection of $(X, \mathcal{J}_X)$ with respect to **PrSULim** ($\cong$ **SUnif**) which is easily checked.

3) a) **SPsU** is a bireflective subconstruct of **SULim**: Let $(X, \mathcal{J}_X) \in |\mathbf{SULim}|$. Then $(X, (\mathcal{J}_X)_{Ps})$ is a semi–pseudouniform space provided that $(\mathcal{J}_X)_{Ps} = \{\mathcal{F} \in F(X \times X) : \mathcal{U} \in \mathcal{J}_X$ for all ultrafilters $\mathcal{U}$ on $X \times X$ with $\mathcal{F} \subset \mathcal{U}\}$ and $1_X : (X, \mathcal{J}_X) \longrightarrow (X, (\mathcal{J}_X)_{Ps})$ is the desired bireflection of $(X, \mathcal{J}_X)$.
b) Since **SULim** is bireflective in **SUConv** (cf. 2.3.2.3.), it follows from a) that **SPsU** is also bireflective in **SUConv**.

6.3.3 Proposition. *Let $(X, \mathcal{J}_X)$ be a semi–pseudouniform space and $A \subset X$ a non–empty precompact subset. Then the filter (Δ_A) on $X \times X$ generated by the diagonal Δ_A in $A \times A$ belongs to $\mathcal{J}_X$.*

Proof. Let $\mathcal{W}$ be an ultrafilter on $X \times X$ finer than (Δ_A). Then there is an ultrafilter $\mathcal{U}$ on X containing A such that $\Delta_{\mathcal{U}} = \mathcal{W}$, where $\Delta_{\mathcal{U}}$ is the filter on $X \times X$ generated by the filter $\Delta'_{\mathcal{U}} = \{\Delta_U : U \in \mathcal{U}\}$ on Δ_X, namely $\mathcal{U} = \{U \subset X : \Delta_U \in \mathcal{W}\}$ (note: since $\mathcal{U}$ is an ultrafilter, $\Delta_{\mathcal{U}} \subset \mathcal{W}$ is an ultrafilter too, because $\Delta'_{\mathcal{U}}$ is an ultrafilter). Since $\mathcal{U} \times \mathcal{U} \subset \Delta_{\mathcal{U}}$ and A is a precompact subset of X, $\Delta_{\mathcal{U}} = \mathcal{W}$ belongs to $\mathcal{J}_X$. Thus, since $(X, \mathcal{J}_X)$ is semi–pseudouniform, (Δ_A) belongs to $\mathcal{J}_X$.

6.3.4 Corollary. *Every precompact semi–pseudouniform space is diagonal.*

6.3.5 Proposition. *Let $\mathbf{X} = (X, \mathcal{J}_X)$ and $\mathbf{Y} = (Y, \mathcal{J}_Y)$ be semiuniform convergence spaces. Then $([\mathbf{X}, \mathbf{Y}], \mathcal{J}_{X,Y})$ is a semi–pseudouniform space provided that $\mathbf{Y}$ is a semi–pseudouniform space. The inverse implication is valid provided that $X \neq \emptyset$.*

Proof. 1) Let $\mathbf{Y}$ be semi–pseudouniform and $\Phi \in F([\mathbf{X}, \mathbf{Y}] \times [\mathbf{X}, \mathbf{Y}])$ such that $\Phi \notin \mathcal{J}_{X,Y}$. Then there is some $\mathcal{F} \in \mathcal{J}_X$ with $\Phi(\mathcal{F}) = (e_{X,Y} \times e_{X,Y})(\mathcal{F} \times \Phi) \notin \mathcal{J}_Y$. Since $\mathcal{J}_Y$ fulfills (Ps) there is an ultrafilter $\mathcal{U}$ on $Y \times Y$ such that $\mathcal{U} \notin \mathcal{J}_Y$ and $\Phi(\mathcal{F}) \subset \mathcal{U}$. Furthermore, there is an ultrafilter $\mathcal{U}'$ on $(X \times [\mathbf{X}, \mathbf{Y}]) \times (X \times [\mathbf{X}, \mathbf{Y}])$

such that $\mathcal{F} \times \Phi \subset \mathcal{U}'$ and $e_{X,Y} \times e_{X,Y}(\mathcal{U}') = \mathcal{U}$ (cf. 0.2.3.26. and note that we do not distinguish between $(X \times Z) \times (X \times Z)$ and $(X \times X) \times (Z \times Z)$ where $Z = [\mathbf{X}, \mathbf{Y}]$). Let $p_X : X \times Z \longrightarrow X$ (resp. $p_Z : X \times Z \longrightarrow Z$) denote the projection and put $p_X \times p_X(\mathcal{U}') = \mathcal{V}$ and $p_Z \times p_Z(\mathcal{U}') = \Omega$. Then $\mathcal{V}$ and Ω are ultrafilters such that $\mathcal{F} \subset \mathcal{V}$, $\Phi \subset \Omega$ and $\mathcal{V} \times \Omega \subset \mathcal{U}'$, i.e. $(e_{X,Y} \times e_{X,Y})(\mathcal{V} \times \Omega) \subset (e_{X,Y} \times e_{X,Y})(\mathcal{U}') = \mathcal{U} \notin \mathcal{J}_Y$ with $\mathcal{V} \in \mathcal{J}_X$. Thus, $\Omega(\mathcal{V}) \notin \mathcal{J}_Y$, i.e. $\Omega \notin \mathcal{J}_{X,Y}$.
2) Let $p_Y : \mathbf{X} \times \mathbf{Y} \longrightarrow \mathbf{Y}$ be the projection. Since **SUConv** is cartesian closed, $\overline{p_Y} : \mathbf{Y} \longrightarrow ([\mathbf{X}, \mathbf{Y}], \mathcal{J}_{X,Y})$ is uniformly continuous. Let $x \in X$. Then $e_{X,Y}(x, \cdot)$: $([\mathbf{X}, \mathbf{Y}], \mathcal{J}_{X,Y}) \longrightarrow \mathbf{Y}$ is uniformly continuous and obviously, $e_{X,Y}(x, \cdot) \circ \overline{p_Y} = 1_Y$ which implies that $\overline{p_Y}$ is an embedding. Thus, it follows from 6.3.2. 3) b) that $\mathbf{Y}$ is semi–pseudouniform.

6.3.6 Remarks. 1) It follows from 6.3.2. 3) b) and 6.3.5. that **SPsU** is cartesian closed and *the natural function spaces in* **SPsU** *are formed as in* **SUConv** (cf. 3.1.8.).
2) If in the above proposition $\mathbf{X}$ and $\mathbf{Y}$ are uniform limit spaces, $([\mathbf{X}, \mathbf{Y}], \mathcal{J}_{X,Y})$ is a uniform limit space (cf. exercise 25)). Thus, it follows from 6.3.5. that **PsU** *is cartesian closed* since it is bireflective in **ULim** (restrict the bireflector $\mathcal{R}$: **SULim** $\longrightarrow$ **SPsU** to **ULim**!).

6.3.7 Definition. Let $\mathbf{X} = (X, \mathcal{J}_X)$ and $\mathbf{Y} = (Y, \mathcal{J}_Y)$ be semiuniform convergence spaces and $M \subset [\mathbf{X}, \mathbf{Y}]$. Then M is called *uniformly equicontinuous* provided that $(\Delta_M) \in \mathcal{J}_{X,Y}$ (i.e. for each $\mathcal{F} \in \mathcal{J}_X$ the filter $(\Delta_M)(\mathcal{F})$ generated by $\{\Delta_M(F) : F \in \mathcal{F}\}$ with $\Delta_M(F) = \{(f(a), f(b)) : f \in M, (a, b) \in F\}$ belongs to $\mathcal{J}_Y$), if the filter (Δ_M) exists.

6.3.8 Remarks. 1) *Let* $\mathbf{X}$ *and* $\mathbf{Y}$ *be uniform spaces* (= principal uniform limit spaces). *Then a subset* $M \subset [\mathbf{X}, \mathbf{Y}]$ *is uniformly equicontinuous iff for each entourage* V *of* $\mathbf{Y}$ *there is an entourage* U *of* $\mathbf{X}$ *such that for each* $f \in M$, $(x, x') \in U$ *implies* $(f(x), f(x')) \in V$.
2) *Let* $\mathbf{X}$ *be a symmetric topological space* (= topological semiuniform convergence space) *and* $\mathbf{Y}$ *a uniform space.* Then $[\mathbf{X}, \mathbf{Y}]$ is equal to $C(\mathbf{X}, \mathbf{Y})$. *A subset* $M \subset C(\mathbf{X}, \mathbf{Y})$ *is uniformly equicontinuous iff it is equicontinuous in the usual sense, i.e. for each* $x_0 \in X$ *and each entourage* V *of* $\mathbf{Y}$ *there is some neighborhood* $U(x_0, V)$ *of* x_0 *in* $\mathbf{X}$ *such that for each* $x \in U(x_0, V)$ *and each* $f \in M$, $(f(x_0), f(x)) \in V$.
3) If $\mathbf{X}$ is a topological space (resp. uniform space) and $\mathbf{Y}$ a uniform space, then *equicontinuity (resp. uniform equicontinuity) of a subset* M *of* $F(X, Y) = Y^X$ can be defined similarly to 2) (resp. 1)) where X (resp. Y) denotes the underlying set of $\mathbf{X}$ (resp. $\mathbf{Y}$). But obviously, if $M \subset F(X, Y)$ is equicontinuous (resp. uniformly equicontinuous), then each $f \in M$ is continuous (resp. uniformly continuous).

6.3.9 Proposition. *Let* $\mathbf{X}$ *be a topological space (resp. uniform space) and* $\mathbf{Y}$ *a uniform space. A subset* M *of* $F(X, Y)$ *is equicontinuous (resp. uniformly* *equicontinuous) iff the map $g : \mathbf{X} \longrightarrow F_u(M,Y)$ defined by $g(x)(f) = f(x)$ is continuous (resp. uniformly continuous).*

Proof. By 6.1.6. the sets $W(V)$, where V is an entourage of $\mathbf{Y}$, form a base for the uniformity of $F_u(M,Y)$. In particular, $(g(x),g(x')) \in W(V)$ iff $(g(x)(f),g(x')(f)) \in V$ for each $f \in M$. Obviously, g is continuous at each $x_0 \in X$ (resp. uniformly continuous) iff for each entourage V of $\mathbf{Y}$ there is a neighborhood $U(x_0,V)$ of x_0 in $\mathbf{X}$ (resp. an entourage U of $\mathbf{X}$) such that $x \in U(x_0,V)$ (resp. $(x,x') \in U$) implies $(g(x_0)(f),\ g(x)(f)) \in V$ for each $f \in M$ [i.e. $g(x) \in W(V)(g(x_0))$] (resp. $(g(x)(f),g(x')(f)) \in V$ for each $f \in M$ [i.e. $(g(x),g(x')) \in W(V)$]). By definition, this is equivalent to the equicontinuity (resp. uniform equicontinuity) of M.

6.3.10 Corollary. *Let $\mathbf{X}$ be a compact Hausdorff space and $\mathbf{Y}$ a uniform space. Then every equicontinuous subset M of $F(X,Y)$ is uniformly equicontinuous.*

Proof. Since $\mathbf{X}$ is compact Hausdorff, each continuous map from $\mathbf{X}$ into $F_u(M,Y)$ is uniformly continuous (cf. 4.3.2.24. and 4.3.2.25.). Thus the corollary follows from the above proposition.

6.3.11 Remark. Obviously, uniform equicontinuity implies equicontinuity. The converse is not true (cf. exercise 72)) unless $\mathbf{X}$ is compact Hausdorff according to the above corollary.

6.3.12 Proposition. *Let $\mathbf{X}$ be a topological space (resp. uniform space) and $\mathbf{Y}$ a uniform space. A subset M of $F(X,Y)$ is equicontinuous (resp. uniformly equicontinuous) iff the closure $\overline{M}$ of M in $F(X,Y)$ supplied with the uniform structure of simple convergence is equicontinuous (resp. uniformly equicontinuous).*

Proof. 1) "$\Longleftarrow$". Obvious.
2) "$\Longrightarrow$". Let V be a closed entourage of $\mathbf{Y}$ and $x_0 \in X$. By assumption there is some neighborhood U_{x_0} of x_0 in $\mathbf{X}$ (resp. an entourage U of $\mathbf{X}$) such that for each $f \in M$, $x \in U_{x_0}$ (resp. $(x,x') \in U$) implies $(f(x_0),f(x)) \in V$ (resp. $(f(x),f(x')) \in V$). Since V is closed, the set $N = \{f \in Y^X\colon\ (f(x_0),f(x)) \in V$ for each $x \in U_{x_0}\}$ (resp. $N = \{f \in Y^X : (f(x),f(x')) \in V$ for each $(x,x') \in U\}$) is closed in the uniform product space $\mathbf{Y}^X$ [note: For each $x \in X$, the x-th projection $p_x : \mathbf{Y}^X \longrightarrow \mathbf{Y}$ is continuous; further $h : \mathbf{Y}^X \longrightarrow \mathbf{Y}^X \times \mathbf{Y}^X$ defined by $h(f) = (f,f)$ is continuous and for any $u,v \in X$, $h^{-1}[(p_u \times p_v)^{-1}[V]] = \{f \in Y^X\colon (f(u),f(v)) \in V\}$ is closed; consequently $N = \bigcap\{\{f \in Y^X : f(x_0),f(x) \in V\} : x \in U_0\}$ (resp. $N = \bigcap\{\{f \in Y^X : (f(x),f(x')) \in V\}\colon (x,x') \in U\}$) is closed.]. Obviously, $M \subset N$, which implies $\overline{M} \subset \overline{N} = N$, i.e. $\overline{M}$ is equicontinuous (resp. uniformly equicontinuous).

6.3.13 Proposition. *Let $\mathbf{X}$ be a topological space and $\mathbf{Y}$ a uniform space. If*

M is an equicontinuous subset of $C(\mathbf{X}, \mathbf{Y})$, then the uniform structure of compact convergence on M coincides with the uniform structure of simple convergence on M.

Proof. By 6.1.15. 3) it suffices to prove that the uniform structure of simple convergence on M is finer than the uniform structure of compact convergence on M. In other words: It suffices to prove that for each entourage V of $\mathbf{Y}$ and each compact subset K of $\mathbf{X}$ there is an entourage W of $\mathbf{Y}$ and a finite subset F of X such that

(*) $(f,g) \in M \times M$ and$(f(x), g(x)) \in W$ for each $x \in F$

implies

(**) $(f(x), g(x)) \in V$ for each $x \in K$.

[i.e. $(\bigcap_{x \in F}(p_x \times p_x)^{-1}[W]) \cap (M \times M) \subset W(K,V) \cap (M \times M)$, where $p_x : \mathbf{Y}^X \longrightarrow \mathbf{Y}$ denotes the x–th projection].

Let W be a symmetric entourage of $\mathbf{Y}$ such that $W^3 \subset V$. Since M is equicontinuous, for each $x \in X$ there is an open neighborhood U_x of x in $\mathbf{X}$ such that for each $f \in M$ and each $x' \in U_x$, $(f(x), f(x')) \in W$. Furthermore, since K is compact, there is a finite subset $F = \{x_1, \dots, x_n\}$ of X such that $K \subset \bigcup_{i=1}^{n} U_{x_i}$. Let (*) be satisfied. For each $x \in K$ there is some $i \in \{1, \dots, n\}$ such that $x \in U_{x_i}$, which implies $(f(x_i), f(x)) \in W$ and $(g(x_i), g(x)) \in W$. Since, by assumption, $(f(x_i), g(x_i)) \in W$, it follows that $(f(x), g(x)) \in W^3 \subset V$.

6.3.14 Corollary. *Let $\mathbf{X}$ be a topological space, $\mathbf{Y}$ a uniform space and $M \subset C(\mathbf{X}, \mathbf{Y})$ equicontinuous. Then the closure $\overline{M}$ of M in $F_s(X,Y)$ coincides with the closure of M in $C_c(\mathbf{X}, \mathbf{Y})$.*

Proof. By 6.3.12., $\overline{M}$ is equicontinuous and consequently, $\overline{M} \subset C(\mathbf{X}, \mathbf{Y})$. By 6.3.13., the topology of simple convergence on $\overline{M}$ coincides with the topology of compact convergence on $\overline{M}$. Thus, the corollary is proved.

6.3.15 Definition. Let **Δ–SUConv** be the construct of diagonal semiuniform convergence spaces (and uniformly continuous maps). If X is a set and $(Y, \mathcal{J}_Y) \in |\mathbf{\Delta{-}SUConv}|$, then the **Δ–SUConv**-structure

$$\mathcal{J}^u_{Y^X} = \{\Phi \in F(Y^X \times Y^X) : \Phi((\Delta_X)) \in \mathcal{J}_Y \text{ if the filter } (\Delta_X) \text{ exists}\}$$

is called the *structure of uniform convergence* on Y^X.

6.3.16 Remarks. 1) Let $(Y, \mathcal{V})$ be a uniform space and $(Y, \mathcal{J}_Y)$ the corresponding principal uniform limit space, i.e. $\mathcal{J}_Y = [\mathcal{V}]$. Then, obviously, $(Y, [\mathcal{V}])$ is diagonal, i.e. a Cook–Fischer space. If X is a set, then the **Δ–SUConv**–structure $\mathcal{J}^u_{Y^X}$ is a **PrULim**–structure which is generated by the (usual) uniform structure $\mathcal{U}$ of uniform convergence on $Y^X = F(X,Y)$ (cf. 6.1.6. for the definition of $\mathcal{U}$); namely, it is easily checked that $\Phi((\Delta_X)) \supset \mathcal{V}$ iff $\Phi \supset \{W(V) : V \in \mathcal{V}\}$, i.e. $\mathcal{J}^u_{Y^X} = [\mathcal{U}]$. Thus, *the* **Δ–SUConv**–*structure of uniform convergence as* *defined above is the natural generalization of the uniform structure of uniform convergence.*

2) *The* $\boldsymbol{\Delta}$**–SUConv**–*structure (including the uniform structure) of uniform convergence can be derived from the natural function space structure in the realm of semiuniform convergence spaces*; namely, if the set X in 6.3.15. is endowed with the discrete $\boldsymbol{\Delta}$**–SUConv**–structure $[(\Delta_X)] = \{\mathcal{F} \in F(X \times X) : \Delta_X \in \mathcal{F}\}$, then $[(X,[(\Delta_X)]),(Y,\mathcal{J}_Y)]_{\boldsymbol{\Delta}\text{-}\mathbf{SUConv}} = [(X,[(\Delta_X)]),(Y,\mathcal{J}_Y)]_{\mathbf{SUConv}} = Y^X$ and the uniformly continuous **SUConv**–structure $\mathcal{J}_{X,Y}$ coincides with the structure of uniform convergence on Y^X.

3) Let X be a set and $\mathcal{J}_X = \{\dot{x} \times \dot{x} : x \in X\}$, i.e. $\mathcal{J}_X$ is the discrete **SUConv**–structure on X. If $(Y,\mathcal{J}_Y)$ is a semiuniform convergence space, then $[(X,\mathcal{J}_X),(Y,\mathcal{J}_Y)] = Y^X$ and for the uniformly continuous **SUConv**–structure $\mathcal{J}_{X,Y}$ on this set one obtains

$$\mathcal{J}_{X,Y} = \{\Phi \in F(Y^X \times Y^X) : (p_x \times p_x)(\Phi) \in \mathcal{J}_Y \quad \text{for each} \quad x \in X\}$$

where $p_x : Y^X \longrightarrow Y$ denotes the x–th projection (note: $\Phi(\dot{x} \times \dot{x}) = (p_x \times p_x)(\Phi)$ for each $\Phi \in F(Y^X \times Y^X)$), i.e. $\mathcal{J}_{X,Y}$ is the product structure (in **SUConv**) on Y^X. It is called the **SUConv**–*structure of simple convergence* and is denoted by $\mathcal{J}^s_{Y^X}$ (note: If (Y^X, q_s) is the underlying Kent convergence of $(Y^X, \mathcal{J}^s_{Y^X})$, then $(\Phi, f) \in q_s$ iff $(p_x(\Phi), f(x)) \in q_{\gamma_{\mathcal{J}_Y}}$ for each $x \in X$). If $(Y,\mathcal{J}_Y)$ is a diagonal semiuniform convergence space, then $(Y^X, \mathcal{J}^s_{Y^X})$ is also diagonal, since, obviously, $\boldsymbol{\Delta}$**–SUConv** is closed under formation of initial structures in **SUConv** (i.e. $\boldsymbol{\Delta}$–**SUConv** is a bireflective subconstruct of **SUConv**). Then

$$\mathcal{J}^u_{Y^X} \subset \mathcal{J}^s_{Y^X} \,,$$

i.e. *the* $\boldsymbol{\Delta}$**–SUConv**–*structure of uniform convergence is finer than the* $\boldsymbol{\Delta}$–**SUConv**–*structure of simple convergence* (note: $(\Delta_X) \subset \dot{x} \times \dot{x}$ for each $x \in X$). *Furthermore,* $\mathcal{J}^s_{Y^X}$ *has been derived from the natural function space structure in the realm of semiuniform convergence spaces. If* $(Y,\mathcal{J}_Y)$ *is a uniform space (= principal uniform limit space), then* $\mathcal{J}^s_{Y^X}$ *is uniform, since* **Unif** *is bireflective in* **SUConv**, *and it is generated by the (usual) uniform structure of simple convergence on* Y^X.

4) If X is a set, $(Y,\mathcal{J}_Y)$ a diagonal semiuniform convergence space and $\mathcal{S} \subset \mathcal{P}(X)$, then the $\boldsymbol{\Delta}$**–SUConv**–structure $\mathcal{J}^{\mathcal{S}}_{Y^X}$ *of uniform convergence on members of* $\mathcal{S}$ can be defined to be the initial **SUConv**–structure on Y^X with respect to $(\varphi_S : Y^X \longrightarrow (Y^S, \mathcal{J}^u_{Y^S}))_{S \in \mathcal{S}}$ where $\varphi_S(f) = f \mid S$. Obviously,

$$\mathcal{J}^u_{Y^X} \subset \mathcal{J}^{\mathcal{S}}_{Y^X} \subset \mathcal{J}^s_{Y^X}$$

provided that $\mathcal{S}$ is a covering of X.

If $(Y,\mathcal{V})$ is a uniform space and $(Y,\mathcal{J}_Y)$ its corresponding semiuniform convergence space, then $\mathcal{J}^{\mathcal{S}}_{Y^X}$ is a **PrULim**–structure which is generated by the usual uniform structure $\mathcal{U} \mid \mathcal{S}$ of uniform convergence on members of $\mathcal{S}$ (cf. 6.1.13. for the definition of $\mathcal{U} \mid \mathcal{S}$).

6.3.17 Proposition. *Let* $\mathbf{X} = (X, \mathcal{J}_X)$ *and* $\mathbf{Y} = (Y, \mathcal{J}_Y)$ *be diagonal semiuniform convergence spaces,* $\mathcal{J}_{X,Y}$ *the uniformly continuous* **SUConv**-*structure on* $[\mathbf{X}, \mathbf{Y}]$ *and* $\mathcal{J}^u_{X,Y}$ *the structure of uniform convergence on* $[\mathbf{X}, \mathbf{Y}]$ *(i.e.* $\mathcal{J}^u_{X,Y}$ *is induced by* $\mathcal{J}^u_{Y^X}$*). Then*

$$1_{[\mathbf{X},\mathbf{Y}]} : ([\mathbf{X}, \mathbf{Y}], \mathcal{J}_{X,Y}) \longrightarrow ([\mathbf{X}, \mathbf{Y}], \mathcal{J}^u_{X,Y})$$

is uniformly continuous, i.e. $\mathcal{J}_{X,Y}$ *is finer than* $\mathcal{J}^u_{X,Y}$.

Proof. Since $\mathbf{X}$ is diagonal, $(\Delta_X) \in \mathcal{J}_X$ and, by definition of $\mathcal{J}_{X,Y}$, $\Phi((\Delta_X)) \in \mathcal{J}_Y$ for each $\Phi \in \mathcal{J}_{X,Y}$. If $i : [\mathbf{X}, \mathbf{Y}] \longrightarrow Y^X$ denotes the inclusion map, then $(i \times i(\Phi))((\Delta_X)) = \Phi((\Delta_X))$, and consequently, $(i \times i)(\Phi) \in \mathcal{J}^u_{Y^X}$. Thus, $\Phi \in \mathcal{J}^u_{X,Y}$ for each $\Phi \in \mathcal{J}_{X,Y}$.

6.3.18 Remarks. 1) Since $\mathbf{\Delta}$-**SUConv** is bireflective in **SUConv** and $\mathcal{J}^u_{Y^X}$ is diagonal, $\mathcal{J}^u_{X,Y}$ is also diagonal. On the other hand, $([\mathbf{X}, \mathbf{Y}], \mathcal{J}_{X,Y})$ is generally not diagonal, e.g. if $\mathbf{X} = \mathbf{Y} = I\!R_u$, then $([I\!R_u, I\!R_u], \mathcal{J}_{I\!R,I\!R})$ is a uniform limit space which is not diagonal (cf. 4.2.2.8.). Thus, $\mathcal{J}_{X,Y} \neq \mathcal{J}^u_{X,Y}$ in general.
2) *If* $\mathbf{X}$ *and* $\mathbf{Y}$ *are Cook-Fischer spaces, then* $q_{\gamma_{\mathcal{J}_{X,Y}}} = q_{\gamma_{\mathcal{J}^u_{X,Y}}}$ *though* $\mathcal{J}_{X,Y}$ *might be unequal to* $\mathcal{J}^u_{X,Y}$. Since $q_{\gamma_{\mathcal{J}_{X,Y}}} \subset q_{\gamma_{\mathcal{J}^u_{X,Y}}}$ follows immediately from 6.3.17., it remains to show that $q_{\gamma_{\mathcal{J}^u_{X,Y}}} \subset q_{\gamma_{\mathcal{J}_{X,Y}}}$. Let $(\mathcal{F}, f) \in q_{\gamma_{\mathcal{J}^u_{X,Y}}}$, i.e. $\mathcal{F} \times \dot{f} \in \mathcal{J}^u_{X,Y}$(($[\mathbf{X}, \mathbf{Y}], \mathcal{J}^u_{X,Y}$) is a Cook-Fischer space!). Thus, $(\mathcal{F} \times \dot{f})((\Delta_X)) \in \mathcal{J}_Y$. In order to prove that $(\mathcal{F}, f) \in q_{\gamma_{\mathcal{J}_{X,Y}}}$, i.e. $\mathcal{F} \times \dot{f} \in \mathcal{J}_{X,Y}$(($[\mathbf{X}, \mathbf{Y}], \mathcal{J}_{X,Y}$) is a uniform limit space!), let $\mathcal{G} \in \mathcal{J}_X$. Since $f : \mathbf{X} \longrightarrow \mathbf{Y}$ is uniformly continuous, $(f \times f)(\mathcal{G}) = (\dot{f} \times \dot{f})(\mathcal{G}) \in \mathcal{J}_Y$. Consequently, it follows from $(\mathcal{F} \times \dot{f})(\mathcal{G}) = (\mathcal{F} \times \dot{f})(\mathcal{G} \circ (\Delta_X)) = ((\dot{f} \times \dot{f}) \circ (\mathcal{F} \times \dot{f}))(\mathcal{G} \circ (\Delta_X)) \supset (\dot{f} \times \dot{f})(\mathcal{G}) \circ (\mathcal{F} \times \dot{f})((\Delta_X))$ that $(\mathcal{F} \times \dot{f})(\mathcal{G}) \in \mathcal{J}_Y$. Hence, $\mathcal{F} \times \dot{f} \in \mathcal{J}_{X,Y}$.

6.3.19 Theorem (Necessary conditions for precompactness in $([\mathbf{X}, \mathbf{Y}], \mathcal{J}_{X,Y})$). *Let* $\mathbf{X} = (X, \mathcal{J}_X)$ *be a semiuniform convergence space,* $\mathbf{Y} = (Y, \mathcal{J}_Y)$ *a semi-pseudouniform space and* M *a precompact subset of* $([\mathbf{X}, \mathbf{Y}], \mathcal{J}_{X,Y})$. *Then the following are satisfied:*
(1) M *is uniformly equicontinuous,*
(2) $M(x) = \{f(x) : f \in M\} \subset Y$ *is precompact for each* $x \in X$.

Proof. Since $\mathbf{Y}$ is semi-pseudouniform, it follows from 6.3.5. that $([\mathbf{X}, \mathbf{Y}], \mathcal{J}_{X,Y})$ is semi-pseudouniform. Thus, by 6.3.3., M is uniformly equicontinuous. For each $x \in X$, the restriction map $e_{X,Y}|_{\{x\} \times M} : \{x\} \times M \longrightarrow Y$ of the evaluation map $e_{X,Y} : X \times [\mathbf{X}, \mathbf{Y}] \longrightarrow Y$ is uniformly continuous. Since$\{x\}$ and M are precompact, $\{x\} \times M$ is precompact (cf. 4.3.2.2). Consequently, $(e_{X,Y}|_{\{x\} \times M})[\{x\} \times M] = M(x) \subset Y$ is precompact (cf. 4.3.2.10.).

6.3.20 Definition. A semi-pseudouniform space (resp. pseudouniform space) $\mathbf{Y} = (Y, \mathcal{J}_Y)$ is called a *semi-Ascoli space* (resp. *Ascoli space*) provided that for every precompact semiuniform convergence space $\mathbf{X} = (X, \mathcal{J}_X)$ and every uniformly equicontinuous set $M \subset [\mathbf{X}, \mathbf{Y}]$ the subspace structure of M with respect to the product structure of Y^X coincides with the subspace structure of M in $([\mathbf{X},\mathbf{Y}],\mathcal{J}_{X,Y})$.

6.3.21 Remark. Obviously, a semi–pseudouniform space (resp. pseudouniform space) $\mathbf{Y}=(Y,\mathcal{J}_Y)$ is a semi-Ascoli space (resp. Ascoli space) iff for every precompact semiuniform convergence space $\mathbf{X}=(X,\mathcal{J}_X)$ and every uniformly equicontinuous set $M\subset[\mathbf{X},\mathbf{Y}]$ the subspace structure of M in $([\mathbf{X},\mathbf{Y}],\mathcal{J}_{X,Y})$ contains the subspace structure of M with respect to the product structure of Y^X (i.e. *the structure of simple convergence on* M).

6.3.22 Proposition. *Let* $\mathbf{Y}=(Y,\mathcal{J}_Y)$ *be a pseudouniform space. Then the following are equivalent:*
(1) $\mathbf{Y}$ *is an Ascoli space,*
(2) *For a map* $f:X\times Z\longrightarrow Y$ *such that the filter* $f\times f((\Delta_X)\times(\mathcal{U}\times\mathcal{U}))\in\mathcal{J}_Y$ *for every ultrafilter* $\mathcal{U}$ *on* Z, *and for a filter* Φ *on* $X\times X$ *such that the filter* $f\times f(\Phi\times(\dot{z}\times\dot{z}))\in\mathcal{J}_Y$ *for every* $z\in Z$, $f\times f(\Phi\times(\Delta_Z))\in\mathcal{J}_Y$ *(whenever* X *and* Z *are non-empty).*

Proof. (1) $\Longrightarrow$ (2). Provide X in (2) with the discrete **PrULim**–structure $\mathcal{J}_X=[(\Delta_X)]$, and provide Z with the initial **PsU**–structure $\mathcal{J}_Z$ with respect to $f^*:Z\longrightarrow[\mathbf{X},\mathbf{Y}]$ defined by $f^*(z)(x)=f(x,z)$ where $\mathbf{X}=(X,\mathcal{J}_X)$ and $\mathbf{Y}=(Y,\mathcal{J}_Y)$, i.e. $\mathcal{J}_Z=\{\mathcal{F}\in F(Z\times Z):f^*\times f^*(\mathcal{F})\in\mathcal{J}_{X,Y}\}$.

Then $\mathbf{Z}=(Z,\mathcal{J}_Z)$ is precompact by assumption (note: for each ultrafilter $\mathcal{U}$ on Z, $((f^*\times f^*)(\mathcal{U}\times\mathcal{U}))(\Delta_X))=f\times f((\Delta_X)\times(\mathcal{U}\times\mathcal{U}))\in\mathcal{J}_Y$, i.e. $\mathcal{U}\times\mathcal{U}\in\mathcal{J}_Z$). Let $h:X\longrightarrow[\mathbf{Z},\mathbf{Y}]$ be defined by $h(x)(z)=f(x,z)$. It is easily checked that $h[X]\subset[\mathbf{Z},\mathbf{Y}]$ is uniformly equicontinuous. Furthermore, $h\times h(\Phi)$ is uniform with respect to the structure of simple convergence (note: $(p_z\times p_z)(h\times h(\Phi))=f\times f(\Phi\times(\dot{z}\times\dot{z}))\in\mathcal{J}_Y$ for each $z\in Z$ by assumption, where $p_z:\mathbf{Y}^Z\longrightarrow\mathbf{Y}$ denotes the z–th projection).
By 6.3.4., $(\Delta_Z)\in\mathcal{J}_Z$. If $\mathbf{Y}$ is an Ascoli space, then $ev\times ev(h\times h(\Phi)\times(\Delta_Z))=f\times f(\Phi\times(\Delta_Z))\in\mathcal{J}_Y$, since the evaluation map $ev:[\mathbf{Z},\mathbf{Y}]\times\mathbf{Z}\longrightarrow\mathbf{Y}$ is uniformly continuous.

(2) $\Longrightarrow$ (1). Let $\mathbf{Y}=(Y,\mathcal{J}_Y)$ be a pseudouniform space, $\mathbf{X}=(X,\mathcal{J}_X)$ a precompact semiuniform convergence space and $M\subset[\mathbf{X},\mathbf{Y}]$ a uniformly equicontinuous set. A filter $\mathcal{F}\in F(M\times M)$ is pointwise uniform (i.e. $\mathcal{F}$ belongs to the subspace structure of M with respect to the product structure of Y^X) iff $p_x\times p_x(\mathcal{F})=ev\times ev(\mathcal{F}\times(\dot{x}\times\dot{x}))\in\mathcal{J}_Y$ for each $x\in X$, where $p_x:Y^X\longrightarrow Y$ denotes the x–th projection and $ev:M\times X\longrightarrow Y$ is the evaluation map. In order to prove that $\mathbf{Y}$ is an Ascoli space, it suffices to verify that each pointwise uniform $\mathcal{F}\in F(M\times M)$ is uniform, i.e. for each $\Phi\in\mathcal{J}_X$, $\mathcal{F}(\Phi)=ev\times ev(\mathcal{F}\times\Phi)\in\mathcal{J}_Y$. Since M is uniformly equicontinuous, $(\Delta_M)\in\mathcal{J}_{X,Y}$, i.e. for each $\mathcal{G}\in\mathcal{J}_X$, $(\Delta_M)(\mathcal{G})=ev\times ev((\Delta_M)\times\mathcal{G})\in\mathcal{J}_Y$. Consequently,

(α) $ev\times ev\,((\Delta_M)\times\Phi)\in\mathcal{J}_Y$

and, since $\mathbf{X}$ is precompact,

(β) $ev \times ev\,((\Delta_M) \times (\mathcal{U} \times \mathcal{U})) \in \mathcal{J}_Y$ for each ultrafilter $\mathcal{U}$ on X.

Furthermore, by assumption (put $f = ev$),

(γ) $ev \times ev\,(\mathcal{F} \times (\Delta_X)) \in \mathcal{J}_Y$.

Since the case that $X = \emptyset$ or $M = \emptyset$ is trivial, it may be assumed (without loss of generality) that $X \cap M \neq \emptyset$. Then $\mathcal{F} \times \Phi = ((\Delta_M) \times \Phi) \circ (\mathcal{F} \times (\Delta_X))$ and one obtains

(δ) $ev \times ev\,((\Delta_M) \times \Phi) \circ (ev \times ev)\,(\mathcal{F} \times (\Delta_X)) \subset ev \times ev\,(\mathcal{F} \times \Phi)$.

By (α) and (γ), it follows from (δ), that $ev \times ev\,(\mathcal{F} \times \Phi) \in \mathcal{J}_Y$ since $\mathbf{Y}$ is a uniform limit space.

6.3.23 Proposition. 1) *The construct* **Asc** *of Ascoli spaces (and uniformly continuous maps) is bireflective in* **SUConv** (resp. **PsU**).
2) *If* $\mathbf{X} = (X, \mathcal{J}_X)$ *is a uniform limit space and* $\mathbf{Y} = (Y, \mathcal{J}_Y)$ *a principal uniform limit space (= uniform space), then* $([\mathbf{X}, \mathbf{Y}], \mathcal{J}_{X,Y})$ *is an Ascoli space.*

Proof. 1) Let $(g_i : (X, \mathcal{J}_X) \longrightarrow (X_i, \mathcal{J}_{X_i}))_{i \in I}$ be an initial source in **SUConv** such that each $(X_i, \mathcal{J}_{X_i})$ is an Ascoli space. Since **PsU** is bireflective in **SUConv** (cf. 6.3.6.2) and note that **ULim** is bireflective in **SUConv**), $(X, \mathcal{J}_X)$ is pseudouniform.
Obviously, the assumptions of 6.3.22. (2) are valid for a map f and a filter Φ iff they are valid for all $g_i \circ f$, and the conclusion is valid for f iff it is valid for all $g_i \circ f$. Thus, $(Y, \mathcal{J}_X)$ is an Ascoli space, since all $(X_i, \mathcal{J}_{X_i})$ are Ascoli spaces.
2) During this proof use the following convention: If $A \subset X \times X$ and $B \subset Y \times Y$, then $A \times B$ is considered to be a subset of $(X \times Y) \times (X \times Y)$ and consequently, for a map $f : X \times Y \longrightarrow Z$, $f \times f\,[A \times B] = \{(f(x,y), f(x',y')) : (x,x') \in A$ and $(y,y') \in B\}$.
Since $\mathbf{Y}$ is uniform, it is pseudouniform. Consequently, $([\mathbf{X}, \mathbf{Y}], \mathcal{J}_{X,Y})$ is pseudouniform. In order to prove that it is an Ascoli space, 6.3.22. (2) must be proved for a map $f : S \times T \longrightarrow [\mathbf{X}, \mathbf{Y}]$ and a filter Φ on $S \times S$. Define $g : S \times T \times X \longrightarrow Y$ by $g(s,t,x) = f(s,t)(x)$.

By definition of $\mathcal{J}_{X,Y}$ it suffices to prove the following: If $\mathcal{W}$ denotes the filter of entourages of $\mathbf{Y}$, i.e. $\mathcal{J}_Y = [\mathcal{W}]$, then

(1) $g \times g\,[(\Delta_S) \times (\mathcal{U} \times \mathcal{U}) \times \Omega] \supset \mathcal{W}$ for all ultrafilter $\mathcal{U}$ on T

and

(2) $g \times g\,[\Phi \times (\dot{t} \times \dot{t}) \times \Omega] \supset \mathcal{W}$ for all $t \in T$

imply

(3) $g \times g\,[\Phi \times (\Delta_T) \times \Omega] \supset \mathcal{W}$

provided that $\Omega = \Omega^{-1} \in \mathcal{J}_X$ (note e.g.: $(f \times f(\Phi \times (\Delta_T))(\Omega) =$
$(ev \times ev)\,((f \times f(\Phi \times (\Delta_T))) \times \Omega) = g \times g\,[\Phi \times (\Delta_T) \times \Omega]$).

If (3) is not true, there is a $W \in \mathcal{W}$ such that $W \not\supset g \times g\,[A \times \Delta_T \times B]$ for all $A \in \Phi$ and $B \in \Omega$. Then $\{F_{A,B} : A \in \Phi,\ B \in \Omega\}$ is a filter base on T provided that $F_{A,B} = \{t \in T : g \times g\,[A \times \{(t,t)\} \times B] \not\subset W\}$ (obviously, $F_{A,B} \neq \emptyset$ for all $A \in \Phi$ and $B \in \Omega$ and $F_{A\cap A',B\cap B'} \subset F_{A,B} \cap F_{A',B'}$ for all $A, A' \in \Phi$ and $B, B' \in \Omega$). Consequently, there is an ultrafilter $\mathcal{U}$ on T such that all $F_{A,B}$ belong to $\mathcal{U}$. Now let $V = V^{-1} \in \mathcal{W}$ such that $V^3 \subset W$. By (1) there are $U \in \mathcal{U}$ and $B_1 \in \Omega$ such that $g \times g\,[\Delta_S \times (U \times U) \times B_1] \subset V$. Choose $t_0 \in U$. Then, by (2), there are $A \in \Phi$ and $B' \in \Omega$, i.e. $B' = B_2^{-1}$ with $B_2 \in \Omega$, such that $g \times g\,[A \times \{(t_0,t_0)\} \times B_2^{-1}] \subset V$. Put $B = B_1 \circ B_2^{-1} \circ B_1$. Then $B \in \Omega$, since $B_1 \cap B_2 \subset B$. Furthermore,

$$(\Delta_S \times (U \times U) \times B_1) \circ (A \times \{(t_0,t_0)\} \times B_2^{-1}) \circ (\Delta_S \times (U \times U) \times B_1) = A \times (U \times U) \times B$$

and

$$\begin{aligned}(g \times g)[A \times (U \times U) \times B] &\subset (g \times g)(\Delta_S \times (U \times U) \times B_1)\\ &\circ (g \times g)(A \times \{(t_0,t_0)\} \times B_2^{-1}) \circ (g \times g)(\Delta_S \times (U \times U) \times B_1)\end{aligned}$$

which implies $g \times g\,[A \times (U \times U) \times B] \subset V^3 \subset W$. Consequently, $U \cap F_{A,B} = \emptyset$, in contrast to $U \in \mathcal{U}$ and $F_{A,B} \in \mathcal{U}$. This contradiction proves that $([\mathbf{X},\mathbf{Y}], \mathcal{J}_{X,Y})$ is an Ascoli space.

6.3.24 Remarks. 1) A semiuniform convergence space $\mathbf{X} = (X, \mathcal{J}_X)$ is called a *u-space* provided that the map $i_X : \mathbf{X} \longrightarrow U^2(\mathbf{X})$ as defined under 4.2.2.4. is initial. By means of the proof of 4.2.2.5. every uniform space X is a u–space.
2) By 6.3.23. *every uniform limit space which is a u–space is an Ascoli space. In particular,* by 1), *every uniform space is an Ascoli space, but there are Ascoli spaces which are not uniform,* e.g. $([\mathbb{R}_u, \mathbb{R}_u], \mathcal{J}_{\mathbb{R},\mathbb{R}}) = U(\mathbb{R}_u)$ (cf. 4.2.2.7. and 4.2.2.8.).

6.3.25 Theorem (Sufficient conditions for precompactness in $([\mathbf{X},\mathbf{Y}], \mathcal{J}_{X,Y})$). *Let $\mathbf{X} = (X, \mathcal{J}_X)$ be a locally precompact semiuniform convergence space and $\mathbf{Y} = (Y, \mathcal{J}_Y)$ a semi–Ascoli space. Then $M \subset [\mathbf{X},\mathbf{Y}]$ is a precompact subset of $([\mathbf{X},\mathbf{Y}], \mathcal{J}_{X,Y})$ provided that the following are satisfied:*
(1) M is uniformly equicontinuous,
(2) $M(x) = \{f(x) : f \in M\} \subset Y$ is precompact for each $x \in X$.

Proof. Let N be the space $(M, (\mathcal{J}_{Y^X})_M)$, where $(\mathcal{J}_{Y^X})_M$ denotes the subspace structure of M in the product space $(Y^X, \mathcal{J}_{Y^X})$. Since $M \subset \prod_{x\in X} M(x) \subset Y^X$, N is precompact (note (2) and cf. 4.3.2.2.). Consider first the case where $\mathbf{X}$ is precompact. Since $\mathbf{Y}$ is a semi–Ascoli space and M is uniformly equicontinuous, $N = (M, (\mathcal{J}_{X,Y})_M)$ where $(\mathcal{J}_{X,Y})_M$ denotes the subspace structure of M with respect to $\mathcal{J}_{X,Y}$. Consider next any locally precompact semiuniform convergence space $\mathbf{X}$. By 6.2.2.18 the family $(j_i : \mathbf{C}_i \longrightarrow \mathbf{X})_{i\in I}$ of the inclusions of all precompact subspaces $\mathbf{C}_i = (C_i, \mathcal{J}_{C_i})$ of $\mathbf{X}$ is a final epi–sink in **SUConv**. Then the source $(h_i : ([\mathbf{X},\mathbf{Y}], \mathcal{J}_{X,Y}) \longrightarrow ([\mathbf{C}_i,\mathbf{Y}], \mathcal{J}_{C_i,Y}))_{i\in I}$ with $h_i(f) = f \circ j_i = f \mid C_i$ is an initial mono–source in **SUConv**. For each $i \in I$, $M_i = h_i[M]$ is uniformly equicontinuous (namely, since h_i is uniformly continuous, $h_i \times h_i((\Delta_M)) = (\Delta_{M_i}) \in \mathcal{J}_{C_i,Y}$ for each $i \in I$). Furthermore, for each $x_i \in C_i$, $M_i(x_i) = M(x_i)$ is precompact by assumption. Thus, using the first case of this proof, M_i is precompact for each $i \in I$. Since

$(h_i \mid M : (M, (\mathcal{J}_{X,Y})_M) \longrightarrow (M_i, (\mathcal{J}_{C_i,Y})_{M_i}))_{i\in I}$ is an initial mono-source in **SUConv**, $(M, (\mathcal{J}_{X,Y})_M)$ is precompact (cf. 4.3.2.2.).

6.3.26 Corollary 1 (Necessary conditions for compactness in $([\mathbf{X},\mathbf{Y}], \mathcal{J}_{X,Y})$). *Let $\mathbf{X} = (X, \mathcal{J}_X)$ be a semiuniform convergence space and $\mathbf{Y} = (Y, \mathcal{J}_Y)$ a semi–pseudouniform T_2–space. If M is a compact subset of $([\mathbf{X},\mathbf{Y}], \mathcal{J}_{X,Y})$, then the following are satisfied:*
(1) M is closed in $([\mathbf{X},\mathbf{Y}], \mathcal{J}_{X,Y})$, i.e. $M = \overline{M}$, where $\overline{M}$ denotes the closure of M with respect to the underlying Kent convergence space of $([\mathbf{X},\mathbf{Y}], \mathcal{J}_{X,Y})$,
(2) M is uniformly equicontinuous,
(3) $M(x) \subset Y$ is compact for each $x \in X$.

Proof. (1) By 4.2.2.3., $([\mathbf{X},\mathbf{Y}], \mathcal{J}_{X,Y})$ is a T_2–space. Since M is compact, M is closed (cf. 4.3.2.14.).
(2) is obvious, since M is precompact and 6.3.19. is valid.
(3) cf. the corresponding proof under 6.3.19. and replace "precompact" by "compact".

6.3.27 Corollary 2 (Sufficient conditions for compactness in $([\mathbf{X},\mathbf{Y}], \mathcal{J}_{X,Y})$). *Let $\mathbf{X} = (X, \mathcal{J}_X)$ be a locally precompact semiuniform convergence space and $\mathbf{Y} = (Y, \mathcal{J}_Y)$ a regular complete semi–Ascoli space (e.g. a complete uniform space). Then $M \subset [\mathbf{X},\mathbf{Y}]$ is a compact subset of $([\mathbf{X},\mathbf{Y}], \mathcal{J}_{X,Y})$ provided that the following are satisfied:*
(1) M is closed in $([\mathbf{X},\mathbf{Y}], \mathcal{J}_{X,Y})$,
(2) M is uniformly equicontinuous,
(3) $M(x) \subset Y$ is compact for each $x \in X$.

Proof. By 4.2.2.1. and 6.3.2. 1), $([\mathbf{X},\mathbf{Y}], \mathcal{J}_{X,Y})$ is complete. By 6.3.25., M is precompact. Since M is closed, M is complete, too (cf. 2.3.3.28.). Thus, M is compact.

6.3.28 Corollary 3. *Let $\mathbf{X} = (X, \mathcal{X})$ be a compact Hausdorff space and $\mathbf{Y} = (Y, \mathcal{V})$ a separated uniform space. Then $M \subset C_u(\mathbf{X},\mathbf{Y})$ is relatively compact, i.e. the closure $\overline{M}$ of M with respect to the topology of uniform convergence is compact, if and only if M is equicontinuous and for each $x \in X$, the set $M(x)$ is relatively compact in $\mathbf{Y}$.*

Proof. $\Longleftarrow$". Let $\mathcal{U}$ be the unique uniformity inducing $\mathcal{X}$ (cf. 4.3.2.25.). It follows from the assumptions on $\mathbf{X}$ and $\mathbf{Y}$ that $C_u(\mathbf{X},\mathbf{Y}) = C_c(\mathbf{X},\mathbf{Y})$ (cf. 6.1.21.). Furthermore, equicontinuity of M means uniform equicontinuity (cf. 6.3.10. and 6.3.11.). Additionally, the closure $\overline{M}$ of M in $C_u(\mathbf{X},\mathbf{Y})$ coincides with the closure of M in $F_s(X,Y)$ (cf. 6.3.14.). Since $p_x[\overline{M}] = \overline{M}(x) \subset \overline{p_x[M]} = \overline{M(x)}$, where $p_x : Y^X \longrightarrow Y$ denotes the x–th projection for each $x \in X$, $\overline{M}(x)$ is relatively compact and consequently precompact for each $x \in X$. Thus, by 6.3.25., $\overline{M}$ is precompact in $([\mathbf{X},\mathbf{Y}], \mathcal{J}_{X,Y})$, where $\mathbf{X} = (X, \mathcal{J}_X)$ and $\mathbf{Y} = (Y, \mathcal{J}_Y)$ with $\mathcal{J}_X = [\mathcal{U}]$ and $\mathcal{J}_Y = [\mathcal{V}]$ respectively (note: According to 6.3.12., $\overline{M}$ is uniformly equicontinuous.). Hence, by 6.3.17., $\overline{M}$ is precompact in $([\mathbf{X},\mathbf{Y}],\mathcal{J}^u_{X,Y}) = C_u(\mathbf{X},\mathbf{Y})$ (note also 4.3.2.24 and 4.3.2.25 and remember that we do not distinguish between principal uniform limit spaces and uniform spaces). Since $\overline{M} \subset \prod_{x\in X}\overline{M}(x) \subset \prod_{x\in X}\overline{M(x)} \subset Y^X$ and $\overline{M(x)}$ is compact and consequently complete, $\overline{M}$ is complete with respect to the structure of simple convergence [as a closed subspace of a complete product space]; thus, $\overline{M}$ is complete with respect to the structure of uniform convergence (cf. 6.1.16. and remember that the structure of uniform convergence is finer than the structure of simple convergence), i.e. as a subspace of $C_u(\mathbf{X},\mathbf{Y})$. Hence, $\overline{M}$ is precompact and complete, i.e. $\overline{M}$ is compact.

"$\Longrightarrow$". Let $N = \overline{M}$. By assumption N is compact in $C_u(\mathbf{X},\mathbf{Y}) = ([\mathbf{X},\mathbf{Y}],\mathcal{J}^u_{X,Y})$, where $\mathbf{X}$ is considered to be a uniform space (= principal uniform limit space) too (as under "$\Longleftarrow$"). Then N is also compact in $([\mathbf{X},\mathbf{Y}],\mathcal{J}_{X,Y})$, since filter convergence with respect to $\mathcal{J}^u_{X,Y}$ coincides with filter convergence with respect to $\mathcal{J}_{X,Y}$ [cf. 6.3.18. 2) and note that $\mathbf{X}$ and $\mathbf{Y}$ are Cook–Fischer spaces]. Thus, by 6.3.26., N is uniformly equicontinuous and $p_x[N] = N(x) \subset Y$ is compact for each $x \in X$. It follows from $M \subset N$ that M is uniformly equicontinuous (= equicontinuous) and, since $\mathbf{Y}$ is a T_2–space, $p_x[M] = M(x)$ is relatively compact in $\mathbf{Y}$ for each $x \in X$.

6.3.29 Remarks. 1) Note, that for the first part of the proof of 6.3.28. ("$\Longleftarrow$") the uniform space $\mathbf{Y}$ need not be separated.

2) If in 6.3.28. the uniform space $\mathbf{Y}$ is metrizable, then it follows from 6.1.8. that $C_u(\mathbf{X},\mathbf{Y})$ is metrizable. Thus, from 6.3.28. results the classical Ascoli theorem:
If $\mathbf{X}$ is a compact Hausdorff space and $\mathbf{Y}$ a metrizable uniform space, then the following are equivalent for each $M \subset C(\mathbf{X},\mathbf{Y})$:
(1) *Each sequence (f_n) in M has a subsequence converging uniformly to some $f \in C(\mathbf{X},\mathbf{Y})$.*
(2) *M is equicontinuous and $M(x)$ is relatively compact in $\mathbf{Y}$ for each $x \in X$.*
If $\mathbf{Y} = \mathbb{R}_u$, then the above conditions are equivalent to
(3) *M is equicontinuous and $M(x)$ is bounded in $\mathbb{R}$ for each $x \in X$.*

3) By definition, every Ascoli space is a semi–Ascoli space. It is unknown whether there are semi–Ascoli spaces which are not Ascoli spaces.

Chapter 7

Relations between Semiuniform Convergence Spaces and Merotopic Spaces (including Nearness Spaces)

M. Katětov [80] originally introduced filter spaces in the realm of his merotopic spaces (studied in the same paper) and called them filter–merotopic spaces. We start the present chapter with this alternative description of filter spaces which have been introduced in chapter 1 and which have also been described in the framework of semiuniform convergence spaces in chapter 2. In other words, a filter space may be regarded as a (filter–)merotopic space or as a **Fil**–determined semiuniform convergence space. Furthermore, the construct **Fil** is bicoreflectively embedded in the construct **Mer** of merotopic spaces, whereas it is bireflectively and bicoreflectively embedded into **SUConv**. As already mentioned in the introduction of this book, the formation of subspaces in **Top** (or $\mathbf{Top_S}$) is not satisfactory. The reason becomes clear, when subspaces of symmetric topological spaces are formed in **SUConv**: they are not topological in general unless they are closed. Since symmetric topological spaces may be regarded as complete filter spaces, subspaces of them, formed in **SUConv**, are filter spaces (regarded as semiuniform convergence spaces). Thus, in order to answer the question how subspaces (in **SUConv**) of symmetric topological spaces, called subtopological spaces, can be characterized axiomatically, we may focus our interest to **Fil**. Such an axiomatic characterization in terms of filters is given in the second part of this chapter. Another characterization due to H.L. Bentley [10] is found, when the description of filter spaces in the realm of merotopic spaces is used, namely a filter space is subtopological iff its corresponding merotopic space is a nearness space. Nearness spaces have been introduced and studied first by H. Herrlich [62]. Since there exist excellent expositions of the theory of nearness spaces, we do not entirely develop this theory, but some basic definitions and results are repeated for the convenience of the reader and hints to the literature are given.
Concerning higher separation axioms in **SUConv**, we have already considered

regular semiuniform convergence spaces in chapter 4. Completely regular topological spaces go back to A. Tychonoff [142]. Completely regular Cauchy spaces have been defined by A. Frič and D.C. Kent [48] as those Cauchy spaces which are subspaces (in the topological construct **Chy** of Cauchy spaces) of completely regular topological spaces. Since subspaces in **Chy** are formed as in **Fil**, this definition is suitable to define completely regular filter spaces correspondingly. Obviously, a completely regular filter space is regular (as a semiuniform convergence space). But if we would define completely regular semiuniform convergence spaces as completely regular filter spaces, we would exclude uniform spaces from consideration. Thus, we might define completely regular semiuniform convergence spaces as those semiuniform convergence spaces whose underlying filter spaces are completely regular. In this case complete regularity would not imply regularity. Thus, a completely regular semiuniform convergence space is a regular semiuniform convergence space such that its underlying filter space is completely regular. It turns out that every uniform space is completely regular and that a symmetric Kent convergence space is completely regular as a semiuniform convergence space iff it is a completely regular topological space (in the usual sense). In other words: complete regularity is such a strong condition that there are no other completely regular convergence stuctures than topological ones. Corresponding to the classical case, complete regularity is closed under formation of initial structures. Finally, an axiomatic description for completely regular filter spaces is given, which can be derived from an axiomatic description of completely regular nearness spaces due to H.L. Bentley, H. Herrlich and G. Ori [14] and whose existence has been observed first by H.L. Bentley and E. Lowen–Colebunders [16]. The definitions of normality and full normality in **SUConv** profit from the corresponding definitions for nearness spaces. A symmetric Kent convergence space is normal (resp. fully normal) as a semiuniform convergence space iff it is a normal (resp. fully normal) topological space in the usual sense, i.e. in the sense of H. Tietze [140] (resp. W. Tukey [141]). Proximity spaces are always fully normal and thus normal, whereas uniform spaces need not be normal. The structural behavior of normal and fully normal symmetric topological spaces is bad, e.g. they are neither closed under formation of subspaces nor products in **Top** (or $\mathbf{Top_S}$). The situation becomes better, when these spaces are regarded as semiuniform convergence spaces, since normality and full normality are hereditary properties in **SUConv**. Furthermore, for normal filter spaces a version of Urysohn's Lemma (resp. Tietze's extension theorem) is satisfied which implies the usual version for normal symmetric topological spaces due to P. Urysohn [144] (resp. H. Tietze [139]). In this context the results of H. Pust [123] concerning normal nearness spaces are used. Fully normal semiuniform convergence spaces which are T_1–spaces (i.e. whose induced topological spaces are T_1–spaces) are called paracompact. A symmetric Kent convergence space is paracompact as a semiuniform convergence space iff it is a paracompact topological space, i.e. a fully normal topological T_1–space. Paracompact topological spaces were originally introduced by J. Dieudonné [35], whereas their equivalence with fully normal T_1–spaces was proved by A.H. Stone [138]. By the hereditariness of full normality in **SUConv**, paracompactness is also hereditary in **SUConv**, which implies that subspaces (in **SUConv**) of paracompact topological spaces (regarded as semiuniform convergence spaces) are paracompact. Furthermore, in the realm of filter spaces, paracompact topological spaces have a nice characterization: they are those topological T_1–spaces which are simultaneously 'topological' and 'm–uniform' as filter spaces (a filter space is topological iff its structure is induced by a symmetric topology, whereas it is m–uniform iff it is uniform as a merotopic space). Besides paracompactness several dimension functions for semiuniform convergence spaces are studied in the fourth section of this chapter. It turns out that the (Lebesgue) covering dimension dim of a symmetric topological space, introduced by E. Čech [28] and much earlier in the more restrictive setting of Euclidean spaces by H. Lebesgue [92], has the property that the dimension of a subspace is less than or equal to the dimension of the original space provided subspaces are formed in **SUConv** and not in **Top** (or $\mathbf{Top}_S$), where this result is true only for closed subspaces. Last but not least, a cohomological characterization of dimension for normal filter spaces (and uniform spaces) of finite dimension is obtained by using corresponding results for nearness spaces due to G. Preuß [110].

The fifth part of the present chapter is devoted to the question how the subspaces, formed in **Fil** (resp. **SUConv**) of compact symmetric topological spaces regarded as filter spaces (resp. semiuniform convergence spaces), called subcompact spaces, can be characterized axiomatically and to the corresponding question for compact Hausdorff spaces. The Herrlich completion of nearness spaces is used for the proof of the characterization of subcompact spaces, whereas the Hausdorff completion of uniform spaces suffices for the proof of the corresponding characterization of the so–called sub–(compact Hausdorff) spaces, since the construct of sub–(compact Hausdorff) spaces is concretely isomorphic to the construct **SepProx** of separated proximity spaces.

7.1 An alternative description of filter spaces in the realm of merotopic spaces

7.1.1. Definitions. Let (X, μ) be a merotopic space.
1) $\mathcal{C} \subset \mathcal{P}(X)$ is called a *Cauchy system* (in (X, μ)) provided that for each $\mathcal{A} \in \mu$ there exist $A \in \mathcal{A}$ and $C \in \mathcal{C}$ with $C \subset A$.
2) A filter on X is called a *Cauchy filter* in (X, μ) provided that it is a Cauchy system.
3) (X, μ) is called a *filter–merotopic space* provided that each Cauchy system in (X, μ) is corefined[1] by some Cauchy filter.
4) (X, μ) is called *contigual* provided that each $\mathcal{A} \in \mu$ contains some finite $\mathcal{B} \in \mu$ (equivalently: each $\mathcal{A} \in \mu$ is refined by some finite $\mathcal{B} \in \mu$).

[1] Let $\mathcal{C}, \mathcal{D} \subset \mathcal{P}(X)$: $\mathcal{C} \ll \mathcal{D} \iff$ For each $C \in \mathcal{C}$ there is some $D \in \mathcal{D}$ such that $D \subset C \iff \mathcal{C}$ *corefines* $\mathcal{D}$.

7.1.2 Remarks. 1) Let (X, μ) be a uniform space, where μ denotes the set of all uniform covers of X. Then (X, μ) is contigual iff it is totally bounded (= precompact) [cf. 4.3.2.15.].

2) a) The construct **Fil–Mer** of filter–merotopic spaces (and uniformly continuous maps) is (concretely) isomorphic to **Fil** (Note: a) If (X, γ) is a filter space, then (X, μ_γ) is a filter–merotopic space provided that $\mu_\gamma = \{\mathcal{A} \subset \mathcal{P}(X)$: for each $\mathcal{F} \in \gamma$ there is some $A \in \mathcal{A}$ with $A \in \mathcal{F}\}$ (It is easily checked that (X, μ_γ) is a merotopic space; if (X, μ_γ) were not a filter–merotopic space, then there were some Cauchy system $\mathcal{C}$ in (X, μ_γ) which is corefined by no $\mathcal{F} \in \gamma$ [cf. d)], i.e. for each $\mathcal{F} \in \gamma$ there were some $F_\mathcal{F} \in \mathcal{F}$ with $F_\mathcal{F} \not\supset C$ for each $C \in \mathcal{C}$, whereas $\mathcal{A} = \{F_\mathcal{F} : \mathcal{F} \in \gamma\} \in \mu_\gamma$ would contain some $F_\mathcal{F}$ such that $F_\mathcal{F} \supset C$ for some $C \in \mathcal{C}$, since $\mathcal{C}$ is a Cauchy system – a contradiction.).
b) If (X, μ) is a merotopic space, then (X, γ_μ) is a filter space provided that γ_μ denotes the set of all Cauchy filters in (X, μ).
c) $\mu = \mu_{\gamma_\mu}$ for each **Fil–Mer**–structure μ on a set X.
[Since $\mu \subset \mu_{\gamma_\mu}$ is obvious, it suffices to verify that $\mu_{\gamma_\mu} \subset \mu$: Let $\mathcal{A} \in \mu_{\gamma_\mu}$. If $\mathcal{A} \notin \mu$, then there is no $\mathcal{B} \in \mu$ such that $\mathcal{B} \prec \mathcal{A}$, i.e. from each $\mathcal{B} \in \mu$ some $B_\mathcal{B}$ can be chosen in such a way that there is no $A \in \mathcal{A}$ with $B_\mathcal{B} \subset A$. Then $\mathcal{C} = \{B_\mathcal{B} : \mathcal{B} \in \mu\}$ were a Cauchy system which is corefined by some $\mathcal{F} \in \gamma_\mu$, since (X, μ) is a filter-merotopic space. Because of $\mathcal{A} \in \mu_{\gamma_\mu}$ there would exist some $A \in \mathcal{A}$ with $A \in \mathcal{F}$ and consequently, there would exist some $B_\mathcal{B} \in \mathcal{C}$ with $B_\mathcal{B} \subset A$ – a contradiction.]
d) $\gamma = \gamma_{\mu_\gamma}$ for each **Fil**–structure γ on a set X.
[Since $\gamma \subset \gamma_{\mu_\gamma}$ is obvious, it suffices to prove $\gamma_{\mu_\gamma} \subset \gamma$: Let $\mathcal{F} \in \gamma_{\mu_\gamma}$. If $\mathcal{F} \notin \gamma$, then there is no $\mathcal{G} \in \gamma$ with $\mathcal{G} \subset \mathcal{F}$, i.e. for each $\mathcal{G} \in \gamma$ there is some $G_\mathcal{G} \in \mathcal{G}$ with $G_\mathcal{G} \notin \mathcal{F}$. Thus, $\mathcal{A} = \{G_\mathcal{G} : \mathcal{G} \in \gamma\} \in \mu_\gamma$. By assumption, $\mathcal{F}$ is a Cauchy filter in (X, μ_γ). Consequently, there would be some $G_\mathcal{G} \in \mathcal{A}$ with $G_\mathcal{G} \in \mathcal{F}$ – a contradiction.]
e) If $f : (X, \gamma) \longrightarrow (X', \gamma')$ is a Cauchy continuous map between filter spaces, then $f : (X, \mu_\gamma) \longrightarrow (X', \mu_{\gamma'})$ is uniformly continuous.
f) If $f : (X, \mu) \longrightarrow (X', \mu')$ is a uniformly continuous map between merotopic spaces, then $f : (X, \gamma_\mu) \longrightarrow (X', \gamma_{\mu'})$ is Cauchy continuous.).
Thus, *we need not distiniguish between filter–merotopic spaces and filter spaces.*

3) a) **Fil–Mer** *is a bicoreflective (full and isomorphism–closed) subconstruct of* **Mer**, where $1_X : (X, \mu_{\gamma_\mu}) \longrightarrow (X, \mu)$ is the bicoreflection of $(X, \mu) \in |\mathbf{Mer}|$ w.r.t. **Fil–Mer**. Thus, final structures in **Fil–Mer** are formed as in **Mer** where they are easy to describe (cf. 4.3.1.3. 2) b)).
b) *Subspaces in* **Fil–Mer** *are formed as in* **Mer**, *in other words:* **Fil–Mer** *is closed under formation of subspaces in* **Mer** (Namely, let (X, μ) be a filter–merotopic space and (Y, η) a subspace of (X, μ) in **Mer**; furthermore, let $\mathcal{C}$ be a Cauchy system in (Y, η) where without loss of generality $\emptyset \notin \mathcal{C}$; since the inclusion map $i : (Y, \eta) \longrightarrow (X, \mu)$ is uniformly continuous, $i\mathcal{C} = \mathcal{C}$ is a Cauchy system in (X, μ) and thus, $\mathcal{C}$ is corefined by some Cauchy filter $\mathcal{F}$ on X; the trace $i^{-1}(\mathcal{F})$ of $\mathcal{F}$ in Y is a Cauchy filter on X which still corefines $\mathcal{C}$, i.e. (Y, η) is filter–merotopic.).

7.1.3 Proposition. *Every contigual merotopic space (X, μ) is a filter space (= filter–merotopic space).*

Proof. Let $\mathcal{C}$ be a Cauchy system in a contigual merotopic space (X, μ). We may assume that $\emptyset \notin \mathcal{C}$. Put

$$\begin{aligned}\Theta &= \{\mathcal{T} \subset \mathcal{P}(X)\backslash\{\emptyset\} : \{C \in \mathcal{C} : C \subset T_1 \cap \cdots \cap T_n\} \\ &\quad \text{is a Cauchy system for any } T_1, \ldots, T_n \in \mathcal{T}\}.\end{aligned}$$

Then $(\Theta, \subset)$ is inductively ordered, i.e. each linearly ordered set Σ in the partially ordered set Θ has an upper bound, namely $\mathcal{T} = \bigcup_{\mathcal{S} \in \Sigma} \mathcal{S} \in \Theta$. By Zorn's lemma, Θ has maximal elements. Let $\mathcal{F}$ be a maximal element of Θ. Then $\mathcal{F}$ is a filter (1. $\emptyset \notin \mathcal{F}$ by the definition of Θ. 2. If $F_1, F_2 \in \mathcal{F}$, then $F_1 \cap F_2 \in \mathcal{F}$, because $\mathcal{F} \subset \mathcal{F} \cup \{F_1 \cap F_2\} \in \Theta$ and $\mathcal{F}$ is maximal. 3. If $F \in \mathcal{F}$ and $X \supset F' \supset F$, then $F' \in \mathcal{F}$ [note: $\mathcal{F} \cup \{F'\} \in \Theta$, since for any $F_1, \ldots, F_n \in \mathcal{F}$, $\{C \in \mathcal{C} : C \subset F_1 \cap \cdots \cap F_n\}$ as well as $\mathcal{B} = \{C \in \mathcal{C} : C \subset F_1 \cap \cdots \cap F_n \cap F\}$ is a Cauchy system and because of $\mathcal{B} \subset \mathcal{D} = \{C \in \mathcal{C} : C \subset F_1 \cap \cdots \cap F_n \cap F'\}$, $\mathcal{D}$ is also a Cauchy system]).
Since $\mathcal{T} \ll \mathcal{C}$ for $\mathcal{T} \in \Theta$, $\mathcal{F} \ll \mathcal{C}$. It remains to verify that $\mathcal{F}$ is a Cauchy system: Let $\mathcal{A} \in \mu$ and $\mathcal{B} = \{B_1, B_2, \ldots, B_n\}$ a finite uniform subcover of $\mathcal{A}$ (the latter one exists by assumption). We assert that $B_\ell \in \mathcal{F}$ for some $\ell \in \{1, \ldots, n\}$. If $B_i \notin \mathcal{F}$ for each $i \in \{1, \ldots, n\}$, then for each $i \in \{1, \ldots, n\}$, there were $F_1^i, \ldots, F_{k_i}^i \in \mathcal{F}$ such that $\mathcal{C}_i = \{C \in \mathcal{C} : C \subset B_i \cap F_1^i \cap \cdots \cap F_{k_i}^i\}$ is not a Cauchy system, because $\mathcal{F} \cup \{B_i\}$ would not belong to Θ. On the other hand, $\mathcal{C}' = \{C \in \mathcal{C} : C \subset \bigcap\{F_j^i : i \in \{1, \ldots, n\}, j \in \{1, \ldots, k_i\}\}\}$ is a Cauchy system ($\mathcal{F} \in \Theta$!). Since $\mathcal{C}_i$ is not a Cauchy system, $\bigcup_{i=1}^n \mathcal{C}_i$ is also not a Cauchy system, i.e. there is some $\mathcal{A}' \in \mu$ such that no element of $\bigcup_{i=1}^n \mathcal{C}_i$ is contained in some $A' \in \mathcal{A}'$. But $\mathcal{C}'' = \mathcal{C}' \backslash \bigcup_{i=1}^n \mathcal{C}_i$ is a Cauchy system (namely, if $\mathcal{C}''$ were not a Cauchy system, there would exist some $\mathcal{A}'' \in \mu$ such that no element of $\mathcal{C}''$ is contained in some $A'' \in \mathcal{A}''$; since $\mathcal{C}'$ is a Cauchy system, there is some $C \in \mathcal{C}'$ and some $M \in \mathcal{A}' \wedge \mathcal{A}'' \in \mu$ with $C \subset M$, where $M = A' \cap A''$ with $A' \in \mathcal{A}'$ and $A'' \in \mathcal{A}''$; because of $C \subset A'$, $C \notin \bigcup_{i=1}^n \mathcal{C}_i$, i.e. $C \in \mathcal{C}''$, and since $C \subset A''$, we obtain a contradiction). Consequently, there are some $B_\ell \in \mathcal{B}$ and some $C \in \mathcal{C}''$ such that $C \subset B_\ell$. Since C is contained in each F_j^i, we have $C \subset B_\ell \cap F_1^\ell \cap \cdots \cap F_{k_\ell}^\ell$, i.e. $C \in \mathcal{C}_\ell$, which is impossible.

7.1.4 Proposition. *Every filter–merotopic uniform space (X, μ) is a Cauchy space.*

Proof. We have to prove that the filter space (X, γ_μ) is a Cauchy space, where γ_μ is the set of all Cauchy filters in (X, μ): Let $\mathcal{F}, \mathcal{G} \in \gamma_\mu$ such that $F \cap G \neq \emptyset$ for each $F \in \mathcal{F}$ and each $G \in \mathcal{G}$, and $\mathcal{U} \in \mu$. Then there exists some $\mathcal{W} \in \mu$ with $\mathcal{W} \, {}^*\!\!\prec \mathcal{U}$. Further, there exist $F \in \mathcal{F}$, $G \in \mathcal{G}$ and $V, W \in \mathcal{W}$ such that $F \subset V$ and $G \subset W$. Since $F \cap G \neq \emptyset$, we obtain $V \cap W \neq \emptyset$. Moreover, there is some $U \in \mathcal{U}$ with $St(V, \mathcal{W}) = \bigcup\{V' \in V : V \cap V' \neq \emptyset\} \subset U$. Because of $V, W \subset St(V, \mathcal{W}) \subset U$ it follows $F \cup G \subset U$. Then, $\mathcal{F} \cap \mathcal{G} \in \gamma_\mu$.

7.1.5 Corollary. *Every proximity space is a Cauchy space.*

Proof. Let **X** be a contigual uniform space, i.e. a proximity space (cf. 7.1.2. 1) and 4.3.2.18.), described by means of uniform covers. By 7.1.3., **X** is a filter–merotopic space. Since **X** is uniform, it follows from 7.1.4. that **X** is a Cauchy space.

7.1.6 Remark. In 7.1.4. and 7.1.5. we have used the following alternative description of Cauchy spaces. A *Cauchy space* is a filter–merotopic space (X,μ) which satisfies the following: If $\mathcal{A}$ and $\mathcal{B}$ are Cauchy systems in (X,μ) with $\emptyset \notin \mathcal{A} \wedge \mathcal{B} = \{A \cap B : A \in \mathcal{A} \text{ and } B \in \mathcal{B}\}$, then $\mathcal{A} \vee \mathcal{B} = \{A \cup B : A \in \mathcal{A}$ and $B \in \mathcal{B}\}$ is a Cauchy system in (X,μ). Obviously, a filter–merotopic space (X,μ) is a Cauchy space iff (X,γ_μ) is a Cauchy space in the usual sense. It is easily checked that the full subconstruct **Chy–Mer** of **Fil–Mer** whose object class consists of all Cauchy spaces (in the sense above) is (concretely) isomorphic to **Chy**. Since **Unif** is bireflective in **Mer** (cf. 4.3.1.3. 3)) and **Prox** is bireflective in **Unif** (cf. 4.3.2.19.), **Prox** is bireflective in **Mer**. This implies that **Prox** *is bireflective in* **Chy–Mer** ($\cong$ **Chy**) [cf. 7.1.5. and restrict the bireflector from **Mer** into **Prox** to **Chy–Mer**].

7.1.7 Proposition. *Every weakly Hausdorff limit space is a complete Cauchy space.*

Proof. Obviously, every weakly Hausdorff limit space (X,q) is symmetric. We have to prove that its corresponding complete filter space (X,γ_q) is a Cauchy space (remember that γ_q is the set of all convergent filters in (X,q)). Let $\mathcal{F},\mathcal{G} \in \gamma_q$ such that $F \cap G \neq \emptyset$ for each $F \in \mathcal{F}$ and each $G \in \mathcal{G}$. Then there exist $x,y \in X$ with $(\mathcal{F},x) \in q$ and $(\mathcal{G},y) \in q$. Furthermore $(\mathcal{H},x) \in q$ and $(\mathcal{H},y) \in q$, where $\mathcal{H} = \{F \cap G : F \in \mathcal{F} \text{ and } G \in \mathcal{G}\}$. Thus, $(\mathcal{F},x) \in q$ and $(\mathcal{G},x) \in q$, which implies $(\mathcal{F} \cap \mathcal{G},x) \in q$, i.e. $\mathcal{F} \cap \mathcal{G} \in \gamma_q$.

7.1.8 Remarks. 1) *The construct* $\mathbf{T_{2W}}$**–Lim** *of weakly Hausdorff limit spaces (and continuous maps) is (concretely) isomorphic to the construct* **CChy** *of complete Cauchy spaces (and Cauchy continuous maps).* [First note that for each Cauchy space (X,γ), (X,q_γ) is a weakly Hausdorff limit space, since $(X,\mathcal{J}^*_\gamma)$ is a uniform limit space with $q_{\gamma_{\mathcal{J}^*_\gamma}} = q_\gamma$ (cf. part 1) a) of the proof of 4.4.18. and use 2.3.3.14. 2) b)). Then use 7.1.7. and apply 2.3.3.11.]
2) **CChy** *is a bicoreflective (full and isomorphism–closed) subconstruct of* **Chy**; namely, the restriction of the bicoreflector $\mathcal{R}_C$: **Fil** $\longrightarrow$ **CFil** (cf. 2.3.3.9.) to **Chy** is easily checked to be the desired bicoreflector from **Chy** into **CChy**.
3) *A limit space* (X,q) *is* T_{2W} *iff it is limit uniformizable*, i.e. iff there is a **ULim**–structure $\mathcal{J}_X$ on X such that $q_{\gamma_{\mathcal{J}_X}} = q$ (If (X,q) is limit uniformizable, then (X,q) is T_{2W} [cf. 2.3.3.14. 2) b)]. Conversely, if (X,q) is T_{2W}, then, by 7.1.7., (X,γ_q) is a Cauchy space and, by part 1) a) of the proof of 4.4.18, $(X,\mathcal{J}^*_{\gamma_q})$ is a uniform limit space such that $\gamma_{\mathcal{J}^*_{\gamma_q}} = \gamma_q$. Thus, $q_{\gamma_{\mathcal{J}^*_{\gamma_q}}} = q_{\gamma_q} = q$, since (X,q) is symmetric [cf. 3) b) of the proof of 2.3.3.11.]). Furthermore, $\mathcal{J}^*_{\gamma_q}$ is the finest

one of all **ULim**–structures $\mathcal{J}_X$ on X with $q_{\gamma_{\mathcal{J}_X}} = q$ provided that (X, q) is a weakly Hausdorff limit space.
4) $\mathbf{T_{2W}}$–**Lim** *is a (full and isomorphism–closed) bireflective subconstruct of* $\mathbf{Lim_S}$ (It is easily verified that $\mathbf{T_{2W}}$–**Lim** is closed under formation of initial structures in **Lim**, i.e. $\mathbf{T_{2W}}$–**Lim** is bireflective in **Lim**. Since every weakly Hausdorff limit space is symmtric, the restriction of the bireflector from **Lim** into $\mathbf{T_{2W}}$–**Lim** to $\mathbf{Lim_S}$ is the desired bireflector $\mathcal{R} : \mathbf{Lim_S} \longrightarrow \mathbf{T_{2W}}$–**Lim**).

7.2 Subtopological spaces

7.2.1 Definitions. A filter space (X, γ) is called
1) *weakly subtopological* provided that its underlying Kent convergence space (X, q_γ) is topological,
2) *subtopological* provided that each $\mathcal{F} \in \gamma$ contains some $\mathcal{G} \in \gamma$ with a q_γ–open base $\mathcal{B}$ (i.e. each $B \in \mathcal{B}$ belongs to $\mathcal{X}_{q_\gamma}$ [cf. the proof of 2.3.1.5]),
3) *topological* provided that it is complete and weakly subtopological.

7.2.2 Remark. The construct **T–Fil** of topological filter spaces (and Cauchy continuous maps) is concretely isomorphic to $\mathbf{Top_S}$ (use 2.3.3.11. 3)), i.e. *we need not distinguish between topological filter spaces and symmetric topological spaces.*

7.2.3 Proposition. *Let (X, γ) be a filter space. Then the following are equivalent:*
(1) *(X, γ) is subtopological,*
(2) *each $\mathcal{F} \in \gamma$ contains some $\mathcal{G} \in \gamma$ such that for each $G \in \mathcal{G}$, $\text{int}_{\mathcal{X}_{q_\gamma}} G \in \mathcal{G}$,*
(3) *for each Cauchy filter $\mathcal{F}$ in (X, γ) the topological neighborhood filter of $\mathcal{F}$, denoted by $\mathcal{U}_{\mathcal{X}_{q_\gamma}}(\mathcal{F})$, is a Cauchy filter in (X, γ), where $\mathcal{U}_{\mathcal{X}_{q_\gamma}}(\mathcal{F}) = \{U \subset X : U \supset O_F \supset F$ for some $F \in \mathcal{F}$ and some $O_F \in \mathcal{X}_{q_\gamma}\}$.*

Proof. (1) $\Longleftrightarrow$ (2) is obvious.
(1) $\Longrightarrow$ (3). Let $\mathcal{F} \in \gamma$. Then there is some $\mathcal{G} \in \gamma$ with $\mathcal{G} \subset \mathcal{F}$ such that $\mathcal{G}$ has a q_γ–open base. Thus, $\mathcal{G} \subset \mathcal{U}_{\mathcal{X}_{q_\gamma}}(\mathcal{F})$ and consequently, $\mathcal{U}_{\mathcal{X}_{q_\gamma}}(\mathcal{F}) \in \gamma$.
(3) $\Longrightarrow$ (1). For each $\mathcal{F} \in F(X)$, $\mathcal{U}_{\mathcal{X}_{q_\gamma}}(\mathcal{F}) \subset \mathcal{F}$ has a q_γ–open base $\mathcal{B} = \{\text{int}_{\mathcal{X}_{q_\gamma}} U : U \in \mathcal{U}_{\mathcal{X}_{q_\gamma}}(\mathcal{F})\}$. By assumption, $\mathcal{U}_{\mathcal{X}_{q_\gamma}}(\mathcal{F}) \in \gamma$ provided that $\mathcal{F} \in \gamma$. Thus, (X, γ) is subtopological.

7.2.4 Proposition. *Every subtopological filter space (X, γ) is weakly subtopological.*

Proof. In order to prove that (X, q_γ) is topological it suffices to prove

$$(*) \qquad (\mathcal{F}, x) \in q_\gamma \Longleftrightarrow \mathcal{F} \supset \mathcal{U}_{\mathcal{X}_{q_\gamma}}(x).$$

Let $(\mathcal{F}, x) \in q_\gamma$. If $U \in \mathcal{U}_{\mathcal{X}_{q_\gamma}}(x)$, then there is some $O \in \mathcal{X}_{q_\gamma}$ with $x \in O \subset U$. Consequently, $O \in \mathcal{F}$ and thus $U \in \mathcal{F}$. Hence $\mathcal{F} \supset \mathcal{U}_{\mathcal{X}_{q_\gamma}}(x)$. Conversely, let $\mathcal{F} \supset \mathcal{U}_{\mathcal{X}_{q_\gamma}}(x)$. In order to prove that $(\mathcal{F}, x) \in q_\gamma$, it suffices to prove

$(\mathcal{U}_{\mathcal{X}_{q_\gamma}}(x), x) \in q_\gamma$. Since $\dot{x} \in \gamma$, $\mathcal{U}_{\mathcal{X}_{q_\gamma}}(\dot{x}) = \mathcal{U}_{\mathcal{X}_{q_\gamma}}(x) = \mathcal{U}_{\mathcal{X}_{q_\gamma}}(x) \cap \dot{x} \in \gamma$, i.e. $(\mathcal{U}_{\mathcal{X}_{q_\gamma}}(x), x) \in q_\gamma$.

7.2.5 Remark. *A weakly subtopological filter space need not be subtopological* as the following *example* shows: Let X be the set $\mathbb{R}\setminus\{0\}$ and $\mathcal{X}$ the topology on X induced by the usual topology on $\mathbb{R}$. Put $\gamma = \gamma_{q_\mathcal{X}} \cup \{\mathcal{F} \in F(X) : \mathcal{F} \supset \mathcal{G}\}$ where $\mathcal{G}$ denotes the elementary filter of the sequence $(\frac{1}{n})_{n\in\mathbb{N}}$ in X. Then (X,γ) is a filter space whose underlying Kent convergence space is the topological space $(X, q_\mathcal{X})$. Thus, (X,γ) is weakly subtopological, but not subtopological, since $\mathcal{G} \in \gamma$ does not have a q_γ–open base.

7.2.6 Proposition. *The construct* **SubTop** *of subtopological filter spaces (and Cauchy continuous maps) is a bireflective (full and isomorphism–closed) subconstruct of* **Fil**.

Proof. 1) **SubTop** is closed under formation of subspaces in **Fil**: Let $(X',\gamma') \in |\mathbf{SubTop}|$ and (X,γ) a subspace in **Fil**, i.e. (without loss of generality) $X \subset X'$ and $\gamma = \{\mathcal{F} \in F(X) : i(\mathcal{F}) \in \gamma'\}$, where $i : X \longrightarrow X'$ denotes the inclusion map. By assumption, for each $\mathcal{F} \in \gamma$, there is some $\mathcal{G} \in \gamma'$ such that $\mathcal{G} \subset i(\mathcal{F})$ and $\mathcal{G}$ has a $q_{\gamma'}$–open base. Thus, $\mathcal{F} \supset i^{-1}(\mathcal{G})$, where $i^{-1}(\mathcal{G})$ is a Cauchy filter in (X,γ) with a q_γ–open base ($i : (X, \mathcal{X}_{q_\gamma}) \longrightarrow (X', \mathcal{X}_{q_{\gamma'}})$ is continuous!).
2) **SubTop** is closed under formation of products in **Fil**. Let $((X_i,\gamma_i))_{i\in I}$ be a family of **SubTop**–objects and $((\prod_{i\in I} X_i, \gamma), (p_i)_{i\in I})$ their product in **Fil**, i.e. $\gamma = \{\mathcal{F} \in F(\prod X_i) : p_i(\mathcal{F}) \in \gamma_i$ for each $i \in I\}$, where $p_i : \prod X_i \longrightarrow X_i$ denotes the i–th projection. If $\mathcal{F} \in \gamma$, then by assumption, for each $i \in I$, there is some $\mathcal{G}_i \in \gamma_i$ such that $\mathcal{G}_i \subset p_i(\mathcal{F})$ and $\mathcal{G}_i$ contains $\operatorname{int}_{\mathcal{X}_{q_{\gamma_i}}} G_i$ for each $G_i \in \mathcal{G}_i$. Hence, the product filter $\prod_{i\in I} p_i(\mathcal{F}) \subset \mathcal{F}$ contains the product filter $\prod_{i\in I} \mathcal{G}_i \in \gamma$ (note: $p_i(\prod \mathcal{G}_i) = \mathcal{G}_i \in \gamma_i$ for each $i \in I$) which has a q_γ–open base $\mathcal{B} = \{\prod_{i\in I} \operatorname{int}_{\mathcal{X}_{q_{\gamma_i}}} G_i : G_i \in \mathcal{G}_i$ for each $i \in I$ and $G_i \neq X_i$ for at most finitely many $i \in I\}$ (note: $\prod_{i\in I} \operatorname{int}_{\mathcal{X}_{q_{\gamma_i}}} G_i \in \mathcal{X}$ provided that $(\prod_{i\in I} X_i, \mathcal{X})$ denotes the product of $((X_i, \mathcal{X}_{q_{\gamma_i}}))_{i\in I}$ in $\mathbf{Top_S}$).
3) All indiscrete **Fil**–objects, i.e. all filter spaces (X,γ) such that $\gamma = F(X)$, belong to $|\mathbf{SubTop}|$, since for each $\mathcal{F} \in F(X)$, $\mathcal{U}_{\mathcal{X}_{q_\gamma}}(\mathcal{F}) \in F(X)$.
It follows from 1), 2) and 3) that **SubTop** is bireflective in **Fil** (cf. the theorem under 2.2.11. 2)).

7.2.7 Theorem *Let (X,γ) be a subtopological filter space. Put $X^* = X \cup \{\mathcal{F} \in \gamma$: $\mathcal{F}$ has a q_γ–open base and is non–convergent$\}$ and $\gamma^* = \{\mathcal{H} \in F(X^*) : \mathcal{H} \supset \mathcal{G}$ for some $\mathcal{G} \in \mathcal{B}^*\}$, where $\mathcal{B}^* = \{i(\mathcal{F}) \cap \dot{x} : (\mathcal{F}, x) \in q_\gamma\} \cup \{i(\mathcal{F}) \cap \dot{\mathcal{F}} : \mathcal{F} \in \gamma$ has a q_γ–open base and is non–convergent$\}$ and $i : X \longrightarrow X^*$ denotes the inclusion map. Then (X^*,γ^*) is a complete filter space containing (X,γ) as a dense subspace.*

Proof. 1) Obviously, (X^*,γ^*) is a filter space.
2) (X,γ) is a subspace of (X^*,γ^*), i.e. $\gamma = \{\mathcal{F} \in F(X) : i(\mathcal{F}) \in \gamma^*\}$: If $\mathcal{F} \in \gamma$, then $i(\mathcal{F}) \in \gamma^*$. Conversely, let $\mathcal{F} \in F(X)$ such that $i(\mathcal{F}) \in \gamma^*$. Then there is

some $(\mathcal{G}, x) \in q_\gamma$ with $i(\mathcal{F}) \supset i(\mathcal{G}) \cap \dot{x}$ or there is a non–convergent $\mathcal{G} \in \gamma$ (with a q_γ–open base) such that $i(\mathcal{F}) \supset i(\mathcal{G}) \cap \dot{\mathcal{G}}$. In the first case $\mathcal{G}' = \mathcal{G} \cap \dot{x} \in \gamma$ and $i(\mathcal{G}') = i(\mathcal{G}) \cap i(\dot{x}) = i(\mathcal{G}) \cap \dot{x} \subset i(\mathcal{F})$; thus, $\mathcal{F} = i^{-1}(i(\mathcal{F})) \supset i^{-1}(i(\mathcal{G}')) = \mathcal{G}'$ and consequently, $\mathcal{F} \in \gamma$. In the second case $\mathcal{F} = i^{-1}(i(\mathcal{F})) \supset i^{-1}(i(\mathcal{G}) \cap \dot{\mathcal{G}}) = \mathcal{G}$, which implies $\mathcal{F} \in \gamma$.
3) X is dense in (X^*, γ^*), i.e. $X^* \subset \overline{X} = \{y \in X : \text{there is some } \mathcal{G} \in F(X^*) \text{ with } (\mathcal{G}, y) \in q_{\gamma^*} \text{ and } X \in \mathcal{G}\}$. If $y \in X^*$, then $y \in X$ or $y = \mathcal{F} \in \gamma$ is non–convergent and has a q_γ–open base. In the first case $y \in \overline{X}$, since $X \in i(\dot{y}) \xrightarrow{q_{\gamma^*}} i(y) = y$. In the second case $y \in \overline{X}$, since $X \in i(\mathcal{F}) \xrightarrow{q_{\gamma^*}} \mathcal{F} = y$.
4) (X^*, γ^*) is complete: If $\mathcal{G} \in \gamma^*$, then $\mathcal{G} \supset i(\mathcal{F}) \cap \dot{x}$ with $(\mathcal{F}, x) \in q_\gamma$ or $\mathcal{G} \supset i(\mathcal{F}) \cap \dot{\mathcal{F}}$ where $\mathcal{F} \in \gamma$ is non–convergent and has a q_γ–open base. In the first case it follows from $i(\mathcal{F}) \cap \dot{x} \xrightarrow{q_{\gamma^*}} x$ that $\mathcal{G} \xrightarrow{q_{\gamma^*}} x$. In the second case it follows that $\mathcal{G} \cap \dot{\mathcal{F}} \in \gamma^*$, i.e. $\mathcal{G} \xrightarrow{q_{\gamma^*}} \mathcal{F}$.

7.2.8 Proposition. *Let (X, γ) be a subtopological filter space and (X^*, γ^*) its above completion (cf. 7.2.7.). If (X^*, q_{γ^*}) is the underlying Kent convergence space of (X^*, γ^*), then the following is valid:*

1. $x \in X$: $\quad (\mathcal{F}^*, x) \in q_{\gamma^*} \iff \mathcal{F}^* \cap \dot{x} \supset i(\mathcal{F}) \cap \dot{x}$ for some $\mathcal{F} \in F(X)$ with $\mathcal{F} \cap \dot{x} \in \gamma$,
2. $\mathcal{F} \in X^* \backslash X$: $\quad (\mathcal{F}^*, \mathcal{F}) \in q_{\gamma^*} \iff \mathcal{F}^* \cap \dot{\mathcal{F}} \supset i(\mathcal{F}) \cap \dot{\mathcal{F}}$.

Proof. 1. "$\Longrightarrow$". The case $\mathcal{F}^* \cap \dot{x} \supset i(\mathcal{F}) \cap \dot{\mathcal{F}}$ for some non–convergent $\mathcal{F} \in \gamma$ with a q_γ–open base cannot occur, since otherwise for each $F \in \mathcal{F}$ there were some $F^* \in \mathcal{F}^*$ such that $F^* \cup \{x\} \subset F \cup \{\mathcal{F}\}$, and consequently, $x \in F$, i.e. $\mathcal{F} \subset \dot{x}$ which would imply $\mathcal{F} \cap \dot{x} = \mathcal{F} \in \gamma$, i.e. $\mathcal{F} \xrightarrow{q_\gamma} x$ – a contradiction. If $\mathcal{F}^* \cap \dot{x} \supset i(\mathcal{F}) \cap \dot{y}$ for some $(\mathcal{F}, y) \in q_\gamma$, then for $\mathcal{G} = \mathcal{F} \cap \dot{y}$, $\mathcal{G} \cap \dot{x} = \mathcal{G} \in \gamma$, i.e. $(\mathcal{G}, x) \in q_\gamma$, and $i(\mathcal{G}) \cap \dot{x} \subset i(\mathcal{G}) \subset \mathcal{F}^* \cap \dot{x}$.
2. "$\Longrightarrow$". The case $\mathcal{F}^* \cap \dot{\mathcal{F}} \supset i(\mathcal{G}) \cap \dot{x}$ for some $(\mathcal{G}, x) \in q_\gamma$ cannot occur, since $X \in \mathcal{G}$ and thus $X \cup \{x\} = X \in i(\mathcal{G}) \cap \dot{x}$ but on the other side $X \notin \mathcal{F}^* \cap \dot{\mathcal{F}}$, i.e. $X \not\supset F^* \cup \{\mathcal{F}\}$ for each $F^* \in \mathcal{F}^*$, because $\mathcal{F} \notin X$. Consequently, it remains the case that there is some non–convergent $\mathcal{G} \in \gamma$ with a q_γ–open base such that $\mathcal{F}^* \cap \dot{\mathcal{F}} \supset i(\mathcal{G}) \cap \dot{\mathcal{G}}$. Let $G \in \mathcal{G} \subset i(\mathcal{G})$. Then $G \cup \{\mathcal{G}\} = F^* \cup \{\mathcal{F}\}$ for some $F^* \in \mathcal{F}^*$. Thus, $\mathcal{F} = \mathcal{G}$, since $\mathcal{F} \notin G \subset X$. Hence, $\mathcal{F}^* \cap \dot{\mathcal{F}} \supset i(\mathcal{F}) \cap \dot{\mathcal{F}}$.
The inverse implications under 1. and 2. are trivial.

7.2.9 Corollary. *Let (X, γ) be a subtopological filter space and (X^*, γ^*) its above completion. The neighborhood filters of the points of (X^*, q_{γ^*}) are obtained as follows:*

1. $x \in X : \mathcal{U}_{q_{\gamma^*}}(x) = i(\mathcal{U}_{q_\gamma}(x))$,
2. $\mathcal{F} \in X^* \backslash X : \mathcal{U}_{q_{\gamma^*}}(\mathcal{F}) = i(\mathcal{F}) \cap \dot{\mathcal{F}}$.

Proof. 1. $i(\mathcal{U}_{q_\gamma}(x)) = i(\bigcap_{\mathcal{F} \cap \dot{x} \in \gamma} \mathcal{F}) = \bigcap_{\mathcal{F} \cap \dot{x} \in \gamma} i(\mathcal{F}) = \bigcap_{\mathcal{F} \cap \dot{x} \in \gamma} i(\mathcal{F} \cap \dot{x}) = \bigcap_{\mathcal{F} \cap \dot{x} \in \gamma} i(\mathcal{F}) \cap \dot{x} = \bigcap_{\mathcal{F}^* \cap \dot{x} \in \gamma^*} \mathcal{F}^* \cap \dot{x} = \bigcap_{\mathcal{F}^* \xrightarrow{q_{\gamma^*}} x} \mathcal{F}^* = \mathcal{U}_{q_{\gamma^*}}(x)$ (cf. 7.2.8.1.)

2. $\mathcal{U}_{q_{\gamma^*}}(\mathcal{F}) = \bigcap_{\mathcal{F}^* \xrightarrow{q_{\gamma^*}} \mathcal{F}} \mathcal{F}^* = \bigcap_{\mathcal{F}^* \xrightarrow{q_{\gamma^*}} \mathcal{F}} \mathcal{F}^* \cap \dot{\mathcal{F}} = i(\mathcal{F}) \cap \dot{\mathcal{F}}$ (cf. 7.2.8.2.).

7.2.10 Proposition. *Every topological filter space is subtopological.*

Proof. Let (X, γ) be a topological filter space. Then (X, γ) is weakly subtopological, i.e. (X, q_γ) is topological, and complete, i.e. from $\mathcal{F} \in \gamma$ follows the existence of some $x \in X$ with $\mathcal{F} \xrightarrow{q_\gamma} x$ or equivalently $\mathcal{F} \supset \mathcal{U}_{q_\gamma}(x)$, where $\mathcal{U}_{q_\gamma}(x)$ has a q_γ-open base. Thus, (X, γ) is subtopological.

7.2.11 Theorem. *Let (X, γ) be a filter space. Then the following are equivalent:*
(1) *(X, γ) is subtopological,*
(2) *(X, γ) is a dense subspace (in* **Fil***) of some topological filter space,*
(3) *(X, γ) is a subspace (in* **Fil***) of some topological filter space.*

Proof. (1) $\Longrightarrow$ (2). It suffices to show that the completion (X^*, γ^*) of (X, γ) (cf. 7.2.7.) is a topological filter space, i.e. (X^*, q_{γ^*}) is topological. Since by 7.2.9., C_6) is valid, C_7) must be proved for (X^*, q_{γ^*}) (cf. 2.3.1.1. 2)):
1. $x \in X$: If $U_x^* \in \mathcal{U}_{q_{\gamma^*}}(x)$, then, by 7.2.9.1., there is some $U_x \in \mathcal{U}_{q_\gamma}(x)$ with $U_x \subset U_x^*$. Since (X, q_γ) is topological, there is some $V_x \in \mathcal{U}_{q_\gamma}(x) \subset \mathcal{U}_{q_{\gamma^*}}(x)$ such that for each $z \in V_x$, $U_x \in \mathcal{U}_{q_\gamma}(z) \subset i(\mathcal{U}_{q_\gamma}(z)) = \mathcal{U}_{q_{\gamma^*}}(z)$.
2. $\mathcal{F} \in X^* \backslash X$: Let $U_{\mathcal{F}}^* \in \mathcal{U}_{q_{\gamma^*}}(\mathcal{F}) = i(\mathcal{F}) \cap \dot{\mathcal{F}}$ (cf. 7.2.9.2.), i.e. $U_{\mathcal{F}}^* \supset F \cup \{\mathcal{F}\}$ for some $F \in \mathcal{F}$. Put $V_{\mathcal{F}}^* = (\text{int}_{\mathcal{X}_{q_\gamma}} F) \cup \{\mathcal{F}\}$. Then $V_{\mathcal{F}}^* \subset U_{\mathcal{F}}^*$ and since by assumption, $\text{int}_{\mathcal{X}_{q_\gamma}} F \in \mathcal{F}$, $V_{\mathcal{F}}^* \in \mathcal{U}_{q_{\gamma^*}}(\mathcal{F})$. Let $z \in V_{\mathcal{F}}^*$. Then a) $z = \mathcal{F}$ or b) $z \in \text{int}_{\mathcal{X}_{q_\gamma}} F$. In case a), $U_{\mathcal{F}}^* \in \mathcal{U}_{q_{\gamma^*}}(\mathcal{F})$ by assumption. In case b), $U_{\mathcal{F}}^* \in \mathcal{U}_{q_{\gamma^*}}(z)$ since $\text{int}_{\mathcal{X}_{q_\gamma}} F \in \mathcal{U}_{q_\gamma}(z) \subset i(\mathcal{U}_{q_\gamma}(z)) = \mathcal{U}_{q_{\gamma^*}}(z)$ and $U_{\mathcal{F}}^* \supset \text{int}_{\mathcal{X}_{q_\gamma}} F$.
(2) $\Longrightarrow$ (3). Obvious.
(3) $\Longrightarrow$ (1). Let (Y, η) be a topological filter space containing (X, γ) as a subspace. Since every topological filter space is subtopological (cf. 7.2.10.) and **SubTop** is bireflective in **Fil** (and thus closed under formation of subspaces), (X, γ) is also subtopological.

7.2.12 Definitions. A semiuniform convergence space $(X, \mathcal{J}_X)$ is called
1) *subtopological* provided that it is **Fil**-determined and its corresponding filter space $(X, \gamma_{\mathcal{J}_X})$ is subtopological,
2) *topological* provided that it is a convergence space and its corresponding Kent convergence space $(X, q_{\gamma_{\mathcal{J}_X}})$ is topological.

7.2.13 Remark. Obviously, the construct **T–SUConv** of topological semiuniform convergence spaces (and uniformly continuous maps) is (concretely) isomorphic to $\mathbf{Top}_S$. Furthermore, **T–SUConv** *is closed under formation of closed subspaces and products in* **SUConv**.

7.2.14 Corollary. *Let $(X, \mathcal{J}_X)$ be a semiuniform convergence space. Then the following are equivalent:*
(1) $(X, \mathcal{J}_X)$ is subtopological,
(2) $(X, \mathcal{J}_X)$ is a subspace (in **SUConv**) of some topological semiuniform convergence space,

(3) $(X, \mathcal{J}_X)$ is a dense subspace (in **SUConv**) of some topological semiuniform convergence space.

Proof. Since **Fil–D–SUConv** $\cong$ **Fil** is bireflective in **SUConv** (and thus subspaces in **Fil** are formed as in **SUConv**) and **T–SUConv** is (concretely) isomorphic to $\mathbf{Top_S}$ as well as to **T–Fil**, the above corollary is an immediate consequence of 7.2.11.

7.2.15 Remark. *Concerning the formation of subspaces, semiuniform convergence spaces are better behaved than topological spaces* as the following *example* shows: Though there is a difference of a topological nature between the removal of the point 0 and the removal of the closed unit interval $[0, 1]$ from the usual topological space $\mathbb{R}_t$ of real numbers, the obtained topological spaces are not distinguishable, i.e. they are homeomorphic. But if we do the same in the realm of semiuniform convergence spaces, we obtain non–isomorphic semiuniform convergence spaces; namely $\mathbb{R} \setminus \{0\}$ is connected as a dense subspace (formed in **SUConv**) of the connected topological space $\mathbb{R}_t$ whereas $\mathbb{R} \setminus [0, 1]$ regarded as a subspace (in **SUConv**) of $\mathbb{R}_t$ has two components. Neither $\mathbb{R} \setminus \{0\}$ nor $\mathbb{R} \setminus [0, 1]$ regarded as subspaces in **SUConv** are topological semiuniform convergence spaces, since they are not complete, but they are subtopological semiuniform convergence spaces according to 7.2.14.

7.2.16 Definition. A merotopic space (X, μ) is called a *nearness space* provided that the following is satisfied:
Near) $\mathcal{A} \in \mu$ implies $\{\mathrm{int}_\mu A : A \in \mathcal{A}\} \in \mu$, where $\mathrm{int}_\mu A = \{x \in X : \{A, X \setminus \{x\}\} \in \mu$.

7.2.17 Remarks. 1) *The construct* **Near** *of nearness spaces (and uniformly continuous maps) is a bireflective (full and isomorphism–closed) subconstruct of* **Mer** (cf. e.g. [111; 3.2.2.5.], where merotopic spaces are called seminearness spaces).
2) a) *Every uniform space (X, μ) described by means of uniform covers is a nearness space*; namely, if $\mathcal{A} \in \mu$, then there is some $\mathcal{B} \in \mu$ with $\mathcal{B} \overset{*}{\prec} \mathcal{A}$ which implies that $\mathcal{B} \prec \{\mathrm{int}_\mu A : A \in \mathcal{A}\}$ ($B \in \mathcal{B}$ implies that there is some $A \in \mathcal{A}$ such that $St\,(B, \mathcal{B}) \subset A$ and consequently, $B \subset \mathrm{int}_\mu A$) and thus, by Mer_1), $\{\mathrm{int}_\mu A : A \in \mathcal{A}\} \in \mu$.
b) Since **Unif** ($\cong$ **U–Mer**) is bireflective in **Mer** (cf. 4.3.1.3. 3)), it follows from a) that **Unif** *is also bireflective in* **Near**.
3) a) For every nearness space (X, μ) there is a symmetric topological space $(X, \mathcal{X}_\mu)$ defined by $\mathcal{X}_\mu = \{O \subset X : \mathrm{int}_\mu O = O\}$, and for each $A \subset X$, $\mathrm{int}_{\mathcal{X}_\mu} A = \mathrm{int}_\mu A$.
b) Every symmetric topological space $(X, \mathcal{X})$ defines a nearness space $(X, \mu_\mathcal{X})$ such that $\mathrm{int}_{\mu_\mathcal{X}} A = \mathrm{int}_\mathcal{X} A$ for each $A \subset X$, where $\mu_\mathcal{X}$ consists of all covers of X which are refined by some open cover of $(X, \mathcal{X})$.
c) A nearness space (X, μ) is called *topological* provided that $\mu = \mu_{\mathcal{X}_\mu}$ or equivalently, $X = \bigcup_{A \in \mathcal{A}} \mathrm{int}_\mu A$ implies $\mathcal{A} \in \mu$.

d) *The construct* **T–Near** *of topological nearness spaces is (concretely) isomorphic to* $\mathbf{Top}_S$ (note: $\mu_{\mathcal{X}_\mu} = \mu$ for every topological **Near**–structure μ, and $\mathcal{X}_{\mu_\mathcal{X}} = \mathcal{X}$ for each symmetric topology $\mathcal{X}$).
e) **T–Near** *is bicoreflective in* **Near**; namely, if (X, μ) is a nearness space, then $1_X : (X, \mu_{\mathcal{X}_\mu}) \longrightarrow (X, \mu)$ is the desired bicoreflection of (X, μ) w.r.t. **T–Near** and $(X, \mu_{\mathcal{X}_\mu})$ (resp. $(X, \mathcal{X}_\mu)$) is called the *underlying topological space* of (X, μ). (cf. [111; 3.1.1.3.] for the proofs of 3) a) – e)).
f) If $(X, \mathcal{X})$ is a symmetric topological space, then

$$\mu_{\gamma_{q\mathcal{X}}} = \mu_\mathcal{X} .$$

(α) Let $\mathcal{U} \in \mu_{\gamma_{q\mathcal{X}}}$. Thus, each convergent filter in $(X, \mathcal{X})$ contains some $U \in \mathcal{U}$. Hence, $\{U^0 : U \in \mathcal{U}\} \prec \mathcal{U}$ and is an open cover of X; namely, if $x \in X$, the neighborhood filter $\mathcal{U}_\mathcal{X}(x)$ of x in $(X, \mathcal{X})$ converges to x and thus there is some $U \in \mathcal{U}$ such that $U \in \mathcal{U}_\mathcal{X}(x)$, i.e. $x \in U^0 \subset U$. Therefore, $\mathcal{U} \in \mu_\mathcal{X}$.
(β) Let $\mathcal{U} \in \mu_\mathcal{X}$, i.e. there is an open cover $\mathcal{O}$ of $(X, \mathcal{X})$ with $\mathcal{O} \prec \mathcal{U}$. If $\mathcal{F}$ is a filter on X such that there is some $x \in X$ with $\mathcal{F} \supset \mathcal{U}_\mathcal{X}(x)$, then there are some $O \in \mathcal{O}$ and some $U \in \mathcal{U}$ such that $x \in O \subset U$. Thus, $U \in \mathcal{U}_\mathcal{X}(x) \subset \mathcal{F}$. Consequently, $\mathcal{U} \in \mu_{\gamma_{q\mathcal{X}}}$.)

7.2.18 Definition. A nearness space (X, μ) is called *subtopological* provided that it is a subspace (in **Near**) of some topological nearness space.

7.2.19 Proposition. *A filter space* (X, γ) *is subtopological iff* (X, μ_γ) *is a subtopological nearness space.*

Proof. "$\Longrightarrow$". Let (X, γ) be a subtopological filter space, i.e. there is some topological filter space (X', γ') such that (X, γ) is a subspace (in **Fil**) of (X', γ'). Since **Fil** $\cong$ **Fil–Mer**, (X, μ_γ) is a subspace (in **Fil–Mer**) of $(X', \mu_{\gamma'})$, where $(X', \mu_{\gamma'})$ is a topological nearness space, since $\mu_{\gamma'} = \mu_{\mathcal{X}_{q_{\gamma'}}}$. By 7.1.2. 3) b), (X, μ_γ) is a subspace in **Mer** of $(X', \mu_{\gamma'})$, and by 7.2.17. 1), (X, μ_γ) is a nearness space, i.e. (X, μ_γ) is a subspace (in **Near**) of the topological nearness space $(X', \mu_{\gamma'})$.
"$\Longleftarrow$". Let (X, γ) be a filter space such that (X, μ_γ) is a subtopological nearness space, i.e. there is a topological nearness space (X', μ') such that (X, μ_γ) is a subspace in **Near** (resp. **Mer**) of (X', μ'). But (X, μ_γ) is also a subspace of (X', μ') in **Fil–Mer**, since (X', μ') is filtermerotopic (this follows from $\mu' = \mu_{\mathcal{X}_{\mu'}}$ and $\mu_{\gamma_{q\mathcal{X}_{\mu'}}} = \mu_{\mathcal{X}_{\mu'}}$) and subspaces in **Fil–Mer** are formed as in **Mer**. Since **Fil–Mer** $\cong$ **Fil**, $(X, \gamma) = (X, \gamma_{\mu_\gamma})$ is a subspace in **Fil** of $(X', \gamma_{\mu'})$, where it is easily checked that $\gamma_{\mu'} = \gamma_{q\mathcal{X}_{\mu'}}$, which implies that $(X, \gamma_{\mu'})$ is topological. Thus, (X, γ) is subtopological.

7.2.20. Since a theorem due to H.L. Bentley says that a nearness space is subtopological iff it is filtermerotopic (cf. e.g. [69; 3.1.9]), one obtains from 7.2.19. the following

Corollary. *A filter space* (X, γ) *is subtopological iff* (X, μ_γ) *is a nearness space.*

7.2.21 Remark. Obviously, the construct $\mathbf{Sub_{SUConv}T\text{–}SUConv}$ of all subspaces (in **SUConv**) of topological semiuniform convergence spaces (and uniformly continuous maps) is (concretely) isomorphic to **SubTop**. Furthermore, the construct $\mathbf{Sub_{Near}T\text{–}Near}$ of all subspaces (in **Near**) of topological nearness spaces (and uniformly continuous maps) is also (concretely) isomorphic to **SubTop** by means of the restriction of the (concrete) isomorphism $\mathbf{Fil\text{–}Mer} \cong \mathbf{Fil}$ to $\mathbf{Sub_{Near}T\text{–}Near}$. In other words: *Subtopological spaces originally defined here in the realm of filter spaces can also be described in the realm of semiuniform convergence spaces or nearness spaces.*

7.3 Complete regularity and normality

7.3.1 Definitions. 1) A filter space (X, γ) is called *completely regular* provided that it is a subspace (in **Fil**) of some completely regular topological space (regarded as a filter space).
2) A semiuniform convergence space $(X, \mathcal{J}_X)$ is called *completely regular* provided that it is regular and its underlying filter space $(X, \gamma_{\mathcal{J}_X})$ is completely regular.

7.3.2 Proposition. *The underlying Kent convergence space of a completely regular semiuniform convergence space is a completely regular topological space (in the usual sense).*

Proof. Let $(X, \mathcal{J}_X)$ be a completely regular semiuniform convergence space. Then $(X, \gamma_{\mathcal{J}_X})$ is a completely regular filter space, i.e. there is a completely regular topological space $(Y, \mathcal{Y})$ such that $(X, \gamma_{\mathcal{J}_X})$ is a subspace (in **Fil**) of $(Y, \gamma_{q_\mathcal{Y}})$, where $\gamma_{q_\mathcal{Y}}$ consists of all convergent filters on $(Y, \mathcal{Y})$. Since initial structures in **Fil** induce initial structures in $\mathbf{KConv_S}$, $(X, q_{\gamma_{\mathcal{J}_X}})$ is a subspace (in $\mathbf{KConv_S}$) of $(Y, q_\mathcal{Y})$ $[(\mathcal{F}, y) \in q_\mathcal{Y}$ iff $\mathcal{F}$ converges to y in $(Y, \mathcal{Y})]$. Since $(Y, q_\mathcal{Y})$ is a completely regular topological space and $\mathbf{Top_S}$ is closed under formation of subspaces in $\mathbf{KConv_S}$, $(X, q_{\gamma_{\mathcal{J}_X}})$ is a completely regular topological space.

7.3.3 Corollary. *A symmetric Kent convergence space (X, q) is completely regular (as a semiuniform convergence space) iff it is a completely regular topological space.*

Proof. "$\Longrightarrow$". If $(X, \mathcal{J}_{\gamma_q})$ is completely regular, then, by 7.3.2., $(X, q_{\gamma_{\mathcal{J}_{\gamma_q}}}) = (X, q)$ is a completely regular topological space.
"$\Longleftarrow$". If (X, q) is a completely regular topological space, then $(X, \mathcal{J}_{\gamma_q})$ is a regular semiuniform convergence space (cf. 4.2.1.3.). Furthermore, $(X, \gamma_{\mathcal{J}_{\gamma_q}}) = (X, \gamma_q)$ is completely regular by 7.3.1. 1).

7.3.4 Remark. Every uniform space $(X, \mathcal{V})$ can be densely embedded into a complete uniform space $(X^*, \mathcal{V}^*)$: Let X^* be the set of all Cauchy filters on $(X, \mathcal{V})$ and let $\mathcal{V}^*$ be generated by $\{\widetilde{V} : V = V^{-1} \in \mathcal{V}\}$ with $\widetilde{V} = \{(\mathcal{F}, \mathcal{G}) \in X^* \times X^*:$

there is some $M \in \mathcal{F} \cap \mathcal{G}$ with $M \times M \subset V\}$. Then $r_X : (X, \mathcal{V}) \longrightarrow (X^*, \mathcal{V}^*)$, defined by $r_X(x) = \dot{x}$ for each $x \in X$, is a dense embedding of $(X, \mathcal{V})$ into $(X^*, \mathcal{V}^*)$, where $(X^*, \mathcal{V}^*)$ is a compete uniform space by construction (cf. e.g. [128; II.3.4, Satz 1]).

7.3.5 Proposition. *Every uniform space is completely regular.*

Proof. By 4.2.1.5., every uniform space is regular. Furthermore, if $(X, \mathcal{V})$ is a uniform space and $(X^*, \mathcal{V}^*)$ a complete uniform space containing $(X, \mathcal{V})$ as a dense subspace (cf. 7.3.4.), then the underlying filter space (= Cauchy space) $(X, \gamma_{\mathcal{V}})$ of $(X, \mathcal{V})$ is a subspace (in **Fil**) of the underlying filter space (= Cauchy space) $(X^*, \gamma_{\mathcal{V}^*})$ of $(X^*, \mathcal{V}^*)$. Since $(X^*, \mathcal{V}^*)$ is complete, $(X^*, \gamma_{\mathcal{V}^*})$ is the corresponding filter space of the underlying competely regular topological space of $(X^*, \mathcal{V}^*)$. Thus, $(X, \gamma_{\mathcal{V}})$ is a completely regular filter space. Therefore, everything is proved.

7.3.6 Proposition. *Every completely regular filter space is regular* (*as a semiuniform convergence space*).

Proof. Let (X, γ) be a completely regular filter space. Since (X, γ) is a subspace (in **Fil**) of a completely regular topological space, which is regular as a semiuniform convergence space, (X, γ) (regarded as a semiuniform convergence space) is also regular because the regular semiuniform convergence spaces form a bireflective subconstruct of **SUConv** and subspaces in **Fil–D–SUConv** ($\cong$ **Fil**) are formed as in **SUConv**.

7.3.7 Remark. The two preceding propositions demonstrate that for important examples of semiuniform convergence spaces, whose underlying filter spaces are completely regular, the regularity is automatically fulfilled. But this is not always the case as the following *example* shows: Let X be the set $\mathbb{R}$ of real numbers and $\mathcal{X}$ the usual topology on $\mathbb{R}$. Define $A \subset \Delta_X$ by $(x, x) \in A$ iff $0 < x < 1$ and let $\mathcal{U}_{\mathcal{X}}(x)$ be the neighborhood filter of $x \in X$ with respect to $\mathcal{X}$. Put $\mathcal{J}_X = \{\mathcal{F} \in F(\mathbb{R}^2) : \mathcal{F} \supset (\{A\})\} \cup \{\mathcal{F} \in F(\mathbb{R}^2) : \mathcal{F} \supset \mathcal{U}_{\mathcal{X}}(x) \times \mathcal{U}_{\mathcal{X}}(x)$ for some $x \in \mathbb{R}\}$. Then $(X, \mathcal{J}_X)$ is a semiuniform convergence space such that the underlying filter space $(X, \gamma_{\mathcal{J}_X})$ is completely regular (obviously, $\gamma_{\mathcal{J}_X} = \{\mathcal{F} \in F(\mathbb{R}) : \mathcal{F}$ converges in $(X, \mathcal{X})\}$), but $(X, \mathcal{J}_X)$ is not regular, since $(\{A\})$ belongs to $\mathcal{J}_X$, whereas the subfilter $\overline{(\{A\})} = (\{\overline{A}\})$ does not belong to $\mathcal{J}_X$.

7.3.8 Proposition. *Every completely regular filter space is a Cauchy space.*

Proof. Since each completely regular topological space is a T_{2W}–space (cf. exercise 76)) [and thus a Cauchy space (cf. 7.1.7.)] and **Chy** is bireflective in **Fil** (cf. 4.4.19. 4)), every subspace in **Fil** of a completely regular topological space, i.e. every completely regular filter space, is a Cauchy space.

7.3.9 Definition. A Cauchy space (X, γ) is called *uniformizable* provided that there is a uniformity $\mathcal{V}$ on X such that γ is the set of all Cauchy filters on $(X, \mathcal{V})$ (the uniformity $\mathcal{V}$ is called a *compatible uniformity*).

7.3.10 Corollary. *Every uniformizable Cauchy space is completely regular.*

Proof. If (X, γ) is a uniformizable Cauchy space, there is a uniformity $\mathcal{V}$ on X such that $\gamma = \gamma_{\mathcal{V}}$, where $\gamma_{\mathcal{V}}$ denotes the set of all Cauchy filters on $(X, \mathcal{V})$. By 7.3.5., (X, γ) is completely regular.

7.3.11 Remarks. 1) *The construct* **UChy** *of all uniformizable Cauchy spaces (and Cauchy continuous maps) is a bireflective (full and isomorphism-closed) subconstruct of* **Fil** [Let $(f_i : (X, \gamma) \longrightarrow (X_i, \gamma_i))_{i \in I}$ be an initial source in **Fil**, where $(X_i, \gamma_i) \in |\mathbf{UChy}|$ for each $i \in I$, i.e. for each $i \in I$, there is a uniformity $\mathcal{V}_i$ on X_i such that $\gamma_{\mathcal{V}_i} = \gamma_i$. Consider the initial source $(f_i : (X, \mathcal{V}) \longrightarrow (X_i, \mathcal{V}_i))_{i \in I}$ in **Unif**. Then $(f_i : (X, [\mathcal{V}]) \longrightarrow (X_i, [\mathcal{V}_i]))_{i \in I}$ is initial in **SUConv** and $(f_i : (X, \gamma_{[\mathcal{V}]}) \longrightarrow (X, \gamma_{[\mathcal{V}_i]}))_{i \in I}$ is initial in **Fil**, which implies $\gamma = \gamma_{[\mathcal{V}]} = \gamma_{\mathcal{V}}$, i.e. (X, γ) is a uniformizable Cauchy space].

2) **UChy** *is a bireflective (full and isomorphism-closed) subconstruct of the construct* **CRegFil** *of all completely regular filter spaces (and Cauchy continuous maps)*. [By 7.3.8. and 7.3.10., the restriction to **CRegFil** of the bireflector from **Chy** to **UChy** is the bireflector from **CRegFil** to **UChy**].

3) **UChy** *is bicoreflectively embedded in* **Unif** [**UChy** is (concretely) isomorphic to a (full and isomorphism-closed) subconstruct of **Unif**; namely, if (X, γ) is a uniformizable Cauchy space, there exists a finest uniformity $\mathcal{V}_\gamma$ with $\gamma_{\mathcal{V}_\gamma} = \gamma$; thus, the desired subconstruct of **Unif** consists of all *Cauchy fine uniform spaces*, where a uniform space $(X, \mathcal{W})$ is called Cauchy fine provided that $\mathcal{V}_{\gamma_\mathcal{W}} = \mathcal{W}$. The desired bicoreflector is then the restriction of the bicoreflector from **ULim** to **Chy–D–ULim** ($\cong$ **Chy**) to **PrULim** $\cong$ **Unif**; in other words: *the bicoreflective* **UChy**–*modification of a uniform space* $(X, \mathcal{W})$ *is its underlying filter space* (= *Cauchy space*) $(X, \gamma_\mathcal{W})$].

7.3.12 Proposition. 1) **CRegFil** *is a bireflective (full and isomorphism–closed) subconstruct of* **Fil**.
2) *The construct* **CReg** *of all completely regular semiuniform convergence spaces (and uniformly continuous maps) is a bireflective (full and isomorphism–closed) subconstruct of* **SUConv**.

Proof. 1) Obviously, **CRegFil** is closed under formation of subspaces in **Fil**. Since products of subspaces are subspaces of products and products in $\mathbf{Top_S}$ are formed as in **Fil**, **CRegFil** is also closed under formation of products in **Fil**. Furthermore, all indiscrete **Fil**–objects, i.e. all filter spaces (X, γ) with $\gamma = F(X)$, are topological and completely regular. Thus, **CRegFil** is bireflective in **Fil** (cf. the theorem under 2.2.11. 2)).
2) Let $(f_i : (X, \mathcal{J}_X) \longrightarrow (X_i, \mathcal{J}_{X_i}))_{i \in I}$ be an initial source in **SUConv** such that all $(X_i, \mathcal{J}_{X_i})$ are completely regular. Then $(X, \mathcal{J}_X)$ is completely regular:

a) By 4.2.1.6., $(X, \mathcal{J}_X)$ is regular.
b) By 2.3.3.17. a), $\gamma_{\mathcal{J}_X}$ is the initial **Fil**–structure w.r.t. (f_i), and, by 1), $(X, \gamma_{\mathcal{J}_X})$ is a completely regular filter space.

7.3.13 Remarks. 1) By 7.3.12., a subspace (in **SUConv**) of a completely regular topological space (regarded as a semiuniform convergence space) is a completely regular semiuniform convergence space. It follows from 7.3.5. that the inverse is not true, namely the uniform space $\mathbb{R}_u$ of real numbers is completely regular but it is not a subspace (in **SUConv**) of a completely regular topological space [otherwise it would be **Fil**–determined, which is impossible since the uniform structure of $\mathbb{R}_u$ is not indiscrete (cf. exercise 57)].

2) a) **CregFil** *is bireflective in* **Chy**, where the bireflector from **Chy** into **CregFil** is the restriction of the bireflector from **Fil** into **CregFil** (cf. 7.3.8. and 7.3.12. 1)).
b) **CregFil** *is bireflective in* **SubTop**, where the bireflector from **SubTop** into **CregFil** is the restriction of the bireflector from **Fil** into **SubTop** (cf. 7.2.6. and note that each completely regular filter space is subtopological by definition).

3) *The construct* **CRegTop** *of completely regular topological spaces is bicoreflective in* **CregFil**, where the bicoreflector from **CregFil** into **CRegTop** is the restriction of the bicoreflector from **Fil** into **KConv**$_S$. Similarly, **Top**$_S$ *is bicoreflective in* **SubTop**. Together with former results we get the following diagram, where r (resp. c) stands for embedding as a bireflective (resp. bicoreflective) [full and isomorphism–closed] subconstruct:

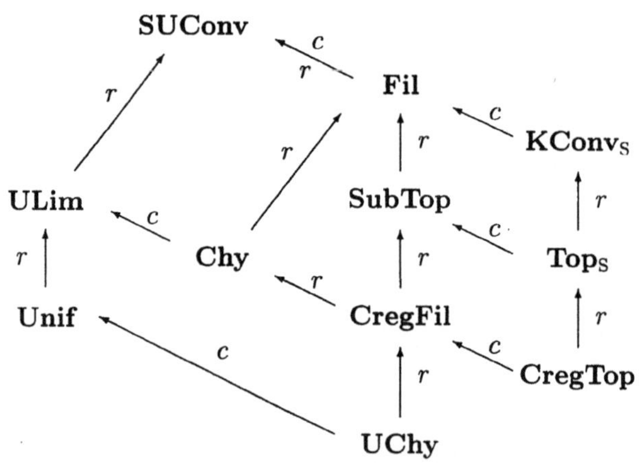

It follows immediately from the above diagram that the underlying Kent convergence space of a uniform space is a completely regular topological space – a result that has been proved earlier in another context.

4) a) Let (X, μ) be a merotopic space and $A, B \subset X$. Then A is *completely within* B iff there is a uniformly continuous map $f : (X, \mu) \longrightarrow ([0,1], \mu_t)$ such that $f[A] \subset \{0\}$ and $f[X \backslash B] \subset \{1\}$, where μ_t is the set of all covers of X which are refined by some open cover of $[0,1]$ w.r.t. the usual topology on $[0,1]$.

b) A merotopic space (X, μ) is called *completely regular* provided that one of the following two equivalent conditions is satisfied:
(1) For each $\mathcal{U} \in \mu$, $\mathcal{V} = \{V \subset X$: V is completely within U for some $U \in \mathcal{U}\} \in \mu$.
(2) For each Cauchy system $\mathcal{C}$ in (X, μ), $\mathcal{D} = \{D \subset X$: C is completely within D for some $C \in \mathcal{C}\}$ is a Cauchy system in (X, μ).

5) By [14] and [16], the following is valid:

Theorem. *Let (X, μ) be a merotopic space. Then the following are equivalent:*
(1) (X, μ) is a completely regular filter–merotopic space,
(2) (X, μ) is a subspace (in **Mer***) of a completely regular topological space (regarded as a nearness space).*
Then, obviously, *a filter space (X, γ) is completely regular iff (X, μ_γ) is a completely regular filter–merotopic space.* Since subspaces (in **Mer**) of nearness spaces are nearness spaces (cf. 7.2.17. 1)), the above theorem is indeed a theorem on nearness spaces.

7.3.14 Definition. Let (X, γ) be a filter space and $A, B \subset X$. Then A is said to be *completely within* B provided that there is a Cauchy continuous map $f : (X, \gamma) \longrightarrow ([0, 1], \gamma_t)$ such that $f[A] \subset \{0\}$ and $f[X \backslash B] \subset \{1\}$, where γ_t denotes the set of all convergent filters on the closed unit interval $[0, 1]$ endowed with the usual topology.

7.3.15 Remark. It is easily checked that for each filter space (X, γ) and subsets A, B, C, D of X the following are valid:
0) A is completely within B implies $A \subset B$.
1) A is completely within B and $B \subset C$ imply that A is completely within C.
2) A is completely within B and C is completely within D imply that $A \cap B$ is completely within $C \cap D$.
3) A is completely within B in (X, γ) iff A is completely within B in (X, μ_γ).

It follows from 0), 1) and 2) that for each filter $\mathcal{F}$ on X, $\mathcal{G} = \{G \subset X$: F is completely within G for some $F \in \mathcal{F}\}$ is a subfilter of $\mathcal{F}$.

7.3.16 Theorem. *A filter space (X, γ) is completely regular iff for each $\mathcal{F} \in \gamma$ the subfilter $\mathcal{G} = \{G \subset X$: F is completely within G for some $F \in \mathcal{F}\}$ belongs to γ.*

Proof. "$\Longrightarrow$". Let $\mathcal{F} \in \gamma$. Then $\mathcal{F}$ is a Cauchy system in (X, μ_γ), where (X, μ_γ) is a completely regular filter–merotopic space by assumption. By 7.3.13. 4), the filter $\mathcal{G} = \{G \subset X$: F is competely within G for some $F \in \mathcal{F}\}$ is a Cauchy system in (X, μ_γ), i.e. $\mathcal{G} \in \gamma$.
"$\Longleftarrow$". It follows from the assumption on (X, γ) that (X, μ_γ) fulfills the condition (2) under 7.3.13. 4), i.e. (X, μ_γ) is a completely regular filter–merotopic space, and, by 7.3.13. 5), (X, γ) is completely regular.

7.3.17 Remark. A filter space (X, γ) is completely regular iff $(X, \mathcal{J}_\gamma)$ is a completely regular semiuniform convergence space (cf. 7.3.6. and remember $\gamma_{\mathcal{J}_\gamma} = \gamma$). Furthermore, the construct $\mathcal{A}$ of all **Fil**–determined completely regular semiuniform convergence spaces (and uniformly continuous maps) is (concretely) isomorphic to **CRegFil**, i.e. **CRegFil** can be embedded into **Creg**. This embedding is bireflective and bicoreflective (the bicoreflector is the restriction of the bicoreflector from **SUConv** into **Fil–D–SUConv** $\cong$ **Fil**; since **Fil–D–SUConv** and **Creg** are both bireflective subconstructs of **SUConv**, $\mathcal{A}$ is bireflective in **SUConv** and thus in **Creg**). Since each uniform space is completely regular (cf. 7.3.3.), we obtain the following diagram:

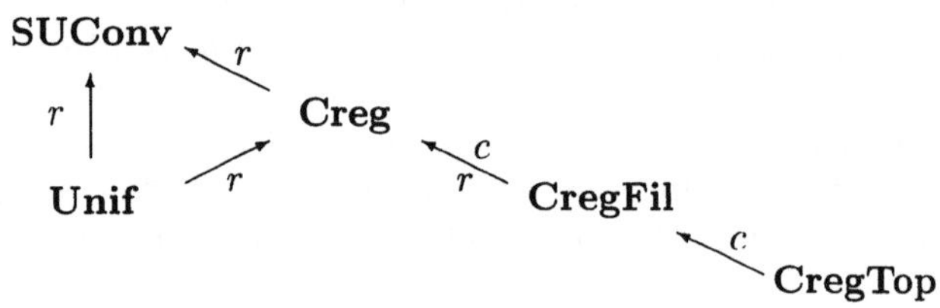

7.3.18 Definitions. 1) A merotopic space (X, μ) is called
a) *regular* provided that the following is satisfied:
(R) For each $\mathcal{U} \in \mu$, there is some (refinement) $\mathcal{V} \in \mu$ such that for each $V \in \mathcal{V}$, there exists some $U \in \mathcal{U}$ with $\{X \backslash V, U\} \in \mu$,
b) *normal* provided that (X, μ) and (X, μ_c) are regular, where $\mu_c = \{\mathcal{A} \in \mu$: there is some finite $\mathcal{V} \in \mu$ with $\mathcal{V} \prec \mathcal{A}\}$.
2) Let (X, γ) be a filter space and (X, μ_γ) its corresponding (filter–)merotopic space. Then (X, γ) is called *merotopically normal* (shortly: m–normal) provided that (X, μ_γ) is normal.
3) A semiuniform convergence space $(X, \mathcal{J}_X)$ is called *normal* provided that it is regular and its underlying filter space $(X, \gamma_{\mathcal{J}_X})$ is m–normal.

7.3.19 Remarks. 1) Every regular merotopic space (X, μ) is a nearness space [Namely, if $\mathcal{U} \in \mu$, then there exists some $\mathcal{V} \in \mu$ such that (R) is satisfied; thus $\mathcal{V} \prec \{\text{int}_\mu U : U \in \mathcal{U}\}$, which implies $\{\text{int}_\mu U : U \in \mathcal{U}\} \in \mu$.].
2) Every normal nearness space (X, μ) is completely regular (cf. [14]).

7.3.20 Proposition. *Every normal semiuniform convergence space is completely regular.*

Proof. Let $(X, \mathcal{J}_X)$ be a normal semiuniform convergence space. Then $(X, \mathcal{J}_X)$ is regular. Furthermore, $(X, \mu_{\gamma_{\mathcal{J}_X}})$ is a nearness space which is normal and thus complelety regular by 7.3.19. 2). Consequently, by 7.3.13. 5), $(X, \gamma_{\mathcal{J}_X})$ is completely regular.

7.3.21 Proposition. *Let* **X** *be a filter space. Then the following are equivalent:*
(1) **X** *is normal (as a semiuniform convergence space),*
(2) **X** *is m–normal.*

Proof. (1) $\Longrightarrow$ (2). This follows immediately from the definitions.
(2) $\Longrightarrow$ (1). Since **X** coincides with its underlying filter space whenever **X** is considered to be a semiuniform convergence space, it suffices to prove that **X** is regular. Since **X** is normal as a filter–merotopic space, it is also completely regular. Consequently, by 7.3.13. 5), the filter space **X** is completely regular. Thus, by 7.3.6., **X** is regular as a semiuniform convergence space.

7.3.22 Remarks. 1) A symmetric topological space $(X, \mathcal{X})$ is regular iff $(X, \mu_{\mathcal{X}})$ is a regular (topological) nearness space (cf. [111; 6.2.7. ② b) and 6.2.7. ③]).
2) If $(X, \mathcal{X})$ is a symmetric topological space, then the following are equivalent:
(1) $(X, \mathcal{X})$ is normal (in the usual sense),
(2) $(X, \mu_{\mathcal{X}})$ is a normal nearness space,
(3) $(X, (\mu_{\mathcal{X}})_c)$ is a regular nearness space.
(cf. [111; 7.2.2.])

7.3.23 Proposition. *A symmetric Kent convergence space is normal (as a semiuniform convergence space) iff it is a normal topological space in the usual sense.*

Proof. "$\Longrightarrow$". Let (X, q) be a symmetric Kent convergence space such that $(X, \mathcal{J}_{\gamma_q})$ is normal. Then $(X, \mathcal{J}_{\gamma_q})$ is completely regular and by 7.3.2., (X, q) is a (completely regular) topological space, i.e. $q = q_{\mathcal{X}_q}$ (cf. b) γ) (2) in the proof of 2.3.1.8.). By assumption, $(X, \gamma_{\mathcal{J}_{\gamma_q}}) = (X, \gamma_q)$ is m–normal, i.e. $(X, \mu_{\mathcal{X}_q})$ is normal, since for each symmetric topological space $(X', \mathcal{X}')$, $\mu_{\gamma_{q_{\mathcal{X}'}}} = \mu_{\mathcal{X}'}$. This implies that $(X, \mathcal{X}_q)$ (resp. (X, q)) is a normal topological space in the usual sense (cf. 7.3.22. 2)).
"$\Longleftarrow$". Let (X, q) be a normal symmetric topological space in the usual sense, i.e. $\mathcal{X}_q$ is a normal symmetric topology on X. In order to prove that $(X, \mathcal{J}_{\gamma_q})$ is normal it suffices to show that (X, μ_{γ_q}) is normal, since (X, γ_q) is a filter space and 7.3.21. is valid. Obviously, $\mu_{\gamma_q} = \mu_{\mathcal{X}_q}$. Since $(X, \mathcal{X}_q)$ is a normal symmetric topological space, it is regular and by 7.3.22. 1), $(X, \mu_{\mathcal{X}_q})$ is regular. Furthermore, $(X, (\mu_{\mathcal{X}_q})_c)$ is regular by 7.3.22. 2). Thus, $(X, \mu_{\mathcal{X}_q})$ is normal and the proof is finished.

7.3.24 Remarks. 1) *The underlying Kent convergence space of a normal semiuniform convergence space need not be normal* as the following *example* shows (whereas it is always a completely regular topological space by 7.3.20. and 7.3.2.): Let $(X, \mathcal{X})$ be a completely regular Hausdorff space, which is not normal, e.g. the Niemytzki plane (cf. [43; 1.2.4., 1.4.5. and 1.5.9.]), and (X, μ_F) the corresponding fine uniform space described by means of uniform covers, i.e. μ_F is the set of all covers of X which are refined by some normally open cover of $(X, \mathcal{X})$. Further, consider $(\mu_F)_c$ and its corresponding Weil uniformity $\mathcal{W}_{(\mu_F)_c}$. Then $(X, \mathcal{W}_{(\mu_F)_c})$ is a totally bounded uniform space which is separated since its underlying topological space is the Hausdorff space $(X, \mathcal{X})$. Consequently, $(X, [\mathcal{W}_{(\mu_F)_c}])$ is a normal semiuniform convergence space (cf. 7.4.11. and 7.4.3.), whose underlying Kent

convergence space (X, q_X) is not normal.
2) *A uniform space need not be normal as a semiuniform convergence space.* In the following an *example* of a complete separated uniform space is given which is not normal: Let $\mathbb{R}_u$ be the uniform space of real numbers and $\mathbb{R}_u^{\mathbb{R}}$ the uniform product space. Let $\mathcal{V}$ be the uniformity of $\mathbb{R}_u^{\mathbb{R}}$ and put $\mathcal{J}_{\mathbb{R}^{\mathbb{R}}} = [\mathcal{V}]$, i.e. $(\mathbb{R}^{\mathbb{R}}, \mathcal{J}_{\mathbb{R}^{\mathbb{R}}})$ is the principal uniform limit space corresponding to $\mathbb{R}_u^{\mathbb{R}}$. Then the set $\gamma_{\mathcal{J}_{\mathbb{R}^{\mathbb{R}}}}$ of its Cauchy filters is given by

$$(*) \qquad \gamma_{\mathcal{J}_{\mathbb{R}^{\mathbb{R}}}} = \{\mathcal{F} \in F(\mathbb{R}^{\mathbb{R}}) : \mathcal{F} \text{ converges in } \mathbb{R}_t^{\mathbb{R}}\},$$

where $\mathbb{R}_t$ denotes the usual topological space of real numbers and the topological product space $\mathbb{R}_t^{\mathbb{R}}$ is the underlying Kent convergence space of $(\mathbb{R}^{\mathbb{R}}, \mathcal{J}_{\mathbb{R}^{\mathbb{R}}})$. Further,

$$(**) \qquad \mu_{\gamma_{\mathcal{J}_{R^{\mathbb{R}}}}} = \mu_{\mathcal{X}_{\mathcal{V}}},$$

where $\mathcal{X}_{\mathcal{V}}$ denotes the topology induced by $\mathcal{V}$, i.e. the topology of $\mathbb{R}_t^{\mathbb{R}}$. Since $\mathbb{R}_t^{\mathbb{R}}$ is not normal in the usual topological sense (cf. [149; 21 C. 5]) $(\mathbb{R}^{\mathbb{R}}, \mu_{\mathcal{X}_{\mathcal{V}}})$ cannot be normal (cf. 7.3.22. 2)). By (**), this implies that the complete separated uniform space $\mathbb{R}_u^{\mathbb{R}}$ is not normal as a semiuniform convergence space.
3) Every subspace (in **Near**) of a normal nearness space is normal (cf. [111; 6.2.7. ⑤ and 7.2.10. ①]).

7.3.25 Proposition. *Every subspace (in* **SUConv***) of a normal semiuniform convergence space is normal.*

Proof. Let $(X, \mathcal{J}_X)$ be a normal semiuniform convergence space and $U \subset X$. If $\mathcal{J}_U$ denotes the initial **SUConv**–structure on U w.r.t. the inclusion map $i : U \longrightarrow X$, then $(U, \mathcal{J}_U)$ is regular (cf. 4.2.1.6.). By 2.3.3.17. a), $(U, \gamma_{\mathcal{J}_U})$ is a subspace (in **Fil**) of $(X, \gamma_{\mathcal{J}_X})$. Since subspaces in **Fil–Mer** ($\cong$ **Fil**) are formed as in **Mer** (cf. 7.1.2. 3) b)), $(U, \mu_{\gamma_{\mathcal{J}_U}})$ is a subspace of $(X, \mu_{\gamma_{\mathcal{J}_X}})$. Then, by 7.3.24. 3), $(U, \mu_{\gamma_{\mathcal{J}_U}})$ is normal since $(X, \mu_{\gamma_{\mathcal{J}_X}})$ is a normal nearness space by assumption.

7.3.26 Remarks. 1) As is well–known subspaces in **Top** (resp. **Top$_\mathbf{S}$**) of normal symmetric topological spaces need not be normal in general. But if subspaces are formed in **SUConv** one obtains from 7.3.25. the following result: *Subspaces (in* **SUConv***) of normal symmetric topological spaces are normal.*
2) The class of all normal semiuniform convergence spaces does not coincide with the class of all subspaces (in **SUConv**) of normal symmetric topological spaces as the following *example* shows:
Modify the example under 7.3.7. by defining $A \subset \Delta_X$ as follows:

$$(x, x) \in A \text{ iff } 0 \leq x \leq 1\,.$$

Then $(X, \mathcal{J}_X)$ is a normal semiuniform convergence space which is not a subspace of a normal topological semiuniform convergence space since it is not **Fil**–determined.
3) Since products of symmetric topological spaces are formed in **Top$_\mathbf{S}$** as in **SU**-

Conv and in **Top**$_{\mathbf{S}}$ normality is not finitely productive, it follows from 7.3.23. that *in* **SUConv** *normality is not finitely productive.*
4) Let (X,μ) be a normal nearness space. Then the following are valid (cf. [111; 7.2.7. and 7.2.13.]):
1. *Urysohn's Lemma.* Whenever $\{A,B\} \subset \mathcal{P}(X)$ is not near (i.e. $\{X\backslash A, X\backslash B\} \in \mu$) there is a uniformly continuous map $f : (X,\mu) \longrightarrow [0,1]$ from (X,μ) into the closed unit interval $[0,1]$ (endowed with its usual uniform structure) such that $f[A] \subset \{0\}$ and $f[B] \subset \{1\}$.
2. *Tietze's extension theorem.* If (A,μ_A) is a nearness subspace of (X,μ), then every uniformly continuous map $f : (A,\mu_A) \longrightarrow [0,1]$ from (A,μ_A) into the closed unit interval $[0,1]$ (endowed with its usual uniform structure) has a uniformly continuous extension $F : (X,\mu) \longrightarrow [0,1]$.

7.3.27 Definition. Let (X,γ) be a filter space. Then $\mathcal{A} \subset \mathcal{P}(X)$ is called *near* provided that there is some $\mathcal{F} \in \gamma$ such that for each $A \in \mathcal{A}$, $X\backslash A \notin \mathcal{F}$.

7.3.28 Theorem (Urysohn's Lemma). *Let (X,γ) be a normal filter space. Whenever $\{A,B\} \subset \mathcal{P}(X)$ is not near, there is a Cauchy continuous map $f : (X,\gamma) \longrightarrow ([0,1],\gamma_t)$ such that $f[A] \subset \{0\}$ and $f[B] \subset \{1\}$, where γ_t denotes the set of all convergent filters on the closed unit interval $[0,1]$ endowed with the usual topology.*

Proof. Apply 7.3.26. 4) 1. and note:
1. $\{A,B\} \subset \mathcal{P}(X)$ is not near in (X,γ) iff $\{A,B\}$ is not near in (X,μ_γ), i.e. $\{X\backslash A, X\backslash B\} \in \mu_\gamma$.
2. $f : (X,\gamma) \longrightarrow ([0,1],\gamma_t)$ is Cauchy continuous iff $f : (X,\mu_\gamma) \longrightarrow ([0,1],\mu_{\gamma_t})$ is uniformly continuous (μ_{γ_t} coincides with the fine uniform structure on $[0,1]$, i.e. it is the usual uniform structure on $[0,1]$).

7.3.29 Theorem (Tietze's extension theorem). *Let (X,γ) be a normal filter space and (A,γ_A) a subspace (in* **Fil***) of (X,γ). Then every Cauchy continuous map $f : (A,\gamma_A) \longrightarrow ([0,1],\gamma_t)$ has a Cauchy continuous extension $F : (X,\gamma) \longrightarrow ([0,1],\gamma_t)$.*

Proof. Apply 7.3.26. 4) 2. and note that subspaces of filter spaces are formed in **Fil** ($\cong$ **Fil–Mer**) as in **Mer** (cf. 7.1.2. 3) b)).

7.3.30 Remark. For normal R_0–spaces, the classical version of Urysohn's Lemma follows from 7.3.28 (note: $A = \overline{A}$, $B = \overline{B}$ and $A \cap B = \emptyset$ imply $\{A,B\}$ is not near) and the classical Tietze extension theorem is an immediate consequence of 7.3.29 (note: every closed subspace of an R_0–space is a subspace in **Fil**).

7.4 Paracompactness and dimension

7.4.1 Definitions. 1) A filter space (X, γ) is called *merotopically uniform* (shortly: *m–uniform*) provided that (X, μ_γ) is uniform.
2) A semiuniform convergence space $(X, \mathcal{J}_X)$ is called *fully normal* provided that it is regular and its underlying filter space $(X, \gamma_{\mathcal{J}_X})$ is m–uniform.
3) A semiuniform convergence space is called *paracompact* provided that it is T_1 and fully normal.

7.4.2 Remark. Every uniform space (X, μ) described by means of uniform covers is a normal nearness space:
1. (X, μ) is regular, since for each $\mathcal{U} \in \mu$ there is some $\mathcal{V} \in \mu$ with $\mathcal{V} \overset{*}{\prec} \mathcal{U}$, i.e. for each $V \in \mathcal{V}$, there exists some $U \in \mathcal{U}$ with $St(V, \mathcal{V}) \subset U$, and consequently, $\mathcal{V} \prec \{X \backslash V, U\}$ which implies $\{X \backslash V, U\} \in \mu$.
2. (X, μ_c) is regular, namely by 4.3.2.19., (X, μ_c) is a proximity space and by 1. it is regular.

7.4.3 Proposition. *Every fully normal semiuniform convergence space is normal.*

Proof. Let $(X, \mathcal{J}_X) \in |\mathbf{SUConv}|$ be fully normal. It suffices to prove that $(X, \gamma_{\mathcal{J}_X})$ is m–normal. But this is obvious, since $(X, \mu_{\gamma_{\mathcal{J}_X}})$ is uniform by assumption and consequently, by 7.4.2. it is normal.

7.4.4 Corollary. *Every paracompact semiuniform convergence space is T_4, i.e. normal and T_1.*

7.4.5 Proposition. *A filter space is fully normal (as a semiuniform convergence space) iff it is m–uniform.*

Proof. "$\Longrightarrow$". If $(X, \gamma) \in |\mathbf{Fil}|$ such that $(X, \mathcal{J}_\gamma)$ is fully normal, then $(X, \gamma_{\mathcal{J}_\gamma}) = (X, \gamma)$ is m–uniform.
"$\Longleftarrow$". If $(X, \gamma) \in |\mathbf{Fil}|$ is m–uniform, then (X, μ_γ) is uniform and thus it is normal. By 7.3.21, $(X, \mathcal{J}_\gamma)$ is normal and consequently it is regular. Furthermore, $(X, \gamma_{\mathcal{J}_\gamma}) = (X, \gamma)$ is m–uniform by assumption.

7.4.6 Remark. It is easily checked that a symmetric topological space $(X, \mathcal{X})$ is fully normal in the usual sense (cf. 4.3.2.26. 2)) iff $(X, \mu_{\mathcal{X}})$ is a uniform space.

7.4.7 Proposition. *A symmetric Kent convergence space is fully normal (as a semiuniform convergence space) iff it is a fully normal topological space in the usual sense* (cf. 4.3.2.26. 2)).

Proof. "$\Longrightarrow$". Let (X, q) be a symmetric Kent convergence space such that $(X, \mathcal{J}_{\gamma_q})$ is fully normal. Then $(X, \mathcal{J}_{\gamma_q})$ is completely regular and by 7.3.2., (X, q) is a (completely regular) topological space, i.e. $q = q_{\mathcal{X}_q}$. By assumption,

$(X, \gamma_{\mathcal{J}_{\gamma_q}}) = (X, \gamma_q)$ is m–uniform, i.e. $(X, \mu_{\gamma_q}) = (X, \mu_{\mathcal{X}_q})$ is uniform. By 7.4.6., $(X, \mathcal{X}_q)$ (resp. (X, q)) is fully normal.
"$\Longleftarrow$". Let (X, q) be a fully normal R_0–space in the usual sense, i.e. $(X, \mathcal{X}_q)$ is a fully normal symmetric topological space. In order to prove that $(X, \mathcal{J}_{\gamma_q})$ is fully normal, it suffices to show that (X, μ_{γ_q}) is uniform since (X, γ_q) is a filter space and 7.4.5. is valid. But this is obvious because $(X, \mu_{\gamma_q}) = (X, \mu_{\mathcal{X}_q})$ and $(X, \mathcal{X}_q)$ is a fully normal R_0–space (cf. 7.4.6.).

7.4.8 Corollary. *A symmetric topological space is paracompact (as a semiuniform convergence space) iff it is paracompact in the usual sense* (cf. 4.3.2.26. 2)).

7.4.9 Proposition. *A symmetric topological space is fully normal in the usual sense iff it is topological and m–uniform as a filter space.*

Proof. "$\Longrightarrow$". Let $(X, \mathcal{X})$ be a fully normal R_0–space and $(X, \gamma_{q_\mathcal{X}})$ its corresponding filter space. Since $(X, \gamma_{q_\mathcal{X}})$ is complete and $(X, q_{\gamma_{q_\mathcal{X}}}) = (X, q_\mathcal{X})$ is topological, $(X, \gamma_{q_\mathcal{X}})$ is topological. Furthermore, $(X, \mu_{\gamma_{q_\mathcal{X}}}) = (X, \mu_\mathcal{X})$ is uniform, since $(X, \mathcal{X})$ is fully normal. Thus, $(X, \gamma_{q_\mathcal{X}})$ is m–uniform.
"$\Longleftarrow$". Let $(X, \mathcal{X})$ be an R_0–space such that $(X, \gamma_{q_\mathcal{X}})$ is (topological and) m–uniform. By 7.4.5., $(X, \mathcal{J}_{\gamma_{q_\mathcal{X}}})$ is fully normal, and by 7.4.7., $(X, q_\mathcal{X})$ (resp. $(X, \mathcal{X})$) is fully normal.

7.4.10 Corollary. *A topological T_1–space is paracompact in the usual sense iff it is topological and m–uniform as a filter space.*

7.4.11 Proposition. *Every proximity space is fully normal (as a semiuniform convergence space).*

Proof. Let $(X, \mathcal{J}_X)$ be a proximity space, i.e. $\mathcal{J}_X = [\mathcal{V}]$ where $\mathcal{V}$ is a totally bounded uniformity on X. Since $(X, \mathcal{J}_X)$ is uniform, it is regular (cf. 4.2.1.5.). It remains to show that $(X, \gamma_{\mathcal{J}_X})$ is m–uniform. Let $(X, \mu_\mathcal{V})$ be the merotopic space corresponding to $(X, \mathcal{V})$, i.e. $\mu_\mathcal{V} = \{\mathcal{A} \subset \mathcal{P}(X)\colon$ there is some $V \in \mathcal{V}$ with $\mathcal{A}_V \prec \mathcal{A}\}$, where $\mathcal{A}_V = \{V(x) : x \in X\}$. If $\gamma_{\mu_\mathcal{V}}$ denotes the set of all Cauchy filters on $(X, \mu_\mathcal{V})$ (resp. $(X, \mathcal{V})$), then

(1) $\gamma_{\mu_\mathcal{V}} = \gamma_{\mathcal{J}_X}$.

Since $(X, \mu_\mathcal{V})$ is a contigual merotopic space, it is filter-merotopic (cf. 7.1.3.) and thus $\mu_{\gamma_{\mu_\mathcal{V}}} = \mu_\mathcal{V}$ (cf. 7.1.2. 2) c)). Consequently, by (1), one obtains:

(2) $\mu_{\gamma_{\mathcal{J}_X}} = \mu_\mathcal{V}$.

Since $(X, \mu_\mathcal{V})$ is uniform, it follows from (2) that $(X, \gamma_{\mathcal{J}_X})$ is m–uniform.

7.4.12 Corollary. *Every separated proximity space is paracompact (as a semiuniform convergence space).*

7.4.13 Proposition. *A complete uniform space is fully normal (as a semiu-*

niform convergence space) iff its underlying topological space is fully normal.

Proof. "$\Longleftarrow$". Let $(X, \mathcal{V})$ be a complete uniform space such that its underlying topological space $(X, \mathcal{X}_{\mathcal{V}})$ is fully normal. Put $\mathcal{J}_X = [\mathcal{V}]$. Since $(X, \mathcal{V})$ is complete, $\mu_{\gamma_{\mathcal{J}_X}} = \mu_{\gamma_{q\mathcal{X}_{\mathcal{V}}}}$ and since $(X, \mathcal{X}_{\mathcal{V}})$ is fully normal, $\mu_{\gamma_{\mathcal{J}_X}} = \mu_{\gamma_{q\mathcal{X}_{\mathcal{V}}}} = \mu_{\mathcal{X}_{\mathcal{V}}}$ is a uniform structure. Furthermore, $(X, \mathcal{J}_X)$ is regular. Thus, $(X, \mathcal{J}_X)$ is fully normal.
"$\Longrightarrow$". Let $(X, \mathcal{V})$ be a complete uniform space such that $(X, [\mathcal{V}])$ is fully normal. Put $\mathcal{J}_X = [\mathcal{V}]$. By assumption, $(X, \mu_{\gamma_{\mathcal{J}_X}})$ is uniform. Since $\mu_{\gamma_{\mathcal{J}_X}} = \mu_{\mathcal{X}_{\mathcal{V}}}$, $(X, \mathcal{X}_{\mathcal{V}})$ is fully normal.

7.4.14 Remarks. 1) By 7.4.13, the uniform space $\mathbb{R}_u$ of real numbers is paracompact as a semiuniform convergence space. Furthermore, the topological space $\mathbb{R}_t$ of real numbers is paracompact as a semiuniform convergence space (cf. 7.4.8.).
2) A complete uniform space need not have an underlying topological space which is fully normal as the example under 7.3.24. 2) shows.
3) *The underlying Kent convergence space of a fully normal semiuniform convergence space need not be fully normal,* namely, there is a separated proximity space whose underlying topological space is not paracompact: Consider a completely regular Hausdorff space which is not paracompact (e.g. the ordinal space Ω_0 of all ordinals less than the first uncountable ordinal ω_1). Continuing as in 7.3.24. 1) leads to the desired example.

7.4.15 Proposition. *Every subspace (in* **SUConv***) of a fully normal (resp. paracompact) semiuniform convergence space is fully normal (resp. paracompact).*

Proof. Let $(X, \mathcal{J}_X)$ be a fully normal semiuniform convergence space, $U \subset X$ and $\mathcal{J}_U$ the initial **SUConv**–structure on X w.r.t. the inclusion map $i : U \longrightarrow X$:
a) $(U, \mathcal{J}_U)$ is regular since $(X, \mathcal{J}_X)$ is regular (cf. 4.2.16.).
b) $(U, \gamma_{\mathcal{J}_U})$ is a subspace (in **Fil**) of $(X, \gamma_{\mathcal{J}_X})$. Since subspaces in **Fil** ($\cong$ **Fil–Mer**) are formed as in **Mer** (cf. 7.1.2. 3) b)), $(U, \mu_{\gamma_{\mathcal{J}_U}})$ is a subspace in (**Mer**) of $(X, \mu_{\gamma_{\mathcal{J}_X}})$. Since **Unif** is bireflective in **Mer** (cf. 4.3.1.3. 3) a)), $(U, \mu_{\gamma_{\mathcal{J}_U}})$ is uniform, i.e. $(U, \gamma_{\mathcal{J}_U})$ is m–uniform. It follows from a) and b) that $(U, \mathcal{J}_U)$ is fully normal.

Since obviously subspaces of T_1–spaces are T_1–spaces, the above proposition is also valid for paracompactness.

7.4.16 Remarks. 1) As is well–known, subspaces in **Top** (or **Top**$_{\mathbf{S}}$) of paracompact topological spaces need not be paracompact. But if subspaces are formed in **SUConv** one obtains from 7.4.15. the following result:
Subspaces of paracompact topological spaces are paracompact.

2) A paracompact semiuniform convergence space need not be a subspace (in **SUConv**) of a paracompact topological space, because the example under 7.3.26. 2) is a counterexample (indeed, $(X, \mathcal{J}_X)$ is paracompact!). Together with 7.4.15. one

obtains that the class of semiuniform convergence spaces which are subspaces of paracompact topological spaces is a proper subclass of the class of paracompact semiuniform convergence spaces.

3) Since *paracompactness* is not finitely productive in $\mathbf{Top_S}$, it *is also not finitely productive in* **SUConv** (cf. the corresponding result for normality under 7.3.26. 3)).

4) a) A semiuniform convergence space $(X, \mathcal{J}_X)$ is called *topologically metrizable* provided that there is a metric d on X such that $\mathcal{J}_X = \mathcal{J}_{\gamma_{q\mathcal{X}_d}}$ where $\mathcal{X}_d$ is the topology on X induced by d. It has already been mentioned under 4.3.2.26. 2) that every metrizable topological space is paracompact. By 7.4.8. this result reads now as follows:
Every topologically metrizable semiuniform convergence space is paracompact.
b) Let (X, d) be a metric space and (X^*, d^*) its metric completion in the usual sense, i.e. (X^*, d^*) is a complete metric space containing (X, d) as a dense subspace. Then $(X^*, \mathcal{W}_{d^*})$ is a complete uniform space containing $(X, \mathcal{W}_d)$ as a dense subspace. Using a) it follows from 7.4.13. that $(X^*, [\mathcal{W}_{d^*}])$ is paracompact. Thus, by 7.4.15., $(X, [\mathcal{W}_d])$ is paracompact, i.e. *every metrizable semiuniform convergence space is paracompact.*

5) Full normality is related to the other higher separation axioms by means of the following implication scheme:

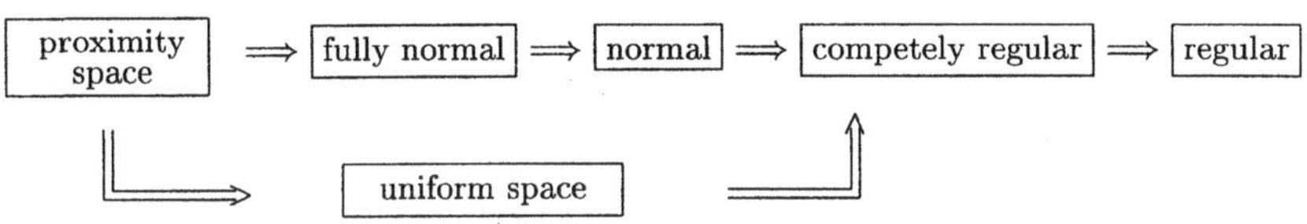

We have already mentioned that a normal T_1–space is called a T_4–*space*. A completely regular (resp. regular) T_1–space is called a $T_{3\frac{1}{2}}$–*space* (resp. T_3–*space*). Thus, adding the T_1–axiom one obtains from the above implication scheme the following one, where 4) is also incorporated:

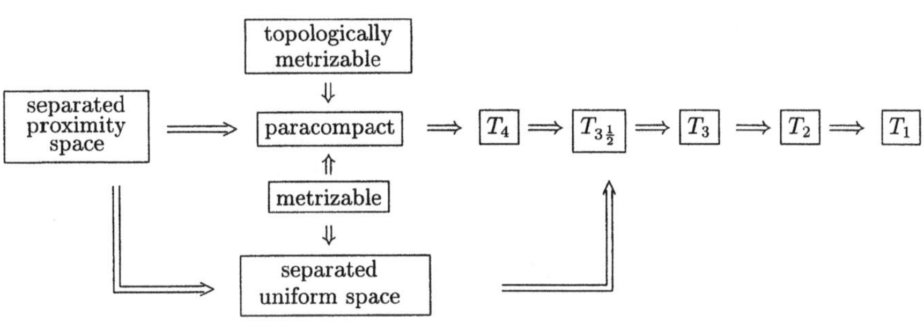

("$T_3 \Longrightarrow T_2$": Let $(X, \mathcal{J}_X) \in |\mathbf{SUConv}|$ be a T_3–space and $\mathcal{F}$ a filter on X converging to $x, y \in X$ in $(X, q_{\gamma_{\mathcal{J}_X}})$. Since $(X, \mathcal{J}_X)$ is regular, $(X, q_{\gamma_{\mathcal{J}_X}})$ is regular by 4.2.1.2. Thus, $\overline{\mathcal{F}}$ converges to x, y in $(X, q_{\gamma_{\mathcal{J}_X}})$. Since $(X, q_{\gamma_{\mathcal{J}_X}})$ is a T_1–space, $\{y\}$ is closed and $X\backslash\{y\}$ is open in $(X, q_{\gamma_{\mathcal{J}_X}})$, i.e. $X\backslash\{y\} \in \mathcal{X}_{q_{\gamma_{\mathcal{J}_X}}}$. Consequently, if x were unequal to y, $X\backslash\{y\}$ would belong to $\overline{\mathcal{F}}$, i.e. there would be some $F \in \mathcal{F}$ such that $\overline{F} = cl_{q_{\gamma_{\mathcal{J}_X}}} F \subset X\backslash\{y\}$, which would imply $y \in X\backslash cl_{q_{\gamma_{\mathcal{J}_X}}} F = \text{int}_{q_{\gamma_{\mathcal{J}_X}}}(X\backslash F) = \{z \in X\colon \text{ for each } \mathcal{G} \in F(X) \text{ with } (\mathcal{G}, z) \in q_{\gamma_{\mathcal{J}_X}}, X\backslash F \in \mathcal{G}\}$ and thus $X\backslash F \in \mathcal{F}$ – a contradiction. Therefore, $(X, q_{\gamma_{\mathcal{J}_X}})$ is a T_2–space, i.e. $(X, \mathcal{J}_X)$ is a T_2–space.)

7.4.17 Definitions. A) Let (X, μ) be a merotopic space.
1) The *large dimension* Dim (X, μ) of (X, μ) is said to be $\leq n$ provided that every uniform cover $\mathcal{U}$ of X has a refinement $\mathcal{V} \in \mu$ of order $\leq n+1$ (i.e. each $x \in X$ is contained in at most $n+1$ elements of $\mathcal{V}$). If Dim $(X, \mu) \leq n$ and the statement Dim $(X, \mu) \leq n-1$ is false, we say Dim $(X, \mu) = n$. If the statement Dim $(X, \mu) \leq n$ is false for all n, then we write Dim $(X, \mu) = \infty$. In case $X = \emptyset$, Dim $(X, \mu) = -1$.
2) The *small dimension* dim (X, μ) of (X, μ) is defined to be the large dimension of (X, μ_c). Especially, dim $(X, \mu) \leq n$ iff every finite uniform cover $\mathcal{U}$ of X has a (finite) refinement $\mathcal{V} \in \mu$ of order $\leq n+1$.
B) Let $(X, \mathcal{J}_X)$ be a semiuniform convergence space.
1) a) The *large filter dimension* $\text{Dim}_f(X, \mathcal{J}_X)$ is defined to be Dim $(X, \mu_{\gamma_{\mathcal{J}_X}})$.
b) The *small filter dimension* $\dim_f(X, \mathcal{J}_X)$ of $(X, \mathcal{J}_X)$ is defined to be dim $(X, \mu_{\gamma_{\mathcal{J}_X}})$.
2) a) The *large uniform dimension* $\text{Dim}_u(X, \mathcal{J}_X)$ of $(X, \mathcal{J}_X)$ is defined to be the large dimension of the (uniformly) merotopic space $(X, \mu_{\mathcal{V}})$ corresponding to the underlying uniform space $(X, \mathcal{V})$ of $(X, \mathcal{J}_X)$.
b) The *small uniform dimension* $\dim_u(X, \mathcal{J}_X)$ of $(X, \mathcal{J}_X)$ is defined to be the small dimension of $(X, \mu_{\mathcal{V}})$ (cf. a)).

7.4.18 Remarks. 1) a) For symmetric topological spaces $(X, \mathcal{X})$, one writes dim $(X, \mathcal{X})$ instead of dim $(X, \mu_{\mathcal{X}})$ (and Dim $(X, \mathcal{X})$ instead of Dim $(X, \mu_{\mathcal{X}})$), where $(X, \mu_{\mathcal{X}})$ is the merotopic space (= topological nearness space) corresponding to $\mathcal{X}$. Then dim $(X, \mathcal{X})$ coincides with *the (Lebesgue) covering dimension* of $(X, \mathcal{X})$ (cf. [103; definition I.4]).
b) For paracompact topological spaces $(X, \mathcal{X})$, dim $(X, \mathcal{X})$ = Dim $(X, \mathcal{X})$ (cf. [102; 9–14]).

2) Let $(X, \mathcal{V})$ be a uniform space. Then the following are valid:
a) dim $(X, \mu_{\mathcal{V}}) = \dim_u(X, [\mathcal{V}])$ coincides with Isbell's uniform dimension $\delta d(X, \mu_{\mathcal{V}})$ and Dim $(X, \mu_{\mathcal{V}}) = \text{Dim}_u(X, [\mathcal{V}])$ is identical with Isbell's large dimension $\Delta d(X, \mu_{\mathcal{V}})$ (cf. [79]).
b) $\dim_u(X, [\mathcal{V}]) = \text{Dim}_u(X, [\mathcal{V}])$ provided that $\text{Dim}_u(X, [\mathcal{V}]) < \infty$ (cf. [79]).

7.4.19 Proposition. 1) *If $(X, \mathcal{X})$ is a symmetric topological space, then*

$\dim_f(X, \mathcal{J}_{\gamma_{q\mathcal{X}}}) = \dim\,(X, \mathcal{X})$ *and* $\mathrm{Dim}_f(X, \mathcal{J}_{\gamma_{q\mathcal{X}}}) = \mathrm{Dim}\,(X, \mathcal{X})$.
2) *If* $(X, \mathcal{X})$ *is a paracompact topological space, then additionally,* $\dim_u(X, \mathcal{J}_{\gamma_{q\mathcal{X}}}) = \dim_f(X, \mathcal{J}_{\gamma_{q\mathcal{X}}}) = \mathrm{Dim}_u(X, \mathcal{J}_{\gamma_{q\mathcal{X}}}) = Dim_f(X, \mathcal{J}_{\gamma_{q\mathcal{X}}})$.

Proof. 1) $(X, \gamma_{q\mathcal{X}})$ is the underlying filter space of $(X, \mathcal{J}_{\gamma_{q\mathcal{X}}})$, where $\gamma_{q\mathcal{X}}$ consists of all convergent filters in $(X, \mathcal{X})$. Hence, $(X, \mu_{\gamma_{q\mathcal{X}}}) = (X, \mu_{\mathcal{X}})$ (cf. 7.2.17. 3) f)) which implies the desired equalities.

2) Since $(X, \mathcal{X})$ is paracompact, there is a finest uniformity $\mathcal{V}$ on X which induces $\mathcal{X}$ and $(X, \mathcal{V})$ is the underlying uniform space of $(X, \mathcal{J}_{\gamma_{q\mathcal{X}}})$ (cf. 4.3.2.26. 3) b)). Then $\mu_{\mathcal{V}} = \mu_{\mathcal{X}}$. This implies (together with 1)) that $\dim_u(X, \mathcal{J}_{\gamma_{q\mathcal{X}}}) = \dim\,(X, \mu_{\mathcal{V}}) = \dim\,(X, \mu_{\mathcal{X}}) = \dim_f(X, \mathcal{J}_{\gamma_{q\mathcal{X}}}) = \dim\,(X, \mathcal{X})$ and $\mathrm{Dim}_u(X, \mathcal{J}_{\gamma_{q\mathcal{X}}}) = \mathrm{Dim}\,(X, \mu_{\mathcal{V}}) = \mathrm{Dim}\,(X, \mu_{\mathcal{X}}) = \mathrm{Dim}_f(X, \mathcal{J}_{\gamma_{q\mathcal{X}}}) = \mathrm{Dim}\,(X, \mathcal{X})$. Consequently, the desired result follows from 7.4.18. 1) b).

7.4.20 Remark. For normal symmetric topological spaces $(X, \mathcal{X})$, $\mathrm{Dim}_f(X, \mathcal{J}_{\gamma_{q\mathcal{X}}})$ may differ from $\dim_f(X, \mathcal{J}_{\gamma_{q\mathcal{X}}})$, e.g. if $(X, \mathcal{X})$ is equal to the ordinal space Ω_0 of all ordinals less than the first uncountable ordinal ω_1, then $\dim \Omega_0 = 0$ and $\mathrm{Dim}\,\Omega_0 = \infty$ (cf. [69; 5.4.8]).

7.4.21 Proposition. *Let* $(X, \mathcal{V})$ *be a proximity space. Then* $\dim_u(X, [\mathcal{V}]) = \dim_f(X, [\mathcal{V}]) = \dim\,(X, \mu_{\mathcal{V}}) = \mathrm{Dim}_u\,(X, [\mathcal{V}]) = \mathrm{Dim}_f(X, [\mathcal{V}]) = \mathrm{Dim}\,(X, \mu_{\mathcal{V}})$.

Proof. Since $\mu_{\gamma_{[\mathcal{V}]}} = \mu_{\mathcal{V}}$ (cf. (2) in the proof of 7.4.11.), $\dim_u(X, [\mathcal{V}]) = \dim\,(X, \mu_{\mathcal{V}}) = \dim\,(X, \mu_{\gamma_{[\mathcal{V}]}}) = \dim_f(X, [\mathcal{V}])$ and $\mathrm{Dim}_u(X, [\mathcal{V}]) = \mathrm{Dim}\,(X, \mu_{\mathcal{V}}) = \mathrm{Dim}_f(X, [\mathcal{V}])$. Furthermore, $\mathrm{Dim}\,(X, \mu_{\mathcal{V}}) = \dim\,(X, \mu_{\mathcal{V}})$ because $(\mu_{\mathcal{V}})_c = \mu_{\mathcal{V}}$.

7.4.22 Remark. For a given proximity space the common value of all dimension functions considered above is known as Smirnov's δ–dimension of it (cf. [135]).

7.4.23 Proposition. *If* $(X, \mathcal{V})$ *is a complete uniform space and* $\mathcal{X}_{\mathcal{V}}$ *the topology induced by* $\mathcal{V}$, *then* $\dim_f(X, [\mathcal{V}]) = \dim\,(X, \mathcal{X}_{\mathcal{V}})$ *and* $\mathrm{Dim}_f(X, [\mathcal{V}]) = \mathrm{Dim}\,(X, \mathcal{X}_{\mathcal{V}})$.

Proof. Since by assumption $\mu_{\gamma_{[\mathcal{V}]}} = \mu_{\mathcal{X}_{\mathcal{V}}}$, the desired result is obvious.

7.4.24 Remarks. 1) If $I\!R^n_t$ denotes the usual topological space with underlying set $I\!R^n$, then $\dim I\!R^n_t = n$ (cf. e.g. [43; 7.3.19]).
2) If $I\!R^n_u$ denotes the usual uniform space with underlying set $I\!R^n$, then $\dim_u I\!R^n_u = \mathrm{Dim}\,I\!R^n_u = n$ (cf. [79; V. 12]).

7.4.25 Corollary. $\mathrm{Dim}_u I\!R^n_t = \mathrm{Dim}_f I\!R^n_t = \mathrm{Dim}\,I\!R^n_t = \mathrm{Dim}_f I\!R^n_u = \mathrm{Dim}_u I\!R^n_u = \dim I\!R^n_t = \dim_f I\!R^n_t = \dim_u I\!R^n_t = \dim_f I\!R^n_u = \dim_u I\!R^n_u = n$.

Proof. Since $I\!R^n_t$ is a paracompact topological space, it follows from 7.4.19. that $\dim_u I\!R^n_t = \dim_f I\!R^n_t = \dim I\!R^n_t = \mathrm{Dim}_u I\!R^n_t = \mathrm{Dim}_f I\!R^n_t = \mathrm{Dim}\,I\!R^n_t$, where $\dim I\!R^n_t = n$ by 7.4.24. 1). Since $I\!R^n_u$ is a complete uniform space, it follows from 7.4.23. that $\dim_f I\!R_u^n = \dim I\!R_t^n$ and $\mathrm{Dim}_f I\!R_u^n = \mathrm{Dim}\, I\!R_t^n$. Thus, by 7.4.24. 2), the corollary is proved.

7.4.26 Remarks. 1) For each merotopic space (X, μ),

$$\dim\,(X, \mu) \leq \mathrm{Dim}\,(X, \mu)$$

(cf. [69; 5.4.7]).
2) For each merotopic space (X, μ) and each subspace (A, μ_A) of (X, μ) (in **Mer**), the following are valid:
a) $\mathrm{Dim}\,(A, \mu_A) \leq \mathrm{Dim}\,(X, \mu)$
b) $\dim\,(A, \mu_A) \leq \dim\,(X, \mu)$
(cf. [69; 5.4.10]).
3) For each regular merotopic space (X, μ) and each dense subspace (A, μ_A) of (X, μ), $\mathrm{Dim}\,(A, \mu_A) = \mathrm{Dim}\,(X, \mu)$ (cf. [69; 5.4.11.]).

7.4.27 Corollary. *If $(X, \mathcal{J}_X)$ is a semiuniform convergence space, then the following are valid:*
a) $\dim_f(X, \mathcal{J}_X) \leq \mathrm{Dim}_f(X, \mathcal{J}_X)$,
b) $\dim_u(X, \mathcal{J}_X) \leq \mathrm{Dim}_u(X, \mathcal{J}_X)$.

Proof. Apply 7.4.26. 1).

7.4.28 Corollary. 1) *Let $(X, \mathcal{J}_X)$ be a semiuniform convergence space and $(A, \mathcal{J}_A)$ a subspace (in **SUConv**) of $(X, \mathcal{J}_X)$. Then the following are satisfied:*
a) $\mathrm{Dim}_f(A, \mathcal{J}_A) \leq \mathrm{Dim}_f(X, \mathcal{J}_X)$,
b) $\dim_f(A, \mathcal{J}_A) \leq \dim_f(X, \mathcal{J}_X)$.

2) *Let $(X, \mathcal{J}_X)$ be a principal uniform limit space (= uniform space) and $(A, \mathcal{J}_A)$ a subspace (in **SUConv**) of $(X, \mathcal{J}_X)$. Then the following are satisfied:*
a) $\mathrm{Dim}_u(A, \mathcal{J}_A) \leq \mathrm{Dim}_u(X, \mathcal{J}_X)$,
b) $\dim_u(A, \mathcal{J}_A) \leq \dim_u(X, \mathcal{J}_X)$.

Proof. Apply 7.4.26. 2).

7.4.29 Corollary. *Let **X** be a symmetric topological space and **A** a closed subspace in **Top** (or **Top$_S$**) of **X**. Then*

$$\dim \mathbf{A} \leq \dim \mathbf{X} \quad \text{and} \quad \mathrm{Dim}\, \mathbf{A} \leq \mathrm{Dim}\, \mathbf{X}\,.$$

Proof. Since closed subspaces in **Top$_S$** are formed as in **SUConv**, the desired result follows from 7.4.28. 1) and 7.4.19. 1).

7.4.30 Corollary. 1) *Let $(X, \mathcal{X})$ be a regular topological space and $(A, \mathcal{J}_A)$ a dense subspace (in **SUConv**) of $(X, \mathcal{J}_{\gamma_{q\mathcal{X}}})$. Then*

$$\mathrm{Dim}_f(X, \mathcal{J}_{\gamma_{q\mathcal{X}}}) = \mathrm{Dim}_f(A, \mathcal{J}_A)\,.$$

2) *Let $(X, \mathcal{J}_X)$ be a principal uniform limit space (= uniform space) and $(A, \mathcal{J}_A)$ a dense subspace (in* **SUConv**) *of $(X, \mathcal{J}_X)$. Then*

$$\mathrm{Dim}_u(X, \mathcal{J}_X) = \mathrm{Dim}_u(A, \mathcal{J}_A) .$$

Proof. Apply 7.4.26. 3).

7.4.31 Remark. If (X, μ) is a normal nearness space, then dim $(X, \mu) \leq n$ iff every uniformly continuous map of any subspace (A, μ_A) from (X, μ) into the n-sphere S^n (endowed with its usual uniform structure) has a uniformly continuous extension over (X, μ) (cf. [111; 7.2.17]).

7.4.32 Theorem. 1) *Let (X, γ) be a normal filter space. Then* $\dim_f(X, \mathcal{J}_\gamma) \leq n$ *iff every Cauchy continuous map from any subspace (A, γ_A) (in* **Fil**) *of (X, γ) into the n-sphere S^n has a Cauchy continuous extension over (X, γ), where S^n is endowed with its usual topological* **Fil**-*structure consisting of all convergent filters w.r.t. the usual topology of S^n.*
2) *Let $(X, \mathcal{J}_X)$ be a principal uniform limit space (= uniform space). Then* $\dim_u(X, \mathcal{J}_X) \leq n$ *iff every uniformly continuous map from any subspace (in* **SUConv**) *of $(X, \mathcal{J}_X)$ into the n-sphere S^n has a uniformly continuous extension over $(X, \mathcal{J}_X)$, where S^n is endowed with its usual uniform structure.*

Proof. For every normal filter space (X, γ), (X, μ_γ) is a normal nearness space, and for every uniform space $(X, [\mathcal{V}])$, $(X, \mu_{\mathcal{V}})$ is a normal nearness space. Furthermore, the corresponding merotopic structure of the usual **Fil**-structure of S^n is the usual uniform structure. Thus, the above theorem follows from the result under 7.4.31.

7.4.33 Corollary. *Let $(X, \mathcal{X})$ be a normal R_0-space. Then* dim $(X, \mathcal{X}) \leq n$ *iff every continuous map from any closed subspace $(A, \mathcal{X}_A)$ (in* **Top**) *of $(X, \mathcal{X})$ into the n-sphere S^n has a continuous extension over $(X, \mathcal{X})$.*

7.4.34 Remarks. 1) In [111] the Čech cohomology theory for pairs of nearness spaces based on finite uniform covers has been studied. If $\mathbf{X}$ is a nearness space and $\mathbf{A}$ a subspace (in **Near**) of $\mathbf{X}$, then the n-dimensional Čech cohomology group (with coefficients $\mathbb{Z}$) of $(\mathbf{X}, \mathbf{A})$ based on finite uniform covers is denoted by $\check{H}^n_f(\mathbf{X}, \mathbf{A})$ (or $\check{H}^n_f(\mathbf{X})$ if A is empty). *For every normal nearness space* $\mathbf{X}$ *of finite small dimension,* dim $\mathbf{X} \leq n$ *iff for every integer $m \geq n+1$ and every subspace* $\mathbf{A}$ *of* $\mathbf{X}$, $\check{H}^m_f(\mathbf{X}, \mathbf{A}) = 0$ (cf. [111; 7.3.3.]).

2) a) If $\mathbf{X} = (X, \gamma)$ is a filter space and $\mathbf{A} = (A, \gamma_A)$ a subspace (in **Fil**) of $\mathbf{X}$, then $\check{H}^n_f(\mathbf{X}, \mathbf{A})$ is defined to be $\check{H}^n_f((X, \mu_\gamma), (A, \mu_{\gamma_A}))$.
b) If $\mathbf{X} = (X, \mathcal{V})$ is uniform space and $\mathbf{A} = (A, \mathcal{V}_A)$ a subspace (in **Unif**) of $\mathbf{X}$, then $\check{H}^n_f(\mathbf{X}, \mathbf{A})$ is defined to be $\check{H}^n_f((X, \mu_{\mathcal{V}}), (A, \mu_{\mathcal{V}_A}))$.

7.4.35 Theorem. 1) *Let* $\mathbf{X}$ *be a normal filter space of finite small filter dimension. Then* $\dim_f \mathbf{X} \leq n$ *iff for every integer $m \geq n+1$ and every subspace* $\mathbf{A}$ *(in* **Fil**) *of* $\mathbf{X}$, $\check{H}^m_f(\mathbf{X}, \mathbf{A}) = 0$.

2) *Let* **X** *be a uniform space of finite small uniform dimension. Then* $\dim_u \mathbf{X} \leq n$ *iff for every integer* $m \geq n+1$ *and every subspace* **A** *(in* **Unif***) of* **X**, $\check{H}_f^m(\mathbf{X}, \mathbf{A}) = 0$.

Proof. Apply the result under 7.4.34. 1) and recall 7.4.34. 2).

7.4.36 Remark. If **X** is a uniform space of finite large uniform dimension, then by 7.4.2.7. 2) b), $\dim_u \mathbf{X} = \operatorname{Dim}_u \mathbf{X}$. Thus, in this case, 7.4.35. 2) gives a cohomological characterization of the large uniform dimension of **X**, too.

7.5 Subcompact and sub–(compact Hausdorff) spaces

7.5.1 Definitions. A filter space (X, γ) is called
a) *merotopically contigual* (shortly: *m–contigual*) provided that its corresponding filter–merotopic space (X, μ_γ) is contigual,
b) *subcompact* provided that it is subtopological and *m*–contigual.

7.5.2 Remark. *The construct* **C–Mer** *of contigual merotopic spaces (and uniformly continuous maps) is bireflective in* **Mer** *and hence in* **Fil–Mer** (cf. 7.1.3): If (X, μ) is a merotopic space and μ_c is defined as under 7.3.18. 1) b), then $1_X : (X, \mu) \longrightarrow (X, \mu_c)$ is the desired bireflection of (X, μ) w.r.t. **C–Mer** (cf. also [111; 3.1.3.3].).

7.5.3 Proposition. *A filter space* (X, γ) *is* m*–contigual iff the following is satisfied: If from each* $\mathcal{F} \in \gamma$ *some* $F_{\mathcal{F}}$ *is chosen, then there are finitely many* $\mathcal{F}_1, \dots, \mathcal{F}_n \in \gamma$ *such that for each* $\mathcal{G} \in \gamma$ *there is some* $i \in \{1, \dots, n\}$ *with* $F_{\mathcal{F}_i} \in \mathcal{G}$.

Proof. 1) Let (X, γ) be m–contigual and choose from each $\mathcal{F} \in \gamma$ some $F_{\mathcal{F}}$. Then $\mathcal{A} = \{F_{\mathcal{F}} : \mathcal{F} \in \gamma\} \in \mu_\gamma$ and by assumption there is some finite $\mathcal{B} \in \mu_\gamma$ such that $\mathcal{B} \subset \mathcal{A}$, i.e. there are finitely many $\mathcal{F}_1 \cdots \mathcal{F}_n \in \gamma$ such that for each $\mathcal{G} \in \gamma$ there is some $i \in \{1, \dots, n\}$ with $F_{\mathcal{F}_i} \in \mathcal{G}$.
2) The inverse implication is proved similarly.

7.5.4 Remark. In the following the *canonical completion* (or *Herrlich completion*) (X^*, μ^*) of a nearness space (X, μ) is needed:

a) If (X, μ) is a nearness space, then $\mathcal{A} \subset \mathcal{P}(X)$ is called *near* provided that $\{X \backslash A : A \in \mathcal{A}\} \notin \mu$. Let ξ_μ denote the set of all near collections in (X, μ). A non–empty subset $\mathcal{A}$ of $\mathcal{P}(X)$ is called a *cluster* provided that $\mathcal{A}$ is a maximal element of ξ_μ, ordered by inclusion. A point $x \in X$ is called an *adherence point* of a subset $\mathcal{A}$ of $\mathcal{P}(X)$ provided that $x \in \bigcap_{A \in \mathcal{A}} \overline{A}$ (where the closure $\overline{A}$ of A is formed in $(X, \mathcal{X}_\mu)$ [cf. 7.2.17. 3) a)]). (X, μ) is called *complete* provided that every cluster has an adherence point.

b) α) A uniform space (X, μ) is complete as a nearness space iff it is complete in the usual sense [i.e. iff each Cauchy filter converges] (cf. [111; 6.2.2]).
β) Every topological nearness space (X, μ) is complete, since $\mathcal{A} \subset \mathcal{P}(X)$ is near iff $\bigcap_{A \in \mathcal{A}} \overline{A} \neq \emptyset$.

c) The canonical completion (X^*, μ^*) of a nearness space (X, μ) can be constructed as follows: Put $X^* = X \cup X'$ where X' is the set of all clusters in (X, μ) without an adherence point. If μ^* denotes the set of all covers $\mathcal{U}^*$ of X^* for which there exists some $\mathcal{U} \in \mu$ with $o(\mathcal{U}) \prec \mathcal{U}^*$, where $o(\mathcal{U}) = \{\text{int}_\mu U \cup \{x^* \in X' : (X \backslash U) \notin x^*\} : U \in \mathcal{U}\}$, then (X^*, μ^*) is a complete nearness space containing (X, μ) as a dense subspace (cf. [111; 6.2.3.]).

d) *If* (X, μ) *is a contigual nearness space, then* (X^*, μ^*) *is a compact topological nearness space* (cf. [111; 6.2.11]).

7.5.5 Theorem. *Let* (X, γ) *be a filter spae. Then the following are equivalent:*
(1) (X, γ) is subcompact,
(2) (X, γ) is a dense subspace (in **Fil**) of a compact symmetric topological space (regarded as a filter space),
(3) (X, γ) is a subspace (in **Fil**) of a compact symmetric topological space (regarded as a filter space).

Proof. (1) $\Longrightarrow$ (2). By 7.2.20 (X, μ_γ) is a contigual nearness space and hence, by 7.5.4. d), (X^*, μ_γ^*) is a compact topological nearness space. Since (X^*, μ_γ^*) is topological, $\mu_\gamma^* = \mu_{\mathcal{X}_{\mu_\gamma^*}}$. Put $\mathcal{X}^* = \mathcal{X}_{\mu_\gamma^*}$. Hence, $\mu_{\gamma_{q_{\mathcal{X}^*}}} = \mu_{\mathcal{X}^*} = \mu_\gamma^*$, i.e. (X^*, μ_γ^*) is filtermerotopic. Consequently, since (X, μ_γ) is a subspace in **Near** of (X^*, μ_γ^*), $(X, \gamma_{\mu_\gamma}) = (X, \gamma)$ is a subspace in **Fil** of $(X^*, \gamma_{\mu_\gamma^*}) = (X^*, \gamma_{q_{\mathcal{X}^*}})$, where $(X^*, \gamma_{q_{\mathcal{X}^*}})$ is a compact symmetric topological space regarded as a filter space. Furthermore, X is dense in $(X^*, \mathcal{X}^*)$, i.e. in $(X^*, \gamma_{q_{\mathcal{X}^*}})$.

(2) $\Longrightarrow$ (3). This implication is obvious.

(3) $\Longrightarrow$ (1). Let (X, γ) be a subspace in **Fil** of a compact topological filter space (X', γ'), i.e. there is a compact symmetric topological space $(X', \mathcal{X}')$ such that $\gamma' = \gamma_{q_{\mathcal{X}'}}$. By 7.2.11., (X, γ) is subtopological. Furthermore, since $\mu_{\mathcal{X}'} = \mu_{\gamma_{q_{\mathcal{X}'}}} = \mu_{\gamma'}$, (X', μ'_γ) is a contigual nearness space and (X, μ_γ) is a subspace of it in **Fil–Mer**. By 7.5.2., (X, μ_γ) is contigual, i.e. (X, γ) is m–contigual. Thus, (X, γ) is subcompact.

7.5.6 Corollary. *The construct* **Sub$_{\mathbf{Fil}}$Comp** *of subcompact filter spaces (and Cauchy continuous maps) is bireflective in* **Fil**.

Proof. 1) **Sub$_{\mathbf{Fil}}$Comp** is closed under formation of subspaces in **Fil** by 7.5.5. since subspaces of subspaces are subspaces.

2) **Sub$_{\mathbf{Fil}}$Comp** is closed under formation of products in **Fil** by 7.5.5., since products of subspaces are subspaces of products and Tychonoff's theorem applies.

3) Since every indiscrete filter space is an indiscrete topological space regarded

as a filter space, it is compact and topological. Consequently, all indiscrete filter spaces belong to $|\mathbf{Sub_{Fil}Comp}|$.

Since $\mathbf{Sub_{Fil}Comp}$ is a full and isomorphism–closed subconstruct of **Fil**, it follows from 1) – 3) that it is bireflective in **Fil**.

7.5.7 Remark. Let $(X, \mathcal{X})$ be a topological T_1–space and $(X, \mu_{\mathcal{X}})$ its corresponding topological nearness space. Then $(X, (\mu_{\mathcal{X}})_c)$ is a contigual nearness space such that $\mathcal{X}_{(\mu_{\mathcal{X}})_c} = \mathcal{X}$. By 7.5.4. d), its Herrlich completion $(X^*, (\mu_{\mathcal{X}})_c^*)$ is a compact topological nearness space which is $T_1[= N_1]$ (cf. [111; 6.2.5. ① 1.], i.e. $(X^*, \mathcal{X}_{(\mu_{\mathcal{X}})_c^*})$ is a compact topological T_1–space containing $(X, \mathcal{X})$ as a dense subspace. Then $(X^*, \mathcal{X}_{(\mu_{\mathcal{X}})_c^*})$ is called the *Wallman compactification* of $(X, \mathcal{X})$, denoted by $w(X)$. It has the following extension property:

For each compact Hausdorff space $(Y, \mathcal{Y})$ and each continuous map $f : (X, \mathcal{X}) \longrightarrow (Y, \mathcal{Y})$ there is a unique continuous map $\overline{f} : w(X) \longrightarrow (Y, \mathcal{Y})$ such that $\overline{f} \circ i = f$, where $i : (X, \mathcal{X}) \longrightarrow w(X)$ denotes the inclusion map.

(*Proof.* Since $(Y, \mathcal{Y})$ is compact, $(Y, \mu_{\mathcal{Y}})$ is a contigual nearness space. By assumption, $f : (X, \mu_{\mathcal{X}}) \longrightarrow (Y, \mu_{\mathcal{Y}})$ is uniformly continuous, and since $1_X : (X, \mu_{\mathcal{X}}) \longrightarrow (X, (\mu_{\mathcal{X}})_c)$ is the contigual bireflection of $(X, \mu_{\mathcal{X}})$, $f : (X, (\mu_{\mathcal{X}})_c) \longrightarrow (Y, \mu_{\mathcal{Y}})$ is uniformly continuous. By means of the extension theorem for nearness spaces (cf. [111; 6.2.9.])), there is a unique uniformly continuous map $\overline{f} : (X^*, (\mu_{\mathcal{X}})_c^*) \longrightarrow (Y, \mu_{\mathcal{Y}})$ such that $\overline{f} \circ i = f$, where $i : (X, (\mu_{\mathcal{X}})_c) \longrightarrow (X^*, (\mu_{\mathcal{X}})_c^*)$ denotes the inclusion map, since $(Y, \mu_{\mathcal{Y}})$ is complete as a topological nearness space and by assumption also regular and T_1 $(= N_1)$. Thus, for the underlying topological spaces, one obtains the desired result).

If $(X, \mathcal{X})$ is a normal Hausdorff space, then $(X, \mu_{\mathcal{X}})$ is a normal topological nearness space which is T_1, i.e. $(X, (\mu_{\mathcal{X}})_c)$ is regular, T_1, and contigual. Put $\mathcal{X}_{(\mu_{\mathcal{X}})_c^*} = \mathcal{X}^*$. Using [111; 6.2.8. and 6.2.7. ③], $w(X) = (X^*, \mathcal{X}^*)$ is a compact Hausdorff space containing $(X, \mathcal{X})$ as a dense subspace and it follows from the above extension property that *for each normal Hausdorff space $(X, \mathcal{X})$ the Wallman compactification $w(X)$ coincides with the Stone–Čech compactification $\beta(X)$ (up to isomorphism).*

7.5.8 Definition. A semiuniform convergence space $(X, \mathcal{J}_X)$ is called *subcompact* provided that it is **Fil**–determined and its corresponding filter space $(X, \gamma_{\mathcal{J}_X})$ is subcompact.

7.5.9 Corollary. *If $(X, \mathcal{J}_X)$ is a semiuniform convergence space, then the following are equivalent:*
(1) $(X, \mathcal{J}_X)$ is subcompact,
(2) $(X, \mathcal{J}_X)$ is a dense subspace (in **SUConv***) of a compact symmetric topological space (regarded as a semiuniform convergence space),*
(3) $(X, \mathcal{J}_X)$ is a subspace (in **SUConv***) of a compact symmetric topological space (regarded as a semiuniform convergence space).*

Proof. Apply 7.5.5.

7.5.10 Proposition. *The construct* $\mathbf{Subs_{SUConv}Comp}$ *of all subcompact semi-uniform convergence spaces (and uniformly continuous maps) is a bireflective (full and isomorphism–closed) subconstruct of* **SUConv**.

Proof. By (3) of the above corollary subcompactness is closed under formation of subspaces and products (Tychonoff!), and all indiscrete **SUConv**–objects are compact and topological and hence subcompact.

7.5.11 Proposition. *Let (X,γ) be a subtopological filter space and (X,μ_γ) its corresponding nearness space. Then, $\mathcal{X}_{q_\gamma} = \mathcal{X}_{\mu_\gamma}$.*

Proof. The bicoreflection of (X,μ_γ) w.r.t. **T–Near** is given by

$(*)\quad 1_X : (X, \mu_{\mathcal{X}_{\mu_\gamma}}) \longrightarrow (X, \mu_\gamma)$.

Since (X,γ) is weakly subtopological, (X,q_γ) is topological, i.e. $q_\gamma = q_{\mathcal{X}_{q_\gamma}}$. By 7.2.17. 3) f), $\mu_{\gamma_{q_{\mathcal{X}_{q_\gamma}}}} = \mu_{\mathcal{X}_{q_\gamma}}$.

On the other hand, $1_X : (X,\gamma_{q_\gamma}) \longrightarrow (X,\gamma)$ is the bicoreflection of (X,γ) w.r.t. **T–Fil** ($\cong$ **T–Near**). Since $\mu_{\gamma_{q_\gamma}} = \mu_{\gamma_{q_{\mathcal{X}_{q_\gamma}}}} = \mu_{\mathcal{X}_{q_\gamma}}$,

$(**)\quad 1_X : (X, \mu_{\mathcal{X}_{q_\gamma}}) \longrightarrow (X, \mu_\gamma)$

is then the bicoreflection of (X,μ_γ) w.r.t. **T–Near**. It follows from $(*)$ and $(**)$ that $\mu_{\mathcal{X}_{\mu_\gamma}} = \mu_{\mathcal{X}_{q_\gamma}}$, which implies $\mathcal{X}_{\mu_\gamma} = \mathcal{X}_{q_\gamma}$.

7.5.12 Corollary. *Let (X,γ) be a subtopological filter space. Then (X,γ) is T_1 (i.e. $(X,\mathcal{X}_{q_\gamma})$ is T_1) iff (X,μ_γ) is T_1 (i.e. $(X,\mathcal{X}_{\mu_\gamma})$ is T_1).*

7.5.13 Definitions. A filter space (X,γ) is called

1) *m–proximal* provided that (X,μ_γ) is a proximity space (= precompact uniform space),

2) *sub–(compact Hausdorff)* provided that it is T_1 and m–proximal.

7.5.14 Theorem. *Let (X,γ) be a filter space. Then the following are equivalent:*

(1) *(X,γ) is sub–(compact Hausdorff),*

(2) *(X,γ) is a dense subspace (in* **Fil**) *of a compact Hausdorff space (regarded as a filter space).*

(3) *(X,γ) is a subspace (in* **Fil**) *of a compact Hausdorff space (regarded as a filter space).*

Proof. (1) $\Longrightarrow$ (2). Since (X,γ) is T_1 and m–proximal, (X,μ_γ) is a separated proximity space (cf. the above corollary and note that by 7.2.20 (X,γ) is subtopological, since (X,μ_γ) is a proximity space and thus a nearness space). Consequently, the Hausdorff completion (X^*,μ^*_γ) of (X,μ_γ) is compact (i.e. $(X^*,\mathcal{X}_{\mu^*_\gamma})$

is compact) and contains (X, μ_γ) as a dense subspace. Put $\mathcal{X}_{\mu_\gamma^*} = \mathcal{X}^*$. Since a compact Hausdorff space is uniquely uniformizable and 7.2.17. 3) f) is valid, $\mu_\gamma^* = \mu_{\mathcal{X}^*} = \mu_{\gamma_{q_{\mathcal{X}^*}}}$ and (X, γ) is a dense subspace of $(X^*, \gamma_{q_{\mathcal{X}^*}})$. q.e.d.

(2) $\Longrightarrow$ (3) is obvious.

(3) $\Longrightarrow$ (1). Let (X, γ) be a subspace (in **Fil**) of (X', γ') with $\gamma' = \gamma_{q_{\mathcal{X}'}}$, where $(X', \mathcal{X}')$ is a compact Hausdorff space. Then (X, γ) is T_1 as a subspace of a T_1–space. Furthermore, (X, μ_γ) is a subspace of $(X', \mu_{\gamma'}) = (X', \mu_{\mathcal{X}'})$ in **Fil–Mer**, and by 7.1.2. 3) b) also a subspace in **Mer**. Since $(X', \mu_{\mathcal{X}'})$ is contigual and uniform, it follows from 7.5.2. and 4.3.1.3. 3) a) that (X, μ_γ) is contigual and uniform, in other words: a proximity space, i.e. (X, γ) is m–proximal.

7.5.15 Definition. A semiuniform convergence space $(X, \mathcal{J}_X)$ is called *sub–(compact Hausdorff)* provided that it is **Fil**–determined and its corresponding filter space $(X, \gamma_{\mathcal{J}_X})$ is sub–(compact Hausdorff).

7.5.16 Corollary. *If $(X, \mathcal{J}_X)$ is a semiuniform convergence space, then the following are equivalent:*

(1) *$(X, \mathcal{J}_X)$ is sub–(compact Hausdorff),*

(2) *$(X, \mathcal{J}_X)$ is a dense subspace (in* **SUConv***) of a compact Hausdorff space (regarded as a semiuniform convergence space),*

(3) *$(X, \mathcal{J}_X)$ is a subspace (in* **SUConv***) of a compact Hausdorff space (regarded as a semiuniform convergence space).*

Proof. Apply 7.5.14.

7.5.17 Proposition. *The construct* $\mathbf{Sub_{SUConv}CompH}$ *of all sub–(compact Hausdorff) semiuniform convergence spaces (and uniformly continuous maps) is an epireflective (full and isomorphism–closed) subconstruct of* **SUConv**.

Proof. $\mathbf{Sub_{SUConv}CompH}$ is closed under formation of subspaces and products (in **SUConv**) by means of (3) of the above corollary (Tychonoff!).

7.5.18 Proposition. *Every sub–(compact Hausdorff) semiuniform convergence space is paracompact.*

Proof. Let $(X, \mathcal{J}_X) \in |\mathbf{SUConv}|$ be sub–(compact Hausdorff). Since each compact Hausdorff space is a paracompact topological space and thus, by 7.4.8., also paracompact as a semiuniform convergence space, it follows from 7.4.15. and 7.5.16. that $(X, \mathcal{J}_X)$ is paracompact.

7.5.19 Remarks. 1) Obviously, the following implication scheme is valid for semiuniform convergence spaces:

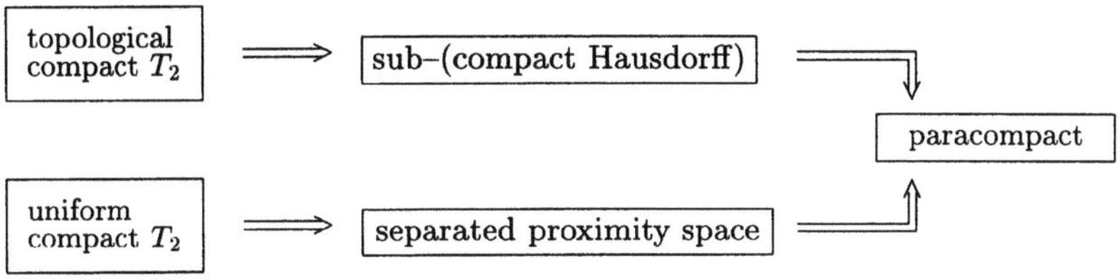

2) a) *The construct* **M–ProxFil** *of m–proximal filter spaces* (*and Cauchy continuous maps*) *is* (*concretely*) *isomorphic to* **Prox** via the restriction of the isomorphism between **Fil** and **Fil–Mer** [cf. 7.1.2. 2) and remember 7.1.3.]. Furthermore, every m–proximal filter space is a uniformizable Cauchy space; namely, if (X, γ) is an m–proximal filter space, (X, μ_γ) is a proximity space and $\gamma = \gamma_{\mu_\gamma} = \gamma_{\mathcal{W}_{\mu_\gamma}}$, whereas the inverse implication is not true (the underlying Cauchy space of $\mathbb{R}_u$ is a counterexample). According to 7.1.6., **Prox** is bireflective in **Chy–Mer**, in other words: **M–ProxFil** is bireflective in **Chy**. By restriction of the bireflector from **Chy** into **M–ProxFil** to **UChy**, we obtain that **M–ProxFil** *is bireflective in* **UChy**.

b) It follows from a) that *the subconstructs* (*of* **SUConv**) $\mathbf{Sub_{SUConv}CompH}$ *and* **SepProx** *of separated proximity spaces* (*and uniformly continuous maps*) *are concretely isomorphic*, but $|\mathbf{Sub_{SUConv}CompH}| \cap |\mathbf{SepProx}|$ does not contain a space with more than one point, since those semiuniform convergence spaces which are simultaneously uniform and **Fil**–determined are exactly the indiscrete uniform spaces (cf. exercise 57)).

Appendix

Some Algebraically Topological Aspects in the Realm of Convenient Topology

A.1 Cohomology for filter spaces

A.1.1. In order to give an exact definition of a cohomology theory for filter spaces we introduce at first the category $\mathbf{Fil}_2$ of pairs of filter spaces:
Objects of $\mathbf{Fil}_2$ are pairs $((X, \gamma_X), (Y, \gamma_Y))$ – shortly (X, Y) – where (X, γ_X) is a filter space and (Y, γ_Y) a subspace of (X, γ_X) in $\mathbf{Fil}$.
Morphisms $f : (X, Y) \to (X', Y')$ are Cauchy continuous maps $f : X \to X'$ such that $f[Y] \subset Y'$.

Definition. Let G be a fixed abelian group. A *cohomology theory* for filter spaces with coefficients G is a pair (H^*, δ^*) where $H^* = (H^q)_{q \in \mathbb{Z}}$ is a family of contravariant functors $H^q : \mathbf{Fil}_2 \to \mathbf{Ab}$ from the category $\mathbf{Fil}_2$ into the construct $\mathbf{Ab}$ of abelian groups (and homomorphisms) for each integer q and $\delta^* = (\delta^q)_{q \in \mathbb{Z}}$ is a family of natural transformations $\delta^q : H^q \circ T \to H^{q+1}$ with a functor $T : \mathbf{Fil}_2 \to \mathbf{Fil}_2$ defined by $T(X, Y) = (Y, \emptyset)$ and $T(f) = f|_Y$ for each $f : (X, Y) \to (X', Y')$ such that the following are satisfied:

1) *Exactness axiom.* For any $(X, Y) \in |\mathbf{Fil}_2|$ with inclusion maps $i : (Y, \emptyset) \to (X, \emptyset)$ and $j : (X, \emptyset) \to (X, Y)$ there is an exact sequence

$$\ldots \xrightarrow{\delta^{q-1}_{(X,Y)}} H^q(X, Y) \xrightarrow{H^q(j)} H^q(X, \emptyset) \xrightarrow{H^q(i)} H^q(Y, \emptyset) \xrightarrow{\delta^q_{(X,Y)}} H^{q+1}(X, Y) \to \ldots$$

2) *Uniform homotopy axiom.* If $f : (X, Y) \to (Z, W)$ and $g : (X, Y) \to (Z, W)$ are uniformly homotopic (i.e. if the pairs (X, Y) and (Z, W) of filter spaces are considered to be pairs of filtermerotopic spaces, there is a uniform homotopy H between f and g, i.e. a uniformly continuous map $H : (X \times I, Y \times I) \to (Z, W)$ such that $H(\cdot, 0) = f$ and $H(\cdot, 1) = g$, where I denotes the unit interval $[0, 1]$ with its usual uniform structure and $\times$ stands for forming products in $\mathbf{Mer}$),

then $H^q(f) = H^q(g)$ for each integer q.

3) *Excision axiom.* If Y and U are subspaces of $(X, \gamma) \in |\mathbf{Fil}|$ such that $X \backslash U \in \mathcal{F}$ or $Y \in \mathcal{F}$ for each $\mathcal{F} \in \gamma$, then the inclusion map $i : (X \backslash U, Y \backslash U) \to (X, Y)$ induces isomorphisms

$$H^q(i) : H^q(X, Y) \to H^q(X \backslash U, Y \backslash U)$$

for each integer q.

4) *Dimension axiom.* If P is a filter space with a single point, then

$$H^q(P, \emptyset) \cong \begin{cases} 0 & \text{if } q \neq 0 \\ G & \text{if } q = 0. \end{cases}$$

A.1.2.1. From now on let G be a fixed abelian group with more than one element. The Alexander cohomology group will be taken to be based on G and explicit denotation of G will be suppressed. Let $(X, Y) \in |\mathbf{Fil_2}|$. Then the Alexander cochain complex

$$0 \to C^0(X, Y) \xrightarrow{\partial^0} C^1(X, Y) \to \cdots \to C^{\nu-1}(X, Y) \xrightarrow{\partial^{\nu-1}} C^{\nu}(X, Y) \to \cdots$$

is defined as follows:

(1) For each non-negative integer q, let $F^q(X)$ denote the abelian group of all functions $f : X^{n+1} \to G$ with the operations being defined pointwise, and $F^q(X) = 0$ for each negative integer q. A function $f \in F^q(X), q \geq 0$, is called *locally zero* provided that for each Cauchy filter $\mathcal{F}$ on X there is some $F \in \mathcal{F}$ such that $f(x_0, \ldots, x_q) = 0$ for any $(q+1)$–tuple of elements of F. The subgroup of $F^q(X)$ consisting of all locally zero functions is denoted by $F_0^q(X)$. Let $C^q(X)$ be the quotient group $F^q(X)/F_0^q(X)$. The inclusion map $i : Y \to X$ gives rise to a homomorphism $F^q(i) : F^q(X) \to F^q(Y)$, defined by $F^q(i)(f)(y_0, \ldots, y_q) = f(y_0, \ldots, y_q)$, which induces a homomorphism $C^q(i) : C^q(X) \to C^q(Y)$, whose kernel is denoted by $C^q(X, Y)$.

(2) Let us define homomorphisms $\delta^q : F^q(X) \to F^{q+1}(X), q \geq 0$, by

$$\delta^q(f)(x_0, \ldots, x_{q+1}) = \sum_{i=0}^{q+1} (-1)^i f(x_0, \ldots, \widehat{x}_i, \ldots, x_{q+1})$$

where $\widehat{x}_i$ means that x_i is to be omitted. δ^q induces a homomorphism $(\delta^q)^* : C^q(X) \to C^{q+1}(X)$. Obviously, $(\delta^q)^*[\operatorname{Ker} C^q(i)] \subset \operatorname{Ker} C^{q+1}(i)$. The induced homorphism $\partial^q : C^q(X, Y) \to C^{q+1}(X, Y)$ is called *coboundary operator.* Furthermore, $\partial^{q+1} \circ \partial^q = 0$ for each integer q. The quotient group $H^q(X, Y) = \operatorname{Ker} \partial^q / \operatorname{Im} \partial^{q-1}$ is called the *q–dimensional Alexander cohomology group* of the pair (X, Y) of filter spaces.
In the following we write $H^q(X)$ instead of $H^q(X, \emptyset)$. Obviously, $H^q(X, X) = 0$.

A.1.2.2. Let $f : (X, \gamma) \to (X', \gamma')$ be a Cauchy continuous map between filter spaces. Then a homomorphism $F^q(f) : F^q(X') \to F^q(X)$ is defined as follows:

$$\begin{cases} F^q(f) = 0 \quad \text{if} \quad q < 0 \\ F^q(f)(h)(x_0, \dots, x_q) = h(f(x_0), \dots, f(x_q)) \quad \text{for each} \quad h \in F^q(X) \\ \text{and each} \quad (x_0, \dots, x_q) \in X^{q+1} \quad \text{if} \quad q \geq 0. \end{cases}$$

Since for each integer q, $F^q(f)[F_0^q(X')] \subset F_0^q(X)$, $F^q(f)$ induces a homomorphism $C^q(f) : C^q(X') \to C^q(X)$.
If $(X,Y), (X',Y') \in |\mathbf{Fil_2}|$ and $f : (X,Y) \to (X',Y')$ is Cauchy continuous, then $C^q(f)[C^q(X',Y')] \subset C^q(X,Y)$. Let us denote the restriction $C^q(f)|C^q(X',Y')$ again by $C^q(f)$. Then $C^q(f) : C^q(X',Y') \to C^q(X,Y)$ is a homomorphism which induces a homomorphism $H^q(f) : H^q(X',Y') \to H^q(X,Y)$. It is easily checked that $H^q : \mathbf{Fil_2} \to \mathbf{Ab}$ is a contravariant functor, called the *q–dimensional Alexander cohomology functor for pairs of filter spaces.*

A.1.2.3. If $(X,Y) \in |\mathbf{Fil_2}|$, then the homomorphism $C^q(i) : C^q(X) \to C^q(Y)$ induced by the inclusion map $i : Y \to X$ is surjective (since $F^q(i) : F^q(X) \to F^q(Y)$ is surjective for each integer q). Now we can proceed as in the topological case (cf.[136; p. 308] or [76; Chapter 3, Section 5]) in defining the *connecting homomorphism*

$$\delta^q : H^q(Y) \to H^{q+1}(X,Y).$$

Furthermore, $\delta^* = (\delta^q)_{q \in \mathbb{Z}}$ is a family of natural transformations $\delta^q : H^q \circ T \to H^{q+1}$, where the functor T is defined as under 3.1.

A.1.3. Examples. 1) Let (X,γ) be a filter space and (X,μ_γ) its corresponding filter–merotopic space. Then a function $f \in F^q(X), q \geq 0$, is locally zero in the sense above iff it is locally zero in the merotopic sense, i.e. iff there is some $\mathcal{A} \in \mu_\gamma$ such that for each $A \in \mathcal{A}$ and each $(q+1)$–tuple $(x_0, \dots, x_q)$ of elements of A, $f(x_0, \dots, x_q) = 0$. Furthermore, subspaces in **Fil-Mer** are formed as in **Mer** (cf. 7.1.3. 3) b)). Thus, *if $(X,Y) \in |\mathbf{Fil_2}|$ is considered to be a pair of filtermerotopic spaces, the Alexander cohomology groups $H^q(X,Y)$ as introduced above are nothing else than the Alexander cohomology groups of the merotopic pair (X,Y) introduced in* [11], *which are isomorphic to the Čech cohomology groups $\check{H}^q(X,Y)$ of the merotopic pair (X,Y)* (cf. also [11]).

2) Let (X,q) be a symmetric topological space and (X,γ_q) its corresponding filter space, i.e. $\gamma_q = \{\mathcal{F} \in F(X)$: there is some $x \in X$ with $\mathcal{F} \supset \mathcal{U}_q(x)\}$. Then a function $f \in F^q(X), q \geq 0$, is locally zero in the sense above iff it is locally zero in the usual topological sense, i.e. for each $x \in X$ there is some $U_x \in \mathcal{U}_q(x)$ such that for each $(q+1)$–tuple $(x_0, \dots, x_q)$ of elements of U_x, $f(x_0, \dots, x_q) = 0$.
Since subspaces in $\mathbf{Top_s}$ are formed as in $\mathbf{KConv_s}$ and closed subspaces in $\mathbf{KConv_s}$ are formed as in **Fil** (cf. 2.3.3.11. 3) and 2.3.3.27. 1)), a closed pair (X,Y) of symmetric topological spaces (i.e. X is a symmetric topological space and Y a closed subspace of X) may be considered to be an element of $|\mathbf{Fil_2}|$. Thus, *for closed pairs of symmetric topological spaces, the Alexander cohomology groups as defined above coincide with the Alexander cohomology groups in the*

usual topological sense (cf. e.g. [76] or [136]).

A.1.4 Theorem. *If H^q denotes the q-dimensional Alexander cohomology functor for pairs of filter spaces and $\delta^q : H^q(Y) \to H^{q+1}(X, Y)$ is the connecting homomorphism, then (H^*, δ^*) is a cohomology theory for filter spaces provided that $H^* = (H^q)_{q\in\mathbb{Z}}$ and $\delta^* = (\delta^q)_{q\in\mathbb{Z}}$.*

Proof. It suffices to prove the four axioms for a cohomology theory (cf. A.1.1.). In this context it is useful to know that Bentley's variant of the Eilenberg–Steenrod axioms is valid for the Alexander cohomology of merotopic pairs (cf. [11]).
1) It follows immediately from A.1.3. 1) that the exactness axiom is fulfilled, since it is fulfilled in the merotopic case.
2) The uniform homotopy axiom is fulfilled, since it its fulfilled in the merotopic case.
3) Let (X, γ) be a filter space and (X, μ_γ) its corresponding filter–merotopic space. If Y and U are subsets of X, then the following are equivalent:
(1) For each $\mathcal{F} \in \gamma, X\backslash U \in \mathcal{F}$ or $Y \in \mathcal{F}$,
(2) $\{X\backslash U, Y\} \in \mu_\gamma$.
This equivalence is an immediate consequence of the definition of μ_γ. Thus, the excision axiom is satisfied, because it is satisfied in the merotopic case (cf. A.1.3.1) and use the above mentioned equivalent description of the excision condition).
The dimension axiom is satisfied, because it is satisfied in the merotopic case (use A.1.3.1)).

A.1.5 Remark. 1) Finite products in **Fil** $\cong$ **Fil–Mer** are not formed as in **Mer** as the following example shows: Let $(X, \mathcal{X})$ and $(Y, \mathcal{Y})$ be paracompact topological spaces such that their topological product $(X, \mathcal{X}) \times (Y, \mathcal{Y})$ is not paracompact. Further, let $(X, \gamma_{q\mathcal{X}})$ and $(Y, \gamma_{q\mathcal{Y}})$ be their corresponding filter spaces and $(X, \gamma_{q\mathcal{X}}) \times (Y, \gamma_{q\mathcal{Y}}) = (X \times Y, \gamma_{X\times Y})$ their product in **Fil**. Then $(X, \mu_{\gamma_{q\mathcal{X}}}) \times (Y, \mu_{\gamma_{q\mathcal{Y}}}) \neq (X \times Y, \mu_{\gamma_{X\times Y}})$, where $\mu_{\gamma_{q\mathcal{X}}}$ (resp. $\mu_{\gamma_{q\mathcal{Y}}}$) is the merotopic structure induced by $\mathcal{X}$ (resp. $\mathcal{Y}$) and $\mu_{\gamma_{X\times Y}}$ is the merotopic structure induced by the product topology of $(X, \mathcal{X}) \times (Y, \mathcal{Y})$. Since the construct **Unif** of uniform spaces is bireflective in **Mer**, $(X, \mu_{\gamma_{q\mathcal{X}}}) \times (Y, \mu_{\gamma_{q\mathcal{Y}}})$ is uniform, but $(X \times Y, \mu_{\gamma_{X\times Y}})$ is not uniform because $(X, \mathcal{X}) \times (Y, \mathcal{Y})$ is not paracompact.

2) Because of 1) it is unknown whether the uniform homotopy axiom for pairs of filter spaces can be replaced by the following *homotopy axiom*: If $f : (X, Y) \to (Z, W)$ and $g : (X, Y) \to (Z, W)$ are homotopic (i.e. there exists a Cauchy continuous map $H : (X \times I, Y \times I) \to (Z, W)$ such that $H(\cdot, 0) = f$ and $H(\cdot, 1) = g$ where I denotes the unit interval $[0, 1]$ with the usual **Fil**–structure consisting of all convergent filters on $[0, 1]$ w.r.t. the usual topology), then $H^q(f) = H^q(g)$ for each integer q.

A.1.6 Definition. Let (X, γ) be a filter space. A map $f : X \to G$ is called *locally constant* provided that for each $\mathcal{F} \in \gamma$ there is some $F \in \mathcal{F}$ such that $f|F$

is constant.

A.1.7 Proposition. *Let (X,γ) be a filter space and G_d the set G endowed with the discrete* **Fil***-structure. A map $f : X \to G$ is locally constant iff $f : (X,\gamma) \to G_d$ is Cauchy continuous.*

Proof. 1) "$\Longrightarrow$". Let $\mathcal{F} \in \gamma$. Then there is some $F \in \gamma$ such that $f|F$ is constant, i.e. $f[F] = \{g\}$ for some $g \in G$. Thus, $\dot{g} = f(\mathcal{F})$, i.e. $f(\mathcal{F})$ is a Cauchy filter in G_d.
2) "$\Longleftarrow$". Let $\mathcal{F} \in \gamma$. Then $f(\mathcal{F}) = \dot{g}$ for some $g \in G$. Hence, $\{g\} \supset f[F]$ for some $F \in \gamma$, i.e. $f|F$ is constant.

A.1.8 Theorem *Let $(X,Y) \in |\mathbf{Fil_2}|$. Then $H^0(X,Y)$ is isomorphic to the abelian group of all locally constant functions from X to G which vanish on Y.*

Proof. If (X,γ) is a filter space and (X,μ_γ) its corresponding filter–merotopic space, then a map $f : X \to G$ is locally constant iff there is some $\mathcal{A} \in \mu_\gamma$ such that f is constant on each $A \in \mathcal{A}$. Thus, A.1.8. follows immediately from the corresponding result for merotopic pairs (cf. [109;4.2.]).

A.1.9 Corollary. *Let (X,γ) be a nonempty filter space. Then (X,γ) is connected (i.e. each Cauchy continuous map $f : (X,\gamma) \to (\{0,1\},\{\dot{0},\dot{1}\})$ is constant) iff $j : G \to H^0(X)$, defined by $j(g) = \overline{g}$ for each $g \in G$ is an isomorphism where $\overline{g}(x) = g$ for each $x \in X$.*

Proof. Use A.1.8. and A.1.7, and note that (X,γ) is connected iff each Cauchy continuous map $f : (X,\gamma) \to G_d$ is constant.

A.1.10 Remark. The question arises whether *there are (absolute) cohomology functors $\overline{H}^q$:* **SUConv** $\to$ **Ab** (*resp.* H^q_u: **SUConv** $\to$ **Ab**) *such that $\overline{H}^0$ (resp. H^0_u) describes connectedness (resp. uniform connectedness)* corresponding to A.1.9. The construct **Fil** as well as the construct **Unif** of uniform spaces are bireflectively embedded into **SUConv**. If R_f : **SUConv** $\to$ **Fil** (resp. R_u : **SUConv** $\to$ **Unif**) denotes the corresponding bireflector, then using 5.1.32. a semiuniform convergence space $(X,\mathcal{J}_X)$ is connected (resp. uniformly connected) iff $R_f((X,\mathcal{J}_X))$ (resp. $R_u((X,\mathcal{J}_X))$ is connected (resp. uniformly connected). Consequently, it makes sense to define

$$\overline{H}^q = H^q \circ R_f \text{ and } H^q_u = \check{H}^q \circ R_u$$

where H^q : **Fil** $\to$ **Ab** denotes the q–dimensional (absolute) cohomology functor as described above and $\check{H}^q$: **Unif** $\to$ **Ab** denotes the q–dimensional (absolute) Čech cohomology functor as defined for merotopic spaces (cf. e.g. [11]) in order to answer the above question positively. Remember that every uniform space (or more generally every diagonal semiuniform convergence space) is connected, but not uniformly connected in general.

A.2 Path connectedness and fundamental groups for limit spaces

A.2.1 Definitions. 1) Let $\mathbf{X} = (X, q)$ be a limit space and $[0,1]_t$ the unit interval $[0,1]$ endowed with the usual topological structure (= limit space structure). Then a *path* in $\mathbf{X}$ is a continuous map $f : [0,1]_t \to \mathbf{X}$. If $f : [0,1]_t \to \mathbf{X}$ is a path in $\mathbf{X}$, then $f(0)$ is called its starting point and $f(1)$ its end point and we say that f runs from $f(0)$ to $f(1)$.
2) A limit space (X, q) is called *path-connected* provided that for any two points $x, y \in X$ there is a path in (X, q) running from x to y.
3) A subset of a limit space is called *path–connected* provided that it is path–connected as a subspace.

A.2.2 Proposition. *Let (X, q) be a pretopological space and (X', q') a limit space. If A_1, A_2 are closed subsets of (X, q) with $X = A_1 \cup A_2$, then a map $f : (X, q) \to (X', q')$ is continuous iff the restrictions $f|A_i$ are continuous for each $i \in \{1, 2\}$.*

Proof. 1) '$\Longrightarrow$". This implication is obvious.
2) "$\Longleftarrow$". Let $(\mathcal{F}, x) \in q$, i.e. $\mathcal{F} \supset \mathcal{U}_q(x)$.
a) Let $x \in X \backslash A_2$. If $i_1 : A_1 \to X$ denotes the inclusion map, then $i_1^{-1}(\mathcal{U}_q(x))$ exists and converges to x, since $i_1(i_1^{-1}(\mathcal{U}_q(x))) \supset \mathcal{U}_q(x)$ and (X, q) is pretopological. Thus, $f|A_1(i_1^{-1}(\mathcal{U}_q(x))$ converges to $f(x)$. But $f|A_1(i_1^{-1}(\mathcal{U}_q(x))) = (f \circ i_1)(i_1^{-1}(\mathcal{U}_q(x)) = f(i_1^{-1}(\mathcal{U}_q(x)) = f(\mathcal{U}_q(x))$ $[f(i_1^{-1}(\mathcal{U}_q(x))) \supset f(\mathcal{U}_q(x))$ is obvious, and $f(\mathcal{U}_q(x)) \supset f(i_1^{-1}(\mathcal{U}_q(x))$ is valid because $X \backslash A_2$ is open$]$. Consequently, $(f(\mathcal{F}), f(x)) \in q'$.
b) Let $x \in X \backslash A_1$: Analogously to a).
c) Let $x \in A_1 \cap A_2$. If $i_2 : A_2 \to X$ denotes the inclusion map, then $i_1^{-1}(\mathcal{U}_q(x))$ and $i_2^{-1}(\mathcal{U}_q(x))$ exist and $\mathcal{U}_q(x) = i_1(i_1^{-1}(\mathcal{U}_q(x)) \cap i_2(i_2^{-1}(\mathcal{U}_q(x))$. Thus, $f(\mathcal{U}_q(x)) = f(i_1(i_1^{-1}(\mathcal{U}_q(x)) \cap f(i_2(i_2^{-1}(\mathcal{U}_q(x))) = f|A_1(i_1^{-1}(\mathcal{U}_q(x)) \cap f|A_2(i_2^{-1}(\mathcal{U}_q(x))$ converges to $f(x)$, since $i_k^{-1}(\mathcal{U}_q(x))$ converges to x in A_k for each $k \in \{1,2\}$, $f|A_k$ is continuous for each $k \in \{1,2\}$ and (X', q') is a limit space. Consequently, $(f(\mathcal{F}), f(x)) \in q$.

A.2.3 Remarks. 1) In the above proposition (X', q') cannot be replaced by a symmetric Kent convergence space which is not a limit space as the following *example* shows:
Let $A_1 = [0, \frac{1}{2}], A_2 = [\frac{1}{2}, 1]$ and $i_k : A_k \to [0,1]$ the inclusion map for each $k \in \{1,2\}$. Furthermore, let q_t be the usual topological structure on $[0,1]$ and q_K a symmetric Kent convergence structure on $[0,1]$ defined as follows:

1. $x \neq \frac{1}{2} : (\mathcal{F}, x) \in q_K \Longleftrightarrow (\mathcal{F}, x) \in q_t$
2. $x = \frac{1}{2}$: Let $\mathcal{F}_1 = i_1(i_1^{-1}(\mathcal{U}_{q_t}(\frac{1}{2}))$ and $\mathcal{F}_2 = i_2(i_2^{-1}(\mathcal{U}_{q_t}(\frac{1}{2}))$. Then $(\mathcal{F}, \frac{1}{2}) \in q_K$ iff $\mathcal{F} \supset \mathcal{F}_1$ or $\mathcal{F} \supset \mathcal{F}_2$ or $\mathcal{F} = \dot{\frac{1}{2}}$.

Obviously, $1_{[0,1]} : ([0,1], q_t) \to ([0,1], q_K)$ is not continuous, but $1_{[0,1]}|A_1$ as well as $1_{[0,1]}|A_2$ are continuous.

2) *In (symmetric) Kent convergence spaces the usual composition of paths (known from topological spaces) is not a path in general* as the following example shows: Let $([0,1],q_t) = [0,1]_t$ and $([0,1],q_K) = [0,1]_K$ (cf. 1)). $f : [0,1]_t \to [0,1]_K$ defined by $f(t) = t/2$ is a path in $[0,1]_K$ running from 0 to 1/2, whereas $g : [0,1]_t \to [0,1]_K$ defined by $g(t) = \frac{t}{2} + \frac{1}{2}$ is a path in $[0,1]_K$ running from $\frac{1}{2}$ to 1. Their composition $h : [0,1]_t \to [0,1]_K$ defined by

$$h(t) = \begin{cases} f(2t) & \text{if } 0 \le t \le \frac{1}{2} \\ g(2t-1) & \text{if } \frac{1}{2} \le t \le 1 \end{cases}$$

is the identitiy map $1_{[0,1]} : [0,1]_t \to [0,1]_K$ which is not a path, since it is not continuous at $\frac{1}{2}$.

A.2.4. By means of A.2.2. the theory of path connectedness for limit spaces can be built up analogously to the theory of path connectedness for topological spaces (cf. e.g. [108]). In particular the following statements are true:
1. Each limit space (X,q) can be decomposed into maximal path–connected subsets, called *path components*, which are the equivalence classes with respect to the following equivalence relation R on X:

$$xRy \iff \text{There is a path } f : [0,1]_t \to (X,q) \text{ running from } x \text{ to } y.$$

2. Continuous images of path–connected limit spaces are path–connected.
3. A product of nonempty limit spaces is path–connected iff each of its factor spaces is path–connected.
4. Every path–connected limit space (X,q) is connected, i.e. each continuous map $f : (X,q) \to D_2$ is constant, where D_2 denotes the two–point discrete topological space.

A.2.5.1. Since **Lim** is cartesian closed, i.e. for any two limit spaces $\mathbf{X}$, $\mathbf{Y}$ there is a natural function space $[\mathbf{X},\mathbf{Y}]_{\mathbf{Lim}}$, *a path H in the natural function space $[\mathbf{X},\mathbf{Y}]_{\mathbf{Lim}}$ running from f to g is a homotopy $H : \mathbf{X} \times [0,1]_t \longrightarrow \mathbf{Y}$ between f and g, i.e. a continuous map $H : \mathbf{X} \times [0,1]_t \longrightarrow \mathbf{Y}$ such that $H(\cdot,0) = f$ and $H(\cdot,1) = g$, and vice versa* (note: $[[0,1]_t,[\mathbf{X},\mathbf{Y}]_{\mathbf{Lim}}] \cong [\mathbf{X} \times [0,1]_t, \mathbf{Y}]_{\mathbf{Lim}}$).
If $\mathbf{X}$ and $\mathbf{Y}$ are topological spaces, then $[\mathbf{X},\mathbf{Y}]_{\mathbf{Lim}}$ need not be a topological space; but if $\mathbf{X}$ is a locally compact topological Hausdorff space, then $[\mathbf{X},\mathbf{Y}]_{\mathbf{Lim}}$ is a topological space and the corresponding topology is the compact–open topology (cf. 6.1.31.).

A.2.5.2. It follows from A.2.5.1. that *the path components in $[\mathbf{X},\mathbf{Y}]_{\mathbf{Lim}}$ are exactly the homotopy classes in $[\mathbf{X},\mathbf{Y}]_{\mathbf{Lim}}$*.

A.2.6. If $\mathbf{X} = (X,q)$ is a limit space and $x_0 \in X$, then by means of A.2.2. the *fundamental group* $\pi_1(X,x_0)$ at the point x_0 can be introduced in the usual way (cf. [74; II.4.], where topological spaces have to be substituted by limit spaces) and if (X, q) is path–connected, one obtains

$$\pi_1(X, x_0) \cong \pi_1(X, x_1)$$

for any two points $x_0, x_1 \in X$, i.e. the fundamental group is independent of the point x_0 and it is denoted by $\pi_1(X)$. By A.2.5.2. the elements of $\pi_1(X, x_0)$ are precisely the path components of the set $\Lambda(X, x_0)$ of all loops with base point x_0 endowed with the subspace structure of the structure of continuous convergence (= natural function space structure) on $[[0,1]_t, \mathbf{X}]_{\mathbf{Lim}}$, where a loop with base point x_0 is a path $f : [0,1]_t \to \mathbf{X}$ such that $f(0) = f(1) = x_0$.

A.2.7 Remark. Let $\mathbf{X} = (X, q)$ be a limit space, $x_0 \in X$. If $Y = \Lambda(X, x_0)$ is endowed with the structure of continuous convergence (cf. A.2.6.) and y_0 is the constant loop defined by $y_0(t) = x_0$ for each $t \in [0,1]$, then for each natural number $n > 1$ the *n–dimensional homotopy group* $\pi_n(X, x_0)$ at the base point x_0 is recurrently defined by $\pi_n(X, x_0) = \pi_{n-1}(Y, y_0)$. This definition coincides with the usual one for topological spaces (cf. [75;p. 200]) provided that $\mathbf{X}$ is a topological space; namely since $[0,1]_t$ is a locally compact topological Hausdorff space, it follows from A.2.5.1. that $[[0,1]_t, \mathbf{X}]_{\mathbf{Lim}}$ is a topological space whose topology is the compact open topology and, since **Top** is a bireflective subconstruct of **Lim**, subspaces in **Top** are formed as in **Lim**.

Exercises

Chapter 1

1) Let $\mathcal{C}$ be a topological construct, X a set and $\mathcal{C}_X$ the set of all $\mathcal{C}$–structures on X. Prove that $(\mathcal{C}_X, \leq)$ is a complete lattice, where $\leq$ denotes the partial order defined in 1.1.4.
2) Show that a partially ordered set $(X, \prec)$ is a complete lattice iff each subset of X has an infimum. What are the consequences of this statement for topological constructs?
3) Let X be a set, $((X_i, \mathcal{J}_{X_i}))_{i\in I}$ a family of semiuniform convergence spaces and $(f_i : X_i \to X)_{i\in I}$ a family of maps. Verify that $\mathcal{J}_X = \{\mathcal{F} \in F(X \times X)$: there is some $i \in I$ and some $\mathcal{F}_i \in \mathcal{J}_{X_i}$ with $(f_i \times f_i)(\mathcal{F}_i) \subset \mathcal{F}\} \cup \{\dot{x} \times \dot{x} : x \in X\}$ is the *final* **SUConv**–*structure* on X w.r.t. the given data. If $X = \bigcup_{i\in I} f_i[X_i]$, then $\mathcal{J}_X = \{\mathcal{F} \in F(X \times X)$: there is some $i \in I$ and some $\mathcal{F}_i \in \mathcal{J}_{X_i}$ with $(f_i \times f_i)(\mathcal{F}_i) \subset \mathcal{F}\}$.
4) Why is the construct **Haus** of Hausdorff spaces (and continuous maps) not topological?
5) Determine the discrete and indiscrete objects of the constructs under 1.1.6.
6) Consider the construct $\mathbf{Top}_1$ of topological T_1–spaces (and continuous maps). Prove that for every set X there is a coarsest $\mathbf{Top}_1$–structure on X, but this one is not the initial $\mathbf{Top}_1$–structure w.r.t. the empty index class provided that $|X| \geq 2$.
7) Let $\mathcal{C}$ be a topological construct. Prove that subspaces and products of indiscrete $\mathcal{C}$–objects are indiscrete, and find the dual assertion.
8) Show that in topological constructs the composition of two embeddings (resp. quotient maps) is an embedding (resp. quotient map).

Chapter 2

9) Let $\mathcal{C}$ be a topological construct. Prove:
a) A source $(f_i : (X, \xi) \to (X_i, \xi_i))_{i\in I}$ in $\mathcal{C}$ is a mono–source iff it *separates points*, i.e. for any two distinct points $x, y \in X$ there is some $i \in I$ with $f_i(x) \neq f_i(y)$.
b) A sink $(f_i : (X_i, \xi_i) \to (X, \xi))_{i\in I}$ in $\mathcal{C}$ is an epi–sink iff it is a *covering sink*, i.e. $X = \bigcup_{i\in I} f_i[X_i]$.

10) Let $\mathcal{F} : \mathcal{A} \to \mathcal{B}$ and $\mathcal{G} : \mathcal{B} \to \mathcal{A}$ be functors. Define functors $\hom(\mathcal{G}(\cdot), \cdot) : \mathcal{B}^* \times \mathcal{A} \to$ **Set** and $\hom(\cdot, \mathcal{F}(\cdot)) : \mathcal{B}^* \times \mathcal{A} \to$ **Set** by

$$\begin{aligned}
\hom(\mathcal{G}(\cdot), \cdot)(B, A) &= [\mathcal{G}(B), A]_{\mathcal{A}} \\
\hom(\mathcal{G}(\cdot), \cdot)(f, g) &= \hom(\mathcal{G}(f), g), \text{where } \hom(\mathcal{G}(f), g)(h) = g \circ h \circ \mathcal{G}(f) \\
\hom(\cdot, \mathcal{F}(\cdot))(B, A) &= [B, \mathcal{F}(A)]_{\mathcal{B}} \\
\hom(\cdot, \mathcal{F}(\cdot))(f, g) &= \hom(f, \mathcal{F}(g)), \text{where } \hom(f, \mathcal{F}(g))(h) = \mathcal{F}(g) \circ h \circ f.
\end{aligned}$$

Prove that $(\mathcal{G}, \mathcal{F})$ is a pair of adjoint functors iff $\hom(\mathcal{G}(\cdot), \cdot)$ and $\hom(\cdot, \mathcal{F}(\cdot))$ are naturally equivalent.
(Hint. " $\Rightarrow$ ". Let $(\mathcal{G}, \mathcal{F}, u, v)$ be an adjoint situation. If $\eta_{(B,A)} : [\mathcal{G}(B), A]_{\mathcal{A}} \to [B, \mathcal{F}(A)]_{\mathcal{B}}$ is defined by $\eta_{(B,A)}(f) = \mathcal{F}(f) \circ u_B$, then $\eta = (\eta_{(B,A)}) : \hom(\mathcal{G}(\cdot), \cdot) \to \hom(\cdot, \mathcal{F}(\cdot))$ is a natural equivalence.
" $\Leftarrow$ ". Let $\eta = (\eta_{(B,A)}) : \hom(\mathcal{G}(\cdot), \cdot) \to \hom(\cdot, \mathcal{F}(\cdot))$ be a natural equivalence. If $u = (u_B)$ is defined by $u_B = \eta_{(B,\mathcal{G}(B))}(1_{\mathcal{G}(B)})$ and $v = (v_A)$ by $v_A = \eta^{-1}_{(\mathcal{F}(A),A)}(1_{\mathcal{F}(A)})$, then $(\mathcal{G}, \mathcal{F}, u, v)$ is an adjoint situation.)
11) Show that there is no subconstruct of **Top** which is simultaneously epireflective and monocoreflective other than **Top**.
12) Determine the coseparators for each topological construct.
13) Let $\mathcal{A}$ be a bireflective (resp. bicoreflective) full subconstruct of a topological construct $\mathcal{C}$. Prove that the following are equivalent:
(1) $\mathcal{A}$ is isomorphism–closed.
(2) For each $(X, \xi) \in |\mathcal{C}|$ there is an $\mathcal{A}$–structure $\xi_{\mathcal{A}}$ on X such that $1_X : (X, \xi) \to (X, \xi_{\mathcal{A}})$ (resp. $1_X : (X, \xi_{\mathcal{A}}) \to (X, \xi)$) is the bireflction (resp. bicoreflection) of (X, ξ) w.r.t. $\mathcal{A}$.
14) Let $(X, \mathcal{X})$ be a symmetric topological space and $(X, \mathcal{J}_{\gamma_{q_\mathcal{X}}})$ its corresponding semiuniform convergence space. Then $(X, \mathcal{J}_{\gamma_{q_\mathcal{X}}})$ fulfills the first axiom of countability in the sense of 2.3.2.7 iff $(X, \mathcal{X})$ fulfills the first axiom countability in the usual sense (cf. 0.2.3.6).
15) a) Prove that the full subconstruct of **SUConv** whose object class consists of all semiuniform convergence spaces fulfilling the first axiom of countability is bicoreflective.
b) Subspaces and countable products of semiuniform convergence spaces fulfilling the first axiom of countability fulfill also the first axiom countability, but arbitrary products need not have this property.
16) Find a Cauchy continuous map which is not uniformly continuous and a continuous map which is not Cauchy continuous.
17) If $(X_i)_{i \in I}$ is a family of sets and $\mathcal{F}_i \in F(X_i)$ for each $i \in I$, then the set $\{\prod_{i \in I} F_i : F_i \in \mathcal{F}_i \text{ for each } i \in I\}$ is a filter base on $\prod_{i \in I} X_i$ generating the so-called *box product filter* $\prod^B_{i \in I} \mathcal{F}_i$. Prove:
a) If $p_i : \prod_{i \in I} X_i \to X_i$ denotes the i–th projection, then $p_i(\prod^B_{i \in I} \mathcal{F}_i) = \mathcal{F}_i$.
b) The procuct filter $\prod_{i \in I} \mathcal{F}_i$ is the coarsest filter $\mathcal{F}$ on $\prod_{i \in I} X_i$ such that $p_i(\mathcal{F}) = \mathcal{F}_i$ for each $i \in I$.
c) If $\mathcal{F}$ is a filter on $\prod_{i \in I} X_i$, then

$$\boxed{\prod_{i \in I} p_i(\mathcal{F}) \subset \mathcal{F}, \text{ where } p_i : \prod_{i \in I} X_i \to X_i \text{ is the } i\text{–th projection}}$$

18) Let $\mathcal{F}, \mathcal{G}, \mathcal{H}$ and $\mathcal{K}$ be filters on a set X. Prove:

a) $(\mathcal{F} \times \mathcal{G}) \circ (\mathcal{H} \times \mathcal{K})$ exists iff $\sup\{\mathcal{F}, \mathcal{K}\}$ exists.
b) $(\mathcal{F} \times \mathcal{G}) \circ (\mathcal{H} \times \mathcal{K}) = \mathcal{H} \times \mathcal{G}$ provided that $\sup\{\mathcal{F}, \mathcal{K}\}$ exists.
c) $(\mathcal{F} \cap \mathcal{G}) \times (\mathcal{F} \cap \mathcal{G}) = (\mathcal{F} \times \mathcal{F}) \cap (\mathcal{F} \times \mathcal{G}) \cap (\mathcal{G} \times \mathcal{F}) \cap (\mathcal{G} \times \mathcal{G})$.

19) Let $(X, \mathcal{J}_X)$ be the coproduct (in **SUConv**) of a family $((X_i, \mathcal{J}_{X_i}))_{i \in I}$ of semiuniform convergence spaces. Prove:
a) $(X, \gamma_{\mathcal{J}_X})$ is the coproduct (in **Fil**) of $((X_i, \gamma_{\mathcal{J}_{X_i}}))_{i \in I}$.
b) $(X, q_{\gamma_{\mathcal{J}_X}})$ is the coproduct (in $\mathbf{KConv}_S$) of $(X_i, q_{\gamma_{\mathcal{J}_{X_i}}})_{i \in I}$.
20) Prove that in **SUConv** coproducts of quotient maps are quotient maps.
21) Verify that in **SUConv** the following are valid:
a) Products of subspaces are subspaces of products.
b) Coproducts of subspaces are subspaces of coproducts.
22) Let $(\boldsymbol{X}_i)_{i \in I}$ be a family of semiuniform convergence spaces. Prove:
a) α) For each $i \in I$, the i–th injection $j_i : \boldsymbol{X}_i \to \coprod_{i \in I} \boldsymbol{X}_i$ is an embedding.
β) Each $\boldsymbol{X}_i$ is an open and closed subspace of $\coprod_{i \in I} \boldsymbol{X}_i$.
b) If all X_i are non–empty, then
α) the i–th projection $p_i : \prod_{i \in I} \boldsymbol{X}_i \to \boldsymbol{X}_i$ is a quotient map for each $i \in I$, and
β) each $\boldsymbol{X}_i$ is a subspace of $\prod_{i \in I} \boldsymbol{X}_i$.

Chapter 3

23) Let $\mathcal{C}$ be a topological construct. Prove: If $(f_i : B_i \to B)_{i \in I}$ is a final epi–sink in $\mathcal{C}$, then there is a set $J \subset I$ such that $(f_j : B_j \to B)_{j \in J}$ is a final epi–sink in $\mathcal{C}$ too.
24) Let $(\mathcal{F}_1, \mathcal{G}_1)$ and $(\mathcal{F}_2, \mathcal{G}_2)$ be pairs of adjoint functors such that $\mathcal{F}_2 \circ \mathcal{F}_1$ is defined. Show that $(\mathcal{F}_2 \circ \mathcal{F}_1, \mathcal{G}_1 \circ \mathcal{G}_2)$ is again an adjoint situation.
25) Let $\boldsymbol{X}$ and $\boldsymbol{Y}$ be semiuniform convergence spaces and $\boldsymbol{Y}^{\mathbf{X}}$ the power object in **SUConv**. Prove:
a) If $\boldsymbol{Y} \in |\mathbf{SULim}|$, then $\boldsymbol{Y}^{\mathbf{X}} \in |\mathbf{SULim}|$.
b) If $\boldsymbol{X}$ and $\boldsymbol{Y}$ are uniform limit spaces, then $\boldsymbol{Y}^{\mathbf{X}}$ is a uniform limit space too.
c) **SULim** and **ULim** are cartesian closed.
26) Prove: **SULim**, **GKonv**, **KConv**, **Lim** and **PsTop** are strong topological universes.

Chapter 4

27) Let $(X, \mathcal{W})$ be a uniform space and $R = \bigcap_{W \in \mathcal{W}} W$. Prove:
a) R is an equivalence relation on X.
b) If $\omega : X \to X/R$ is the natural map and $\mathcal{V} = (\omega \times \omega)(\mathcal{W})$, then $(X/R, \mathcal{V})$ is

a separated uniform space such that $\omega : (X, \mathcal{W}) \to (X/R, \mathcal{V})$ is 1° initial and 2° the epireflection of $(X, \mathcal{W})$ w.r.t. **SepUnif**.
28) Let $(X, \mathcal{X})$ be a topological space, $x \in X$ and $\mathcal{U}(x)$ the neighborhood filter of x. Prove:
a) $\mathcal{U}(x) = \bigcup\{\bigcap_{y \in U_x} \mathcal{U}(y) : U_x \in \mathcal{U}(x)\}$,
b) $\mathcal{U}(x) = \sup\{\bigcap_{y \in U_x} \mathcal{U}(y) : U_x \in \mathcal{U}(x)\}$.
29) Show that in the extension theorem for regular generalized convergence spaces (cf. 4.2.1.11.) the topological space (X, q) may be substituted by a generalized convergence space fulfilling the diagonal property under 4.2.1.2.
30) Prove that coproducts of regular semiuniform convergence spaces are regular.
31) Give an example of a regular semiuniform convergence space whose quotient (space) in **SUConv** is not regular.
(Hint. Construct a paracompact topological space whose quotient in **Fil** is T_1, but not T_2: Consider the closed subspace of $\mathbb{R}^2_t$ whose underlying set is the union of the lines $y = 0$ and $y = 1$ in $\mathbb{R}^2$ and let Y be the quotient set of X obtained by identifying each point $(x, 0)$, for $x \neq 0$, with the corresponding point $(x, 1)$).
32) a) Prove that coproducts of complete semiuniform convergence spaces are complete.
b) Prove that a quotient (in **SUConv**) of a complete semiuniform convergence space need not be weakly complete.
(Hint: Consider $\mathbb{R}^2_u = (\mathbb{R}^2, [j(\mathcal{V} \times \mathcal{V})])$, where $\mathcal{V}$ denotes the usual uniformity on $\mathbb{R}$, and $j : \mathbb{R}^2 \times \mathbb{R}^2 \to \mathbb{R}^2 \times \mathbb{R}^2$ is defined by $j((x, y), (x', y')) = ((x, x'), (y, y'))$. Then the closed subspace $\boldsymbol{A}$ of $\mathbb{R}^2_u$ with underlying set $A = \{(x, y) \in \mathbb{R}^2 : x > 0$ and $y \geq \frac{1}{x}\}$ is a complete uniform space, where its uniformity $\mathcal{W}_A$ has the base $\{j[V_\varepsilon \times V_\varepsilon] \cap (A \times A) : \varepsilon > 0\}$ with $V_\varepsilon = \{(x, y) \in \mathbb{R}^2 : |x - y| < \varepsilon :$

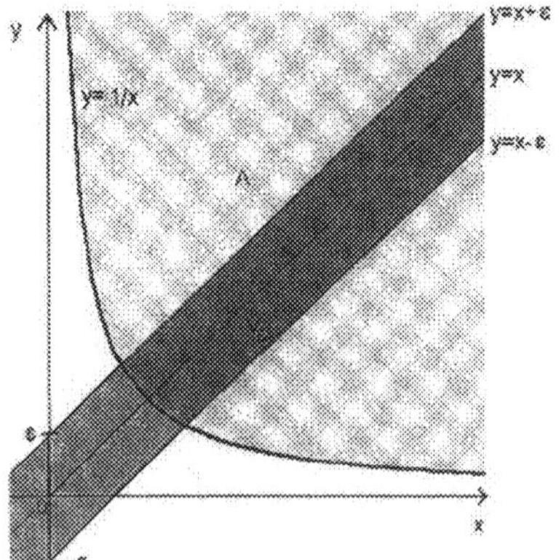

If $\mathbb{R}^+ = \{x \in \mathbb{R} : x > 0\}$, then the map $\omega : A \to \mathbb{R}^+$ defined by $\omega((x, y)) = x$ for each $(x, y) \in A$ is surjective. Let $\mathcal{J}_{\mathbb{R}^+}$ be the final **SUConv**-structure on $\mathbb{R}^+$ w.r.t. ω, i.e. $\mathcal{J}_{\mathbb{R}^+} = \{\mathcal{H} \in F(\mathbb{R}^+ \times \mathbb{R}^+) : \omega \times \omega(\mathcal{W}_A) \subset \mathcal{H}\}$, and let

$X = (I\!R^+, \mathcal{J}_{I\!R^+})$. Then $\omega : A \to X$ is a quotient map in **SUConv**, but X is not weakly complete: Prove that $\omega \times \omega[j[V_\varepsilon \times V_\varepsilon] \cap (A \times A)] = V_\varepsilon \cap (I\!R^+ \times I\!R^+)$ for each $\varepsilon > 0$, i.e. X is a subspace of $I\!R_u = (I\!R, [\mathcal{V}])$, which is not complete [= weakly complete], since the elementary filter of $(\frac{1}{n})_{n \in I\!N}$ is a Cauchy filter on $I\!R^+$ without an adherence point.)

33) Let $(X, \mathcal{J}_X)$ be a uniform limit space satisfying the first axiom of countability. Prove that $(X, \mathcal{J}_X)$ is complete iff each Cauchy sequence (in $(X, \gamma_{\mathcal{J}_X})$) converges (in $(X, q_{\gamma_{\mathcal{J}_X}})$).

34) Verify the following statements:

a) A semiuniform convergence space X is a T_1–space iff all singletons are closed in X.

b) Coproducts (in **SUConv**) of T_1–spaces are T_1–spaces.

c) A quotient space Y (in **SUConv**) of a T_1–space $X \in |\mathbf{SUConv}|$ need not be a T_1–space (note: If a T_1–space Y is a quotient space of a T_1–space X by an equivalence relation R, then the equivalence classes w.r.t. R are closed subsets of X.).

35) Prove: a) A semiuniform convergence space $(X, \mathcal{J}_X)$ is a T_2–space iff for any two distinct elements $x, y \in X$ and arbitrary filters $\mathcal{F}$ and $\mathcal{G}$ on X with $(\mathcal{F}, x) \in q_{\gamma_{\mathcal{J}_X}}$ and $(\mathcal{G}, y) \in q_{\gamma_{\mathcal{J}_X}}$ there are $F \in \mathcal{F}$ and $G \in \mathcal{G}$ such that $F \cap G = \emptyset$.

b) Coproducts (in **SUConv**) of T_2–spaces are T_2–spaces.

(Hint. See exercise 19)b).)

c) Quotients (in **SUConv**) of metrizable semiuniform convergence spaces need not be T_2.

(Hint. Consider the metrizable uniform space $I\!R_u$ of real numbers and define an equivalence relation π on $I\!R$ as follows:

$x \pi y$ iff $x = y$ or ($y = -x$ for $x > 1$ or $x < -1$).)

Chapter 5

36) Let (X, γ) be a filter space. Prove that the following are equivalent:

(a) (X, γ) is connected, i.e. each Cauchy continuous map $f : (X, \gamma) \to (\{0,1\}, \{\dot{0}, \dot{1}\})$ from (X, γ) into the two–point discrete filter space $(\{0,1\}, \{\dot{0}, \dot{1}\})$ is constant.

(b) The merotopic space (X, μ_γ) with $\mu_\gamma = \{\mathcal{A} \subset \mathcal{P}(X) :$ for each $\mathcal{F} \in \gamma$ there is some $A \in \mathcal{A}$ with $A \in \mathcal{F}\}$ is connected, i.e. each uniformly continuous map $f : (X, \mu_\gamma) \to (\{0,1\}, \mu_d)$ is constant, where μ_d consists of all covers of $\{0,1\}$ (i.e. $(\{0,1\}, \mu_d)$ is a two–point discrete merotopic space).

37) Let $(X, \mathcal{W})$ be a proximity space (=totally bounded uniform space). Show that the following are equivalent:

(a) The principal uniform limit space $(X, [\mathcal{W}])$ is uniformly connected.

(b) The underlying filter space (=Cauchy space) $(X, \gamma_{[\mathcal{W}]})$ of $(X, [\mathcal{W}])$ is connected in the sense of 36)(a).

38) Prove that a semiuniform convergence space $(X, \mathcal{J}_X)$ is uniformly connected

iff for each uniformly continuous map $f : (X, \mathcal{J}_X) \to \mathbb{R}_u, \overline{f[X]}$ is an interval.
(A subset I of $\mathbb{R}$ is called an *interval* provided that for $x \in I, y \in I$ and $z \in \mathbb{R}$, $x < z < y$ implies $z \in I$.)
39) Verify that a semiuniform convergence space $(X, \mathcal{J}_X)$ is connected iff for each uniformly continuous map $f : (X, \mathcal{J}_X) \to \mathbb{R}_t, \overline{f[X]}$ is an interval.
40) Prove: A uniform space $(X, \mathcal{W})$ is uniformly connected iff one of the following two equivalent conditions is satisfied:
(1) For each $V \in \mathcal{W}$ and any pair of points $x, y \in X$ there is some $n \in \mathbb{N}$ such that $(x, y) \in V^n$.
(2) For each $V \in \mathcal{W}$ and any pair of points $x, y \in X$ there are finitely many points $x_0, \ldots, x_n \in X$ such that $x_0 = x, x_n = y$ and $(x_{i-1}, x_i) \in V$ for each $i \in \{1, \ldots, n\}$.
41) Let $(X, \mathcal{X})$ be a compact Hausdorff space and $\mathcal{V}$ the unique uniformity inducing $\mathcal{X}$. Prove that $(X, \mathcal{V})$ is uniformly connected iff $(X, \mathcal{X})$ is connected.
42) Let $(X, \mathcal{X})$ be an R_0–space. Show the equivalence of the following statements:
(1) $(X, \mathcal{X})$ is totally disconnected (in the usual sense).
(2) $(X, \mathcal{J}_{\gamma_{q\mathcal{X}}})$ is totally disconnected.

Chapter 6

43) Prove: **Unif** is bireflective in **SUnif**.
44) Let X be a set with $|X| > \aleph_0$ and $\boldsymbol{Y} = (Y, \mathcal{Y})$ a non–indiscrete topological space. Prove that $F_s(X, Y)$ is not metrizable.
(Hint. Prove that $F_s(X, Y)$ does not fulfill the first axiom of countability.)
45) Let $\boldsymbol{X} = (X, \mathcal{X})$ be a topological space, $\boldsymbol{Y} = (Y, \mathcal{V})$ a uniform space, and $\mathcal{B}$ a base for $\mathcal{V}$. Prove that the sets $W(A, V)$, where A is the union of finitely many elements of $\mathcal{S} \subset \mathcal{P}(X)$ and V belongs to $\mathcal{B}$, form a base for $\mathcal{U}|\mathcal{S}$.
46) A filter $\mathcal{F}$ on $F_{\mathcal{S}}(X, Y)$ converges iff $\mathcal{F}$ converges uniformly on each $S \in \mathcal{S}$, where $\mathcal{S}$ need not be a cover of X.
(Hint. Prove first that $\mathcal{F}$ converges on $F_{\mathcal{S}}(X, Y)$ iff $\mathcal{F}$ is a Cauchy filter and $p_x(\mathcal{F})$ converges in Y for all $x \in X$ belonging to at least one $S \in \mathcal{S}$.)
47) Let $\boldsymbol{Y} = (Y, \mathcal{V})$ be a uniform space with $|Y| \geq 2$. Then $F_{\mathcal{S}}(X, Y)$ is separated iff $\boldsymbol{Y}$ is separated and $\mathcal{S}$ is a cover of X.
(Hint. If $\mathcal{S}$ contains a non–empty set, then the subspace $\boldsymbol{Y}'$ of all constant maps of $F_{\mathcal{S}}(X, Y)$ is isomorphic to $\boldsymbol{Y}$.)
48) Let $\boldsymbol{X} = (X, \mathcal{X})$ be a topological space and $\boldsymbol{Y} = (Y, \mathcal{V})$ a uniform space. If $\mathcal{S} \subset \mathcal{P}(X)$ contains a non–empty set, then $F_{\mathcal{S}}(X, Y)$ is complete iff $\boldsymbol{Y}$ is complete.
(Hint. Use 46) and the hint under 47).)
49) Let $\mathcal{S}$ be a set of subsets of a topological space $\boldsymbol{X}$ such that each point of $\boldsymbol{X}$ is an interior point of some $S \in \mathcal{S}$. If $\boldsymbol{Y}$ is a uniform space, then $C_{\mathcal{S}}(\boldsymbol{X}, \boldsymbol{Y})$ is closed in $F_{\mathcal{S}}(X, Y)$.

50) Let $\boldsymbol{Y}$ be a uniform space and $\mathcal{S}$ a cover of a topological space $\boldsymbol{X}$ such that each point of $\boldsymbol{X}$ is an interior point of some $S \in \mathcal{S}$. Then $C_{\mathcal{S}}(\boldsymbol{X}, \boldsymbol{Y})$ is complete (resp. separated) iff $\boldsymbol{Y}$ is complete (resp. separated).
(Hint. Use 47), 48) and 49).)
51) Prove that the Hilbert space introduced in 0.2.1.3.3) is not locally compact (as a topological space).
52) Prove that complex power series converge continuously within their circle of convergence, where you should use only the definition of continuous convergence, i.e. no equivalent formulations.
53) Let $\boldsymbol{X}$ be a topological space satisfying the first axiom of countability, $\boldsymbol{Y}$ an arbitrary topological space, $f \in C(\boldsymbol{X}, \boldsymbol{Y})$ and (f_n) a sequence in $C(\boldsymbol{X}, \boldsymbol{Y})$. Then the elementary filter of (f_n) converges continuously to f iff for each $x \in X$ and each sequence (x_n) in $\boldsymbol{X}$ converging to x the sequence $(f_n(x_n))$ converges to $f(x)$ in $\boldsymbol{Y}$.
54) **PsTop** and $\mathbf{PsTop}_S$ are cartesian closed, where the natural function space structure is the structure of continuous convergence.
(Hint. Use the following statement: If $f : X_1 \times X_2 \to Y$ is a map, $\mathcal{F}_i \in F(X_i)$ for each $i \in \{1, 2\}$ and $\mathcal{V}$ is an ultrafilter on Y with $f(\mathcal{F}_1 \times \mathcal{F}_2) \subset \mathcal{V}$, then there is some ultrafilter $\mathcal{W}$ on X_1 with $\mathcal{F}_1 \subset \mathcal{W}$ and $f(\mathcal{W} \times \mathcal{F}_2) \subset \mathcal{V}$.)
55) If $\boldsymbol{X}$ and $\boldsymbol{Y}$ are compact topological spaces, then $C_{co}(\boldsymbol{X}, \boldsymbol{Y})$ need not be locally compact.
(Hint. Use $X = Y = [0, 1]$.)
56) Let $(X, \mathcal{X})$ be a symmetric topological space and $(X, \mathcal{J}_{\gamma_{q\mathcal{X}}})$ its corresponding semiuniform convergence space. Verify that $(X, \mathcal{J}_{\gamma_{q\mathcal{X}}})$ is locally compact iff $(X, \mathcal{X})$ is locally compact in the usual sense, i.e. each point $x \in X$ has a compact neighborhood.
57) Prove: A semiuniform convergence space is **Fil**–determined and diagonal iff it is indiscrete.
58) Show that the construct of diagonal semiuniform convergence spaces (and uniformly continuous maps) is closed under formation of subspaces, quotients and products. Give an example of two diagonal semiuniform convergence spaces whose coproduct is not diagonal.
59) Let $(X, \mathcal{J}_X)$ be a diagonal semiuniform convergence space. Verify:
a) $(X, \mathcal{J}_X)$ is connected.
b) If $(X, \mathcal{J}_X)$ is locally $\mathcal{E}$–connected, then $(X, \mathcal{J}_X)$ is $\mathcal{E}$–connected.
60) Let $(X, \mathcal{J}_X)$ be a semiuniform convergence space. Prove:
a) If $K \subset X \times X$ is compact, then K^{-1} is also compact.
b) If $K_1, \dots, K_n$ are finitely many compact subsets of X, then their union $\bigcup_{i=1}^{n} K_i$ is compact.
61) Compactly generated topological spaces form a bicoreflective (full and isomorphism–closed) subconstruct of **Top**.
62) The category of compactly generated semiuniform convergence spaces is the smallest bicoreflective (full and isomorphism–closed) subconstruct of **SUConv** which contains the construct of compact semiuniform convergence spaces.
(Note: If $\mathcal{A}$ and $\mathcal{B}$ are full subconstructs of a topological construct $\mathcal{C}$, than $\mathcal{A}$ is *smaller* than $\mathcal{B}$ [or $\mathcal{B}$ *contains* $\mathcal{A}$] iff $|\mathcal{A}| \subset |\mathcal{B}|$.)

63) The {compact Hausdorff}–generated topological spaces form a cartesian closed topological subconstruct of **Top**, where a topological space $(X, \mathcal{X})$ is called *{compact Hausdorff}-generated* provided that $\mathcal{X}$ is the final topology w.r.t. the family of all continuous maps from compact Hausdorff spaces to $(X, \mathcal{X})$.
64) Let $(X, \mathcal{X})$ be a symmetric topological space and $(X, \mathcal{J}_{\gamma_{q\mathcal{X}}})$ its corresponding semiuniform convergence space. Then $(X, \mathcal{X})$ is regular (in the usual sense) iff $(X, \mathcal{J}_{\gamma_{q\mathcal{X}}})$ is t–regular.
65) The construct **SLC-SUConv** of strongly locally compact semiuniform convergence spaces (and uniformly continuous maps) is a bicoreflective cartesian closed topological subconstruct of **SUConv**.
66) The t–regular semiuniform convergence spaces form a bireflective (full and isomorphism closed) subconstruct of **SUConv**.
67) Prove that in exercise 60) 'compact' may be substituted by 'precompact'.
68) Let $\boldsymbol{X}$ be a semiuniform convergence space and $\boldsymbol{X}^*$ its one–point extension in **SUConv**. Prove:
a) $\boldsymbol{X}^*$ is semi–pseudouniform iff $\boldsymbol{X}$ is semi–pseudouniform.
b) **SPsU** is a topological universe, where the natural function spaces and one–point extensions are formed as in **SUConv**.
69) **PsU** is bireflective in **ULim**.
(Hint. Use the following statement: If $\mathcal{F}, \mathcal{G} \in F(X \times X)$ and $\mathcal{U}$ is an ultrafilter on $X \times X$ such that $\mathcal{F} \circ \mathcal{G} \subset \mathcal{U}$, then there are ultrafilters $\mathcal{V}$ and $\mathcal{W}$ on $X \times X$ with $\mathcal{F} \subset \mathcal{V}, \mathcal{G} \subset \mathcal{W}$ and $\mathcal{V} \circ \mathcal{W} \subset \mathcal{U}$.)
70) **PsU** is not extensional.
(Hint. Use 3.2.7.②)
71) Let $\boldsymbol{X}$ be a semiuniform convergence space and $\boldsymbol{Y}$ a semiuniform limit space. Then every non-empty finite set M of uniformly continuous maps from $\boldsymbol{X}$ to $\boldsymbol{Y}$ is uniformly equicontinuous.
72) Define a map $f : I\!R \to I\!R$ by

$$f(x) = \begin{cases} 0 & \text{if } x \leq 0 \\ x & \text{if } 0 \leq x \leq 1 \\ 1 & \text{if } 1 \leq x \end{cases}$$

and prove the following:
a) For each $n \in I\!N, f_n(x) = f(nx - n^2)$ is uniformly continuous with respect to the uniformity on $I\!R$ induced by the Euclidean metric.
b) $\{f_n : n \in I\!N\}$ is equicontinuous but not uniformly equicontinuous.
73) Let k and α be positive numbers. Prove:
a) If (X, d) and (X', d') are metric spaces, then the set $\{f \in F(X, X') : d'(f(x), f(x')) \leq k(d(x, x'))^\alpha\}$ is uniformly equicontinuous.
b) If $I \subset I\!R$ is an interval, then $M = \{f \in F(I, I\!R) : f$ is differentiable on I and $|f'(x)| \leq k$ for each $x \in I\}$ is uniformly equicontinuous.
(Hint. For each $f \in M, |f(x_1) - f(x_2)| \leq k|x_1 - x_2|$ whenever $x_1, x_2 \in I$.)
74) Let X be a set, $(Y, \mathcal{J}_Y)$ a diagonal semiuniform convergence space and $\mathcal{S}$ a non–empty set of subsets of X containing a non–empty subset of X. Prove:
a) The subspace of all constant maps of $(Y^X, \mathcal{J}^{\mathcal{S}}_{YX})$ is isomorphic to $(Y, \mathcal{J}_Y)$.
b) $(Y^X, \mathcal{J}^{\mathcal{S}}_{YX})$ is T_1 iff $(Y, \mathcal{J}_Y)$ is T_1 and $\mathcal{S}$ is a cover of X.

c) If $(Y, \mathcal{J}_Y)$ is a Cook–Fischer space, then b) is also valid for T_2–spaces instead of T_1–spaces.
d) If $(Y^X, \mathcal{J}^{\mathcal{S}}_{YX})$ is T_1 and $(Y, q_{\gamma_{\mathcal{J}_Y}})$ is T_2, then the subspace of all constant maps is closed in $(Y^X, \mathcal{J}^{\mathcal{S}}_{YX})$.
e) Let $\boldsymbol{Y} = (Y, \mathcal{J}_Y)$ be a Cook–Fischer space and $\mathcal{S}$ a cover of X. A filter $\mathcal{F}$ on $(Y^X, \mathcal{J}^{\mathcal{S}}_{YX})$ converges iff $\mathcal{F}$ converges pointwise and $\mathcal{F}$ is a $\mathcal{J}^{\mathcal{S}}_{YX}$–Cauchy filter.
f) Let the assumptions under e) be fulfilled. If $\boldsymbol{X} = (X, q)$ is a generalized convergence space, then the set $C(\boldsymbol{X}, \boldsymbol{Y})$ of all continuous maps from $\boldsymbol{X}$ into $\boldsymbol{Y}$ is closed in $(Y^X, \mathcal{J}^{\mathcal{S}}_{YX})$.

Chapter 7

75) Prove the following statements:
a) A filter space (X, γ) is precompact iff one of the following two equivalent conditions is satisfied:
(1) If from each Cauchy filter $\mathcal{F} \in \gamma$ some $F_{\mathcal{F}}$ is chosen, then there are finitely many Cauchy filters $\mathcal{F}_1, \dots, \mathcal{F}_n \in \gamma$ such that $\bigcup_{i=1}^n F_{\mathcal{F}_i} = X$.
(2) Each uniform cover of (X, μ_γ) contains a finite subcover.
b) A symmetric Kent convergence space (X, q) is compact (=precompact) iff the following condition is satisfied: If from each convergent filter $\mathcal{F}$ in (X, q) some $F_{\mathcal{F}}$ is chosen, then there are finitely many convergent filters $\mathcal{F}_1, \dots, \mathcal{F}_n$ in (X, q) such that $\bigcup_{i=1}^n F_{\mathcal{F}_i} = X$.
76) Prove that every regular topological space is weakly Hausdorff.
77) Let $(X, \mathcal{J}_X)$ be a connected suptopological semiuniform convergence space and A a dense subset of it. Verify that A is connected.
78) Let $(X, \mathcal{J}_X)$ be a subtopological semiuniform convergence space. Prove that each uniformly continuous map $f : (A, \mathcal{J}_A) \to (Y, \mathcal{J}_Y)$ from a dense subspace $(A, \mathcal{J}_A)$ of $(X, \mathcal{J}_X)$ (formed in **SUConv**) into a complete, regular and separated semiuniform convergence space $(Y, \mathcal{J}_Y)$ has a unique uniformly continuous extension $\overline{f} : (X, \mathcal{J}_X) \to (Y, \mathcal{J}_Y)$.
(Hint. Use 5.1.29.)
79) Prove: Coproducts of topological semiuniform convergence spaces are topological.
80) Show the following:
a) Each convergence space $\boldsymbol{X}$ is a quotient space $\boldsymbol{Y}$ (in **SUConv**) of some topological semiuniform convergence space and vice versa.
(Hint. Construct from each convergent filter in $\boldsymbol{X}$ a topological semiuniform convergence space and let $\boldsymbol{Y}$ be their coproduct.)
b) Quotients of topological (subtopological) semiuniform convergence spaces need not be topological (subtopological).
81) Prove: Coproducts of completely regular semiuniform convergence spaces are completely regular.
(Hint. Use 21)b), 79) and 30).)

82) Verify that quotients of completely regular semiuniform convergence spaces are not completely regular in general.
(Hint. Use the example under 31).)
83) Show that uniformizable Cauchy spaces are exactly the subspaces (in **Fil**) of the completely uniformizable topological spaces.
84) Verify: Every normal symmetric topological space is regular.
85) Show: a) Coproducts of normal (fully normal) semiuniform convergence spaces are normal (fully normal).
b) Quotients of normal (fully normal) semiuniform convergence spaces need not be normal (fully normal).
(Hint. Use the example under 31).)
86) Prove that countable products of metrizable (topologically metrizable) semiuniform convergence spaces are metrizable (topologically metrizable) and verify that arbitrary products do not have this property in general.
87) Let $(X, \mathcal{W})$ be a uniform space. Prove that $(X, [\mathcal{W}])$ is uniformly zerodimensional iff $\mathrm{Dim}_u(X, [\mathcal{W}]) = 0$.
88) Let $\boldsymbol{X} = (X, \mathcal{X})$ be a topological T_1–space. Prove the equivalence of the following conditions:
(1) $\boldsymbol{X}$ is zerodimensional.
(2) $\mathrm{ind}\boldsymbol{X} = 0$, i.e. $\mathcal{X}$ has a base consisting of open–closed subsets in $\boldsymbol{X}$.
Show that each of the above conditions is equivalent to each of the following provided that $\boldsymbol{X}$ has a countable base:
(3) $\dim \boldsymbol{X} = 0$.
(4) $\mathrm{Dim}\boldsymbol{X} = 0$.
89) Give an example of a quotient (in **SUConv**) of a sub–(compact Hausdorff) space which is not sub–(compact Hausdorff).
(Hint. Consider the usual compact Hausdorff space $[-2, +2]$ and define an equivalence relation ρ on $[-2, +2]$ as follows:
$x\rho y$ iff $x = y$ or ($y = -x$ for $x > 1$ or $x < 1$). Verify that $[-2, +2]/\rho$ is not T_2 as a quotient in **SUConv**.)
90) Prove: a) Every m–proximal filter space is m–uniform.
b) Every m–uniform filter space is a uniformizable Cauchy space.
Verify that none of the above statements is reversible.

Implication scheme for various SUConv–invariants

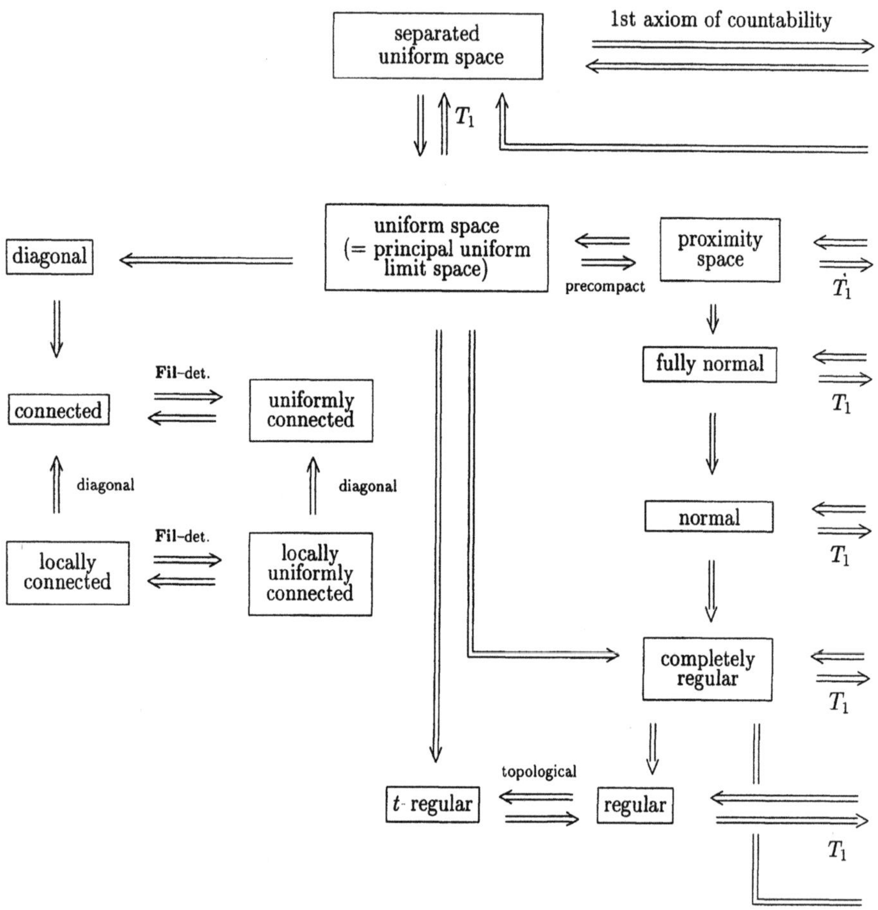

separated uniform space
1st axiom of countability
T_1
diagonal
uniform space (= principal uniform limit space)
proximity space
precompact
T_1
fully normal
T_1
Fil-det.
connected
uniformly connected
diagonal
diagonal
normal
T_1
Fil-det.
locally connected
locally uniformly connected
completely regular
T_1
topological
t- regular
regular
T_1

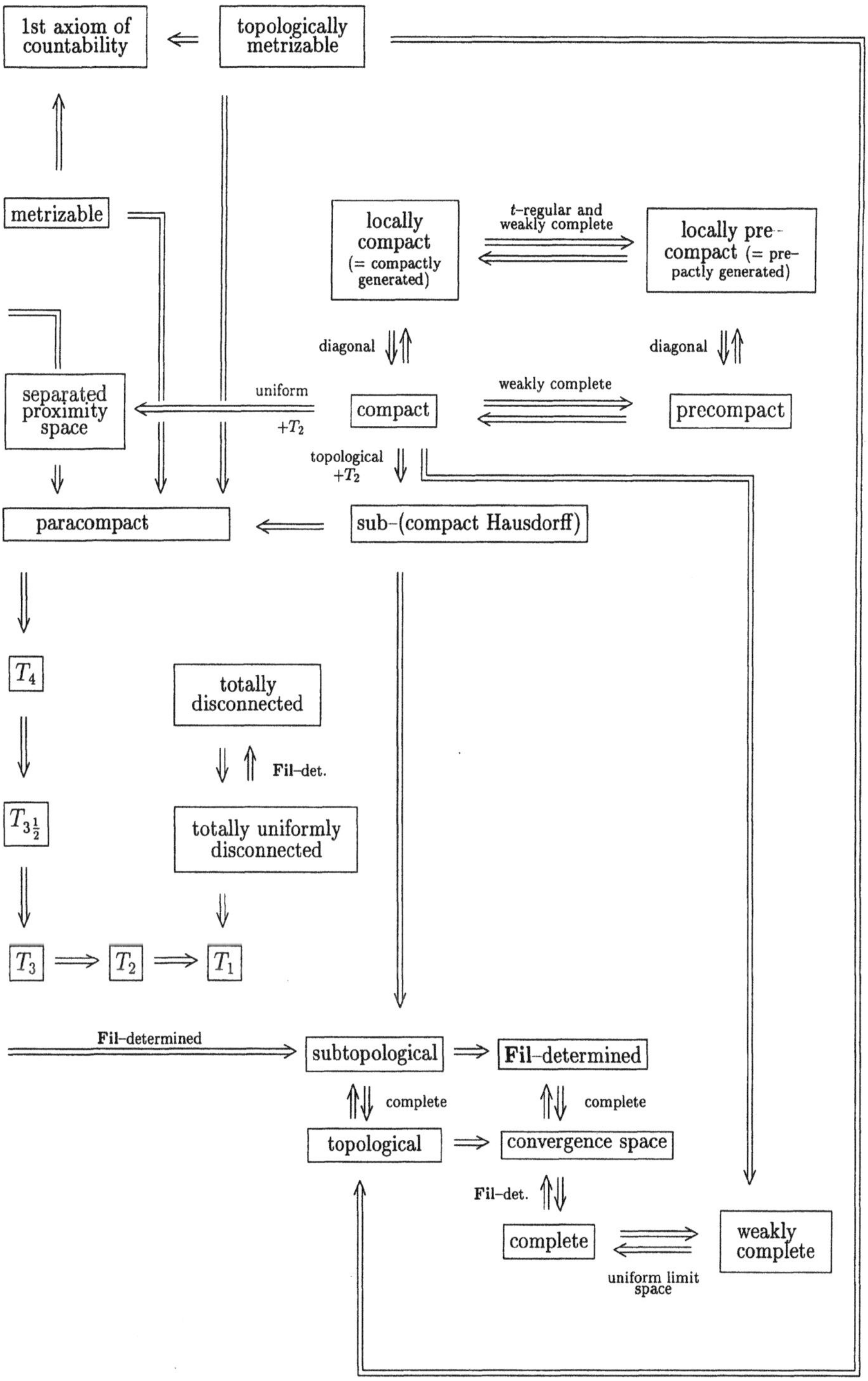

1st axiom of countability
topologically metrizable
metrizable
locally compact (= compactly generated)
t–regular and weakly complete
locally pre-compact (= pre-pactly generated)
diagonal
diagonal
separated proximity space
uniform
$+T_2$
compact
weakly complete
precompact
topological $+T_2$
paracompact
sub–(compact Hausdorff)
T_4
totally disconnected
Fil–det.
$T_{3\frac{1}{2}}$
totally uniformly disconnected
T_3
T_2
T_1
Fil–determined
subtopological
Fil–determined
complete
complete
topological
convergence space
Fil–det.
complete
weakly complete
uniform limit space

Table: Preservation properties of some SUConv–invariants

The following preservation properties are defined similarly to the corresponding properties of topological invariants (cf. 0.2.4.3. 3)).

SUConv–invariants	hereditary	productive	summable	divisible	initially closed	finally closed
T_1	+	+	+	−	−	−
T_2	+	+	+	−	−	−
regular (or t–regular)	+	+	+	−	+	−
completely regular	+	+	+	−	+	−
normal	+	−	+	−	−	−
fully normal (or paracompact)	+	−	+	−	−	−
topological	−	+	+	−	−	−
subtopological	+	+	+	−	+	−
sub–(compact Hausdorff)	+	+	−	−	−	−
diagonal	+	+	−	+	+	−
uniform	+	+	−	−	+	−
compact	−	+	−	+	−	−
locally compact (=compactly generated)	−	−	+	+	−	+
precompact	+	+	−	+	+	−
locally precompact (=precompactly generated)	+	−	+	+	−	+
connected (or uniformly connected)	−	+	−	+	−	−
locally connected (or locally uniformly connected)	−	−	+	+	−	+
totally disconnected (or totally uniformly disconnected)	+	+	+	−	−	−
1st axiom of countability	+	−	+	+	−	+
metrizable	+	−	−	−	−	−
topologically metrizable	−	−	+	−	−	−
Fil–determined	+	+	+	+	+	+
convergence space	−	+	+	+	−	+
complete (or weakly complete)	−	+	+	−	−	−

Concerning the statements in the above table not proved in the text, see the exercises.

Diagram of relations between various subconstructs of SUConv (including their relations to merotopic and nearness spaces)

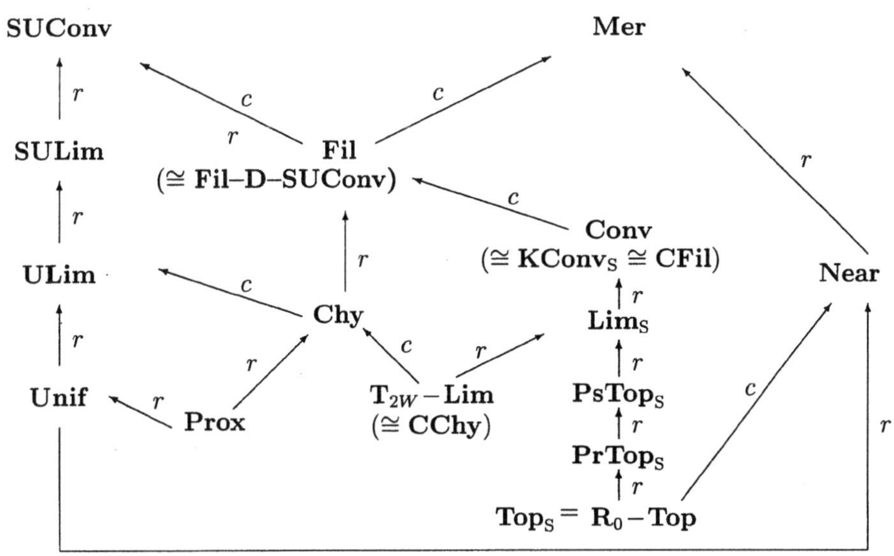

In the above diagram r (resp. c) stands for embedding as a bireflective (resp. bicoreflective) subconstruct.

Bibliography

[1] Adámek, J., H. Herrlich and G.E. Strecker: 1990, *Abstract and Concrete Categories*, Wiley–Intersci. Publ., New York.

[2] Alexandroff, A.: 1924, 'Über die Metrisation der im Kleinen kompakten topologischen Räume', *Math. Ann.* **92**, 294–301.

[3] Alfsen, E.M. and J.E. Fenstad: 1959, 'On the equivalence between proximity structures and totally bounded uniform structures', *Math. Scand.* **7**, 353–360.

[4] Arens, R.: 1946, 'A Topology for Spaces of Transformations', *Ann. of Math.* (2) **47**, 480–495.

[5] Arhangel'skii, A.V.: 1963, 'Some types of factor mappings, and the relations between classes of topological spaces', *Dokl. Akad. Nauk SSSR* **153**, 743–746 (= *Soviet Math. Dokl.* **4**, 1726–1729).

[6] Arhangel'skii, A.V. and R. Wiegandt: 1975, 'Connectednesses and disconnectednesses in topology', *Gen. Top. Appl.* **5**, 9–33.

[7] Arzelà, C.: 1895, 'Sull'Integrabilita Delle Equazioni Differenziali Ordinarie', *Mem. Accad. Bologna* (5) **5**, 257–270.

[8] Ascoli, G.: 1883, 'Le Curve Limite di una Varietà Data di Curve', *Mem. Accad. Lincei* (3) **18**, 521–586.

[9] Behling, A.: 1992, *Einbettung uniformer Räume in topologische Universen,* PhD. thesis, Free University, Berlin.

[10] Bentley, H.L.: 1975, 'Nearness spaces and extensions of topological spaces', In: *Studies in Topology*, pp. 47–66, Academic Press, New York.

[11] Bentley, H.L.: 1983, 'Homology and cohomology for merotopic and nearness spaces', *Quaest. Math.* **6**, 27–47.

[12] Bentley, H.L., H. Herrlich and E. Lowen–Colebunders: 1987, 'The category of Cauchy spaces is cartesian closed', *Topology Appl.* **27**, 105–112.

[13] Bentley, H.L., H. Herrlich and E. Lowen–Colebunders: 1990, 'Convergence', *J. Pure and Applied Algebra* **68**, 27–40.

[14] Bentley, H.L., H. Herrlich and R. Ori: 1989, 'Zero sets and complete regularity for nearness spaces', *Categorical Topology and its relations to Analysis, Algebra and Combinatorics*, World Scient. Publ., Singapore, 446–461.

[15] Bentley, H.L., H. Herrlich and W.A. Robertson: 1976, 'Convenient categories for topologists', *Comm. Math. Univ. Carolinae* **17**, 207–227.

[16] Bentley, H.L. and E. Lowen–Colebunders: 1992, 'The cartesian closed topological hull of the category of completely regular filter spaces', *Cahiers Topol. Géom. Diff.* **23**, 345–360.

[17] Binz, E. and H.H. Keller: 1966, Funktionenräume in der Kategorie der Limesräume', *Ann. Acad. Fenn. Ser. A* I *Mathematica* **383**, 1–21.

[18] Bourbaki, N.: 1957, *Théorie des ensembles*, Chapitre 4: Structures, Hermann, Paris.

[19] Bourbaki, N.: 1961, *Topologie générale*, Chapitres 1 et 2, Hermann, Paris.

[20] Breger, H.: 1971, *Die Kategorie der kompakterzeugten Räume als in TOP coreflective Kategorie mit Exponentialgesetz*, Ms Thesis, Univ. Heidelberg.

[21] Brown, R.: 1963, 'Ten topologies for $X \times Y$', *Quart. J. Math.*, *Oxford* (2) **14**, 303–319.

[22] Brown, R.: 1964, 'Function spaces and product topologies', *Quart. J. Math. Oxford* **15**, 238–250.

[23] Cantor, G.: 1872, 'Über die Ausdehnung eines Satzes aus der Theorie der trigonometrischen Reihen', *Math. Ann.* **5**, 123–132.

[24] Cantor, G.: 1883, 'Über unendliche lineare Punktmannigfaltigkeiten', *Math. Ann.* **21**, 545–591.

[25] Carathéodory, C.: 1929, 'Stetige Konvergenz und normale Familien von Funktionen', *Math. Ann.* **101**, 515–533.

[26] Cartan, H.: 1937, 'Théorie des filtres', *C.R. Acad. Sci. Paris* **205**, 595–598.

[27] Cartan, H.: 1937, 'Filtres et ultrafiltres', *C.R. Acad. Sci. Paris* **205**, 777–779.

[28] Čech, E.: 1933, 'Příspěvek k theorii dimense', *Časopis Pěst. Mat. Fys.* **62**, 277–291. (French translation: Čech, E., *Topological papers*, Prague 1968, 129–142).

[29] Čech, E.: 1966, *Topological Spaces*, Revised edition by M. Frolik and M. Katětov, Interscience, London.

[30] Choquet, G.: 1948, 'Convergences', *Ann. Univ. Grenoble Sect. Math. Phys.* (NS) **23**, 57–112.

[31] Činčura, J.: 1991, 'Cartesian closed coreflective subcategories of the category of topological spaces', *Topol. Appl.* **41**, 205–212.

[32] Cook, C.H. and H.R. Fischer: 1965, 'On equicontinuity and continuous convergence', *Math. Ann.* **159**, 94–104.

[33] Cook, C.H. and H.R. Fischer: 1967, 'Uniform convergence spaces', *Math. Ann.* **173**, 290–306.

[34] Császár, A.: 1996, 'λ–complete filter spaces', *Acta Math. Hungar.* **70**, 75–87.

[35] Dieudonné, J.A.: 1944, 'Une généralisation des espaces compacts', *J. Math. Pures Appl.* **23**, 65–76.

[36] Dowker, C.H.: 1952, 'Topology of metric complexes', *Amer. J. Math.* **74**, 555–577.

[37] Dubuc, E.J. and H. Porta: 1971, 'Convenient Categories of Topological Algebras, and their Duality Theory', *J. Pure and Appl. Algebra* **1** (3), 281–316.

[38] Dugundji, J.: 1966, *Topology*, Allyn and Bacon, Boston.

[39] Efremovič, V.A.: 1949, 'Nonequimorphism of the Euclidean and Lobačevski spaces', *Uspekhi Mat. Nauk* **4**, **2**, 178–179.

[40] Efremovič, V.A.: 1951, 'Infinitesimal spaces', *Doklady Akad. Nauk SSSR* **76**, 341–343.

[41] Efremovič, V.A.: 1951, 'Infinitesimal spaces', *Mat. Sbornik* **6**, **4**, 203–204.

[42] Efremovič, V.A.: 1952, '*The geometry of proximity*', *Mat. Sbornik* **31** (**73**), 189–200.

[43] Engelking, R.: 1977, *General Topology*. Polish Scient. Publ., Warszawa.

[44] Fischer, H.R.: 1959, 'Limesräume', *Math. Ann.* **137**, 269–303.

[45] Fox, R.H.: 1945, 'On Topologies for Function Spaces', *Bull. Amer. Math. Soc.* **51**, 429–432.

[46] Franke, D.: 1975, *Funktionenalgebren in kartesisch abgeschlossenen Kategorien*, PhD. thesis, Free University, Berlin.

[47] Fréchet, M.: 1906, 'Sur quelques points du calcul fonctionnel', *Rend. Palermo* **22**, 1–74.

[48] Frič, A. and D.C. Kent: 1978, 'On the natural completion functor for Cauchy spaces', *Bull Austral. Math. Soc.* **18**, 335–343.

[49] Frölicher, A.: 1972, 'Kompakt erzeugte Räume und Limesräume', *Math. Z.* **129**, 57–63.

[50] Gähler, W.: 1977, *Grundstrukturen der Analysis* I, Birkhäuser, Basel.

[51] Gähler, W.: 1978, *Grundstrukturen der Analysis* II, Birkhäuser, Basel.

[52] Gál, I.S.: 1959, 'Proximity relations and precompact structures', *Nederl. Akad. Wetensch. Proc., Ser.* A **62**, 304–326.

[53] Gale, D.: 1950, 'Compact sets of functions and function rings', *Proc. Amer. Math. Soc.* **1**, 303–308.

[54] Gazik, R.J., D.C. Kent and G.D. Richardson: 1974, 'Regular completions of uniform convergence spaces', *Bull. Austral. Math. Soc.* **11**, 413–424.

[55] Hahn, H.: 1914, 'Über die allgemeinste ebene Punktmenge, die stetiges Bild einer Strecke ist', *Jahresber. Deut. Math. Ver.* **23**, 318–322.

[56] Hahn, H.: 1921, *Theorie der reellen Funktionen*, Berlin.

[57] Hahn, H.: 1921, 'Über die Komponenten offener Mengen', *Fund. Math.* **2**, 189–192.

[58] Hausdorff, F.: 1914, *Grundzüge der Mengenlehre*, Veit, Leipzig. (Reprint: 1949, 1965, Chelsea, New York).

[59] Hausdorff, F.: 1935, 'Gestufte Räume', *Fund. Math.* **25**, 486–502.

[60] Hermes, H.: 1955, *Einführung in die Verbandstheorie*, Springer, Berlin.

[61] Herrlich, H.: 1968, *Topologische Reflexionen und Coreflexionen*, Lecture Notes in Mathematics **78**, Springer, Berlin.

[62] Herrlich, H.: 1974, 'Topological structures', In: *Math. Centre Tracts* **52**, pp. 59–122, Math. Centrum, Amsterdam.

[63] Herrlich, H.: 1974, 'Cartesian closed topological categories', *Math. Coll. Univ. Cape Town* **9**, 1–16.

[64] Herrlich, H.: 1982, 'Categorical Topology 1971–1981', In: J. Novák (ed.), *General Topology and its Relations to Modern Analysis and Algebra V* (Proc. Prague 1981), Heldermann, Berlin, 279–383.

[65] Herrlich, H.: 1983, 'Are there convenient subcategories of Top?', *Top. Appl.* **15**, 263–271.

[66] Herrlich, H.: 1986, *Einführung in die Topologie*, Heldermann, Berlin.

[67] Herrlich, H.: 1986, *Topologie* I: *Topologische Räume*, Heldermann, Berlin.

[68] Herrlich, H.: 1987, 'Topological improvements of categories of structured sets', *Top. Appl.* **27**, 145–155.

[69] Herrlich, H.: 1988, *Topologie* II: *Uniforme Räume*, Heldermann, Berlin.

[70] Herrlich, H.: 1988, 'Hereditary topological constructs', In: *General Topology and its Relations to Modern Analysis and Algebra VI* (Z. Folik, ed.), pp. 249–262, Heldermann, Berlin.

[71] Herrlich, H. and M. Hušek: 1992, 'Categorical Topology', In: *Recent Progress in General Topology* (M. Hušek and J. von Mill, eds.), Elsevier, Amsterdam, 371–403.

[72] H. Herrlich and H.E. Porst (eds.): 1991, *Category Theory at Work,* Heldermann, Berlin.

[73] Herrlich, H. and G.E. Strecker: 1997, 'Categorical Topology – its Origins, as exemplified by the Unfolding of the Theory of Topological Reflections and Coreflections before 1971', In: *Handbook of the History of General Topology Vol.* 1 (eds. C.E. Aull and R. Lowen), pp. 255–341, Kluwer, Dordrecht.

[74] Hu, S.–T.: 1959, *Homotopy Theory,* Academic Press, New York.

[75] Hu, S.–T.: 1964, *Elements of General Topology,* Holden–Day, San Francisco.

[76] Hu, S.–T.: 1968, *Cohomology Theory,* Markham, Chicago.

[77] Hušek, M. and D. Pumplün: 1990, 'Disconnectedness', *Quaest. Math.* **13**, 449–459.

[78] Hušek, M. and M.D. Rice: 1978, 'Productivity of coreflective subcategories of uniform spaces', *Gen. Top. Appl.* **9**, 295–306.

[79] Isbell, J.R.: 1964, *Uniform spaces,* Math. Surveys No. 12, Amer. Math. Soc., Providence.

[80] Katětov, M.: 1965, 'On continuity structures and spaces of mappings', *Comm. Math. Univ. Carolinae* **6**, 257–278.

[81] Keller, H.H.: 1968, 'Die Limes–Uniformierbarkeit der Limesräume', *Math. Ann.* **176**, 334–341.

[82] Kelley, J.L.: 1955, *General Topology,* Van Nostrand, Princeton.

[83] Kent, D.C.: 1964, 'Convergence functions and their related topologies', *Fund. Math.* **54**, 125–133.

[84] Kent, D.C.: 1967, 'On convergence groups and convergence uniformities', *Fund. Math.* **60**, 213–222.

[85] Kent, D.C. and N. Rath: 1993, 'Filter spaces', *Appl. Categorical Structures* **1**, 297–309.

[86] Kent, D.C. and G. Richardson: 1974, 'Regular completion of Cauchy spaces', *Pacific J. Math.* **51**, 483–490.

[87] Kent, D.C. and G. Richardson: 1984, 'Cauchy spaces with regular completions', *Pacific J. Math.* **111**, 105–116.

[88] Kneis, G.: 1986, 'Completion Functors for Categories of Convergence Spaces, I. Acceptability Functors', *Math. Nachr.* **129**, 283–312.

[89] Kneis, G.: 1988, 'Completion Functors for Categories of Convergence Spaces, II. Embedding of Separated Acceptable Spaces into their Completion', *Math. Nachr.* **135**, 181–211.

[90] Kowalsky, H.-J.: 1954, 'Limesräume und Komplettierung', *Math. Nachrichten* **12**, 301–340.

[91] Kuratowski, C.: 1922, 'Sur l'opération $\overline{A}$ de l'Analysis Situs', *Fund. Math.* **3**, 182–199.

[92] Lebesgue, H.: 1911, 'Sur la non applicabilité de deux domaines appartenant respectivement à des espaces, de n et $n + p$ dimensions', *Math. Ann.* **70**, 166–168.

[93] Lee, R.: 1976, 'The category of uniform convergence spaces is cartesian closed', *Bull. Austral. Math. Soc.* **15**, 461–465.

[94] Lennes, N.J.: 1911, 'Curves in non-metrical analysis situs etc.', *Am. J. Math.* **33**, 287–326.

[95] Lowen-Colebunders, E.: 1989, *Function Classes of Cauchy Continuous Maps*, Marcel Dekker, New York.

[96] Mac Lane, S.: 1971, *Categories for the Working Mathematician*, Springer, New York.

[97] Méray, Ch.: 1869, 'Remarques sur la nature des quantités définies par la condition de servir de limite à des variables données', *Revue Soc. Savantes, Sci. Math.* **4**, 280–289.

[98] Méray, Ch.: 1872, *Nouveau précis d'analyse infinitésimal*, Paris.

[99] Michael, E.: 1968, 'Local compactness and cartesian products of quotient maps and k-spaces', *Ann. Inst. Fourier* **18**, 281–286.

[100] Moore, R.L.: 1922, 'Concerning connectedness im kleinen and a related property', *Fund. Math.* **3**, 232–237.

[101] Mrowka, S. and W. Pervin: 1964, 'On uniform connectedness', *Proc. Am. Math. Soc.* **15**, 446–449.

[102] Nagami, K.: 1970, *Dimension Theory*, Academic Press, New York.

[103] Nagata, J.: 1965, *Modern Dimension Theory*, Noordhoff and North-Holland, Groningen and Amsterdam.

[104] Nel, L.D.: 1984, 'Topological universes and smooth Gelfand-Naimark duality', *Contemporary Math.* **30**, 244–276.

[105] Nusser, H.: 1998, *Probabilistic Semiuniform Convergence Spaces*, PhD. thesis, Free University, Berlin.

[106] Penon, M.J.: 1973, 'Quasitopos', *C.R. Acad. Sci. Paris, Sér.* A. **276**, 237–240.

[107] Poppe, H.: 1974, *Compactness in General Function Spaces*, VEB Deutscher Verlag der Wissenschaften, Berlin.

[108] Preuß, G.: 1975, *Allgemeine Topologie*, 2. korr. Aufl., Springer, Berlin.

[109] Preuß, G.: 1983, 'Some point set properties of merotopic spaces and their cohomology groups', *Mh. Math.* **96**, 41–56.

[110] Preuß, G.: 1984, 'A cohomological characterization of dimension for normal nearness spaces, In: Categorical Topology, pp. 441–452, *Sigma Ser. Pure Math.* **5**, Heldermann, Berlin.

[111] Preuß, G.: 1988, *Theory of Topological Structures*, Reidel (Kluwer), Dordrecht.

[112] Preuß, G.: 1991, 'Improvement of Cauchy spaces', *Q & A in General Topology* **9**, 159–165.

[113] Preuß, G.: 1992, 'Semiuniform limit spaces and semi-Cauchy spaces', In: Recent Developments of General Topology and its Applications (eds. W. Gähler et al.), pp. 268–278, *Mathematical Research* **67**, Akademie Verlag, Berlin.

[114] Preuß, G.: 1993, 'Cauchy spaces and generalizations', *Math. Japonica* **38**, 803–812.

[115] Preuß, G.: 1995, 'Semiuniform convergence spaces', *Math. Japonica* **41**, 465–491.

[116] Preuß, G.: 1996, 'Local compactness in semi-uniform convergence spaces', *Quaest. Math.* **19**, 453–466.

[117] Preuß, G.: 1997, 'Precompactness and compactness in the natural function spaces of the category of semiuniform convergence spaces', *Q & A in General Topology* **15**, 173–185.

[118] Preuß, G.: 1997, 'Completion of semiuniform convergence spaces', In: Categorical Methods in Algebra and Topology, *Mathematik-Arbeitspapiere* **48**, Univ. Bremen, 353–361 (= *Applied Categorical Structures* **8** (2000), 463–473).

[119] Preuß, G.: 1998, 'The topological universe of locally precompact semiuniform convergence spaces', *Top. Appl.* **82**, 387–396.

[120] Preuß, G.: 1999, 'Higher separation axioms, paracompactness and dimension for semiuniform convergence spaces', *Scientiae Mathematicae* **2**, 321–335.

[121] Preuß, G.: 2000, 'Some algebraically topological aspects in the realm of Convenient Topology', *Quaest. Math.* **23**, 99-112.

[122] Preuß, G.: 2001, 'Characterization of subspaces of important types of convergence spaces', in: Catgegorical Perspectives (eds. J. Koslowski and A. Melton), pp. 177–190, *Trends in Mathematics*, Birkhäuser, Boston.

[123] Pust, H.: 1977, *Normalität und Dimension for Nearness–Räume*, PhD. thesis, Free University, Berlin.

[124] Reed, E.: 1971, 'Completion of uniform convergence spaces', *Math. Ann.* **194**, 83–108.

[125] Riesz, F.: 1907, 'Die Genesis des Raumbegriffes', *Math. Naturwiss. Ber. Ungarn* **24**, 309–353.

[126] Riesz, F.: 1908, 'Stetigkeitsbegriff und abstrakte Mengenlehre', *Atti* IV *Congr. Intern. Mat.*, *Roma,* **II**, 18–24.

[127] Robertson, W.A.: 1975, *Convergence as a nearness concept,* PhD. thesis, Carleton University, Ottawa.

[128] Schubert, H.: 1964, *Topologie,* Teubner, Stuttgart.

[129] Schwarz, F.: 1983, *Funktionenräume und exponentiale Objekte in punkttrennend initialen Kategorien,* PhD. thesis, Univ. Bremen.

[130] Schwarz, F.: 1986, 'Hereditary topological categories and topological universes', *Quaest. Math.* **10**, 197–216.

[131] Schwarz, F.: 1991, 'Representability of partial morphisms in topological and monotopological constructs', *Quaest. Math.* **14**, 1–10.

[132] Sjöberg, B.: 1970, 'Über die Fortsetzbarkeit gleichmäßig stetiger Abbildungen in uniformen Limesräumen', *Soc. Sci. Fennica, Comment. phys.-math.* **40**, 41–46.

[133] Smirnov, Ju.M.: 1952, 'On proximity spaces in the sense of V.A. Efremovič', *Doklady Akad. Nauk SSSR* (*N.S.*) **84**, 895–898.

[134] Smirnov, Ju.M.: 1952, 'On proximity spaces', *Mat. Sbornik* (*N.S.*) **31**, 543–574.

[135] Smirnov, Ju.M.: 1956 (1962), 'On the dimension of proximity spaces', *Math. Sb.* **38**, 283–302 (= *Amer. Math. Soc.* Transl. Sr. 2, **21**, 1–20).

[136] Spanier, E.H.: 1966, *Algebraic Topology*, Springer, New York.

[137] Steenrod, N.E.: 1967, 'A convenient category of topological spaces', *Mich. Math. J.* **14**, 133–152.

[138] Stone, A.H.: 1948, 'Paracompactness and product spaces', *Bull. American Math. Soc.* **54**, 977–982.

[139] Tietze, H.: 1915, 'Über Funktionen, die auf einer abgeschlossenen Menge stetig sind', *J. für die R. and Ang. Math.* **145**, 9–14.

[140] Tietze, H.: 1923, 'Beiträge zur allgemeinen Topologie I', *Math. Ann.* **88**, 290–312.

[141] Tukey, J.W.: 1940, *Convergence and Uniformity in Topology*, Ann. of Math. Studies 2, Princeton Univ. Press, Princeton.

[142] Tychonoff, A.: 1930, 'Über die topologische Erweiterung von Räumen', *Math. Ann.* **102**, 544–561.

[143] Tychonoff, A.: 1935, 'Über einen Funktionenraum', *Math. Ann.* **111**, 762–766.

[144] Urysohn, P.: 1925, 'Über die Mächtigkeit der zusammenhängenden Mengen', *Math. Ann.* **94**, 262–295.

[145] Vietoris, L.: 1921, 'Stetige Mengen', *Monatshefte f. Math.* **31**, 173–204.

[146] Vogt, R.M.: 1971, 'Convenient categories of topological spaces for homotopy theory', *Archiv Math.* **22**, 545–555.

[147] Weil, A.: 1937, *'Sur les espaces à structures uniformes et sur la topologie générale* , Hermann, Paris.

[148] Whitehead, J.H.C.: 1948, 'A note on a theorem due to Borsuk', *Bull. Amer. Math. Soc.* **54**, 1125–1132.

[149] Willard, S.: 1970, *General Topology*, Addison–Wesley, Reading.

[150] Wyler, O.: 1970, 'Ein Komplettierungsfunktor für uniforme Limesräume', *Math. Nachr.* **46**, 1–12.

[151] Wyler, O.: 1974, 'Filter space monads, regularity, completions', In: Topo 72 – General Topology and its Applications, pp. 591–637, *Lecture Notes Math.* **378**, Springer, Berlin.

[152] Wyler, O.: 1979, 'Function spaces in topological categories', In: Categorical Topology (Proc. Conf. Berlin, 1978), pp. 411–419, *Lecture Notes in Mathematics* **719**, Springer, Berlin.

[153] Wyler, O.: 1984, 'Ascoli theorems for cartesian closed categories', In: Categorical Topology (Proc. Conf. Toledo, Ohio 1983), pp. 599–617, *Sigma Series in Pure Mathematics* **5**, Heldermann, Berlin.

List of Axioms

List of Symbols

Index

MIX
Papier aus verantwortungsvollen Quellen
Paper from responsible sources
FSC® C105338

If you have any concerns about our products,
you can contact us on
ProductSafety@springernature.com

In case Publisher is established outside the EU,
the EU authorized representative is:
Springer Nature Customer Service Center GmbH
Europaplatz 3, 69115 Heidelberg, Germany

Printed by Libri Plureos GmbH
in Hamburg, Germany